INTERNATIONAL SERIES IN EARTH SCIENCES

Editor: DEAN EARL INGERSON
Department of Geology, University of Texas, Austin, Texas, USA

VOLUME 35

THE ORE MINERALS AND THEIR INTERGROWTHS

SECOND EDITION

VOLUME 1

Other Pergamon Titles of Interest

BOOKS

See page 441 for a list of other titles in the series

CARTER, ROWAN & HUNTINGTON:
Remote Sensing and Mineral Exploration

JEFFERY & HUTCHISON:
Chemical Methods of Rock Analysis 3rd edition

RIDGE:
Annotated Bibliographies of Mineral Deposits in Africa, Asia (exclusive of the USSR) and Australasia

SAND & MUMPTON:
Natural Zeolites (Occurence, Properties, Use)

SIMPSON:
Rocks and Minerals

VAN OLPHEN & FRIPIAT:
Data Handbook for Clay Materials and other Non-Metallic Minerals

WHITTAKER:
Crystallography for Mineralogists (and Other Solid State Scientists)

JOURNALS

Computers & Geosciences

Geochimica et Cosmochimica Acta

Journal of Structural Geology

Organic Geochemistry

Full details of all Pergamon publications and a free specimen copy of any Pergamon journal available on request from your nearest Pergamon office.

THE ORE MINERALS AND THEIR INTERGROWTHS

SECOND EDITION

by

PAUL RAMDOHR

Heidelberg

English Translation of the 4th Edition
(with Additions and Corrections by the Author)

IN TWO VOLUMES

VOLUME 1

PERGAMON PRESS

OXFORD · NEW YORK · TORONTO · SYDNEY · PARIS · FRANKFURT

UK	Pergamon Press Ltd., Headington Hill Hall, Oxford OX3 0BW, England
USA	Pergamon Press Inc., Maxwell House, Fairview Park, Elmsford, New York 10523, USA
CANADA	Pergamon of Canada, Suite 104, 150 Consumers Road, Willowdale, Ontario M2J 1P9, Canada
AUSTRALIA	Pergamon Press (Aust.) Pty. Ltd., P. O. Box 544, Potts Point, N. S. W. 2011, Australia
FRANCE	Pergamon Press SARL, 24 rue des Ecoles, 75240 Paris, Cedex 05, France
FEDERAL REPUBLIC OF GERMANY	Pergamon Press GmbH, 6242 Kronberg/Taunus, Hammerweg 6, Federal Republic of Germany

First edition 1980

British Library Cataloguing in Publication Data

Ramdohr, Paul
The ore minerals and their intergrowths.
— 2nd ed. — (International series in
earth sciences; vol. 35).
1. Ores
I. Title II. Series
553 TN 265 79-40745

ISBN 0-08-0238017

Printed in GDR

TABLE OF CONTENTS

Volume 1

PREFACE

TO THE SECOND ENGLISH EDITION

The present book will, with all probability, be the last edition revised by the author. Therefore, he asks permission to give some more remarks than those which are to a certain extent technical necessities for the preface of a textbook.

My books on oremicroscopy have obviously inspired many interested scientists to investigate and to clear many questions. Certainly it turned out that not everything, which I thought to have obtained by my work and which I explained, was in all details correct. But that is a natural consequence of the progress in science. I, myself, have in such cases often taken up these questions again and reexamined, with three different results:

1. My statements were simply wrong — due to very varied reasons — or they could, many years ago, not be right due to the then imperfect methods and/or the too sparse material. The latter refers e.g. to the distinction between valleriite and mackinawite or to the complications in the system Ag–Sb between silver and dyscrasite and many others.

2. Other cases show that nature is often much more complicated than we assumed. One example are the Pb–Sb- or Pb–As-sulfosalts, where not only the actual compounds turned out to be much more numerous than anybody expected, but also the number of the components was greater. In addition — or perhaps in a certain opposition — there is now often an occurrence of solid solution with a broad spectrum of different properties complicating the definitions as well as the clear descriptions. Another problem is the "formula", traditionally accepted to be stoichiometric, which they are often not at all. Many members of the niccolite group, e.g. are marked by the vacancy in "cation" sites, in the case of melonite varying between NiTe to $NiTe_2$ — here apparently continuously — while in other cases — I doubt whether it is always justified — to assume a sudden alteration of the stoichiometry with often rather unlikely complicated members. One of these extreme cases is e.g. pyrrhotite. I hope, the reader will understand when I am not willing to agree to the "trend" of some colleagues to always create new names. Tiny splittings in the X-ray powder-diagrams can have essential meaning but it is not necessarily so.

A similar problem is the "anion deficiencies", e.g. frequently occurring in minerals of the sphalerite and wurtzite structures and their relatives, even when not due to low partial pressure of S. To give a new name (or x names!) for a high-temperature chalcopyrite, which became cubic with principally the same lattice, and having lost a trace of S, and perhaps stabilized by a minimum content of Ni, is in my opinion completely unnecessary and in the consequence dangerous. Just in this group there exists a possibility of replacement or a further addition of other metal atoms as a result of the very loosely packed ZnS lattice. Here we should not wonder, when as a result the lattice is distorted from cubic to tetragonal, to o'rhombic, to hexagonal etc. and/or it shows variations in space groups or polytypes. We should try first — even when we find it hard — to emphasize more the common properties of this group, and also many other more than the differences, since small variations in composition can go parallel with very distinct color deviations.

3. There is also a great number of new reports in literature, which apparently contradict my data or really do. The author could, of course, neither investigate all the deposits of the world, nor could he always get complete or in each case authentic material of all minerals — however in the course of his long life he worked hard in order to get at least a good general view. But it astonishes when the author is reproached for oversights, e.g. when somebody states that sphalerite stars in chalcopyrite are not a result of exsolution! I do not at all deny, that for one or the other reason these stars may occur somehow in similar form. However that exsolution is the rule can be proved by correct observations from more than 1000 deposits and laboratory experiments which can be carried out in a few hours! Some beginners put much more weight on statistics of two observations rather than on conclusions drawn from reliable statistics from 50 or 200!

Some discrepancy may derive from the overestimation of the experiment. — "Daraus schließt er messerscharf, daß nicht ist, was nicht sein darf" (From this he concludes "knife-sharp" that something cannot be which ought not to be!) But, nevertheless, it often is so! In dozens or perhaps hundreds of cases low-temperature compounds are known, which experimentally do not occur at all or only at normally intolerably long times of reaction. Vice versa, there are high-temperature minerals, which, in the experiments are absolutely instable below a certain temperature, become durable by tiny additions, and can also occur in nature. In metallurgy these are absolute trivialities. Expressions, such as "do not exist" should not be used at all, "could not be proved experimentally" only with the suggestion "in due time", or "with the experimental apparatus which was at our disposal". The author remembers a time when it was said that pyrite "could not be produced experimentally". But once you know how to make it, it is not at all difficult to make pyrite! A long list of examples could be given.

The trend to give "data" which deliver foolproof right determinations has been developed e.g. since MURDOCH, DAVY and FARNHAM and SHORT, who tried in vain to get it by systematic etching, works now especially in two directions, indeed seeming to be extremely easy to recommend to the experienced observer:

1. The determination of hardness, at first as scratching-hardness (Talmage) and polishing-hardness, now especially as micro-hardness because it seemed to lead to quantitative accessible values.

2. The reflection-behaviour, already subjectively the most obvious characteristic, seems, regarding the modern highly developed methods, to be especially useful for quantitative data.

But both disappointed! The difficulties are here not caused by the technique of measurements but by the material itself: In the chemism where tiny, often almost trace-like components or likewise the pre-treatment can change the hardness completely. The same (in some cases at least) may happen very quickly with the reflection-behavior. In the polishing-technique e.g. already polishing in water or oil, polishing under high or low pressure can cause varying micro-hardness but also the reflection behavior (mostly by differently strong or quick tarnishing) can be influenced considerably. The variability may also be caused by the difference between "real-crystal" and "idealcrystal", where it results in various but not at the first glance always visible properties: in the hardness getting higher or lower (compare the behavior of technically pure (99.5%) Zn with the so-called 5-nines Zn (99·999%). The reflectivity is mostly higher the nearer it comes to the ideal crystal. If e.g. out of these reasons UYTENBOGAARDT & BURKE give in their tables for a surely in hardness not strongly anisotropic and besides that in its chemism rather simple mineral, such as rammelsbergite, a Vicker's-hardness of 368–1048, then this proves clearly that such statements are not very useful and that it is not possible to call them "quantitative". A remark "differs in surprisingly wide data limits" would express much more. The figures of the measurement values might be alright — but when from these a "mean" is taken, this is really rather "risky".

In the reflection behavior we have the same problem. The measurements can be carried out much more accurately than they are significant for the object. We do not know all reasons, why the values vary so strongly already in the same section with excellent fresh polish and exactly the same method. Chances play perhaps the same part as the natural pre-treatment (shearings, recrystallizations) or lattice deficiencies or minute admixtures etc. When it is said from standards which have been used for a long time (not any more), e.g. galena or pyrite, that they have always and everywhere the same reflectivity (of course, only as long as they were not distinctly tarnished) then this was surely a mild self-delusion. With ideal conditions, from the same deposit and with material not being zoned, differences of 4 units in an intermediate reflectivity, i.e. $> 8\%$ are not at all unusual. This seems to be small compared with the hardness-values but concerns in this medium range minerals showing strongly overlapping properties. Anyway, I think, the measurements of reflectivity may have a genuine chance in future if we compare statistical broadness (across the whole spectrum) — and if we are extremely careful! But today we are still very far from this goal.

For very uncommon minerals we should go again and again back to the powder-diagram and the microprobe. However, these too have their tricks. I mention powder-diagram on page 305, here only a few words on the microprobe. The microprobe or the microsonde, as it was first named by its inventor CASTAING, has given many valuable results, sometimes it has simplified the work greatly, but in many cases also shown that the facts are far more complicated than we at first assumed. — In spite of its invaluability it should not be forgotten, that the careful visual microscopic observation is still the primary! First of all, we must observe, that there is something to be seen at one place of the section, which is worthwhile for further investigation, then we can work with the microprobe!

Cases like the discovery of wairauite, CoFe, having been found accidently during the investigation of awaruite and without any suspicion microscopically will, due to time and cost, remain rare exceptions. — Some publications of careful work with the microprobe has only proved things which were already known to BERZELIUS and GUSTAV ROSE some 150 or 120 years ago. On the other hand, there still remain many things much more worthwhile to be investigated.

Now some remarks on literature. It seems to be surprising that my list of literature still mentions very ancient papers. First of all, I think, that this is a matter of gratitude to mention papers which were done with much care and delicacy some 50, even 100 years ago, and with apparatus nowadays looking rather primitive — with rather complete results which are today published as "brandnew"! Second, because it seems to be necessary to refer to the original paper. At the time being, papers have been published which mention no literature older than 1960 — concerning things of which Bd. I by HINTZE (1904) or DANA (1894) give detailed, sometimes better information. Of course, the cited references — after 1960 — had been taken from quotation of other papers. But it can be imagined what becomes of the really original statement of e.g. 1930, when it appears after 4 or 5 relays again in 1968. I could give examples, where the "new statement" differs exactly diametrally from the original one.

Now some comments to the reproaches, that I did not always use the names or the spelling given by HEY. Well, there are several reasons: Schapbachite e.g. has had priority to Matildite for about 30 years. Lievrite, which I use as well as ilvaite, is nearly exactly as old (both nearly 165 years), and the name was given by A. G. WERNER Heta*i*rolite, not heta*e*rolite, as it is nowadays mostly written, derives from ἑταιϱοζ, the companion, and is explained by the name giving author with the Greek word. Written with ae it suggests the Latin form hetaere, which means "girl friend" in a somewhat dubious sense. Jakobsite derives from the Swedish village Jakobsberg. As in Swedish a *c* does not exist (except *ck* for a double *kk* or in proper names) we have to write it with *k*, etc.

Finally, I should like to express my thanks to all those who, from the beginning of the forerunner of this book (SCHNEIDERHÖHN & RAMDOHR, 1932) until today, assisted me with advice and deed, with material, information, and support in questions of apparatus (I only mention the late Dr. H. FREUND), with proofreading, photographs and translations. Altogether, there will be several hundreds all over the world. — Special thanks I should like to say to HEINRICH LÄMMLER, who made during the last 28 years the excellent preparates of unsurmounted quality, and to Mrs. SOFIE-MARIE SCHINDLER, my secretary, who assisted me assiduously during the last 20 years.

Special thanks I should also like to express to the Deutsche Forschungsgemeinschaft for the generous support. Without this help my work would not have been possible, especially not in the time of being emeritus.

During the corrections of my new edition appeared a "Quantitative Data File" of N. F. Henry for IMA/Comm. in form of a 404 card-file. It came too late to use it in my book. I acknowledge highly the lot of new data, but I do not think that my preface should be changed.

PAUL RAMDOHR

For the English edition I must say my thanks to Professor AMSTUTZ who organised the first translation and selected the translators.

TRANSLATORS

K. A. BIEGMAN, Delft, Holland
E. N. CAMERON, Madison, Wisconsin
C. D. CAMPBELL, Pullman, Washington
G. S. DISLER, Toronto, Canada

A. B. EDWARDS, Parkside, Australia (†)
G. M. FRIEDMAN, Troy, New York, USA
G. FRIEDRICH, Aachen, Germany
H. FROHBERG, Toronto, Canada (†)
R. LA GANZA, North Adelaide, S. Australia
W. F. HAEDERLE, La Oroya, Peru
H. D. HOLLAND, Princeton, N.J.
H. E. KAPP, Toronto, Canada
H. KOBE, Auckland, New Zealand
L. KOCH, Sydney-Kensington, Australia
R. KOSER, Pullman, Washington
G. KULLERUD, Lafayette, Indiana
H. VAN DER LAAN, Delft, Holland
B. F. LEONARD, Denver, Colorado
G. J. NEUERBURG, Conifer, Colorado 80433
E. H. NICKEL, Ottawa, Canada (now Perth)
F. W. OSTERWALD, Denver, Colorado
U. PETERSEN, Cambridge, Mass. U.S.A.
G. M. RADISICS, Toronto, Canada
J. RIMSAITE, Ottawa, Canada
H. J. ROORDA, Delft, Holland
C. B. SCLAR, Bethlehem, Penn.
R. K. SOREM, Pullman, Washington
R. G. WAYLAND, Arlington, Virginia
G. WESTNER, Toronto, Canada
A. W. G. WHITTLE, Parkside, S. Australia
H. ZANTOP, Pullman, Washington (now Dartmouth College, Hanover, N. Hamp.)
R. A. ZIMMERMANN, Heidelberg, Germany

To all these gentlemen I owe my sincere gratitude. Professor G. C. AMSTUTZ gave also some suggestions for the new edition. My friend and old pupil, Professor A. EL GORESY was so kind as to read the proofs with me and gave some helpful advices.

ABBREVIATIONS

Abbreviations were avoided wherever possible, even despite the possibility of criticism. For physical and a few crystallographic data the conventional symbols are, of course, used.

n_ω or n_O, n_ε or n_E	— main indices of refraction in uniaxial crystals (ordinary and extraordinary directions).
n_α, n_β, n_γ	— main indices of refraction in biaxial crystals.
R_O, R_E, R_ω, R_ε	— reflectivity of uniaxial crystals.
R_g, R_m, R_p	— reflectivity for biaxial crystals ("grand, moyen, petit"!).
$\varkappa$, or $\varkappa_O$, $\varkappa_E$	— absorption index kappa.
#	— cleavage, or cleavage after . . .
$<$	— smaller than
$>$	— larger than
$\sim$	— approximately or similar
$\gtrsim$	— similar, but somewhat larger
∅	— on the average (or diameter)

INTRODUCTION TO THE GENERAL SECTION

INTERGROWTHS OF THE ORE MINERALS

The study of rocks and other mineral associations, especially of ore deposits, starts out from the purely descriptive. The composition of these materials, their geologic position in the most comprehensive sense, and their gross and fine structures and textures, etc., must be established. Further research must explain, first the "How"', then the "Why". The latter must comprise the *interpretation* of the association of materials and geologic position as well as each detail of texture and structure, even the smallest. On the knowledge obtained, one must build further and must generalize. Things of which the "Why" can be explained directly will serve in the explanation of others. In doing this the danger of circular reasoning will exist; criticism is therefore invariably necessary, self-criticism most of all. The interpretations must, if possible, be reached along several paths, and the validity of the various paths must be carefully considered.

Once these interpretations have been reached, it will be possible to generalize and draw further conclusions that will serve technology and industry. Minute details of mineral composition and texture which appear unimportant can then become of substantial significance.

Although in this volume the "general part" on intergrowths appears before the "special part" that treats the individual minerals, this is in part not justified; there is much which cannot be understood correctly without knowledge of the individual minerals. On the other hand, the textures are fundamentally independent of the components, so that to me the sequence followed appears logical.

The "general part" itself is further divided into four major sections. The first of these serves to facilitate understanding of all the other sections, and indeed of the special part also, and gives a *systematic presentation of ore deposits* as it has developed in the last 50 years. There is no doubt that these divisions, in spite of their physicochemical adaptation to the development of magmatic rocks, are not yet final, since the true sources of material are not yet known and the magma, in very many cases at least, is only the transporting medium. The metamorphic series has likewise become especially fundamental in connection with all relationships seen in ore-microscopy. This seemed important because until now the metamorphism of ores has in many cases been taken partly for granted, although strictly speaking, no concrete support for these conclusions existed. I think that this is true especially of deposits in the "deeper basement rocks". In spite of appreciation of the fact that high-rank metamorphism in ores easily produces primary looking compositions and textures, it must still be said that the sole fact that the associated rocks are metamorphic does not necessarily mean that the deposits are also metamorphic. They may be considerably younger. I have consciously avoided applying further here the nomenclature and lines of thought of B. SANDER (1930) and W. SCHMIDT (1932). Instead I have emphatically given precedence to the empirical.

The rather voluminous section on *ore intergrowths* is not anywhere near complete, and could not be even with very substantial expansion. Perhaps the works of the writer on Rammelsberg (1953a) or Broken Hill (1950a) alone show how diverse in form the textures of a single ore deposit can be and how differently they can be interpreted. Moreover, the complete description of a paragenetic group such as the Co–Ni–Ag–As deposits, which are fundamentally complete in themselves and seemingly have originated within a relatively small range of temperature and pressure, could fill volumes. It is impossible to achieve the desired division into *formal* and *genetic* in a consequent way! In order to avoid too much repetition, reference is often made to the genetic part while dealing with the formal, and even more often vice versa. However, it may be well to mention them at this time in view of the criticism that a strict order is missing.

It would have been desirable to have many more photographic illustrations. This, however, was not possible. References to figures in other parts of the book must compensate for this lack to some extent. Many of the *captions* of the *figures* were for this reason made rather detailed. They are an essential part of the contents of the book. I have acquired experience that this hint is necessary!

Also the subsection on the technical significance of microscopic investigations of ore intergrowths is not only not complete but is moreover only an indication of work already accomplished, of possible means by which problems in this field are to be handled, and of some aids to the work. In view of the scope of work of the writer, he is under the impression that, in this respect, he is not competent to furnish an unobjectionable and error-free account.

The last large section, *recognition of the genetic position of ore deposits* through the observations of ore-microscopy, contains, naturally side by side, things which result from macroscopic as well as microscopic investigations. The section must often refer to well known and acknowledged methods of investigation in transmitted light, to methods of physical chemistry, of geology, etc. Even though, according to a persuasive presentation by NIGGLI, the T and P conditions of the formation of a mineral deposit are not sufficient to characterize it (naturally apart from its composition!), but instead place and temperature of the separation of the ore fluids are also necessary for a description, still the immense importance of "geologic thermometers" is through these facts not questioned. The mineral association [Paragenesis], the host rock, and all the geologic circumstances will often furnish the additional criteria demanded by NIGGLI.

In the present state of knowledge, the section as a whole is incapable of mature presentation in every respect. The questions under consideration have too many aspects to permit them to be answered today, and in part they cannot even be correctly presented. One has only to think of how much has been written on the habit, structure and occurrence of quartz alone in just one group of ore deposits (the gold-quartz deposits), and he will not expect that he will find here all the answers that may possibly be found at a later time, for each mineral, each mineral association and each texture. The fact that the section repeats many things said earlier in the sections on the genetic scheme, or in the section on genetically related textures, requires no explanation.

GENETIC SYSTEMATICS OF ORE DEPOSITS

In the systematics of the science of ore deposits a certain unified scheme has been worked out in the twenties, especially through the classic works of GOLDSCHMIDT, LINDGREN, NIGGLI, SCHNEIDERHÖHN, and others. This scheme is firmly grounded in physical and chemical conditions and can be generally accepted.

Of course, we are still far from being certain and agreeing in genetic explanations of all ore deposits. In addition, further scientific investigations will lead to small alterations of relationships and sequences, and to expansions and condensations. Finally, the arrangement of many transitional types into the system is basically a matter of choice of the individual. However, all these qualifications will not influence the scheme as such in any significant way!

A greater problem is, where the ores come from! Were the ores first deposited in sediments as "an ancient mud" and then mobilized during mountain folding to magmatic or hydrothermal temperatures, or had they the origin from plate tectonics and therefore from the upper mantle, or were they products of magmatic differenciation of basaltic or intermediate or acidic melts — as accepted often as a proven fact in the first half of our century — or did the temperature only come from thick layer of overburden and were so actually a product of gigantic laterial-secretionary processes — all those possibilities and perhaps still more different ones may happen — the physicochemical scheme: Magmatic — pneumatolytic — hydrothermal temperatures, and all at high, medium or low pressures remains valid. Reworking of originally sedimentary deposits by hot solutions formed during folding, faulting, magmatisation becomes a more and more trivial experience.

Whereas in this sense there is far-reaching agreement (crisp ideas of outsiders have to be ignored here as well as elsewhere in science!), there is no such agreement on the question of the ultimate derivation of the metals finally concentrated in ore deposits. Derivation "out of the magma" or "out of the batholith" is often absolutely clear and undisputed; but whence comes the batholith, the magma as such, and whence does it obtain its contents, say of lead or bismuth or uranium ? It is a local, continuous, comparatively superficial cycle, that allows the metals, which perhaps were first fixed in the primeval solidification crust, to migrate into sediments and there to precipitate ? Does it then allow the sediments in the course of folding to be converted to magma which again produces ore mineral deposits, and so on ? Is the magma, the granite, a formation of a resurgent nature, of the kind associated with folded regions and its metal content only material that has been imported out of the inexhaustible depths of the interior of the earth and only guided through the activity of intrusions ? Finally, is the granite, or at least the larger part of its content of the "usual" elements, just as juvenile as the heavier metals brought with it? — We do not know! The questions are difficult to answer also because many times one answer, many times the other, may prove correct. Small ore deposits are surely often to be interpreted in the sense of the first question. Giant occurrences of definite metals in definite localities (e.g., the geochemical tin provinces of the East Indies or Bolivia) force an answer in the terms of the second question. There are peculiar associations that are recorded geochemically as facts but have yet to be explained by crystal chemistry or otherwise. Such are the numerous associations of lead + zinc, of cobalt + nickel + silver + arsenic + bismuth + uranium, and, to mention a rarity, selenium + lead + silver + copper + mercury + palladium. These associations argue that many superficial deposits formed near the surface have a deep source, very probably in the "upper mantle". We cannot pursue these problems further but must certainly pose the question, though otherwise we hold ourselves to tangible tasks!

When we reflect on the very numerous aspects of texture, mineral content, and total mechanism, e.g., of just the eruptive rocks alone, rocks which genetically correspond to only a small part of our system of ore deposits, and on how much had to be worked out in order that we could draw conclusions from their characteristics as to the origin and migration of materials, we will naturally not expect to be able to affirm, with the aid of ore-microscopy, nearly as much regarding ore deposits as a whole as we can regarding magmatic rocks with the help of silicate microscopy. The following text is therefore a first attempt, and surely its various parts are very different in degree of "maturity" and completeness.

The geological processes leading to the establishment of the great classification groups are briefly sketched. In some cases the corresponding mineral associations [Paragenesis], textures, and other characteristics and special features observed with the ore-microscope are taken up here, in other cases, these matters are discussed in later sections.

A. Meteorites

The separation processes, on the grandest scale, which played a role in the cosmic pregeological history of the earth, and by which the earth core, consisting mainly of metallic iron together with a thin sulphide-rich, and different intermediate layers of high-pressure oxides and a silicate outer shell were formed, we know only through investigation of the density of the earth as a whole, through earthquake waves, and through conclusions, by analogy, from the behavior of metal–sulphide–silicate melts in smelting processes. Only the uppermost kilometers of the solid crust of the earth, i.e. the outermost part of the silicate shell, are accessible to us. Of the earth's interior, we have only possible model samples in the form of *meteorites*, surely broken pieces of *foreign* celestial bodies.

The study and knowledge of meteorites and of their individual constituents is one far beyond the scope of this text, and an extensive and difficult special field in which the methods of ore-microscopy and metallography have long since been extensively applied, with regard to the majority of the opaque minerals taking part in the structure of meteorites.

B. Magmatic sequence

This uppermost part of the silicate shell is the scene and the direct source of materials for almost all ore-forming processes. The chemical composition of this shell will not differ greatly from the earth's crust, which has been more extensively investigated by CLARKE and WASHINGTON, GOLDSCHMIDT, BERG, SAHAMA & RANKAMA, WEDEPOHL and others through geochemical and statistical studies. How far it has also been the indirect source is, as I have explained above, impossible to say at the present.

The processes of formation of mineral deposits may be divided, in a manner quite analogous to processes of rock formation, into three great groups: *the magmatic sequence* (*B*), *the sedimentary sequence* (*C*), *and the metamorphic sequence* (*D*). On the whole, petrography and ore genesis become even more difficult to separate; indeed, we must learn to regard rocks unreservedly as a special group of mineral deposits.

I. PLUTONIC ROCK SERIES

If a body of magma of considerable size solidifies at considerable depth, thus with slow cooling and high pressure the formation of plutonic rocks takes place and with it goes the possibility of extensive processes of differentiation. These processes can be based on the separation of an originally homogeneous melt into several melts no longer miscible, on the separation of differently soluble crystal phases through crystallization, and on concentration of the volatile components. These processes broadly overlap one another in time so that separation into several molten phases, e.g., need only take place when the composition is already substantially altered through partial crystallization. Since the latter will be very different in individual cases, errors in interpretation or other difficulties can enter into a classification. E.g., I have come to the conclusion that quite similar titanomagnetite differentiates have in one case arisen through crystallization (especially in acid rocks), in other cases, surely the more frequent, (in basic rocks) through liquid unmixing.

If one makes a rather far-reaching physicochemical simplification, the processes involved in the sequence of rocks formed at depth can be presented through the t–x diagram of a system consisting of an easily volatile component (A) and a difficultly volatile component (B) under a constant high pressure (fig. 1).*

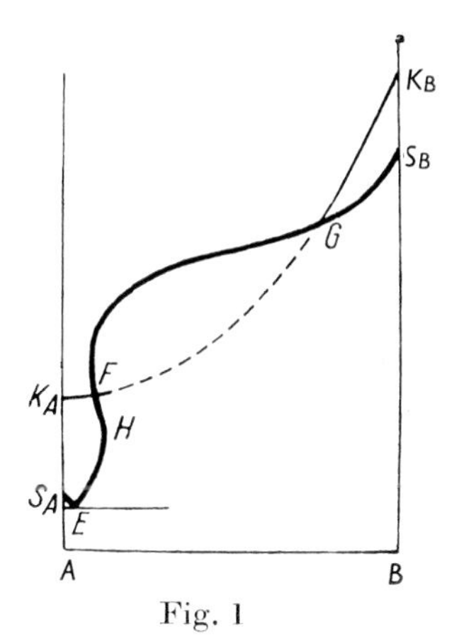

Fig. 1

Out of a melt rich in the refractory component B, as is normally the case, B begins to separate on cooling as soon as the melting curve S_BGFE is reached between S_B and G. The rest melt is therefore enriched in A. The temperature of crystallization decreases simultaneously until at G the melting curve (heavy line) intersects the critical curve K_B–K_A (thin line). The processes playing a part to this point are usually designated as "*intramagmatic stage*" or with a similar term. From point G, the melting curve is very flat; that is, with a slight decrease in temperature, much B separates out, and correspondingly the proportion of A grows at first very rapidly. Then, however, the curve bends back and the dissolving capacity of the liquid for B remains about the same; indeed, even increases somewhat. At F, the critical curve is again reached. The part of the curve G–F indicates the "*Pegmatitic-pneumatolytic stage*". The dashed por-

* Figures, diagrams and drawings are numbered consecutively.

tion of the critical curve between G and F indicates that here melt and gaseous state are principally no longer to be distinguished, and that one can with equal correctness designate the part G–F of the melting curve as the solubility curve of the gas for B. From F–E and S_A, then exists again a melt which, however, is generally called a *solution*. This portion of the curve designates the *hydrothermal stage*. Next, the solubility for B increases, then the "eutectic point", E, which for all practical purposes coincides mostly with the A-ordinate, is reached.

(a) *Intra-magmatic stage*

1. Magmatic differentiation through unmixing of fluids. At very high temperatures the magma dissolves large quantities of sulphides, alomst unlimited amounts of oxides, and traces of metals. With decreasing temperature, these separate out more and more as emulsions and later as small drops. Since the drops are heavier than the rest melt, they tend to be concentrated at the base of the magma body, along the borders through convection currents, or on xenoliths.

Thus we find enrichment:

Of *metals*: native platinum and iridosmium in dunites, whereby it must be noted that solidification points of ultrabasites with rather small K_2O-contents as "melt-breccia" can be lowered till 700°–800° (accord. to a paper by SEIFERT and SCHREYER (1966).

Of *sulphides*: pyrrhotite, pentlandite, and chalcopyrite ("nickeliferous pyrrhotite") in norites and gabbros, also in diabases, diorites, and pyroxenites. Small proportions of these are, by the way, contained in every fresh plutonic rock. Niccolite also occurs occasionally in dunites, along with the species just named and also alone.

Of *oxides*: chromite in very basic rocks, "titanomagnetite" (mostly a mixture of unmixed titanomagnetite with little ilmenite) in some anorthosites and norites; and perhaps also spinel.

Many of these types of deposits possess transitional forms, as could be observed especially by microscopic studies. Besides that often they appear, however, influenced by outside processes, so that explanation and classification are difficult. The segregation of platinum, e.g., is sometimes contemporaneous with the crystallization of chromite, where in the beginning, small crystals of chromite seem to have been the nuclei for small drops of platinum (fig. 301).

It is often even more difficult to classify the sulphide mixtures which belong to this group. These possess very low crystallization temperatures, so that they often are distinctly younger than the silicates and fill cataclastic cracks and cavities in them. Thereby they extend still far into the range in which pneumatolytic and even hydrothermal processes play a part, e.g., in the formation of hornblende out of augites. This detracts *not at all*, however, from their interpretation as being liquid magmatic differentiates, which is valid in most cases. A handsome illustration of an incompletely unmixed sulphide–silicate rock (taken from a transition zone a few centimeters thick), in which the consolidation of the sulphides closely followed that of the silicates, is shown in figs. 4 and 6. In other cases, however, there is no reason to doubt that there has been an important contribution on the part of pneumatolytic processes, in fact, even hydrothermal processes. Thus, we find in Sudbury and in several places in the Bushveld very coarse-grained parts which are unusually rich in pentlandite and chalcopyrite.

They carry their important content of platinum in the form of independent platinum minerals like stibiopalladinite, sperrylite, and cooperite; occasionally they cause contact metasomatic effects, and possess even higher contents of lead and galena, sphalerite, silver, gold, and palladium. In places of extreme differentiation (e.g., at the Frood Mine and at Insizwa) bismuth and tellurium minerals are also found, and occasionally also molybdenite, graphite, gold, and others. At Magnet Heights, Transvaal, the sulphide content of a later titaniferous magnetite intrusion has migrated into an earlier underlying layer, but certainly not truly as a melt, and has there caused replacement of the oxydic iron ores by forming the sulphides pyrrhotite, pentlandite, and cubanite. Skeletal ilmenite proves the nature of the process.

The peculiar *bornite* occurrences in the Ookiep area, Cape Province, are on one hand typical transition deposits; on the other hand they are, according to the work of LATSKY (1942), positively of intramagmatic rather than of pneumatolytic or hydrothermal origin.

Chromite has for a long time been regarded as a model example of crystallization differentiation, in many cases certainly quite correctly. However, more recent investigations, especially of the occurrences in the Bushveld, show that chromite has been intruded almost entirely pure into already solidified silicate rocks, doubtlessly in a molten or pasty condition. This implies a differentiation in the liquid state. In a certain sense, these occurrences can be included in the filter-pressed magmatic deposits.

Occurrences of *titanomagnetite* have been considered usually as crystallization differentiates, on the basis of old (but usually false!) observations that in the majority of rocks magnetite and ilmenite belong to the earliest products of crystallization. However, this view was occasionally contested (e.g., by ZAVARITZKY) and microscopic investigations show that the objections raised are indeed valid in most cases. That is, e.g., the case at Smålands Taberg, where surely olivine and plagioclase were already crystallized while titanomagnetite was still fluid (figs. 2, 3). Presumably there exists a sharp distinction between basic to intermediate rocks, the principal constituents of which (olivine, augite, and plagioclase) crystallize early and *before* the ore minerals, and more acid magmas, the main constituents of which solidify at temperatures several hundred degrees lower, and thus *after* the ore minerals.

2. *Magmatic differentiation through crystallization.* During the cooling of magmas, compounds crystallize out at a stage at which a considerable remainder is still fluid and easily mobile. This takes place naturally at very different temperatures, depending on the bulk chemistry and the external conditions. At this stage the crystals may separate from the rest-magma according to their specific gravity, generally by sinking, and so become concentrated. With further cooling, the proportion of the fluid rest melt becomes smaller and smaller, the magma mushy, and the possibility of segregation less. Even so, it can still be important over very long periods of times. Segregation naturally is more important in large magma chambers than in small ones. Concentrations of this type are very widely distributed. Every small lens of mica in a granite is an example. Most chromite occurrences, many deposits of corundum (fig. 631a), zircon, and monazite, and several, according to earlier views, most occurrences of titanomagnetite and ilmenite are among the economic deposits formed by segregation. Especially widespread are the peridotites and dunites, which as such, rarely have significance themselves as economic deposits but contain such deposits.

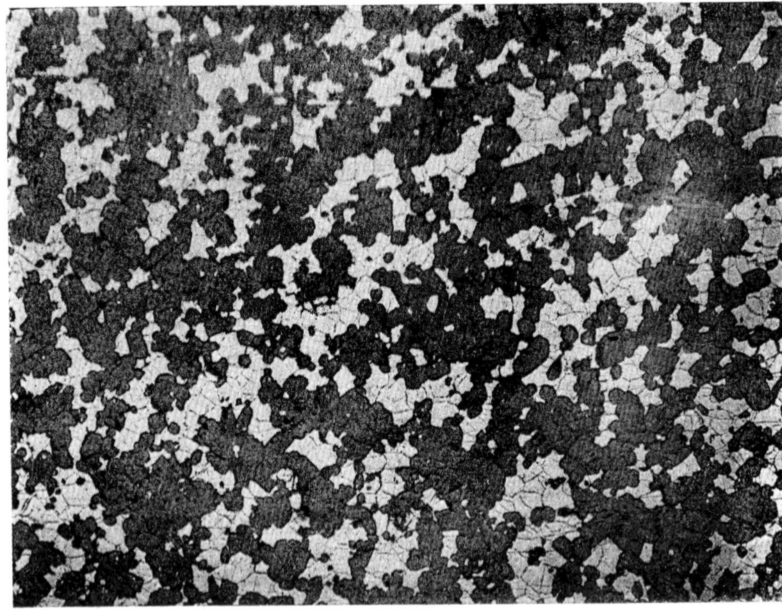

Fig. 2 8 × RAMDOHR

Smålands Taberg, Sweden

Titanomagnetite-olivinite with the characteristic, relatively small ore-mineral content of this type of deposits (cf. fig. 3)

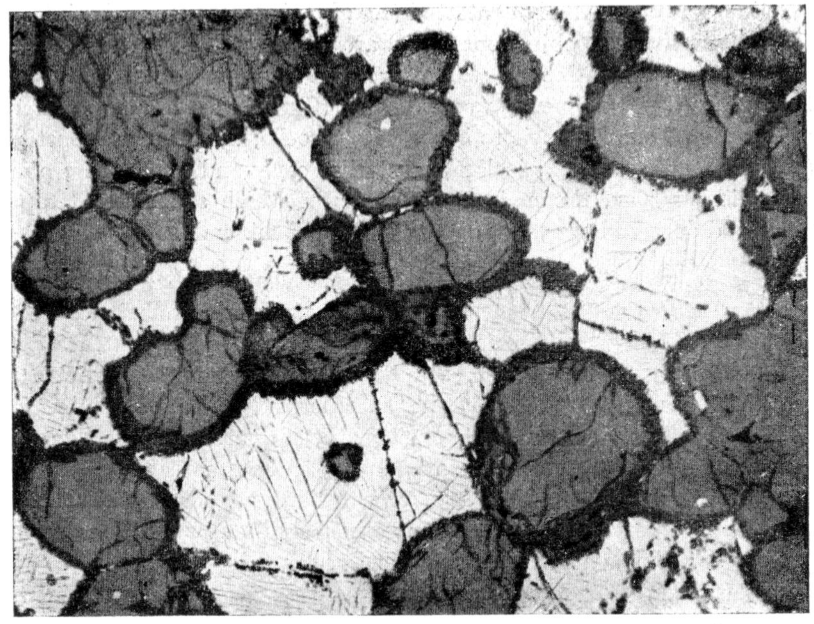

Fig. 3 80 × RAMDOHR

Smålands Taberg, Sweden

Titanomagnetite (light) with *spinel* lamellae // (100), cements idiomorphic *olivine* grains, all of which show reaction rims of *hornblende*

The textures of the ores in these groups are quite variable. Where the proportion of crystals was not very great, they often show an exceptional degree of idiomorphism. Chromite crystals, of course, like all spinel minerals, have strongly rounded corners and edges. Where, on the other hand, the crystals make up the main mass, they form an aggregate of polygonal to rounded grains with individual idiomorphic faces developed against small areas occupied by younger minerals. A widespread cataclasis is extremely common, especially in the case of chromite. In many crystallization differentiates, different minerals of about the same age and specific gravity are associated, so that the term titanomagnetite–spinellite or similar ones are used. It is impossible to give a generally valid age sequence of these minerals. Idiomorphism or xenomorphism gives an indication, but it can be misleading.

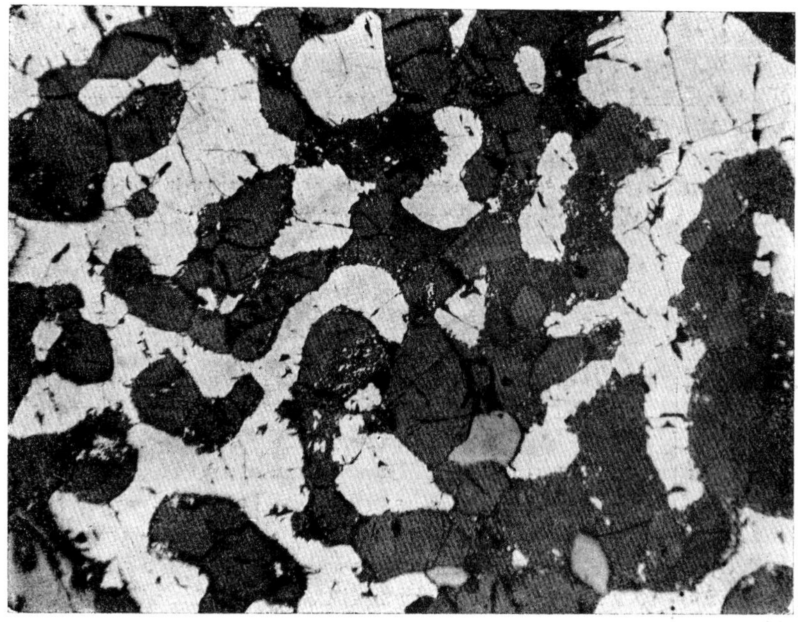

Fig. 4 80 × RAMDOHR

Mouat Mine, Stillwater Valley, Montana

From a border zone, about 1 cm thick, of the former silicate and sulphide melts. — White is *pyrrhotite* (with isolated "flames" of pentlandite); 2 or 3 grains of *magnetite* (light grey). Various shades of dark grey are silicates (augite, olivine, plagioclase). Both immiscible phases are markedly interlaced

3. *Main crystallization of the silicates.* Through the processes described above, the following are withdrawn from the magma: platinum, chromium, nickel (in olivine and in sulphides), a part of the iron, titanium, vanadium, and the bulk of the sulphur, also some magnesia and alumina. The main separation of the silicates now begins, the realm of formation of plutonic rocks in the strict sense. It is not the task of this book to deal with this subject in detail. Geochemically, by far the greatest portion of Al_2O_3, MgO, alkalis, (except Li and Cs), alkaline earths (except Be), SiO_2 as well as large quantities of Fe, Mn, and Ti are withdrawn (fig. 7), together with a series of elements camouflaged in all of the elements mentioned.

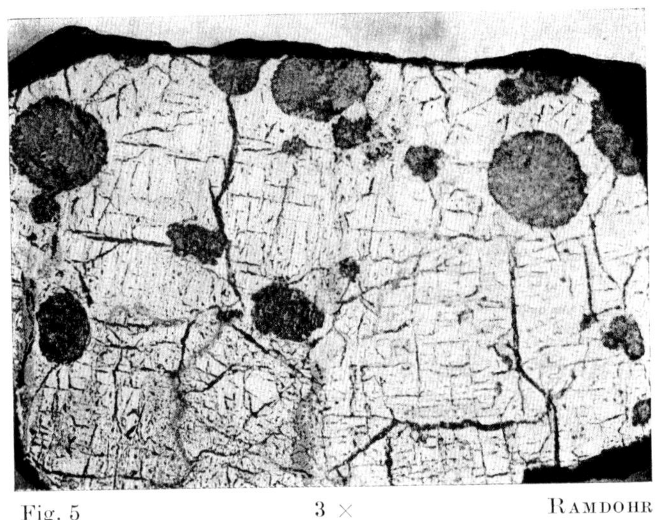

Fig. 5 3 × RAMDOHR
Nysteen, Norway

Nickeliferous pyrrhotite with gangue in drop-shape. The gangue in this case (unusual) remained molten longer than the sulphides

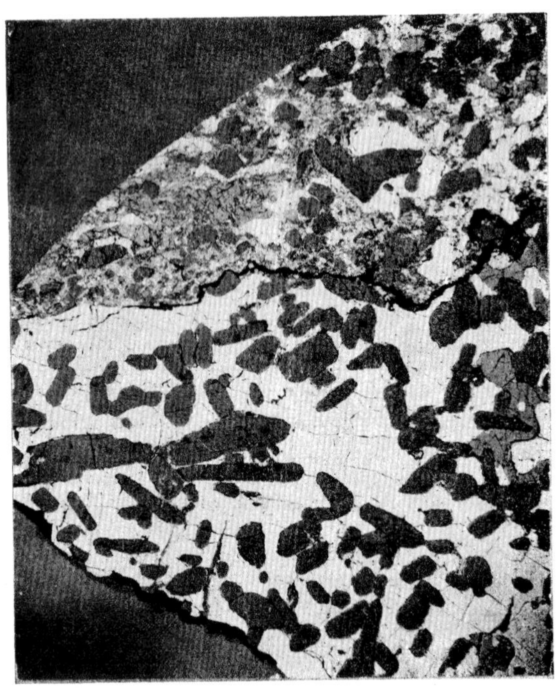

Fig. 5a 4 × RAMDOHR
Matooster Mine, East Bushveld, Transvaal

Pyrrhotite, white, coarse- and finegrained, *magnetite,* medium grey. *silicate,* dark grey, in part *olivine,* in part *orthopyroxene.* The section is just at the border of the sulfide-rich and a silicate- and magnetite-rich zone

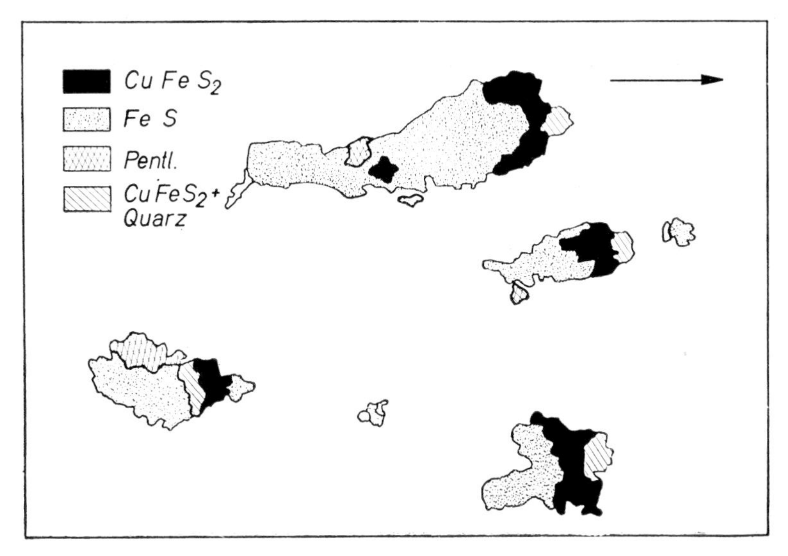

Fig. 6 $2^1/_2 \times$, drawing RAMDOHR
Sohland on the Spree, Saxony

So-called "pipe ore". The drops of ore minerals that were differentiated out of mineralized "diabase" are arranged in vertical tubes (perhaps following old strings of amygdules). In each one of these tubes chalcopyrite, which composes about half the contents, has uniformly migrated upwards, apparently in consequence of its greater mobility

Fig. 7 $8 \times$ RAMDOHR
Mariehamn, Finland

Gabbro, consisting of *augite* (light grey), *plagioclase* (dark grey), *magnetite* (white) and traces of other minerals. The parallel texture seen in the photograph is only a very local feature

The sulphide and oxide fraction is here in general very small but is of special interest for many reasons. The reader may refer to the works of NEWHOUSE (1936) and the writer (1940).

4. Deposits formed by filter pressing. Crystallization differentiates which sink into great depths can be remelted and intruded as independent magmas in the same way as segregations due to liquid immiscibility. It is easy to vizualize these processes. Nevertheless, the deposits included here exhibit certain peculiarities that are difficult to explain in simple terms. By far the most important group are the magnetic deposits of the Kiruna type. Derivation from titanomagnetite differentiates presents difficulties

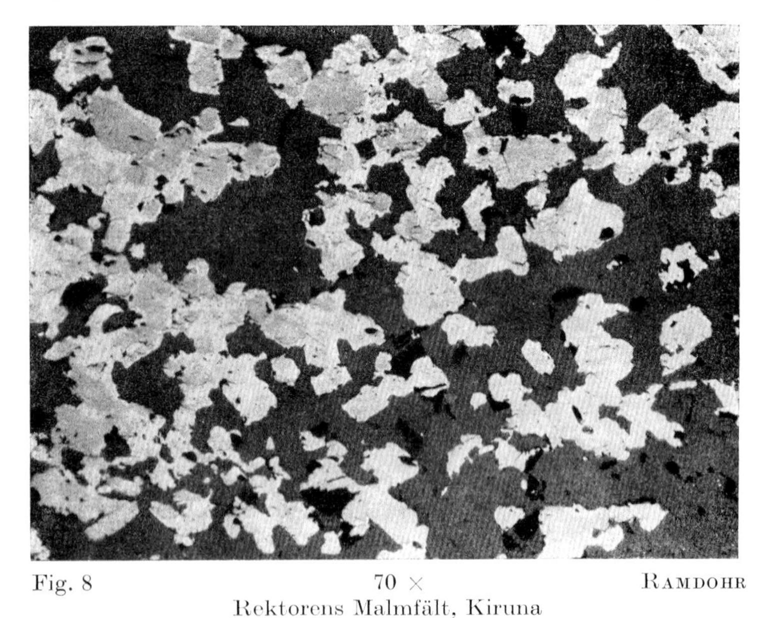

Fig. 8 70 × RAMDOHR
Rektorens Malmfält, Kiruna

Granular aggregate of *magnetite* (grey white) and *apatite* (dark grey). The white fringes around magnetite are *hematite*. The texture is of rather random orientation

because of the low content of Ti and the high content of P. Striking features are the almost unvarying association of such deposits with syenitic rocks, which for the most part show no pronounced characteristics of deep-seated rocks, and the introduction of chlorine and soda into the country rocks in the course of the intrusion of the ore. Texturally the deposits are featured by polygonally-grained masses of almost monomineralic ore (fig. 8), sometimes showing a somewhat fluidal texture in the arrangement of apatite. Larger quantities of specular hematite are in all probability due mostly to hydrothermal or metamorphic activity.

Experimental investigations of the Kiruna-type, proving the occurrence of immiscible melts of magnetite + apatite on one side and Na-rich silicate-melts on the other side, are published by R. FISCHER (1950) and A. R. PHILPOTS (1967).

An argument at least for the possibility of such deposits gives — against many objections and other explanations — the occurrence of El Laco in Chile, where in a sequence of very young lava flows, real lavas, composed actually exclusively of iron ores (martitized magnetite, primary goethite, and younger haematite), can be observed. Probably the original melt was mobile by the presence of a very high content of iron chloride.

As already indicated above, many chromite occurrences are best regarded as belonging to the filter-pressed deposits. The titanomagnetites of the Bushveld also belong here in large part or entirely. In contrast to crystallization differentiates in dunite, in which "chromite" is actually a chrome spinel, (Mg, Fe) Cr_2O_4, the chromite in filter-pressed deposits is very nearly $FeCr_2O_4$ (fig. 546).

(b) *Pegmatitic-pneumatolytic stage*

This stage extends over a very large range of temperature, and accordingly there exist extraordinary differences among deposits of this type. Variations both in mineral content and in type of country rock are, of course, very broad. Variations in country rock are often strongly marked both microscopically and macroscopically where actually only narrow physico-chemical deviations occur. The variations are so marked that one appropriately calls on them as a basis of major subdivision. We therefore distinguish pegmatitic deposits, pneumatolytic veins, contact-pneumatolytic alterations, and pneumatolytic impregnations. Between the pegmatites and the remaining groups there exists still a more definite difference, inasmuch as 1. pegmatites are the bridges between products of this stage and dikes of the plutonic rocks; 2. the content of the normal rock is still decisive; and 3. the "solutions", at time of intrusion, may have had still more the character of a melt than in the case of the remaining groups ("solutions", "melts", and "vapors" in this region are not sharply distinguishable). However, this does not at all mean that pegmatites on the average must have solidified at higher temperatures. In contrast to the pegmatites, deposits of the other three groups are influenced decisively by the behavior of the country rock. Exactly the same fluid solutions can produce a pneumatolytic vein (fig. 9) in a compact chemically unreactive rock, a replacement deposit in an easily replaced limestone, and an impregnation in porous tuffs. What appears chemically unreactive depends of course on the character of the solutions, which here were mostly neutral to strongly acid. It depends also, however, on the existing hydrostatic pressure and on the duration of the inter action, and also on whether easy access of solutions was provided through zones of tension or compression. Thus, especially in the geologically ancient deposits of sphalerite, galena, and chalcopyrite in Central Sweden, which were developed or reworked under high pressure, many different rocks are rather unselectively replaced, where transected by zones of disturbance. Naturally, even if we assume that the introduced solutions are the same, the products precipitated differ according to the reaction with the country rock. Pegmatites, pneumatolytic veins, replacements, and impregnations that are exactly the same in derivation will therefore show a different chemical composition. These differences are especially striking in the "gangue minerals" but are also quite pronounced in the ore minerals. E.g., there are certainly very large occurrences of magnetite among the contact-replacement deposits but none in pegmatites or pneumatolytic veins.

The use of the term "pneumatolytic" varies in some countries and differs always from its original meaning: depositions of volcanic exhalations. Here it should represent the highly mobile supercritical fluids, which may have transitions to melts of may exist independently — so having in part more the properties of melts, in part of hydrothermal solutions.

The terminology is in itself not always very appropriate, since the commercially most important mineral often gives the name even if present only in very small quantities. There are many types of transitions, and many deposits can quite readily be classified at will in different places in the chart.

A further problem here is the somewhat deviating position of the pegmatites in their character as rest melts. Geochemically speaking, they carry especially the elements, the incorporation of which into the minerals of the first and main crystallization stages is difficult because ionic radii are too small or too large (B, Be, Li–Y, Nb, Cs, and rare earths), whereas the remaining groups carry more such elements which are very mobile in the form of easily volatile or easily soluble compounds.

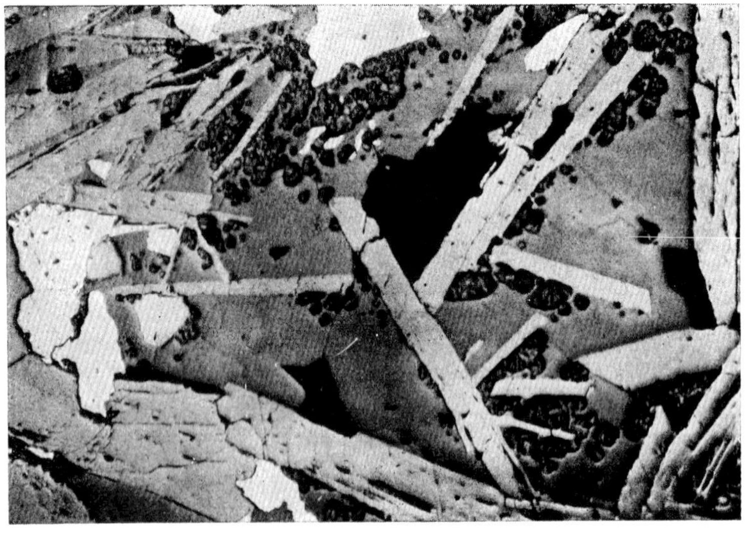

Fig. 9 8 × RAMDOHR

Colavi, Bolivia

Wolframite (hübnerite), in typical lath-shaped crystals with slanted end faces in *quartz* (medium grey), also *arsenopyrite* (white) and "*sericite*" (dark grey)

The *impregnations* can mostly not be distinguished from the "stockwork-like" networks of small veinlets; all stages of transition are recognized. Impregnations have been investigated, relatively rarely as yet, since most of them carry the commercially important minerals only in such low concentrations that exploitation is out of the present. They represent great reserves for the future; their thorough study and exploration is a task of the next decades.

On the way through the porous rocks, gaseous ore solutions are cooled rapidly where formation does not take place at great depths. A transition to hydrothermal impregnations is quite usual. That is the case, e.g., with the "protore" of the "disseminated porphyry copper ores", which are mostly assigned, not quite correctly, to the hydrothermal deposits. Where in the deepest basement rocks the temperature gradient is small, on the contrary, the character of mineralization can remain the same for a long period of time.

TABLE 1. Classification of pegmatitic-pneumatolytic mineral deposits

Pegmatites	Pneumatolytic veins	Contact-pneumatolytic replacements	Impregnations (Disseminations)
cassiterite	cassiterite + topaz ± tourmaline	cassiterite + borates	cassiterite + topaz + fluorite
wolframite + tin minerals (partly scheelite)	*wolframite*	*scheelite*	molybdoscheelite (widespread according to recent work)
molybdenite	*molybdenite*	molybdenite ± powellite	*molybdenite* (partly hydrotherm.)
gold + scheelite	gold + quartz + tourmaline	gold + quartz ± bismuth	gold + quartz + much pyrite
copper minerals	chalcopyrite + tourmaline	*chalcopyrite* ± pyrrhotite (and many others)	*chalcopyrite* tourmaline
sphalerite (rare)		*sphalerite* + *galena* ± chalcop. ± pyrrhot.	sphalerite + galena
magnetite + ilmenite	hematite + magnetite	*magnetite* + contact-silicates	magnetite + hematite
graphite	graphite	*graphite*	graphite
		platinum	
kyanite ± corundum	*sillimanite*		corundum
niobates + *tantalates*		pyrochlore	columbite
Li + *Rb* + *Cs*	lepidolite		
mica			mica
apatite	*apatite*		

(c) *Plutonic-hydrothermal deposits (veins, replacements, impregnations)*

At greater distances from the source of the ore, the temperature of the "solutions" decreases more and more, of course again to varying extents according to the local conditions. The solutions, up to now fluid (supercritical), pass below ~ 400 °C into the region of liquid aqueous solitions. At the same time their transporting capacity for the various constituents changes markedly in connection with the gradual transition of the transporting medium from acid to basic. Very frequently the temperature decrease is determined not only by the distance from the source of the ore, but also through decrease in its activity. Thus, we explain in many cases the change of mineral content in the youngest parts of a deposit, and the partial resorption and pseudomorphism of minerals that had separated earlier. If the solutions become warmer and more active through a revival of the magmatic activity, one can also observe the reverse. The term "rejuvenation" of the vein is here used. The "redspar" ("Rotspat") and bornite occurrences of the Siegerland described by SCHNEIDERHÖHN (1924) and others are excellent examples. Examples are also very widespread elsewhere, and are also summarized by SCHNEIDERHÖHN (1942) in diagrams. In borderline cases they can become similar in content or genetically to metamorphic deposits.

Other things being equal (type of country rock, thickness of cover, etc.) the further decrease in temperature becomes slower and slower. This causes a much more sluggish change of the conditions of mineral formation and correspondingly of the mineral content; in other words, the "primary" differences due to depth become noticeable only over much greater distances. Naturally, this applies especially to deposits formed at greater depths (e.g. the gold–quartz veins of Mysore, India, where the conditions have been exactly studied in the extraordinarily deep mines). In deposits that exhibit the transition to the "epithermal "or "sub-volcanic" types of deposits the effect is more rapid (Schemnitz, the silver-tin ores of Bolivia). There is no lower temperature limit of hydrothermal deposits. They merge imperceptibly with deposits of the vadose surface waters and "ground waters" of every kind. Such deposits have been termed "hydrous deposits" by many. The controversy about the nature and derivation of hydrothermal ore deposits far distant from magmas ("apomagmatic") is well known. In this connection, the type and quantity of the dissolved substances must, in my opinion, be considered. If excessively large aggregations of geochemically otherwise rare elements, like zinc and lead, occur at many places in the world, this fact argues decidedly for ascendent origin, even if all signs point to quite low temperature of formation and no ore-furnishing igneous rock is known. If on the other hand deposits carry, quite predominantly, minerals derived from the immediate country rock, then the rare occurrence of a single crystal of galena, as in the alpine veins, will not serve as any evidence for ascendent origin. Resurgent water, that is, formerly vadose water, can cause mineral associations with characteristics of rather high temperature of formation, if it is heated at great depth and then rises again.

In *form*, the hydrothermal occurrences are again veins, replacement deposits, and impregnations, according to the chemical behavior of the country rock, according to the presence or absence of cracks and pores, and the nature of the *ore-forming* solutions. Every solution is inherently capable of furnishing deposits of these various forms, but individual solutions show tendencies to lead to one or the other form of deposits; e.g., the *gold–quartz–pyrite* occurrences are mostly veins, the hematite deposits are mainly replacements, the pyrite-chalcopyrite-molybdenite associations are often impregnations. The alterations in the country rock are caused primarily through the action of hot alkaline waters. Introduction and removal of material are comparatively slight, even though there are occasionally very strong local alterations. The delimitation of the various groups, "formations" ("Formationen") to use the expression so much applied in early days, is again arbitrary to a great extent. In part it rests on commercial considerations, and it is complicated by transitions in various directions.

Gold–quartz formation: Quartz, pyrite, arsenopyrite, and some gold. In many cases abundant chalcopyrite, sphalerite, galena, pyrrhotite, and tellurides, locally also albite, apatite and others. Mostly as veins, rarely replacements; impregnations were already known and mined in ancient times (e.g. in Spain) and are now becoming important again locally.

Copper–arsenic–iron formation: Two main groups, iron-rich and iron-poor; they are connected, however, through transitions that are treated somewhat unimportant in the literature.

a) Chalcopyrite-pyrite group; in veins, replacements, and impregnations, of which all are of great commercial significance.

The large collective group of the *"instrusive kies-deposits"*- characterized by the association pyrite, pyrrhotite, chalcopyrite, sphalerite, arsenopyrite, by stock work, ore-shoot, and "drag-fold" forms, and by occurrence especially in basement rocks as deep-seated deposits, are sureley of very different origin and belong to the highly metamorphosed deposits. But their temperatures of formation or alteration cause many similarities to this group — in any case they are *not* intrusive. Very high temperature forms are frequent, however, and are closely linked with katametamorphic rocks. Many deposits originally of sedimentary origin have been so stronlgy altered through *high-grade* metamorphism that they are very difficult to classify. Strong alteration of silicate rocks is commonplace. Idioblastic forms and high-temperature index minerals are present everywhere.

b) Chalcopyrite–enargite–tennantite group. Characterized by marked decrease in Fe and therewith in chalcopyrite, and also by a great variety of species in the mineral assemblage.

Lead–zinc–silver formation: In general, galena, sphalerite, chalcopyrite (although sparingly), quartz, and calcite occur quite consistently. The silver is contained mostly in tetrahedrite, freibergite, and pyrargyrite. Several types especially rich in minerals (Freiberg, Přibram, Andreasberg) start to display transitional properties to the subvolcanic formation (fig. 9) or to the cobalt–nickel–arsenic–silver formation. Occurrences in veins are the most common, replacements and impregnations, however, are also of enormous importance locally.

Low-temperature lead–zinc replacements: They are mostly quite isolated, but nonetheless there are several transitions to the formations just discussed. More precise investigation may show such transitions to be rather numerous. Commercially important are almost always the replacement deposits but there exist also veins and possibly also impregnations. The type minerals are galena, sphalerite, wurtzite, pyrite, and marcasite, also lead sulpharsenides. SCHNEIDERHÖHN interprets many of these as "mobilisates" of older deposits ("secondary-hydrothermal"), surely often with good reason.

Cobalt–nickel–silver–uranium–bismuth–arsenic formations: Apparently a rather consistent group which occurs mostly in veins. They are exceptionally variable in mineral content, and commercially very important as a source of silver, cobalt, and nickel, also, especially before the great uranium boom for uranium. This formation is probably the richest in minerals of all, and especially exciting and troublesome from the standpoint of ore-microscopy (e.g., figs. 10, 281, 599ff.).*

* The Ag–Ni–Co-vein deposits have sometimes about the same general sequence of mineralisation: 1. — if present — $UO_2 + Fe_2O_3 + SiO_2$, 2. Silver and SiO_2, 3. Silver and carbonates. 4. Ni–Co-Arsenides, 5. Co–Fe-arsenides and sulfarsenides + Bi + Carbonates, 6. Cu, Fe, Pb, Zn-Sulfides + carbonates. But there are often recurrences and especially the silver compounds may occur in a rather broad field.

The temperature of formation is low, surely lower than the melting point of Bi, but higher than the inversion point α-Ag_2S–β–Ag_2S.

A remark of SHELELSKI & SCOTT (1975) that according fluid inclusions the temperature could go up to 510° is surely a gross mistake.

Tin–silver–zinc formation: This interesting and economically important group is proved recently as a composite. An older tin mineralisation, distinctly related to subvulcanic deposits, is in time followed by mineralisations with lead, silver, zinc, antimony, bismuth and copper. It is remarkable, that the younger mineralisation connected with sulphidisation alters the older one and forms species-rich associations of tin–lead, tin–lead–antimony, complex silver minerals, tin sulphides and so on. The deposits are known mainly from Bolivia and are there of Tertiary age, but they were surely also present elsewhere as near-surface deposits that have mostly been destroyed by erosion. The mineral content is quite variable within individual deposits and assumes many forms. Some veins also carry abundant bismuth minerals (bismuthinite and native bismuth).

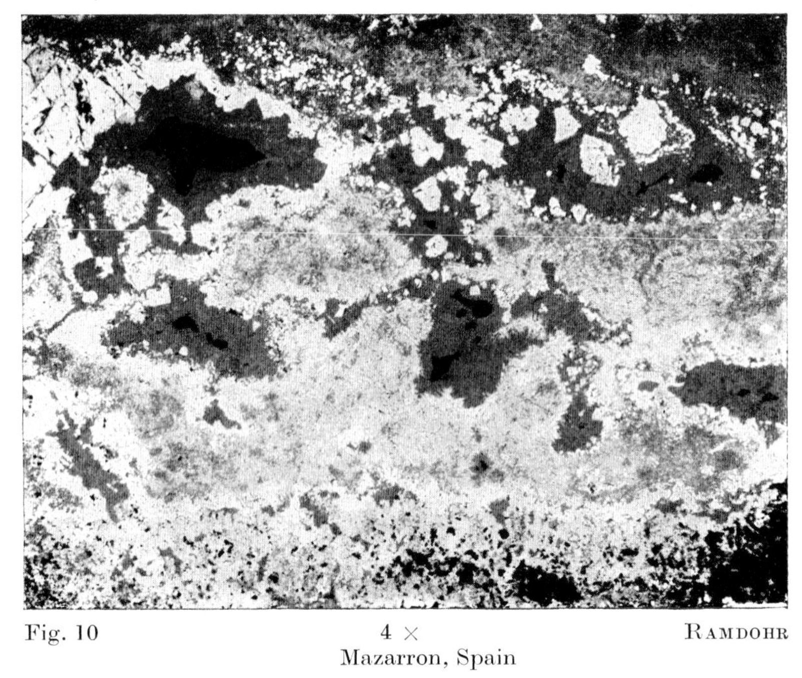

Fig. 10 4 × RAMDOHR
Mazarron, Spain

Concentric [shell-like] texture of an epithermal ore. Main constituent (besides quartz) is *magnetite*, which has formed from hematite. The large grains near top and the light rims are *galena*; below is a broad band of *sphalerite*

Chalcopyrite–siderite formation: Mainly notably lower in temperature of formation than the related chalcopyrite–quartz–pyrite occurrences. Veins can be important as sources of iron and copper, the replacements are copper-poor but are often very large sources of iron. Impregnations are common, but on account of their insignificance from the standpoint of mining, scarcely investigated. The mineral content is mostly monotonous, locally however, rather varied. Sometimes they go back to mobilized strata-bound deposits.

Sulphide–lean formations are important as sources of barite, fluorite, calcite, and celestite. Hematite veins and replacements retain only local significance, but became recently important through high contents of uranium.

II. SUBVOLCANIC SERIES

Parallel to the plutonic hydrothermal series is one in which the formation of the minerals has not taken place at depths of hundreds or thousands of meters but at depths of dozens of metres only. Fundamentally, they cannot be separated sharply from the first and in some mineral associations not at all. The series is marked, however, by certain distinguishing features, and by very rapid variation in mineral content with time and space as well as by the development of the individual minerals. Cases of "telescoping" are common in this series.

An absolute genetic connection with extrusive rocks is not a requirement of this series, nor is the series limited to veins. As observed in *nature*, however, a spatial connection with volcanic rocks and their tuffs is generally present, and metasomatism is unimportant. The occurrences are, of course, "epithermal" in the sense of the American terminology, but herewith nothing is said about the *temperature* of formation.

The concept of the subvolcanic series was introduced by SCHNEIDERHÖHN (1942) to eliminate a contradiction, troublesome over a long period of time, between European and American usage. In this matter, both of the following aspects received due recognition, the theory as developed in Europe by NIGGLI, and the greater amount of empirical data and practice chiefly accumulated in America. The "extrusive rock sequence" discussed in the next section is often linked according to its appearance with the subvolcanic series and in some cases is practically inseparable. The principal difference is, however, that for the subvolcanic deposits the source of ore is a deep body of magma — mostly that from which the volcanic rock itself originated — whereas in the volcanic series the source is the extrusive rock itself.

The multiplicity of forms of this type of deposit is decidedly less than in the typically plutonic suite. Many combinations of elements are missing entirely, others are very rare. The rapid variation in mineral content means that the content of valuable metals in the veins often increases rapidly, then quite suddenly diminishes. The life time of the mines is usually short at the present rate of mining.

Of special commercial significance are the *gold–silver deposits*, which in mineral content are a very inhomogeneous group. The names Siebenbürgen, Comstock Lode, Cripple Creek, and Guanajuato, with their mineral associations are names well known to every investigator of mineral deposits.

The *base metals* are relatively unimportant in this series, if we further indicate clearly that many gold–silver deposits owe their assignment into this series only to the commercial value of their products, whereas chalcopyrite, galena, and sphalerite are by far the predominant components of the ores. The Japanese "black ores" (Kurôko) carry predominantly iron sulphides. Parts of the deposits of Bor and Maidanpek are also to be grouped here. Chalcopyrite predominates, e.g., at Nacozari, whereas Pulacayo, Schemnitz, and several of the veins of the Freiberg region are distinguished by predominant lead–zinc ores. Very strong "telescoping" and formation of pseudomorphs are especially striking.

The *tin–silver–zinc* occurrences of Bolivia are in part to be included here. Cerro Rico de Potosi furnishes the classic example. Hundreds of additional mines show a variable series transitional to very characteristic members of the plutonic sequence in the narrow sense.

Antimony and *quicksilver ores* are precipitated at temperatures often so low that convergence of the plutonic, sub-volcanic, and volcanic-exhalative sequences occurs, and only a very exact and critical evaluation of the geologic relationships and mineralogic relationships permits a definite classification. Mobilisates of former sedimentary deposits are especially here sometimes well developed.

III. EXTRUSIVE SEQUENCE

If a body of magma freezes, not at depth but only after it has achieved a path to the surface of the earth or nearly to the surface, through a volcanic explosion or some other process, then the temperature drop takes place rapidly, and the outside confining pressure is slight. Freezing therefore takes place very quickly, there will be no time for gravitative magmatic differentiation, and finally the slight confining pressure will permit a violent, often explosive boiling off of the readily volatile components. Owing to this escape of gases, the freezing temperature on the average lies several hundred degrees higher than in the case of plutonic rocks. Everything operates, therefore, in such a way that there is substantially less opportunity for the formation of commercially important deposits. In the aftermath of the eruptive activity, at temperatures of about 100 °C, transporting capacity again becomes greater with the appearance of a liquid water phase capable of forming drops. The resulting springs will mostly rearrange the substances of the country rock. Introduction from below is slight. The diagram given on page 7 is altered very importantly through the boiling process (fig. 11).

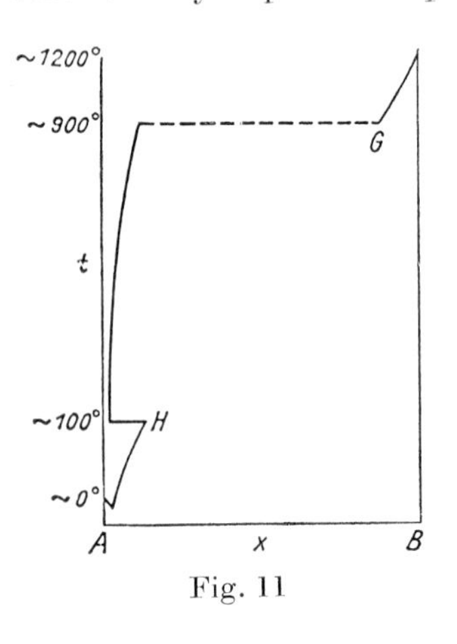

Fig. 11

The *intra-magmatic stage* begins as in the diagram of fig. 1, but very soon, at the instant of boiling, almost the whole portion of *B* separates out by rapid crystallization. The grain size is small — apart from phenocrysts brought up from depth — and, since opportunity for the formation of new minerals is lacking, disequilibria are often preserved in the mineral association as well as in individual components. A *pegmatitic pneumatolytic* range in the sense of the curve for deep-seated rocks does not exist at all, but hot exhalations of volcanoes can appear as mineral-formers in the corresponding temperature range. These minerals are not, however, products of a continuous process of freezing, but substances which crystallize directly out of gases, either as difficulty volatile reaction products of several readily volatile components (like hematite), or as condensation products of substances which crystallize at low pressures without passage through the molten stage (like sal ammoniac or iron chloride). They are, by the way, precisely the minerals for which Bunsen originally coined the expression "pneumatolytic", a term which is used today in a largely figurative sense. At high temperatures the quantity and number of these "volcanic-exhalative" minerals can be large; later they are limited to few species in smaller quantities. Temperatures at the beginning of this stage (*G*) are very high; in the "nuées ardentes" of Mt. Pelée, e.g., gold was melted, and in the exhalations of Katmai temperatures of 650 °C were measured directly. The end of this range, condensation (*H*), lies at the earth's surface, where the processes typically play their role at temperatures only very little above 100 °C and at extremely shallow depths, e.g., in the collecting basins of geysers at about 120° to 130 °C.

Volcanic exhalative deposits are only rarely of direct commercial significance. They are, however, important where exhalations go into the sea and thereby furnish material for large iron deposits and also many heavy metal sulphide deposits; as, e.g., suggested by the occurrences of pyrite, galenobismuthite and bismuthinite in Vulcano, Italy, described by BERNAUER (1939) and many others more recently. In the hydrothermal range, below *H*, ease of solubility is at first very much increased but decreases rapidly. This also brings about the formation of commercially important deposits. At the very lowest temperatures occurs the convergence with other sequences already mentioned.

Intramagmatic stage. Differentiation through gravitative separation of 1. crystalline, or 2. fluid phases plays no role. Although occasionally fist-sized aggregates of magnetite, ilmenite, and augite occur, as well as also small sulphide drops, there is no possibility of concentrations to render the deposits commercial (but — an exception — on page 14 the El Laco deposit).

Main crystallization of the silicates starts, in consequence of the small gas content, at temperatures which on the average are much higher than in the plutonic rocks, and ceases rapidly the moment the boiling off of volatiles occurs. The same mineral often occurs in several generations, which need not correspond to the contrast between phenocrysts and eutectic rest liquid in the usual eutectic diagram, but rather to different stages of boiling off. Of the opaque ore-minerals, titanomagnetite and ilmenite are predominant by far. Sulphides are regularly present in small amounts and in rare cases are somewhat more aboundant.

Volcanic-exhalative stage. The minerals occurring at this stage are very numerous and varied. Of the somewhat more abundant opaque minerals, magnetite, ilmenite, and hematite, all high-temperature products, should be mentioned; of those of medium temperature, e.g., tenorite, and rarely covellite, sphalerite, and galena may be mentioned.

Extrusive-hydrothermal deposits. With rapid crystallization at the surface, "replacement" is lacking and "impregnation" occurs only to quite a small extent. However, the country rocks of the deposits are often volcanic, either little cemented loose material or stronlgy fractured rocks, both easily accessible to impregnation, so that impregnations can extend farther and be more common than one would think.

The quantities of substances actually introduced are small, and where they occasionally may be greater, there is often a transition to "sub-volcanic" types. The most abundant mineral, pyrite, has drawn most of its iron from the country rock and its sulphur from emanations. Rutile or anatase are almost always developed from ilmenite.

Most of the deposits to be mentioned here are uninteresting from the view-point of ore-microscopy; they comprise solfataric sulphur and alunite deposits, occurrences of geyserite, boric acid springs, etc. Deposits of mercury and antimony ores as well as submarine exhalative-sedimentary deposits and gold contents in geysirites have greater significance to ore-microscopy.

Mercury and antimony ore deposits of the volcanic type are noteworthy in that they are almost the only magmatic deposits whose formation can be observed today in hot springs and geysers in regions of extinct and extinguishing volcanism. The very low temperatures of formation explain the difficulties of defining the group (see above), in part also towards spring deposits which bear no relation to the magmatic sequence. The deposits clearly belonging here are mostly unimportant economically.

The submarine exhalations already mentioned can furnish especially favorable environments for microorganisms and therewith create transitions to organic sedimentary deposits, as is discussed in the following section.

Mixed exhalative-sedimentary members. There is a series of deposits the history of which is rather accurately known, but the classification of which is difficult. These are the mixed exhalative-biochemical-sedimentary deposits already mentioned briefly in the preceding section. To mention an example, many types of diatoms build deposits of the so-called kieselgur (diatomaceous earth). These single-celled organisms live actually in every body of warm fresh water, and kieselgur can very well be formed in such bodies out of their siliceous tests. Actually kieselgur is formed in large amounts only where abnormally favorable environments exist. The prefered situation is attained in the warm ponds of water formed on the surfaces of recently chilled basalt flows, where dissolved silica in large amount is present. The extruded rock has therefore furnished the substance and the favorable environment. The life process causes the precipitation. Mobilisates of former sedimentary deposits are especially here sometimes well developed.

Into this group belong especially deposits of hematite associated with diabase and diabasic tuffs, where submarine exhalations from the diabase introduce the iron, and algae and bacteria in the seawater cause precipitation (fig. 140, 560). Since algae produce the same precipitation when the iron is introduced through swamp waters, misinterpretation is a possibility.

In recent years, the writer has come more and more to the conviction that in the formation of many so far controversial types of sulphide deposits these mixed exhalative-sedimentary processes played a decisive part. Recent deposits have been described especially by BERNAUER from Vulcano, Italy (1939). A late Tertiary deposit of this type is El Kebir at Cape Cavallo in Algeria, and a metamorphosed equivalent is the Rammelsberg deposit (1953a). It is clear without further discussion that in a given case close relationships to base-metal occurrences of the sub-volcanic type can exist.

Mixed members, especially "empty thermal springs". The discussion of the magmatic sequences is not to be concluded without mentioning that the physico-chemical conditions described here can also occur without the presence of a "juvenile" magma, so that mineral associations can arise which are in no way magmatic in origin. During great orogenies, parts of the crust formerly near the surface are brought to great depths and thereby into the realm of higher temperatures and pressures. The alterations that rocks and mineral deposits undergo thereby are discussed on pages 39—80 under "metamorphic deposits". However, large quantities of water are set free from the hydrated minerals of the former surface rocks by these processes, and these waters can operate in exactly the same way as the juveline thermal waters of the intrusive sequence. There are even pegmatites of this origin; T. BARTH has shown that, e.g., a large number of pegmatites in southern Norway belongs most probably to this type. At lower temperatures the number of "resurgent" thermal springs becomes greater proportionately to the truly "juvenile" springs. They all have in common, however, from the pegmatites of this origin on downward — that their mineral content is rather monotonous, and that the material is derived from the immediately adjacent country rock. These are "empty springs", economically and from the standpoint of knowledge of ore deposits. They can, of course, contain crystals of magnetite, ilmenite, and hematite, together with small amounts of chalcopyrite, sphalerite, and galena. The most important forms, ornamenting every mineral collection, are those of the alpine fissure minerals.

For the sake of completeness, it must be mentioned that an increasing number of investigators of ore deposits, beginning with LOCKE and BACKLUND, no longer recognize juvenile magma but only "resurgent" magma and derive all magmas* out of the remelting of sediments. According to these workers, the introduced materials of the metallic fraction of the magmatic sequence in the last analysis go back to organically enriched heavy metal contents of remelted swamp sediments. It seems to the writer just as wrong to ignore the *possibility* of the interpretation accepted by the two investigators and later many others as to consider this interpretation to be the *general* rule.

C. Sedimentary sequence

It is difficult to find a general classification for the magmatic sequence of deposits, because transitional deposits occur everywhere and in every direction, nonetheless some basic principles are available that are well founded in physical chemistry. These are given, first, through the decrease in temperature in the consolidation of eruptive rocks and, second, through the important pressure differences between plutonic and effusive rock sequences. In the sedimentary sequence these basic principles are looked for in vain. The temperature range here is determined by the narrow climatic changes, the pressure range by the atmospheric pressure or the small hydrostatic pressure in shallow bodies of water. Much more effective agents here are "fortuitous variables", like the amounts and the seasonal distribution of precipitation, type of flora and fauna, strength of wave action, gradients of streams, and so on and so on! Classification will therefore be left, to a great extent, up to the choice of the individual, and still more frequently than before, deposits will be found that can be classified only forcibly if at all — these will in most cases not be considered.

The source of material of the sedimentary deposits is the weathering rock of the continents, in part also material of the oceans and the atmosphere. The process of weathering is in its own way extremely complicated and influenced by very many factors. However, it permits an easy distinction between pure or prevailingly mechanical alterations and those of a chemical nature. In both cases, enrichment can take place in such a way that some material is removed and the rest becomes a pure concentrate;or the material which is carried away, becomes enriched often by many indirect processes. Thereby one possibility for classification is provided which leads to a series of groups very unequal in size.

I. CONCENTRATION OF MECHANICALLY WEATHERED MATERIALS

Rocks are mechanically broken and diminished in size by rain and drought, strong frost and heat, and by the grinding action of gravel and sand; the latter are set in

* Besides the basic ones.

motion by wave action, running streams and glaciers or wind. The product can become cemented into a sediment without essential changes, but this is rare. Much more frequently the separation of grains of about equal size takes place according to specific gravity and according to grain shape. Heavy mineral grains will be transported less far by surface water or wind than light ones; round grains less far than tabular ones, and large mineral grains less far than small ones.

In this way the following sedimentary rocks are formed, which are distinguished according to grain size as conglomerates and breccias, coarse sandstones, fine sandstones, clays, and so forth. Economically and geochemically more important is the separation of heavy and light constituents. This is true, naturally, only for minerals which are chemically resistant, hard or at least tough, and different in specific gravity from accompanying minerals. Minerals heavier than common rock minerals are gold, platinum, cassiterite, chromite, magnetite, monazite, garnet, and zircon — the typical "*placer minerals*" (figs. 12, 13, 14, 262). Lighter than other common rock minerals is, e.g., the enormously widespread quartz. It is therefore transported farther and furnishes clean quartz sands or sandstones. From the standpoint of ore-microscopy, all these occurrences, are of relatively little interest apart from the cemented and altered placer deposits (Witwatersrand!). Where unconsolidated materials are to be treated for polished sections, the processes are the same as for beneficiation products.

II. ENRICHMENT PROCESSES IN CHEMICALLY WEATHERED MATERIAL

While it is true that mechanical disintegration without a certain degree of contemporaneous chemical decomposition hardly ever takes place, the predominantly chemical alterations are especially important. Many subdivisions are possible.

a) Chemical decomposition consists of the taking up of water of hydration and colloidal water, of carbon dioxide, of nitrates, and of other substances, without marked removal of material. "Soils" originate out of ordinary rocks. The study of these is a special science, soil science.

b) Weathering proceeds in the same way with the taking up of water, etc., and with the formation of colloids, but of the compounds formed some are soluble in water only with extreme difficulty, others are very soluble and are carried away. The fraction remaining behind can form "residual deposits" very important in size and value. Classified as such deposits would be mainly the clays and kaolins, the tropical laterites, the bauxites, iron ores of the Mayari type, nickel ores of the New Caledonian type, residual manganese ores, and many phosphate and apatite deposits.

III. PRECIPITATION OF DISSOLVED SUBSTANCES ON THE CONTINENTS

The "soluble" compounds formed through weathering are soluble to very much different extent. Some are very easily soluble in pure water, others only with difficulty, others yet are transportable only in the presence of other substances in solution, either as double salts or as colloids. This produces a great variety in the type and

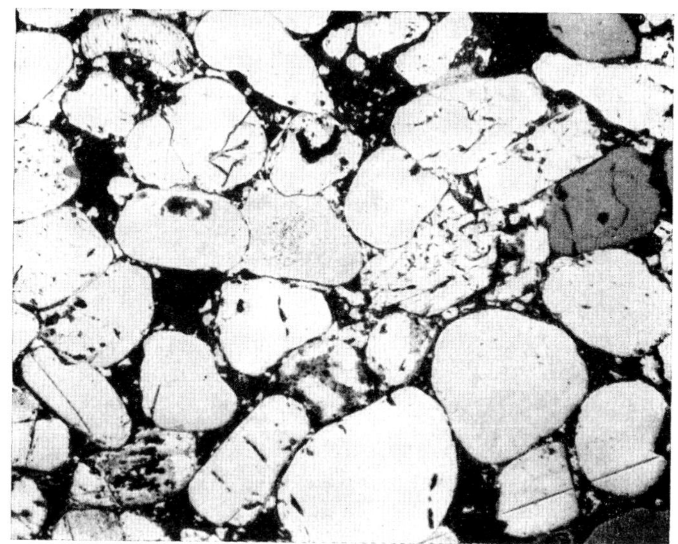

Fig. 12 150 × RAMDOHR

Bothaville, Orange Free State

Ilmenite sandstone (Ecca beds of the Karoo formation). Even though practically no tectonic forces have operated here, the thick disc-shaped grains (oval cross sections) often crush each other. This rock contains > 70% ilmenite, as well as zircon, monazite, rutile, chromite, but no magnetite or hematite (cf. figs. 144a, b)

Fig. 12a 35 × RAMDOHR

Ventersport-Mine, West. Witwatersrand, Transvaal

"Pyrite-sandstone", tiny pebbles of *pyrite* (white) of different, but nearly the same size, beside *chromite*, light grey, *zircon*, darker grey, and other minerals. The matrix, nearly black, is quartz

Fig. 13 80 × RAMDOHR

Pronto mine, Blind River, Ontario

Grains of *uraninite* (not well rounded), in part with traces of crystal forms, accumulated in basal conglomerate of the Huronian with rounded *pyrite*. The pyrite grains have been in part corroded after disturbance of their lattices by radiation

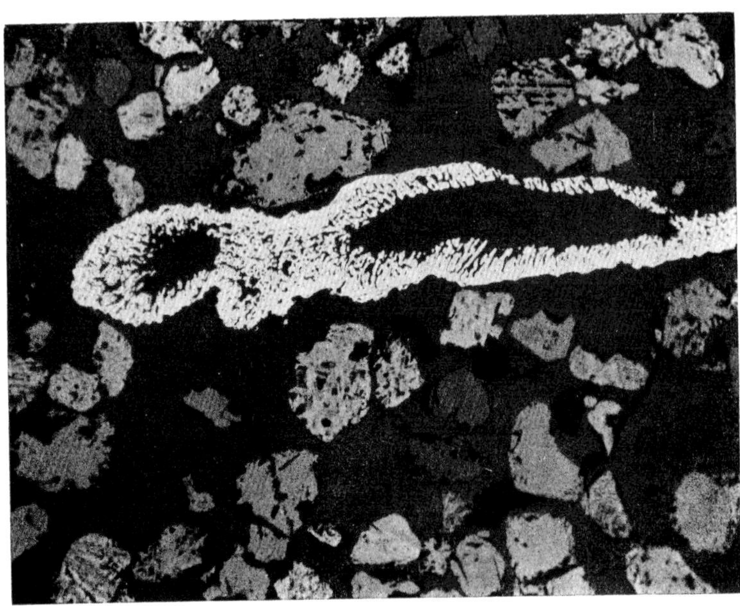

Fig. 14 80 × RAMDOHR

Ohlapian, Siebenbürgen

Section of a concentrate from a gold placer. Abundant grains of the "*black sand*" assemblage: magnetite, ilmenite, titanomagnetite in all forms of development, *monazite*, and *garnet*. One large grain of *gold*, the peculiar shape of which, occurring many times in the same section, probably to be explained as the product of decomposition of a gold telluride

place of precipitation. The water that falls in desert regions and is soon evaporated occupies a quite different position from waters which carry their dissolved mineral substances into the ocean or great inland lakes. The waters of the deserts — water of the humid regions at some time finally does reach the ocean — are apart from the times of rare rainfalls, very rich in dissolved salts and can therefore attack and carry away compounds otherwise insoluble. Manifold conditions (e.g., hydrolysis, exchange reactions, and complete evaporation of water, effects of vegetation or their decayed remains) can cause precipitation. Two groups belong here: 1. The *heavy metal occurrences of arid alluvial basins*, among these especially the copper ore deposits of the "redbed type", similar occurrences of argentite, the vanadium and, by far the most important, the uranium deposits in desert sandstones, and others. Also, a part of the lead and zinc nodule ores may belong here. 2. *Terrestrial salt deposits* like the saltpetre efflorescences of India, perhaps also parts of the saltpetre deposits of Chile, and many gypsum occurrences, etc.

IV. PRECIPITATION OF DISSOLVED SUBSTANCES IN THE SEA OR IN LAKES AND SWAMPS

4a. The salts that reach the ocean or the larger lakes are either so easily soluble or present in such slight concentration that special processes are necessary for their precipitation. In the sea itself, *inorganic chemical* precipitation is slight. However, it becomes large, and is the cause of the development of magnificent salt deposits in lagoons which either are gradually closed off completely from the open sea, or have temporary, or only weak perennial influx. In *inland lakes* in desert regions, these inorganic chemical precipitates can attain *comparatively* greater extent and much more variable composition than oceanic salt deposits. The manifold borate, soda, and mirabilite lakes are examples.

4b. Precipitates caused by activity of organisms have greater significance and endlessly greater variability. These are often naturally combined with inorganic chemical precipitates and clastic sedimentary deposits.

1. The most widespread members are the *marine limestones and dolomites*, in the origin of which innumerable animal and plant groups are involved; also the siliceous rocks, e.g., those formed by radiolaria and sponges, and the marine *phosphate deposits* belong here.

2. Of special economic importance are the *marine occurrences of iron ore* and rather similar *manganese ores*. These, too, occur almost never as pure types and have therefore been interpreted and classified from different viewpoints, which is perfectly within their nature. First of all to be mentioned are the oolitic ores, which consist essentially of limonite and Fe-silicates, many times also of hematite. Either associated with these or alone are oolitic ores of siderite and iron bisulphide if the deposits were formed under reducing conditions. The corresponding oolitic manganese ores consist almost always of "pyrolusite" (figs. 15, 15a, 15b).

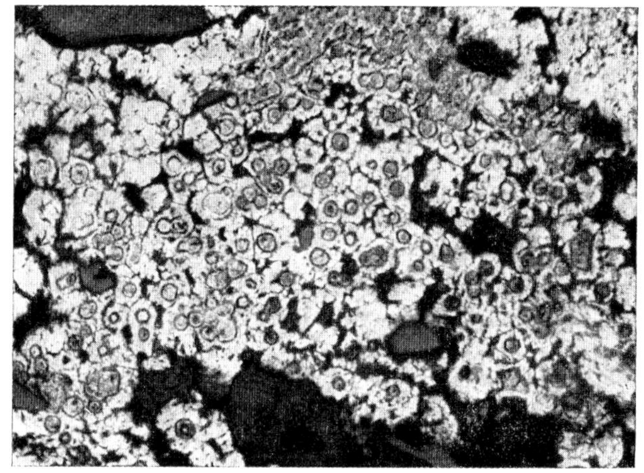

Fig. 15 170 × RAMDOHR
Nicopol, U.S.S.R.

Oolitic texture in porous *pyrolusite* ore. These "small-scale oolites" are certainly not true oolites but strongly resemble the large oolites from the same locality

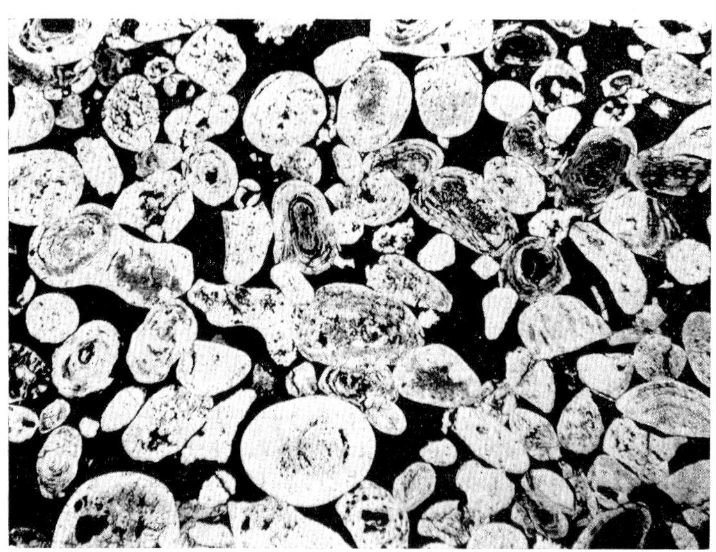

Fig. 15a 5 × RAMDOHR
Wabana (Pyrite layer), New Foundland, Canada

Aggregate of surely primary oolites of *pyrite* in very different forms. The explanation occasionally given, that the forms are pseudomorphosed hematite- or limonite-oolites is wrong, because definite "pyrite bacteria" (framboids) often form the starting inclusion

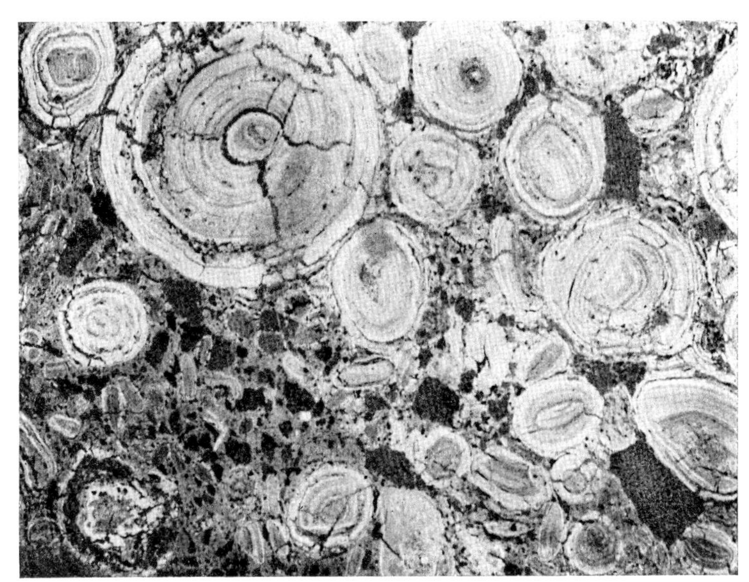

Fig. 15b 8 × RAMDOHR

Nullagine, W. Australia

Pisolitic iron ore composed by *hematite* (brightest), a little *lepidocrocite* (darker) and *goethite* (medium grey). A few grains of gangue (dark grey)

Fig. 15c 30 × RAMDOHR

Rammelsberg, Harz

"Banderz", prototype of a strata-bound sulfide deposit. The ore is essentially a rhythmical sequence of bands of argillaceous- and ore-rich strata. The last one contains especially pyrite, white, and sphalerite, dark grey. The argillaceous material contains partly coaly substance. Each layer contains actually all components but in very different quantity

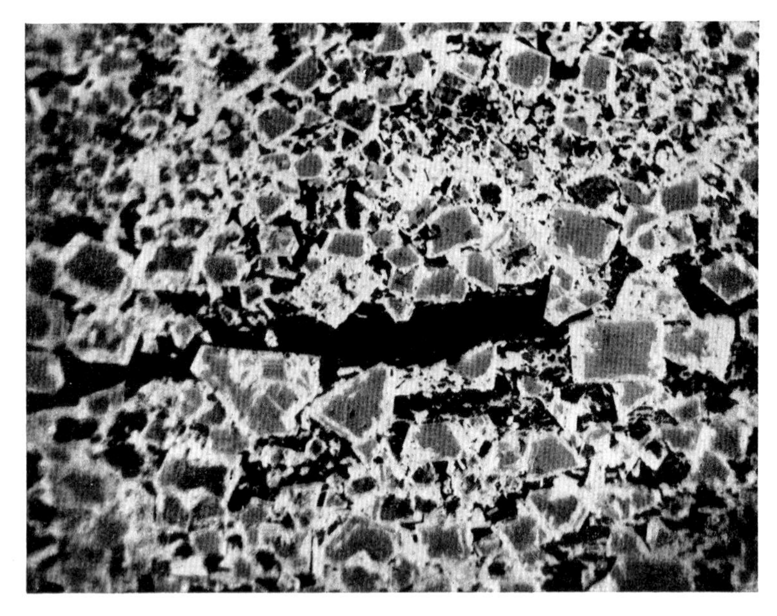

Fig. 15d 250 × Imm. RAMDOHR

Niningara, W.-Australia

Sedimentary but metamorphosed magnetite ore. The idioblastic *magnetite* is covered with a crust of *maghemite*

Manganese and iron ore sediments in very old formations hold a special position. One cannot express an opinion on the origin of these with full assurance, for want of similarities to recent deposits. In part, however, these deposits have an extraordinary significance. Both show alternating layers, often very fine interlayering with quartz and chert. In accordance with their age, they naturally are often very strongly metamorphosed.

3. *Sulphur bacteria* live in waters which are poorly aerated, or in those in which lack of oxygen and reducing conditions prevail through abundantly present decaying organic substances. They can produce sulphur through their life processes, partly as elemental sulphur and partly as hydrogen sulphide. The sulphur which was originally derived from the sulphides of eruptive rocks reaches the sea water in the form of sulphates or is present as primary sulphate of the sea water, and is thus brought into a form in which it can react with solutions of heavy metals; predominantly iron with subordinate zinc, lead, copper and other metals. There it can form sulphides again. Deposits of the sulphur cycle of this type are the bituminous pyrite-rich alum shales, the Mansfeld Kupferschiefer and similar deposits, and finally a part of the occurrences of native sulphur. Other also economically important deposits of elemental sulphur are derived from sulphates through reduction with organic substances, however, without direct operation of life processes of organisms, but instead through the influence of warm water, mostly together with bituminous material (fig. 15a).

4. Diatomaceous earth, infusorial earth (Kieselgur), and tuffaceous limestones are formed in lakes and in spring pools under the direct influence of lower or higher plants.

5. Plants often play the deciding role, also in precipitation of sea, pond, and bog iron ores just as in their solution and transport.

6. Temporally and locally limited are cases, where heavy-metals, esp. Pb, Cu, Zn, Sb, Hg and W are exhalatively or hydrothermally (both being at high water-pressure not more distinguishable) taken up by sea-water during of other plate migrations can be precipitated strongly following one horizon or rhythmically. A long since known example, at first appearing to be mysterious, is the "PbS-Bank" in the Keuper of the German Triassic. In the meantime, e.g. many scheelite-occurrences have become known in the Silurian of the Alps, in the Precambrium of East-Africa, in many PbS-occurrences with limestone, and in stibnite-ores in Bolivia of similar formations. The sulfides of the rift-valley of the Red-Sea are the most important recent examples. Since such precipitates are often strongly metamorphosed and redeposited, their recognition is sometimes difficult and disputed. An open question is still the provenance of the metals (upper-mantle or reworking of old deposits).

V. DEPOSITS OF COAL AND PETROLEUM AND OF THE MATERIALS GENETICALLY RELATED TO THEM

Treatment of the occurrences of peat, bituminous coal and anthracite, of the variable occurrences of petroleum and natural gas, of the bituminous shales, oil shales, and oil sands, is far beyond the scope of this review. The same is true, obviously, of the discussion of the transformation products like asphalt and other substances. On the other hand, the separation of iron carbonate as "white iron ore" is especially mentioned — it is closely related genetically with peat and coal formation — because these deposits in the coal-bearing facies of the carboniferous and other geologic periods have had considerable significance and furnish the "black band" and sphaerosiderite occurrences. They also contain small, and sometimes considerable amounts of sulphides.

VI. THE ZONE OF OXIDATION AND CEMENTATION

It is appropriate to treat the weathering and enrichment zones of sulphide ores separately, although they are closely related to group 2b of this section. This is actually done in probably all books on mineral deposits.

a) In the *zone of oxidation*, i.e., in that depth range in which infiltrating oxygen-bearing waters or, along open channels, air has direct access, the sulphides are destroyed at very different rates. Finally, however, all are destroyed with the formation of water-soluble sulphates (Cu, Ag, Zn, Fe) or insoluble sulphates (Pb), often with free sulphuric acid. These processes are strongly modified by climate, the permeability of the country rock, and its content of acid-neutralizing components. Limonite is almost always produced, often copper carbonates, cerussite, and anglesite. Through armoring with difficulty soluble products, relicts of considerable size can remain. The oxidation zones usually exhibit highly porous masses, in which one can infer the former mineral composition from the shape and kind of cavities and the microscopic relicts. The dissolved sulphates percolate away and migrate vertically or laterally, often quite far. In the *zone of cementation*, that part of the primary deposit where groundwater prevents access of atmospheric oxygen, the chemically nobler elements are precipitated out of the dissolved sulphates by the less noble metals of the primary sulphides. Especially common is precipitation of copper by iron and zinc, of silver by copper, iron, zinc, lead, etc. Phenomena of replacement are produced – the enriched or cementation ores. The range of the cementation ores is limited. It can amount to a few tens of centimeters or to hundreds of meters. For the ore-microscopist, the knowledge and correct judgment of these ores in contrast to the rich primary ores is of very special importance. Gold, cobalt–nickel, and uranium ores, etc., also show characteristic oxidation phenomena, yet their enrichment zones, although present, are not immediately to be compared with those of the elements just named.

b) Many descendent deposits are closely related to the cementation zone. In these, also, substances taken into solution through weathering are precipitated again after short transport. The external forms of these deposits can have the full character of hydrothermal vein deposits. Inward, the deposits show close relationships to eluvial-chemical enrichments. Descendent veins of manganese ore minerals, strontianite, and barite, and some gypsum deposits, may be mentioned as examples.

The metamorphic sequence is treated in detail in a special section of this book, so that it is excluded from the present discussion, although other points of view are particularly emphasized in the special section (cf. p. 34 etc.).

For the explanation of the tables of mineral occurrence, the following remarks are pertinent:

1. Only the more common or paragenetically important ores are included, not those that are quite rare or those that are known only from one or very few deposits and are perhaps of uncertain position. Minerals always occurring together are frequently combined.

2. The symbols ...., ____, ▬▬▬, █████, indicate the relative distribution of the individual minerals. Nothing at all is said about the absolute quantities. Where no indications are given in any column as to whether the mineral occurs in greater or smaller amounts than other minerals, this means that one cannot say anything yet about the principal place of formation of the mineral.

The symbol ____ indicates occasional, mostly also sparse occurrence; the symbol ▬▬▬ means widespread, but present in small amounts or present in larger amounts only in a few of the deposits involved; the symbol █████ means occurring in many or all deposits, and in larger amounts; the symbol .... indicates very rare although not entirely missing.

In order not to endanger the clarity of arrangement of the table too much, many deposits standing somewhat apart had to remain unconsidered (Franklin, Långban) or had to be combined with others.

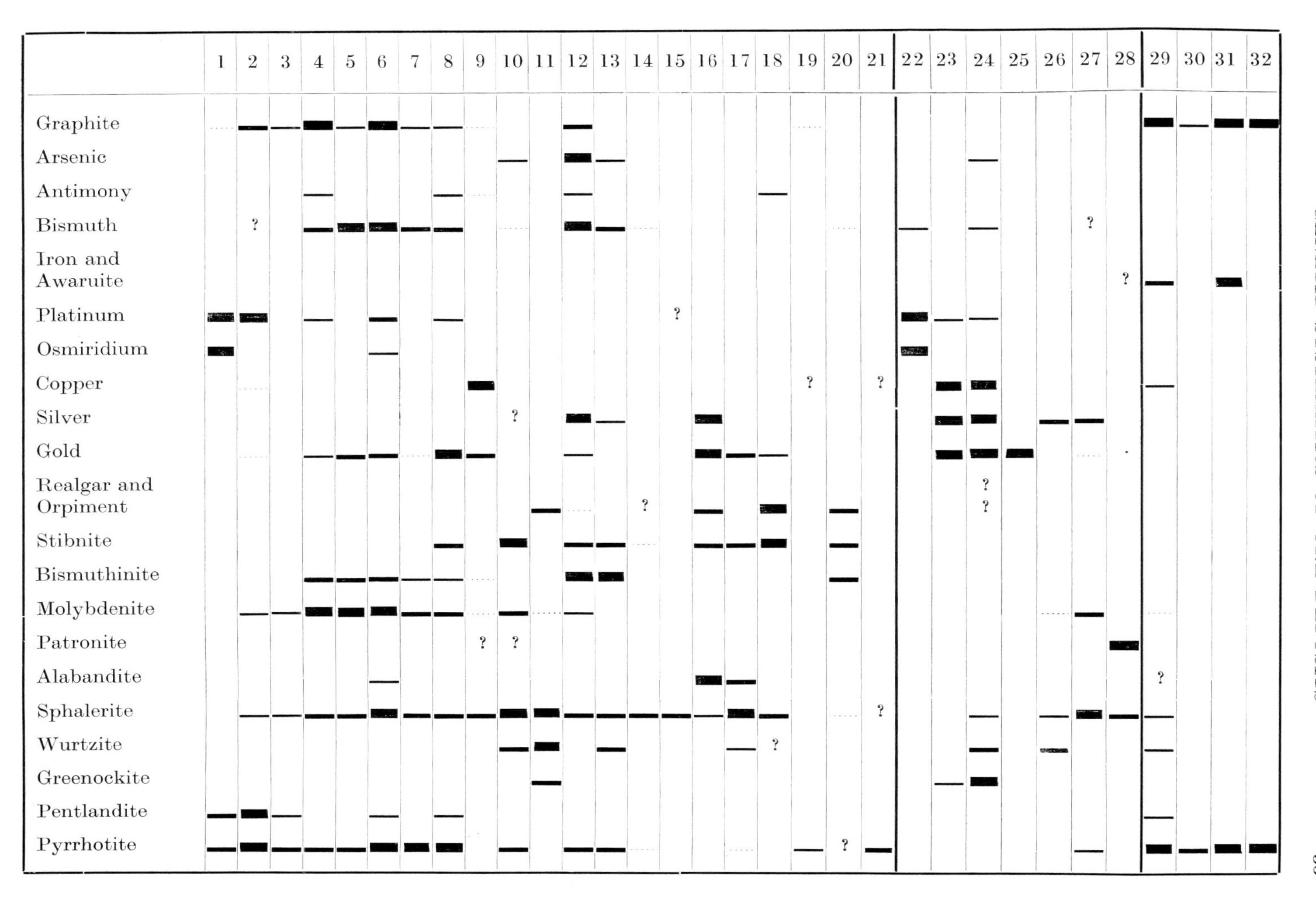
1
2
3
4
5
6
7
8
9
10
11
12
13
14
15
16
17
18
19
20
21
22
23
24
25
26
27
28
29
30
31
32
Graphite
Arsenic
Antimony
Bismuth
Iron and Awaruite
Platinum
Osmiridium
Copper
Silver
Gold
Realgar and Orpiment
Stibnite
Bismuthinite
Molybdenite
Patronite
Alabandite
Sphalerite
Wurtzite
Greenockite
Pentlandite
Pyrrhotite

	1	2	3	4	5	6	7	8	9	10	11	12	13	14	15	16	17	18	19	20	21	22	23	24	25	26	27	28	29	30	31	32
Millerite		⋮	│					│		⋮		│		│										■				■		⋮		
Niccolite and Breithauptite	│	│				│				│		■	│																			
Hauerite																											■					
Pyrite	│	│	│	│	│	│	■	■	■	│	■	│	│	│	│	│	│	■	■		│	⋮		│	│	│	■	│		│		
Sperrylite and Stibiopalladinite		■		│		│	⋮	│													⋮		│	│								
Cubanite					│	■	│	■				■																			│	│
Marcasite										│	■	│	│	│			│	│			│			│	│	│	■	■				
Arsenopyrite	⋮			│	│	│	│	■	■	│		■	│	│	⋮		│	│	│							│	│	⋮				
Löllingite				│	│	■	│	│	■			■	│			│	?															
Safflorite and Rammelsbergite												■	│																│	│		
Skutterudite } Chloanthite }						│	│					■	│	│																	│	
Maucherite		│										■																				
Domeykite } Algodonite } Whitneyite }									■	│							│			?	│											
Dyscrasite						│			│	■		■	■																			│
Galena	⋮	│		│	│	■	│	│	│	■	■	│	│	│	│	│	■	│		│	│	⋮		│		│	■	│				
Clausthalite and Naumannite								│	⋮			│	│	│		■																

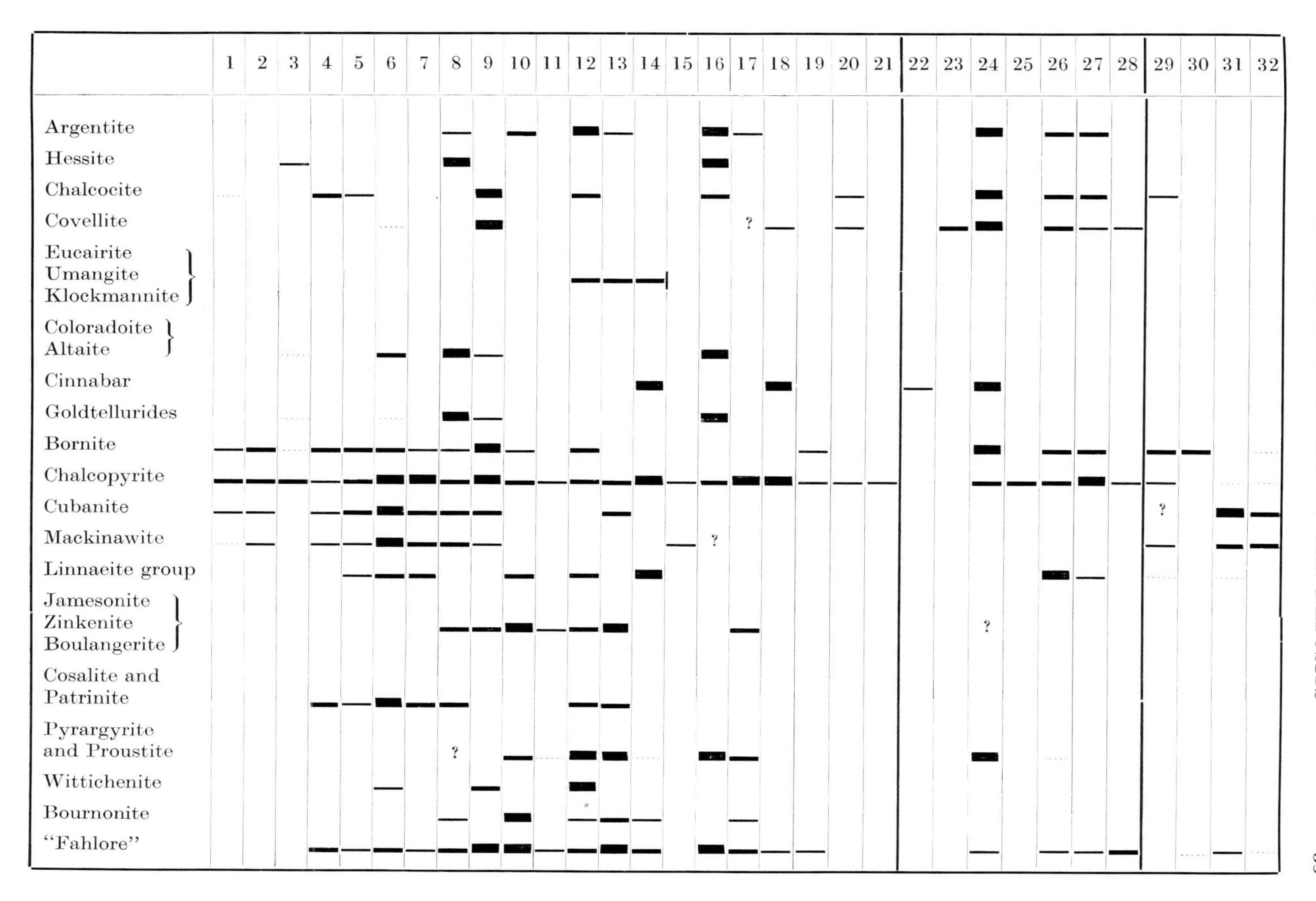
1
2
3
4
5
6
7
8
9
10
11
12
13
14
15
16
17
18
19
20
21
22
23
24
25
26
27
28
29
30
31
32
Argentite
Hessite
Chalcocite
Covellite
Eucairite
Umangite
Klockmannite
Coloradoite
Altaite
Cinnabar
Goldtellurides
Bornite
Chalcopyrite
Cubanite
Mackinawite
Linnaeite group
Jamesonite
Zinkenite
Boulangerite
Cosalite and
Patrinite
Pyrargyrite
and Proustite
Wittichenite
Bournonite
"Fahlore"

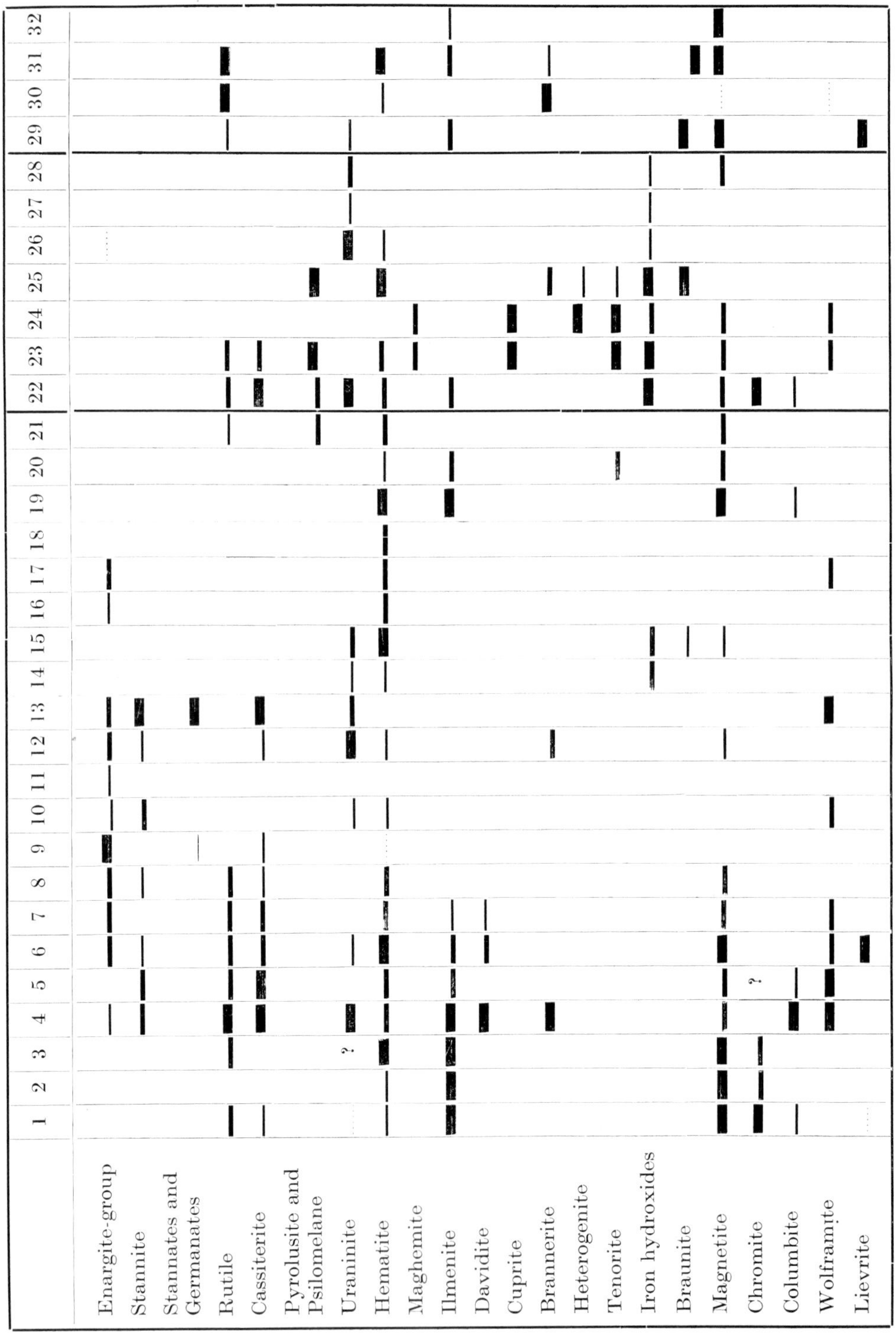

	1	2	3	4	5	6	7	8	9	10	11	12	13	14	15	16	17	18	19	20	21	22	23	24	25	26	27	28	29	30	31	32
Enargite-group																																
Stannite																																
Stannates and Germanates																																
Rutile																																
Cassiterite																																
Pyrolusite and Psilomelane																																
Uraninite			?																													
Hematite																																
Maghemite																																
Ilmenite																																
Davidite																																
Cuprite																																
Brannerite																																
Heterogenite																																
Tenorite																																
Iron hydroxides																																
Braunite																																
Magnetite																																
Chromite					?																											
Columbite																																
Wolframite																																
Lievrite																																

3. Difficulties arose in completing the columns for the metamorphosed deposits. The attempt was made, without any hope whatsoever of completeness to distinguish between those ores *newly* formed through metamorphism, designated by solid lines, and those ores simply retained from the original deposits, indicated by dotted lines, without consideration of the amounts. The table here is certainly rather incomplete.

4. The consecutive numbering of the groups 1–32 is only to facilitate finding one's way.

D. Metamorphic sequence

At present no comprehensive treatment of the metamorphism* of ore minerals exists. It is desirable, therefore, not only to report on single studies in greater detail than in sections A to C, but thoroughly to discuss this material as a whole.

The metamorphism of ore minerals

1. Introduction and general aspects.
2. Metamorphism through change in temperature and in part through change of confining pressure.
   a) Metamorphism in contact with plutonic rocks.
   b) Metamorphism in contact with extrusive rocks.
   c) Load metamorphism.
   d) Diaphthoresis.
3. Metamorphism under the influence of directed pressure, with or without essential effects of temperature change.
   a) General considerations.
      α Concepts of deformation.
      β Aspects of crystallography and structure in deformation.
      γ Concept of recrystallization and formation of porphyroblasts.
      δ Changes in mineral content.
   b) Application of the concept of "depth zones" of rock metamorphism to ores and ore deposits.
   e) Extrahigh pressure alterations.

I. GENERAL ASPECTS

Rocks and mineral deposits which, either as a whole or only in individual parts, have been essentially and characteristically re-formed through processes that are neither related to the original process of formation nor to the effects of atmospheric influences, are called *metamorphosed*.

The metamorphism of rocks has been thoroughly studied, and up to a certain point it is already possible to evaluate the phenomena conclusively in many details. Discussion of this subject lies outside the scope of this book, yet in treating the

* Of course, apart from the earlier editions of this book!

metamorphism of ores and ore deposits which in the broadest sense are also rocks, one must refer to the knowledge already won in connection with rocks and bodies of rocks and apply the concepts used in that field.

The general necessity of a separate section on metamorphism of ores stems from the fact that in their investigation other methods are to be applied, and the fact that in view of the newness of the subject, many things well known for the transparent minerals must be discussed with respect to examples of opaque ore minerals.

Investigations concerning metamorphosed ore deposits exist in great number and are considered in the descriptive part of this book. Many are scarcely more than occasional "aperçus"; other works treat individual cases with great accuracy. In this connection investigations of the following ought to be mentioned: UGLOW, LINDGREN & IRVING, FREBOLD, HUTTENLOCHER, RAMDOHR and SCHNEIDERHÖHN, also DRESCHER and MAUCHER. A summary of all or even the most essential phenomena belonging here, however, has not yet appeared, although SCHNEIDERHÖHN (1958) has already accentuated the problem from the standpoint of ore-microscopy.

If we refer to the definition at the beginning of this section, difficulties of where to draw the boundaries, and other difficulties will be found, — as is common in regard to such definitions. Such cases will be noted at appropriate places. It is also important that "autometamorphic" processes, the aftereffects of the processes of formation, are not considered here. Excluded also, in contrast to the practice employed occasionally in America, is everything which deposits undergo in the way of mechanical alteration through the influence of frost and heat and rain, wind, ice, or vegetation, and in the way of chemical alteration through surface water, oxygen of the air, carbon dioxide, etc., and perhaps also those of mere "aging".

If one disregards, to begin with, the influences of multiple metamorphism, or "poly-metamorphism", which make a clear understanding difficult, then the processes of metamorphism in the strict sense, as seen in the following, are easily divided into two groups very different from each other in the typical case:

First, metamorphism alone through change in temperature and, in part, of the confining pressure. Here belongs the metamorphism of plutonic and extrusive rocks and changes caused through subsidence into greater depths of the earth's crust. Conversely, retrograde metamorphism due to lowering of temperature and relaxation of pressure is also included. With the latter is closely related the diaphthoresis during metamorphism due to directed pressure. Only brief reference can be made to the influence of slight but often certainly very effective amounts of volatile substances furnished by the magma or mobilized out of the neighboring rocks. This is a subject that is both theoretically and factually difficult to treat.

Second, metamorphism in which directed pressure is a decisive factor along with temperature and confining pressure. Here it is necessary to deal with deformation in general, with the influence of deformation on the crystal lattice, with recrystallization, and with other subjects. The basic principles in regard to these factors are treated in the works of SANDER and SCHMIDT (1930; 1932), certainly superbly and in very great detail, but both concern themselves almost exclusively with the silicates and besides are not easy to read. A short easily comprehensible summary has been given by EARL INGERSON.

Among all the examples cited in the present work, cases investigated by the writer himself predominate, and the confusing data of the literature are relied upon only where it was absolutely neccessary. This indeed is designed only to serve simplicity and clarity!

II. METAMORPHISM THROUGH CHANGE IN TEMPERATURE AND CONFINING PRESSURE

Each rock and each mineral deposit is governed as to form and mineralogy by the temperature and pressure prevailing during formation, as well as by many other factors. If temperature and pressure are changed during the later history of a deposit, then as soon as the changes exceed certain limits, the mineral associations, the individual minerals, and many textures are not longer stable, but are altered or at least show a tendency toward alteration.

Only a few minerals are, e.g., stable both at the temperatures and pressures of consolidation of plutonic rocks and at temperatures and pressures prevailing at the surface of the earth. Plutonic rocks, exposed by erosion, have as much tendency for transformation as surface rocks which come into the temperature and pressure range for the formation of plutonic rocks. The general rule that temperature decrease slows the rates of reactions, whereas increase of temperature speeds them up, is the reason that rocks formed at high temperature can maintain themselves (apart from weathering) at the surface for practically unlimited periods, whereas surface rocks under the conditions of plutonic rocks become altered very rapidly and more or less completely.

Plutonic rocks are formed at high but not extreme pressures which hinder partly or completely the escape of gases, and solidification proceeds slowly. Extrusive rocks, on the contrary, solidify rapidly at low pressures; the gases can thereby escape. This determines a fundamental difference which operates in an analogous manner both in rocks and in mineral deposits that are influenced through contact with plutonic or effusive rocks.*

We must therefore distinguish between *metamorphism* at contacts with *plutonic rocks* and at contacts with *effusive rocks*. In the region of the geosynclines rocks and mineral deposits, mostly of sedimentary origin, are often covered by many hundreds, indeed thousands of meters, and therefore are placed under temperature-pressure ranges which produce "*load metamorphism*" that is noticeable and occasionally pronounced. In many respects it has characteristics of the outer regions of contact aureoles of plutonic rocks, but it is distinguished especially by its very great regional extent, and its small and gross structures sometimes indicate that transformation occurred very slowly.

a) *Metamorphism of economic mineral deposits in contact with plutonic rocks* has not yet been observed to be too common. In some instances it is inferred but not proven.

The probability of finding metamorphosed members is naturally largest for iron ore deposits because of their widespread distribution. A good example has been des-

* To bring a mineral association into the physico-chemical conditions of a plutonic or extrusive rock it is normally necessary to contact it with an intruding magma. Exceptionally the same conditions, e.g., of an extrusive rock contact can also be attained in mine fires.

cribed by the writer from the Harz, where the sedimentary (exhalative) hematite occurrences of the "Osteröder Diabaszug" may be followed through all stages of metamorphism! Hematite and siderite are first changed to magnetite, which gradually becomes coarser in grain. Pyrite changes to pyrrhotite, and iron silicates of the thuringite type lead to fayalite and other minerals. Where quartz is present, it reacts in the inner contact halo with magnetite, which leads to the formation of fayalite. Figs. 16 and 17 show exceptionally well the increase in grain size with approach to the contact without any essential metasomatism. Small amounts of spinel were dissolved in magnetite and separated out again in the course of cooling in the form of the usual exsolution bodies. On the other hand, arsenic apparently included by the colloidal pyrite of the original deposit appears as arsenopyrite, and chalcopyrite previously present only in extremely fine dispersion is collected into larger grains. The pressure conditions were such that in the immediate vicinity of the contact pyrite is only partly dissociated and still forms large porphyroblasts. Iron sulphide persists here also into the hybrid rock portions. The sulphur liberated by the dissociation of pyrite has transformed Fe present in other compounds into pyrrhotite.

Fully similar phenomena occur in other places. Thus SCHWARTZ and BRODERICK (1930, 1927), e.g., have described the contact effects on the Gunflint iron formation and the Mesabi iron formation by the Duluth gabbro. Here, however, the formation of fayalite described by ZAPFFE has only a slight extent. Large parts of the deposit are uneconomic (39% Fe) when unmetamorphosed, but have become economical through contact metamorphism., because magnetite has been produced and is easily concentrated to 60% Fe by magnetic separation (cf. COOK, 1936, p. 263). Similar relationships occur in the Mansjö Mountains, where the iron ores and fayalite rocks have been described by ECKERMANN.

Not all at rarely the stage of migmatite formation has been reached, which explains, e.g., the peculiar behavior of a part of the magnetite of the Duluth gabbro (BRODERICK, 1927). Some "orthomagmatic" titanium-lean titaniferous magnetite deposits may likewise belong here.

Manganese ores have also been described at various times in contact aureoles. The occurrences of the Nagpur-Kodur region in Bengal are imposing. Owing to the differences already existing in the source ores, and owing to variable or differing intensity of contact metamorphism, it is quite difficult to see the relationships. Braunite, vredenburgite, sitaparite, and hollandite are widespread, along with a number of manganese-rich silicates. Where the point of melting was reached, very peculiar mixed rocks occur (cf. L. L. FERMOR). — Clearer but very monotonous is a tiny occurrence near Darmstadt, where sedimentary manganese ores of probable Kulm age have been altered to braunite in the immediate vicinity of granite. Viridin, a manganese muscovite, and piedmontite were formed at the same time. A newly discovered impressive deposit of braunite at Otjosondu, South-West-Africa, was perhaps formed at a somewhat higher temperature. Similar occurrences have been described many times. The occurrences of Långban, Jakobsberg, and other places in Värmland, Sweden, belong *perhaps* here.

Rocks of the *lateritic* and *bauxitic* types must have been the source materials of the emery of Smyrna and Naxos. From the standpoint of ore-microscopy, the magnetite and ilmenite are of interest here. Very small occurrences of emery in the gabbro of the Frankenstein in the Odenwald are a type of mixed rock; the magnetite

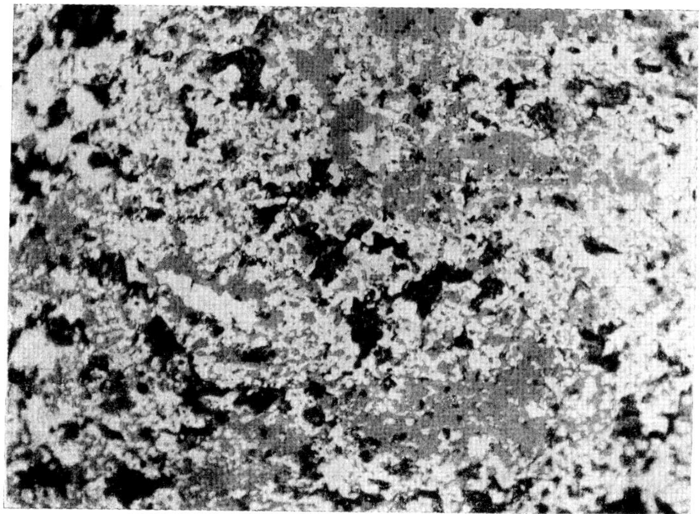

Fig. 16 170 × RAMDOHR

Spitzenberg near Altenau, Harz

Very fine-grained contact product consisting of *pyrrhotite* (white), *magnetite* (grey) and *fayalite* (dark grey to black). Rather far from the contact (~ 100 m)

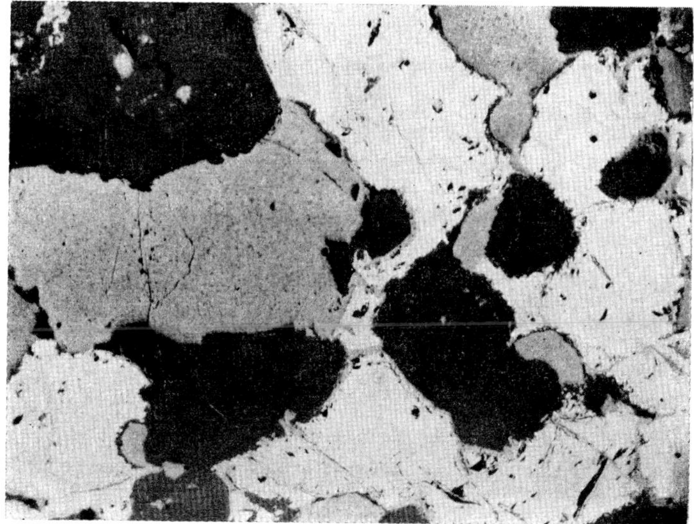

Fig. 17 170 × RAMDOHR

Pinge Riekensglück near Harzburg

Similar in mineral composition and proportions to fig. 16. — In consequence of long duration of contact action and immediate proximity to the contact, the grain size has become much larger and much more uniform

has become a true titanomagnetite by taking up of titanium. Contact metamorphism of a deep-seated rock by another deepseated rock (nepheline syenite of the Pilansberg), leading to complete remelting of the sulphide contents of the rock and to separate intrusion of the sulphides, is assumed by HOFMAN for the pentlandite deposits of Matooster near Rustenburg, Transvaal. I must decline to accept this interpretation, on grounds that cannot be discussed here. Larger sedimentary sulphide occurrences have apparently not yet been described from plutonic rock contacts; yet the first beginnings thereof — alum shales — are well known in contact occurrences through the work of V. M. GOLDSCHMIDT. In these the pyrite is almost always converted to pyrrhotite, the content of bituminous material to graphite. Graphite is generally the product of contact metamorphism of coal and hydrocarbons by plutonic rocks and has often been mentioned but is still scarcely described in the field of ore-microscopy.

In some cases emery deposits of this type have been magmatized completely and became by filterpressing real "corundites" (e.g. Peekskill).

b) *Metamorphism by extrusive rocks.* Here again, metamorphism can be observed mainly on deposits of iron ores. The very slight extent of contact aureoles associated with extrusive rocks naturally lessens the possibilities of recognition and study and also lessens their economic significance. From a theoretical standpoint, and because of similarities with metallurgical processes and greater ease of experimental comprehension, these changes will be, however, of great interest in the future. At present, relatively little is known about them!

The alterations of *siderite veins* of the Siegerland caused by Tertiary basalt vents were observed long ago. They were investigated in detail in the Institute at Aachen by methods of ore-microscopy. Siderite is calcined to magnetite, the change naturally proceeding preferentially from the grain boundaries, fractures, and cleavage cracks. The zone of influence is very narrow, only 10 to 20 centimeters. The magnetite thus formed is exceptionally finegrained and often follows a faint zoning in siderite. In the immediate vicinity of the contact, the siderite is completely destroyed and transformed into powdery and sooty masses of magnetite fully similar to earthy pyrolusite. Of the greatest interest is the alteration of chalcopyrite, which is always present in siderite in small amounts. Towards the contact, a lamellar aggregate of intergrown chalcopyrite and chalcopyrrhotite forms first; the amount of chalcopyrrhotite then increases relative to chalcopyrite. Sometimes mackinawite is also formed perhaps by digestion of pyrrhotite. Still nearer to the contact, pores and magnetite secretions are formed; that is, a partial oxidation begins. As more and more iron in the form of magnetite is withdrawn, bornite appears, perhaps as a transitional product.* The final result is an aggregate of chalcocite plus magnetite.

Basalt and other effusive rocks often have incorporated pyrite or siderite out of the traversed rock units and roasted them to pyrrhotite or magnetite. At Bühl near Kassel, at Ovifak, Greenland, and at several other places, coal was cut at the same

* The formation of bornite is not yet observed in Siegerland reference material, but is formed occasionally in calcined siderite and as a natural product under similar conditions in other localities. Very similar are the alterations which siderite of the same veins shows in the course of its artificial calcining: instead of magnetite, however, partly a very Fe_2O_3-rich Mn–Fe-spinel, partly hematite is so formed. The chalcopyrite fraction behaves similarly.

time, so that the conditions of a natural blast furnace process existed. Iron and cohenite (figs. 19, 309) were formed. — Sphalerite is also known from inclusions in basalt. In contrast to pyrrhotite, it was not melted but, as the writer was able to show in one case (1931 b), was altered to wurtzite (fig. 18) instead. In the process, the sphalerite absorbed FeS, partly from the country rock, partly from associated chalcopyrite and therefore became much darker. Chalcopyrite was thereby converted to bornite. In the descriptive section of this book (figs. 309, 310), the voluminous literature on the Bühl iron EITEL and IRMER (1920, 1920), SCHWARZ and the Ovifak iron LÖFQUIST & BENEDICKS (1941) very recently MEDENBACH (1974) are discussed.

Lignitic coals in the same fashion as hard coals, are converted into columnar, anthracite-like masses through a loss of H, O, and N. The best known example for over a hundred and fifty years is the Hohe Meissner. Natural cokes have formed locally on every continent through contact of basalts with coal seams. Microscopically they are entirely similar to artificial coke.

Many materials formed due to mine fires are closely similar to those at contacts of effusive rocks. The materials thus formed at the United Verde copper mine are examples. There the alterations were often quite varied over distances of a few centimeters. Pyrite was in part decomposed with formation of FeS, which to a rather large extent is dissolved in sphalerite, thereby becoming very dark brown. Chalcopyrite was also decomposed in part with the formation of bornite plus magnetite, etc.

SARKAR & DEB (1974) describe a similar example of contactmetamorphism by extrusive rocks (thick dolerite dykes) at sulfides in the Singhbhum Copper Belt.

The experiments of C. F. PARK (1931) at very low pressures show that the action of hot hydrothermal solutions can give rise to similar mineral associations. The experiments produced associations like those described many years ago by SCHNEIDERHÖHN (1923) from the Siegerland. Concerning the genetic interpretation of the Siegerland occurrence, it is difficult to say whether a revival of the original hydrothermal mineralization is the cause, or whether an action of heated waters in connection with the younger volcanic activity, and independent of the original mineralization, led to effects which have to be classified as metamorphism.

c) *Load metamorphism.* Every deposit, in particular every sedimentary deposit, can be brought into an environment in which temperature-pressure conditions are changed simply by being covered by more and more sediment. This happens particularly in geosynclines. A cover of 2,000 to 3,000 meters is something quite common, and a cover of 10,000 meters is not actually rare, at least for strata geologically very old. Thicknesses substantially greater than this are proved only in folded terrains where directed forces can contribute to the changes. According to the recent ideas and observations on plate-tectonics that overburden may reach several 100 km and cause easily magmatisation. — A cover of 3,000 meters means an increase of about 100 °C in temperature; a cover of 10,000 meters means an increase of 300 to 400 °C. Even in the first case, many delicate gel or zonal structures are lost in certain minerals, many transformations take place, and mixed crystals earlier not possible may be formed. In the second case, on the other hand, extremely strong changes are produced, so that without knowledge of the relationships one would classify the ores as "high-hydrothermal" or something similar. To cite some examples, sphalerite can dissolve considerable amounts of iron; chalcopyrite also enters into sphalerite and can later be detected in the form of small ovoids; chalcocite and bornite dissolve in each other

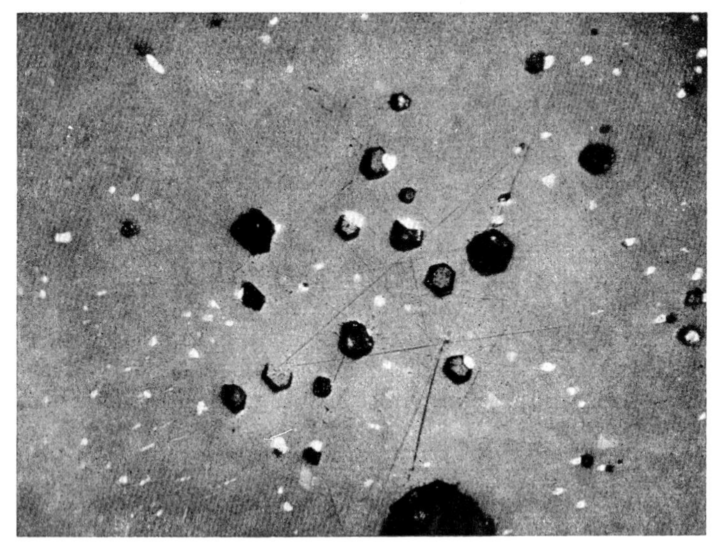

Fig. 18 550 ×, oil. imm. RAMDOHR
Finkenberg near Bonn, Germany

Sphalerite from a basalt inclusion. The sphalerite contains hexagonal pore spaces from former uniform large crystals. They reveal that this sphalerite was heated above the inversion point sphalerite → 1020° → wurtzite. The white inclusions are *chalcopyrite*

Fig. 19 80 ×, nicols almost crossed, immersion RAMDOHR
Ovifak, Greenland

Ferrite, round grains (dark) with a few cementite "needles", embedded in coarse network of *cementite* (light), which shows cracks and indications of cleavage

and exhibit the well known, very varied unmixing textures, like those shown, e.g., in the ores of Zambia. SCHNEIDERHÖHN and BRUMMER regard these ores as originally sedimentary but now metamorphosed in the manner just described.

Minerals and mineral associations which formed as gels undergo a far reaching collection-crystallization [Sammelkristallisation] and become granular, often even coarsely. At the same time, other typical sedimentary structures can be markedly obscured. SCHNEIDERHÖHN (1931) cites very good examples from the manganese ores of Postmasburg, where in addition there has been considerable formation of new minerals from the orginal material, which consisted probably of pyrolusite and psilomelane; some iron ores of Brazil show similar features.

In the formation of the very peculiar mineral assemblage of the Witwatersrand load metamorphism has also been an important factor, indeed apart from supplementary but mostly local heating through intrusive dikes. Magnetite and also ilmenite (except for sparse remnants), and indeed still other iron minerals, transformed into pyrite and pyrrhotite. Pyrrhotite is much more widespread than one might gather from the literature. The titanium of the ilmenite and the titanomagnetite is in rutile, which, again quite in contradiction to the older literature, is the most abundant mineral of the "Banket" after quartz, pyrite, and perhaps sericite and pyrophyllite. The writer has depicted the relationships in two extensive works (1954b; 1958b) Further investigations surely will furnish many new examples.

d) *Diaphthoresis* (with metamorphism due to relief of pressure and cooling).* Every deposit formed in depth and at elevated temperature, at the moment at which it is seen by man, exists under temperature-pressure conditions quite different to those of its origin. At first sight, therefore, one can never say with full certainty whether the present texture and mineral composition were the original ones. In general, the changes with decreasing temperature are not so striking as those with increasing temperature, because certainly the mobility of the molecules decreases markedly with falling temperature; however, at least the high temperature formations must in cooling pass through a considerable range in which the rate of reaction is still great. Reaction is strongly promoted by solutions and above all by diaphthoresis in the narrower sense, which is an inversion of metamorphic conditions. The effects of this consists, e.g., in the appearance, of hydrated and not especially dense minerals, and of phenomena of decomposition of high-temperature minerals, etc.

With further cooling, any given mineral enters into a range in which mobility decreases rapidly and beyond which no further changes occur during geologic times. This boundary is very different for different minerals and equilibria. The boundary for magnetite, e.g., is high, for chalcopyrite low. If boundaries lie high, then the possibility of finding high-temperature minerals preserved through natural or artificial "quenching" or freezing conditions is much greater than if boundaries are low. An example among mixed crystals will make this clear: Titanomagnetites which are not unmixed are quite common in basalts, even in regions which surely cooled rather slowly. On the contrary, mixed crystals of $CuFeS_2$–FeS that are not unmixed are very rare and preserved almost solely as "chalcopyrrhotite", if there additional components effect stabilization. Experimental production of such mixed crystals is

* SCHNEIDERHÖHN (1958) takes the concept of diaphthoresis in a broader sense than is done here, by including the sequences due to relaxation of pressure and cooling.

also successful only with cooling rates very improbable in nature and often not at all. We have to think of the possibility to find in one and the same mineral assemblage, three minerals, of which one has conserved a condition deriving from formation at 600 °C, another a state acquired at 300 °C, the third a state corresponding to 100 °C. One should never forget such considerations in the evaluation of textures that are present or, likewise, those that are lacking in mineral associations.

All exsolutions, the decomposition of many sulphosalts, many so-called replacements, and many lamellar textures due to inversions, etc., are therefore actually signs of metamorphism during cooling, partly also signs of true diaphthoresis. Hydration, so important in the silicates, plays a role here only in magnetite, goethite, olivine, chalcopyrite and in rare instances valleriite.

In practice, it is impossible to draw a sharp boundary between weathering on the one hand and later stages of the original process ("paulopost" processes) on the other, easy though it may be to draw them in theory. Thus, e.g., the formation of pyrite, or pyrite plus magnetite, or pyrite plus marcasite, from pyrrhotite may be the beginning of weathering at one place, metamorphism due to decrease of pressure at another, true diaphthoresis at another yet, and finally a hydrothermal after-effect at still another.

The phenomena mentioned here can have the closest relationship also with effects of directed pressure, treated in the following section.

By way of examples cited already in another connection, we may further mention that the following transformations can take place in a manner quite analogous to those caused by the action of circulating water: pyrrhotite into pyrite plus magnetite, magnetite into hematite, ilmenite into rutile plus hematite possibly into titanite, and pentlandite into millerite. Many sulphosalts decompose during introduction of material. Thus falkmanite is transformed into galena plus bournonite ± native arsenic. Quite similarly geocronite, or members of the fahlore group with much Fe and As, is transformed into ordinary fahlore plus chalcopyrite plus arsenopyrite (Jakobsbakken, Sulitelma). The textural results of such strong alterations of minerals are often myrmekites. Among silicates which directly play a large role in ore deposits may be mentioned the transformation of augite to hornblende, and of the latter in turn to chlorite plus calcite. Similarly, olivine becomes serpentine and serpentine changes to magnesite plus chalcedony.

e) *Effects by extreme pressures.* Highest pressure associations, in which closely packed structures and parageneses, occasionally even a change of coordination number can be formed, became recently important in the study of rocks. In the ore mineralogy only a little is known; interesting data gave the late NEUHAUS in his last papers. Some mixed crystals seem to become more stable, e.g. Ni in pyrite in the association of the kimberlites.

Recently much is done in experimental investigations of ordinary high pressure effects going back to the observations in upper mantle rocks (like kimberlites) or rocks formed through meteoritic impacts. It is without doubt that many minerals with looslely packed structures can undergo by pressure alterations to closer packed structures e. g. graphite — chaoite — diamond, quartz — coesite — stichovite and so on, that minerals with components normally too voluminous to enter into certain lattices e. g. CuS_2 into the pyrite structure, that ore minerals become denser by breakdown (as chalcopyrite forming pyrite + copper) and so on.

III. METAMORPHISM AT HIGH AND HIGHEST PRESSURES

High pressure favors the formation of minerals or associations with higher density (or smaller volume). That becam eimportant since we try to examine rock specimens fo very high depths origin, e. g. the griquaites of Kimberlite pipes or the rocks altered by the impact of meteorites on earth or on the moon. — A new type (or many types) of metamorphism became interesting. — The loss in volume can be reached in many different ways. The mineral can form increasingly denser phases as: quartz — coesite — stishovite or graphite — chaoite — diamond. Or different minerals can be combined and form a new mineral of high density or an assemblage of minerals like eclogites. Furthermore a mineral may break down if one (or more) components can form denser associations. At least different possibilities can be combined. Often the coordination number e. g. against O can increase as in the case of quartz (4) into stishovite (6).

Those effects have been tested by experiments, especially in silicates, in part in oxides. In sulphidic ores not many data of experiments or observations exist. The late A. NEUHAUS gave in his last papers some interesting experiments. RAMDOHR could show that in the high pressure associations of Broken Hill (N.S.W.) alabandite was altered to a hexagonal high pressure phase, which later — in the present state — became by diaphtoresis normal alabandite again. Another observation from a Basutoland pipe did show that there a normal pyrite-contains about 30% Ni, by far more than possible in low pressure pyrite, and at last again in a Basutoland pipe, a breakdown of chalcopyrite into pyrite and native copper, so saving more than 10% of volume. Occasional observations may be mentioned in literature, but seem to be never worked out systematically. Till now the data are too sparse to be treated further, the problems must wait!

IV. METAMORPHISM UNDER THE INFLUENCE OF DIRECTED PRESSURE WITH OR WITHOUT ESSENTIAL INFLUENCE OF TEMPERATURE CHANGES

(a) *General*

α. *Concepts of deformation and the nomenclature of petrofabrics.* The deformation of rocks and ore mineral assemblages is like that of all other substances, in particular the metals used in technology, to be treated with the aid of theoretical mechanics and requires therefore an unavoidable but not always very simple mathematical foundation. Its introduction into petrology, apart from preliminary work by G. F. BECKER, is the merit of B. SANDER (1930) and W. SCHMIDT (1932). We must content ourselves here with quite a brief presentation of their ideas and of the explanation of the principles they have introduced or taken over from mechanics. Reference is made particularly to the detailed report by INGERSON.

A body can be influenced by pressure in ways that are varied to a great extent. 1. Through hydrostatic pressure — all dimensions become reduced in the same way. A sphere remains a sphere. 2. Through directional pressure — in the case of an isotropic substance, a sphere becomes a rotation ellipsoid, the axis of which is parallel to the pressure. 3. By one-sided pressure, in the case of an anisotropic substance, or through shearing stress a sphere becomes a triaxial ellipsoid. Moreover, the pressure body ("strain ellipsoid") can exhibit either a lasting deformation or one that is elastic, so that the body eventually returns to the original state. Only in the last case, that is, the case of elastic deformation, can one speak correctly of "*stress*", and then only if the strains produced in the body are considered, or the strained condition in which the body is placed is pronounced. On the other hand "strain"-ellipsoid signifies only the change of *shape*, without regard to the existing elastic force or the permanent change.*

Finally, deformation can be non-rotational. Mostly, however, it is rotational; indeed "monoclinic rotational" deformation is especially common.

Monoclinic movement shapes. These very ordinary figures are of great significance with regard to rocks and perhaps more so in ores. Among the most striking to follow are those in the water of a waterfall or of a brook flowing over a dam. Among ores, "slickensides" and most formations of "gneissic galena" ["Bleischweif"] are examples of such patterns. The example of the water cannot be applied quite directly, since the cylinder of water forming at a dam *remains* there, or continually forms anew, whereas a "galena-tail" forced over a quartz rib, e.g., will move farther along with the structure impressed upon it. The flow "vortices" are *moved*, although, of course, new "vortices" may arise at the old positions. (This happens in water, by the way, to a certain extent.) The rib or dam can also be carried along, although at a different speed. "Enveloping folds" ["Wickelfalten"] around inclusions furnish especially fine examples, especially those around porphyroblasts. In many of these arrangements of inclusions indicates the "shift", that is the rotation against the source position (fig. 21). "Dead spaces" or pressure-protected positions ("pressure-shadows") are striking features that also result from other kinds of deformation (see above, page 48). They are often filled with minerals occurring as common constituents in the mineral assemblage, but in a textural pattern that deviates from the common texture, which has formed without influence of pressure. Their later development, which often involves an essential influence of solutions, can be recognized by the frequent presence of relatively low-temperature minerals and by the especial purity of all components.

Fine examples of monoclinic movement figures are exhibited by the ores of Rammelsberg (figs. 20–22) and by the ores of Goppenstein and many other places of the south central Alps.

Relationships in the titanomagnetite deposits of Routivare, Lapland, which the writer (1945b) has discussed, are ideally simple. Apart from local special features, they are not complicated, as in the cases named above, by excessive size and hardness differences between the minerals of the original fabric. The basis of the deformation is here a stress, according to the accompanying sketch (fig. 23). A very rigid mass of coarse-grained ore is sandwiched between relatively yielding rocks, which have been stressed along directions corresponding to the arrows and moved relative to the ore. The ore itself has yielded to this stress only along the borders. The effects die out

* In the literature both concepts are often somewhat mixed up, in particular, the "directed pressure", e.g., in the course of mountain building, is designated as "stress".

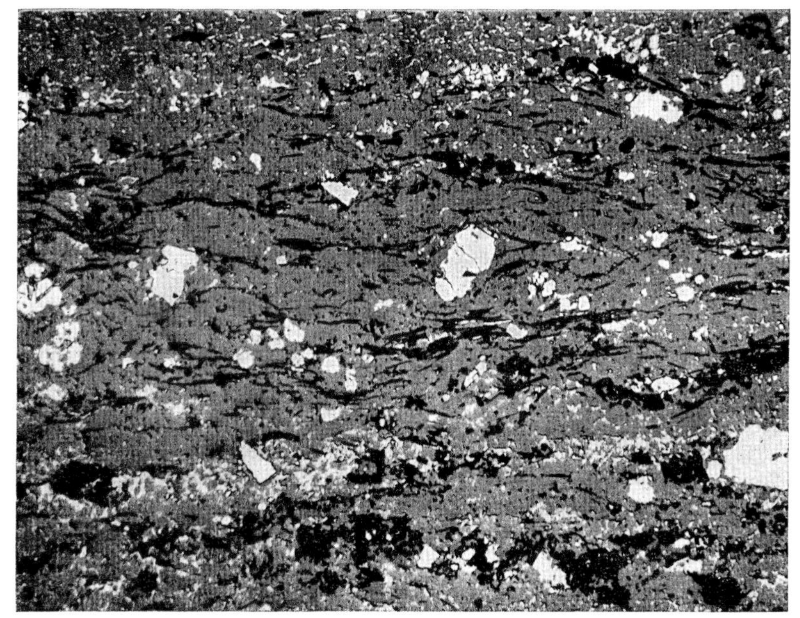

Fig. 20 80 × RAMDOHR

Rammelsberg, Harz

Typical sheared [durchbewegt] zinc ore with strongly oriented flakes of *sericite*. The parallel texture bends distinctly around the "hard bodics" of *pyrite* (white)

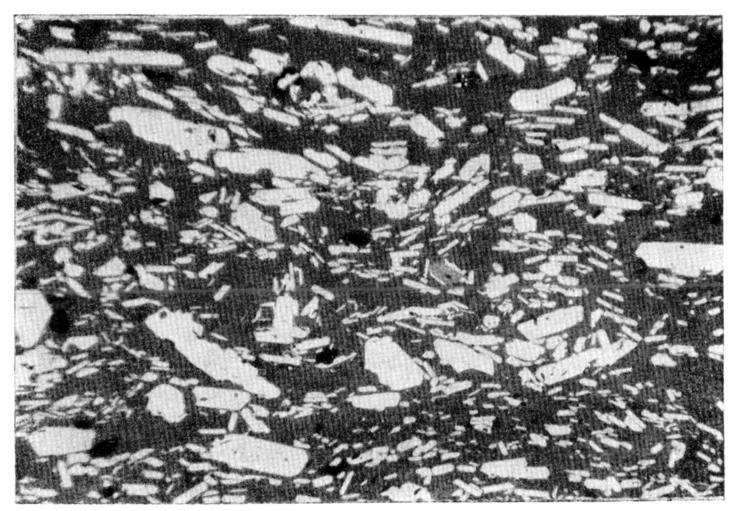

Fig. 20a 35 × RAMDOHR

Yampi Sound, Koolan Island, West Australia

Well oriented rock consisting of *hematite*, white, tiny flakes, very little titanite, grey, and a matrix of quartz, dark-grey

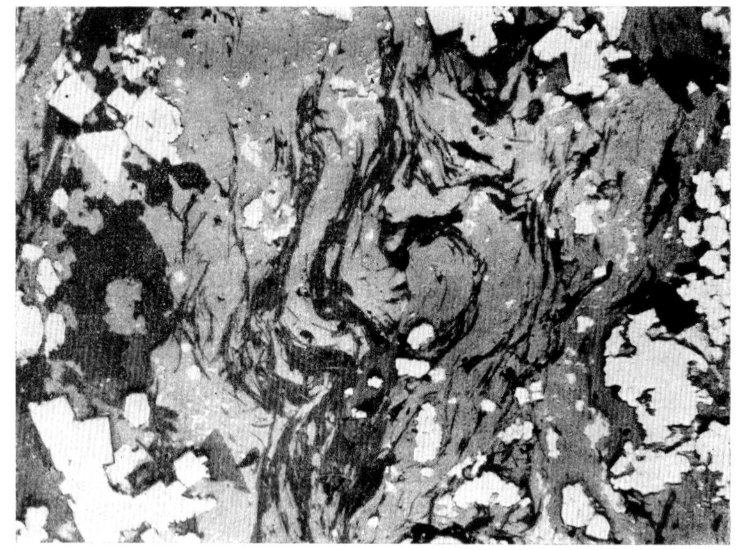

Fig. 21 125 × RAMDOHR

Rammelsberg, Harz

Vortices of compression of sericite in sheared "mottled ore" ["Meliererz"]. *Sericite* and some carbonate are dark and light grey. *Sphalerite*, *pyrite*, and very minor *chalcopyrite* are white

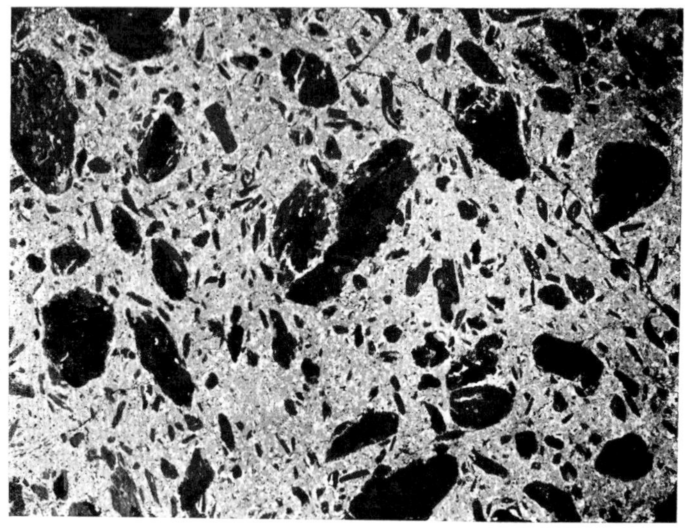

Fig. 22 3 × RAMDOHR

Saxberget, Sweden

Very strongly "tectonized" *galena-sphalerite* ore with interspersed gangue and especially host rock fragments

more to the middle part. Step by step, in the outermost strongly moved portions forms arise that are partly of the laminate eye gneiss ("Augengneiß") type, partly of the pencil gneiss type. The latter have apparently developed in the neighborhood of larger vortex producing obstacles. The laminate portions show a "tailing out" of the "coarse-grained" spinel, which is present as relicts of the motion, in all directions parallel to the foliation. The pencilshaped forms show "tailing out" only in both directions of the b-axis (figs. 24 and 25). The symmetry in the first case is therefore *statically* that of a rotation ellipsoid, in the second case of a triaxial ellipsoid (therefore rhombic!). *Dynamically*, it obviously remains monoclinic. In the part showing pencil structure, ilmenite is arranged in a pronounced "belt fabric"; in the laminate parts, ilmenite shows a simple parallel arrangement.

Triclinic movement shapes are extremely common: on a small scale they occur with the slightest disturbance of the situations described above. On a large scale they are to be expected wherever older stresses exist or also where there occur inclusions with acute angles. To continue with the example of the water dam, such movement pictures occur where water runs over a dam that trends obliquely to its direction of flow. They may also occur at the walls of an ore vein where the shearing motion encounters an oblique band of harder rock. Immediately the relations become more complex. Since to my knowledge such things, however, have not yet been systematically pursued with respect to ores, the reader is referred to the discussions by SANDER, SCHMIDT, and INGERSON.

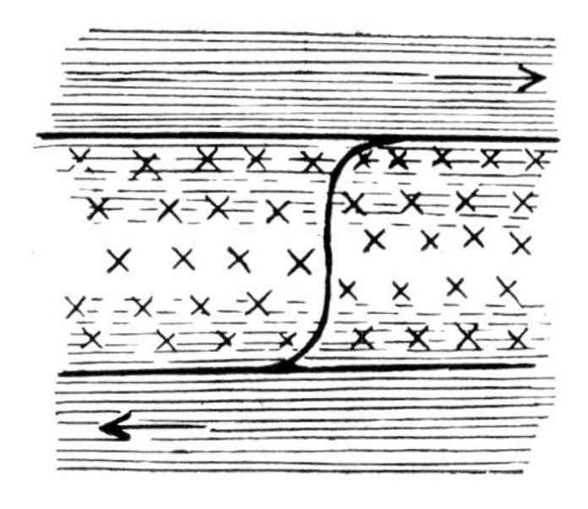

Fig. 23

Competent rock (here it is ore), along its margins against incompetent (very mobile) material, subjected to stress

β. Crystallographic and textural aspects of deformation. In the foregoing it was said that deformation can proceed in part irreversibly, i.e., under pressure or ruptureless "flow", partly reversibly, i.e., without exceeding the elastic limit. Sharp as this difference may be from the standpoint of definition it does not occur without exceptions and transtions. It is, e.g., quite conceivable that at the elevated temperatures (or lower temperatures, also) at which deformation takes place, the elastic limit is not reached, but that with decreasing

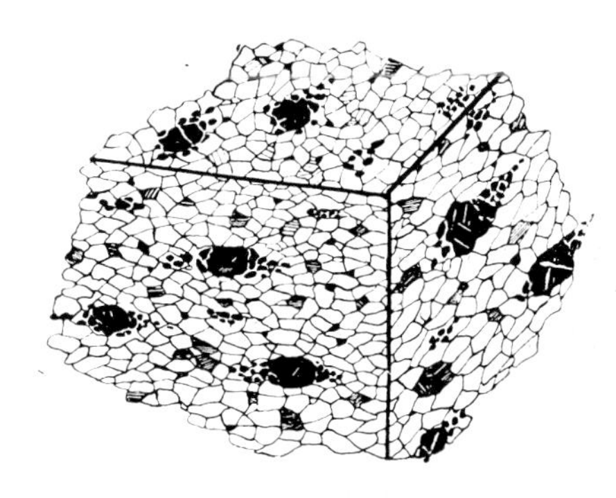

Fig. 24

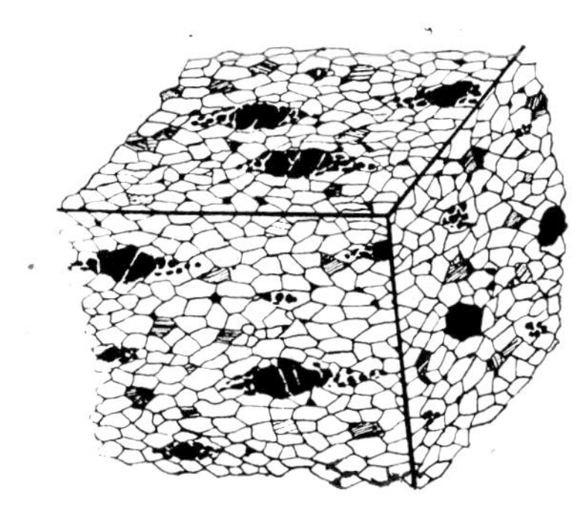

Fig. 25

(or increasing) temperature and maintenance of strain the elastic limit can be far exceeded. Many "latent strains" which express themselves on a large scale, e.g., in "rock bursts" are thus to be explained. Quartz with undulose extinction and the anisotropy shown by many minerals normally isotropic* under normal conditions are also partly explainable in the same way.

Fig. 25a $2^1/_2\times$ RAMDOHR
Mt. Isa, Queensland

Folding of a galena-sphalerite-chert-ore. Caused by subaquated gliding in the time preceding the diagenetic hardening. The whole fold is a member of one layer only. The sequence in broader view remains undisturbed

Such "*latent strains*" are sometimes easily recognizable and very distinct, e.g., in partly cataclastic "fahlore". Occasionally, they can serve as criteria of mechanical stress of a mineral or mixture. Many ores that fly apart in almost an explosive manner on being struck, belong here, as galena, which bursts on the slightest heating, in contrast to other ores of equal grain size and similar occurrence that bear heating without effect. In the same category are several striking deviations and abnormalities in polishing behavior, also, finally, the polishing difficulties which some brittle ores give after careless polishing with emery that is too coarse.

Elastic strain is easily traceable in all stages in substances that can be investigated in transmitted light. The experiment with the glass cube in the jaw press is well known. In reflected light, no experiments of this subject seem to have been conducted either on pressed metal or ore samples. In view of the very great delicacy of many phenomena of anisotropy, they are perhaps not hopeless.

SCHIEBOLD and many others following him have applied Laue photographs to such investigations in metallography with great success. Brief reference may be made to

* Naturally there are indeed very many other causes of strains of this type. E.g., undercooling of mixed crystals, zonal structure in which the zones have different coefficients of expansion, the same in expansion of intergrown minerals, etc.

these works, the conclusions and methods of which could be adapted, with reasonable adjustments, to ores.

Of much greater importance are the *remaining deformations* under fracture or "flow". For the concept of deformation and elasticity no very great difference exists between the two. Crystallographically, however, they are quite different! It is unnecessary to go into the nature of "*fracture*" more exactly. It is well known that many minerals — mostly hard and "brittle" — tend to fracture, others do not. This statement must be kept quite general, however, in accordance with its nature, because temperature, direction of stress in relation to the positions of certain lattice planes,

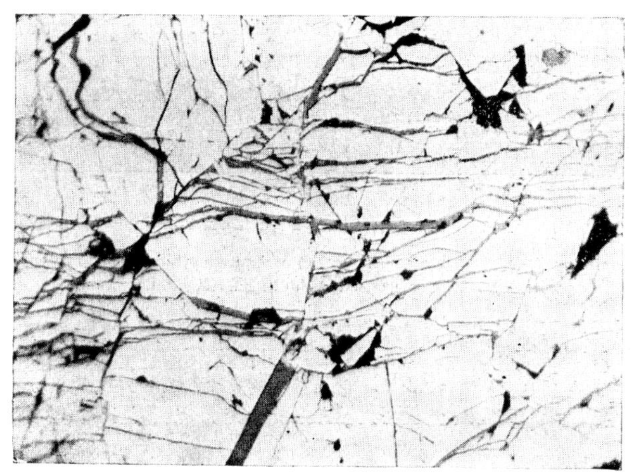

Fig. 26 350 ×, imm. RAMDOHR
Schafalm near Turrach, Austria

Fahlore as cementing medium of a strongly cataclastic *pyrite*

abrupt or gradual action of stress, and quite particularly the manner of embedding in other media and the properties of these influence the "tendency" toward fracturing or toward fractureless deformation quite essentially. Whether fracture takes place with immediate disintegration into small fragments, or with formation of only a few "ruptures", and whether the shape of the fragments or the position of the ruptures follows the "cleavage", can depend from case to case on various influences, e.g., on whether pressure or tensile stress gives rise to the "fractures"*. This, however, is difficult to decipher, since most stresses are accomplished under simultaneous compression *and* tension (figs. 26–31).

The *cleavage* in monomineralic or polymineralic aggregates is frequently quite different from that to be seen macroscopically. "Distinct" or "perfect" cleavages can become very indistinct. "Imperfect" cleavages or cleavages in general scarcely noticeable can play an essential role. E.g., the very perfect cleavage of galena on (100) is scarcely to be noticed as a rupture surface in intense dynamo-metamorphic aggregates, whereas a rather imperfect cleavage of pyrite on (100) plays a marked

* It may be remarked in this connection, as an example, that it is almost impossible to break an argentite crystal through striking it, but with quick tearing a rather good cleavage becomes apparent.

Fig. 27 80 × RAMDOHR
Waldsassen, Bavaria, Germany

Arsenopyrite crystals in *magnetite* (grey) and gangue (black). The arsenopyrite crystals have crushed each other during compression

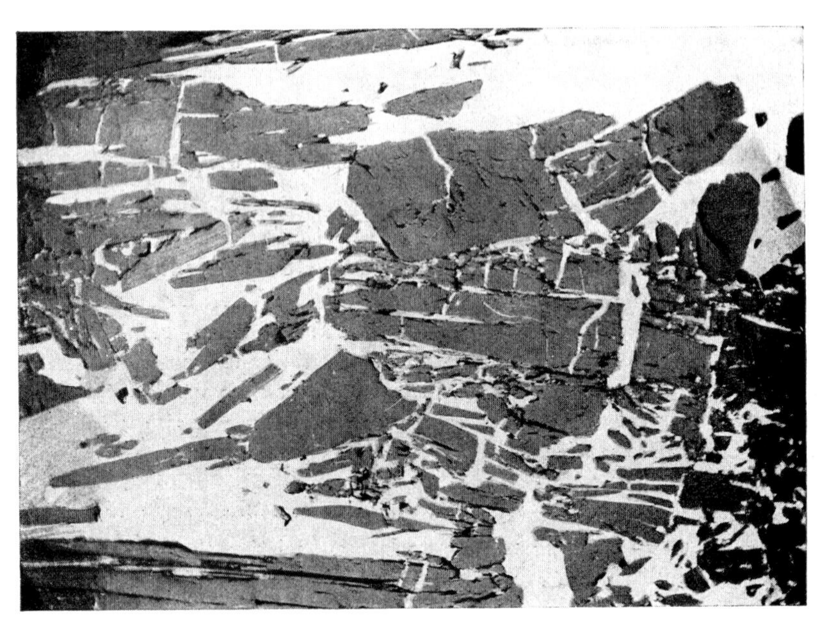

Fig. 28 60 × RAMDOHR
Saxbergets mine, Saxdalen, Central Sweden

Actinolite, strongly cataclastic and cemented with *galena* (white) and *chalcopyrite* (light grey.) The metal content of the fine crack-fillings is easily lost in concentration

Fig. 29 300 ×, imm. RAMDOHR
Ölsnitz, Vogtland, Germany

Arsenopyrite with cataclasis along a zone of shearing; cementation by *quartz* (black)

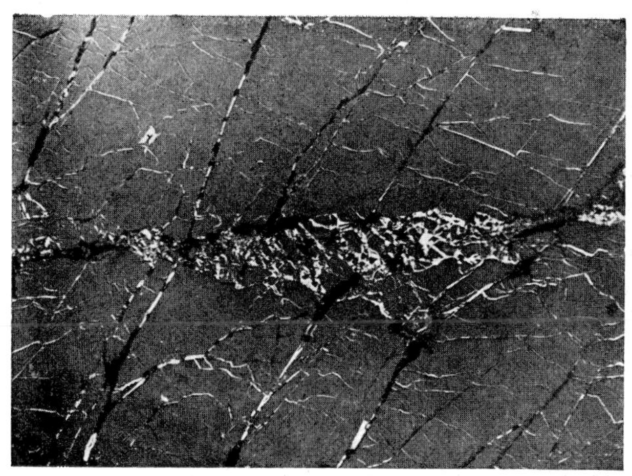

Fig. 30 85 × RAMDOHR
Sudbury, Canada

Cataclastic silicate cemented by sulphides, especially pyrrhotite

role, and the cleavage on (110), scarcely noticed elsewhere, rather often becomes quite distinct. The number of examples could be extended at will.

Distinction between "cleavage" and "parting along glide planes", as, e.g., in hematite along ($10\bar{1}1$) or magnetite along (111), or translation surfaces, as, e.g., also in hematite along (0001), which is often difficult even in individual crystals, is especially difficult or even impractical in aggregates.

Fig. 30a 70 × Ramdohr
Meteorite Norton Co.

Shock induced brecciation in nearly pure enstatite. It is cemented by a melt of different sulfides, not distinguishable with low magnification (troilite, niningerite, alabandite). The large white grain is niningerite.

"*Fractureless deformation*" demands more detailed treatment. It often resembles in its phenomena a complete conversion to the plastic state, indeed to complete melting, but its explanation is quite different, and it is appropriate to emphasize the similarities with melting or "plastification" *less* than happens many times.

Deformation in crystals is effected through the crystallographic phenomena of "*translation*" and "twin gliding".* Cases exist in which the separation of the two phenomena is not at all easy; they are very important and will be discussed specially.

Translation means the phenomenon that a part of a crystal is displaced by mechanical stress along certain planes, the translation planes (*T*). These are lattice planes

* I apply these expressions in the same way as O. Mügge, who first discovered and explained the difference between the two. They may not be quite fortunate linguistically but are in any case correct and preferable to vague and often falsely stated terms. "Simple slip" ["Einfache Schiebung"] is now generally replaced by "twin gliding" ["Zwillingsgleitung"].

that are densely occupied, and mostly by atoms of different ionic state. The movement may be of *every amount* such that there is not disturbance of the unity of the crystal. The direction of displacement in the translation plane (t) is in some cases sharply defined. Under certain conditions it takes place only in one sense of the direction, so that the direction of movement and the opposite direction are not interchangeable! In other cases, displacement is especially easy in particular directions but is likewise possible in other directions. Finally — especially in some crystals of higher symmetry, e.g., where (0001) of a hexagonal crystal is the translation plane — all directions (crystallographically rational as well as irrational) appear to figure almost equally well as translation directions.

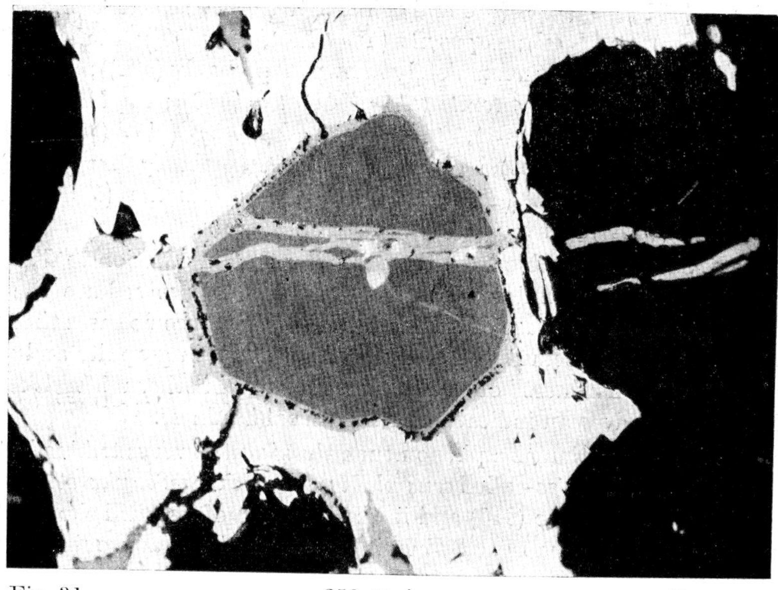

Fig. 31 250 ×, imm. RAMDOHR
Raglan Mine, Ungava, Labrador

Chromspinel (dark grey) shows a fracture healed by *magnetite* (light grey). The vein disappears in the surrounding *pyrrhotite* (white), but starts again in the gangue (black). Magnetite invades likewise the boundary pyrrhotite-chromspinel

For the case in which the amount of gliding of the individual lattice planes or bundles of lattice planes parallel to the translation plane is a whole multiple of the cell dimension in the displacement direction, the crystal after gliding is as complete and undistorted as before,* — at least in theory! In practice, however, this happens seldom or not at all, for reasons which are not always evident. In part, the reason may be incipient separation along translation planes, which are also operative in most cases as cleavages. In part, it may rest on "exfoliations" of certain lattice fragments, in part on the disturbances that cause the contemporaneous appearance of several translations, or a translation together with simple gliding or fracture deformation.

* This assumption is directly defensible and indeed also correct as long as the bonding forces between the individual lattice planes in the "correct" position are considerably greater than in any random intermediate position. In the well defined "layer lattices" these bonding forces are generally so small, that intermediate positions are also conceivable.

In any case the displaced parts are very often easily recognizable because, e.g., they can be etched much more easily, they are affected by preferential replacement and often show strong recrystallization. In many cases translation as such is the reason for recrystallization.

Examples of ore minerals showing translations are — without attempting to offer a complete list! — *graphite* ($T = (0001)$, $t = [10\bar{1}0]$), but intermediate directions are almost as good; *bismuth* ($T = (0001)$, $t = [11\bar{2}0]$); *copper, silver, gold* ($T = (111)$, $t = [211]$, also intermediate directions); *orpiment* ($T = (010)$, $t = [\bullet\bullet\bullet]$); *stibnite* ($T = (010)$, $t = [001]$); *molybdenite* (like graphite); *sphalerite* ($T = (111)$, $t = [112]$); *pyrrhotite* ($T = (0001)$, $t = [10\bar{1}0]$); *arsenopyrite* ($T = (\bullet\bullet\bullet)$*, $t = [\bullet\bullet\bullet]$); *galena* ($T = (100)$, $t = [101]$); *argentite* (probably similar to galena); *covellite* (like graphite); *sylvanite* ($T = (\bullet\bullet\bullet)$, $t = [\bullet\bullet\bullet]$); *nagyagite* ($T = (010)$, $t = [000]$); *chalcopyrite* ($T = (111)$, $t = [\bullet\bullet\bullet]$); *cubanite* (incompletely known, T probably (001)); *valleriite* (probably similar to graphite); *livingstonite* ($T = (001)$, $t = [010]$); *emplectite* ($T = (\bullet\bullet\bullet)$, $t = [\bullet\bullet\bullet]$); *aramayoite* ($T = (\bullet\bullet\bullet)$, $t = [\bullet\bullet\bullet]$); *plagionite* ($T = (\bullet\bullet\bullet)$, $t = [\bullet\bullet\bullet]$); *meneghinite* ($T$ probably $= (100)$, $t = [\bullet\bullet\bullet]$); *sulvanite* (probably like galena); *teallite* ($T = (001)$, $t = [100]$); *rutile* ($T = (100)$, $t = [001]$); *hematite* ($T = (0001)$, $t = [10\bar{1}0]$); *tenorite* (not yet investigated), and many others.

It is immediately recognizable from the examples that in addition to T, t also possesses very simple rational indices, and that the minerals involved are almost all of low "scratch hardness". The ease of translation is so great in some cases that "scratch hardness" *as such* is not accurately determinable, as, e.g., in graphite and valleriite, and only the "scratch hardness" of small grains included in other minerals and firmly surrounded by them gives an indication of the actual hardness.

Connected with translation as a phenomenon that is often very striking but not yet completely explained, is the so-called *crumpled lamellar* texture. E·g., a stibnite crystal bent or waving about [100] to (010) and in the direction of its t will show not a fixed deflection or undulation (like fig. 32), but it remains in itself approximately unbent and optically continuous within a variable range of angles and the flexure phenomenon is brought out only through the overall impression (figs. 32—34).

Besides occurring in stibnite, such crumpling lamellae are displayed well in graphite and molybdenite (figs. 517a, b show a good illustration). The lamellae occur also in pyrrhotite, franckeite and a few other minerals, mostly minerals with rather pronounced layer lattices. These crumpling lamellae often are distinguishable from lamellae caused by pressure twinning only with exceptional difficulty, and several errors of interpretation may have found their way into the literature. The wedge-shaped or spindle-shaped lamellar cross-sections of the crumpling lamellae, which are not quite plane-parallel, are evidence aiding in the correct recognition. The most useful evidence of all, however, is the impossibility of analyzing them definitely from a crystallographic standpoint, since each measurement leads to variable and indeed irrational composition planes.

In the translation planes, a slight torsion around t is often possible. In aggregates especially, this will lead to many striking strains, much more than does pure translation.

---

* Indices not filled in indicate that either no reliable data exist as yet or are not presently available.

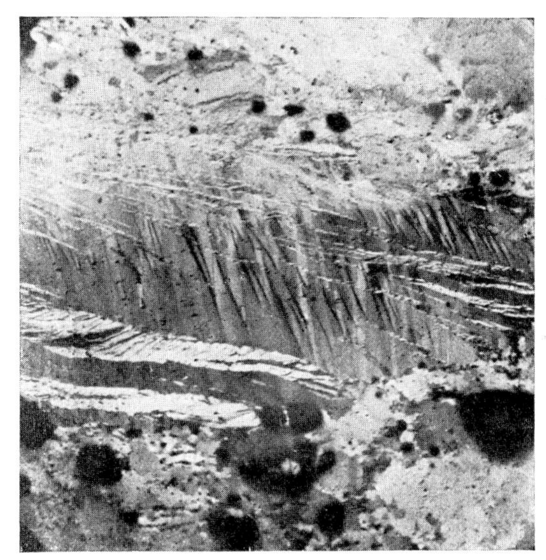

Fig. 32 60 ×, + nic. RAMDOHR
Eidsvold, Queensland

Stibnite with pressure lamellae ("corrugated lamellae") ["Zerknitterungslamellen")

Fig. 33 85 +, + nic. RAMDOHR
Eidsvold, Queensland

Stibnite with pressure lamellae ("corrugated lamellae") ["Zerknitterungslamellen"] and incipient recrystallization

Fig. 34 150 ×, + nic., imm. RAMDOHR
Jakobsbakken mine, Sulitelma, Norway

Wolfsbergite with undulose extinction. (Against *galena* a reaction rim of *bournonite* — both, however, practically invisible between + nic.)

"*Twin gliding*" is very similar to translation. It is different, however, insofar as a crystallographically fixed description can be given not only for the gliding face and the displacement direction in it (in the case of translations T and t, in the case of simple gliding K_1 and σ_1), but also for the second circular cross section of the ellipsoid of deformation and the direction of the principal section in it (K_2 and σ_2). In other words, the *displaced part is in twinned position with respect to the position from which it moved* (fig. 35).

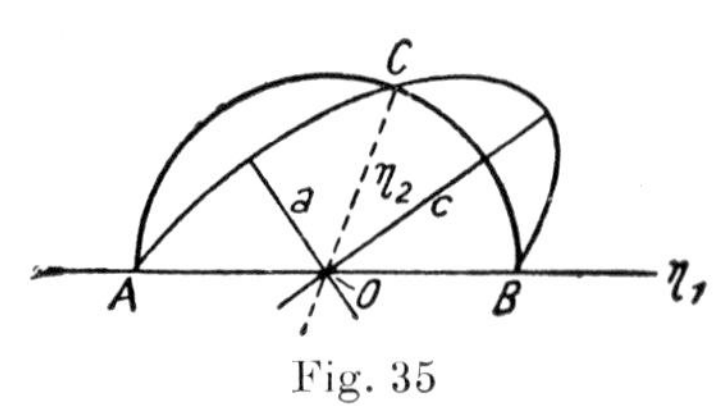

Fig. 35

The various complications, which arise through the fact that, apart from certain high symmetry crystals, in general only either K_1 or K_2 can be crystallographically rational planes and σ_2 or σ_1 resp. such edges and hence pressure twinning corresponds fully to twins of crystals of lower symmetry, are treated in detail by O. MÜGGE (1898, 1920), to whom reference is made.

The most striking crystallographic characteristic of "twin gliding" is in general the appearance of "polysynthetic twinning lamellae" such as those well known in calcite (fig. 630), rutile, hematite (fig. 557), pyrargyrite, and many others. These are always easy to recognize under the ore-microscope, mostly through reflection pleochroism and anisotropy. In isotropic minerals or very weakly anisotropic minerals, lamellae are recognized through etching. Cases are well known, many of them very widespread, in which such twin lamellae (as, e.g., in many plagioclase crystals) result from primary growth or from a transformation with falling temperature, hence no sign of outside mechanical deformation need to be present. In general, however, one must consider deformation as a possible cause of every polysynthetic lamellar twinning. Moreover, inversion lamellae also are signs of stress, although a stress from within the mineral itself, not an outside stress.

A consequence of *higher symmetry* is that in all cases where the pressure does not, by chance, operate *only* in the directions σ_1 of one particular surface out of all the equally possible surfaces K_1, several groups of lamellae always occur. This causes large complications not easily recognized. The different groups of lamellae influence and disturb each other, indeed can completely block each other. At the crossing positions straight rectilinear "canals" arise, which in the case of calcite, where they have been longest known, are often hollow. Similar canals are to be expected in some ores like hematite but are not observable in transmitted light. Microscopic investigation shows in the case of the latter, and especially well in the case of calcite that in their primary state the canals are not all empty. Rather they represent, as L. WEBER and later A. HOLMQUIST showed, positions in the crystal that were originally compressed especially strongly. By reason of the fact that this compression represents an overstressing of the cell lattice, perhaps with complete change to an isotropic condition, these "canals" are attacked by solutions with particular ease and so in fact are often "hollow canals".

For recrystallization processes, also, these "canals" often form the starting point. Accordingly the lamellae crossings are frequently the places where granular recrystallization begins along the twinning lamellae.

The following examples of "simple gliding", again, are only a small selection of the most important, most striking, or (in part) the more thoroughly investigated cases:

Bismuth (*arsenic* and *antimony similarly*) $K_1 = (\bar{1}012)$, $K_2 = (10\bar{1}1)$; *iron* $K_1 = (112)$, $K_2 = (11\bar{1})$; *sphalerite* $K_1 = (111)$, $K_2 = (112)$; *niccolite* $K_1 = (10\bar{1}2)$, $\sigma_2 = [\bullet\bullet\bullet]$*); *marcasite-arsenopyrite* $K_1 = (110)$, $\sigma_2 = [\bullet\bullet\bullet]$; *umangite* $K_1 = (101)$, $\sigma_2 = [\bullet\bullet\bullet]$; *chalcopyrite* $K_1 = (110)$, $\sigma_2 = [\bullet\bullet\bullet]$; $K_1 = (101)$, $\sigma_2 = [\bullet\bullet\bullet]$; *ruby silvers* $K_1 = (10\bar{1}4)$, $\sigma_2 = [\bullet\bullet\bullet]$; *bournonite* $K_1 = (110)$, $\sigma_2 = [110]$ *stephanite* $K_1 = (\bullet\bullet\bullet)$, $\sigma_2 = [\bullet\bullet\bullet]$; *stannite* $K_1 = (111)$, $\sigma_2 = [\bullet\bullet\bullet]$; *rutile, cassiterite* $K_1 = (101)$, $\sigma_2 = [\bar{1}01]$; *magnetite* $K_1 = (111)$, $K_2 = (111)$, *specularite, ilmenite, corundum* $K_1 = (10\bar{1}1)$, $\sigma_2 = [01\bar{1}1]$; *hausmannite* $K_1 = (101)$, $\sigma_2 = [101]$. In *galena* and *chalcocite* special relationships occur (see below).

It is easily shown that among the examples of gliding cited, there are those in which each atom (or ion) in the lattice undergoes one of the macroscopically recognizable displacements in a fully analogous manner (so-called "lattice gliding"). There are also those in which such a shift takes place only for individual atoms or groups of these, whereas other atoms have to move along rather complicated paths ("non-lattice gliding"). The second type is much more widespread than one supposes at first (e.g., calcite, most laws for galena, etc.). The obvious conjecture that *non-lattice* translations are more difficultly produced than lattice-gliding cannot be established through observation.

Some cases of gliding occupy a quite special position. These occur mostly in the very soft "plastic" ores. They were described first by O. MÜGGE in chalcocite (1920) later by H. SEIFERT (1928) in galena. The characteristic feature is that two glide surfaces K_1 can be linked with the same basal zone σ_2. Therefore: 1. In chalcocite with $o_2 = [100]$, $K_1 = (201)$, and $K_{11} = (20\bar{1})$. 2. In galena with $\sigma_2 = [110]$, $K_1 = (131)$ and $K_{11} = (13\bar{1})$. The plane of slip remains thereby the same. This has peculiar consequences, because a crystal will undergo both translations alternating with one another many

* Cf. footnote p. 58. 56

times. As brought out by MÜGGE, the indices of a given face are thereby altered in such a way that the second index remains the same, whereas the fist always varies between the original value and the opposite equal value. The third, however, always becomes negatively larger (fig. 36).

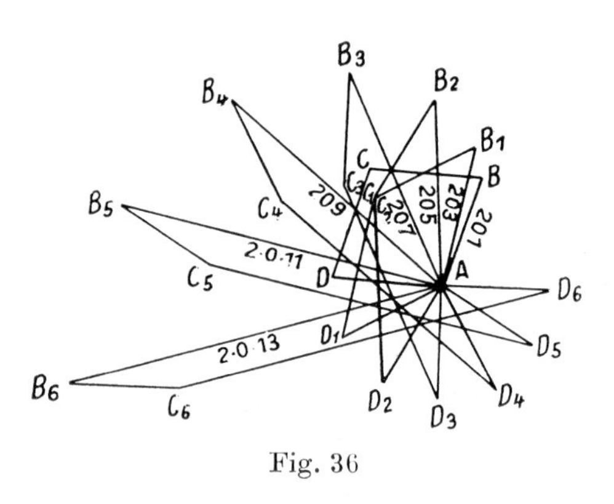

Fig. 36

As the figure taken from MÜGGE shows, in end effect quite thinly tabular grains can apparently be created continuously out of almost isometric grains, and there can arisea very far-reaching similarity to translation distortion and translation in general. Even though gliding with an effect of this kind has as yet been investigated only in galena and chalcocite, a similar relation is to be expected in many other minerals.

Each deformation of a mineral aggregate is, of course, the sum of very many complicated and quite varied deformations of single mineral grains, deformations that are quite complicated even for single mineral grains. One will be able to arrive at interpretation and incontestable description of the deformation *as a whole,* however, only through knowledge and investigation of these *individual deformations.* There is still a great deal to be done in this respect.

γ. *Recrystallization and formation of porphyroblasts.* The concept of "recrystallization" shall hold here only in the sense in which it was first applied. Metals, when rolled or otherwise strongly stressed over the elastic limit, take on properties which are no longer directly to be allied with their usual crystalline condition. Metallographers gave the name recrystallization to the phenomenon that such metals regain their "normal" properties spontaneously, or after longer or shorter periods of heating above a certain temperature. The original hypothesis that the metals become completely or in part "amorphous" during the stress and become "crystalline" by heating again (so is the name recrystallization to be understood) is certainly often untenable, but at the instant after stress, the fabric is so strongly altered, the individual "crystallites" are so traversed by gliding twin lamellae and displaced along translation planes, and the lattices of the crystallites are so strongly deformed and strained that the properties become quite different from the original properties. These conditions of strain as well as the large internal surface resulting from the twin lamellae are "disagreeable" to the crystal. It seeks, therefore, to rebuild its lattice free of strain and its surfaces to a more favorable ratio of surface to volume. This is achieved through recrystallization.

In metallography, and in part in petrography, quite different processes have also been designated as recrystallization, e.g., ordinary "collection" crystallization [Sammelkristallisation] in the course of long periods or during increasing of temperature, the enlargement of grain size in the course of slow circulation of a solvent, etc. Although naturally border cases are conceivable, such applications of the concept are misleading and should be rejected.

Relationships of recrystallization in iron, aluminum, and other metals have been accurately studied. Technology long ago learned that through choice of the degree of rolling and of recystallization temperature quite definite fabric properties could be obtained in the recrystallized product. About recrystallization in ores, on the other hand, we know very little as yet and are dependent on observations which are indeed very numerous but very fragmentary. Their interpretation is often rather hypothetical. Nevertheless a few reliable and interesting statements can be made qualitatively. The examples here adduced should serve only to stimulate further observations and experiments. They are, moreover, only a very small part of the available material.

Fig. 37 Natural size "galena-tail" (steely galena) Landskrone Mine near Siegen, Germany RAMDOHR

Of the three principal variables effective in recrystallization, *degree of stress, temperature*, and *duration* of recrystallization, none is reasonably known. In most cases, however, one will accept the period of time as optionally great and will be able to draw one's own conclusions accordingly. The upper limit of temperature can be set by means of "geologic thermometers", at least in many cases. The degree of stress can occasionally be estimated rather exactly, in other cases, however, not at all. Thus in schists and similar rocks, very perfectly spherical portions of ores (or tiny size up to that of a cherry) are known, which are pressed to rather regular rotation ellipsoids or rolled out into triaxial ellipsoids. For these, and only for these (and not for the remaining constituents of the rocks), the degree of "rolling-out" can therefore be estimated quite well. The other components can show quite differing relationships according to whether they are "more rigid" or "more plastic". The statement: *The stronger the "rolling-out" the greater the inclination to recrystallization*, is easily explained from the discussion given above. Its correctness may be seen at a glance in many "galena-tails" in fine granular sphalerite, in chalcopyrite, etc. The writer has described a good example (1928,d); his illustration is here repeated (fig. 37). The parts next to the selvage along which the motion took place are especially strongly stressed. They were originally rolled out into very thin lamellae; farther away from the selvage, the lamellae become coarser and coarser. Finally, there is the coarse granular material which is hardly influenced at all. The fineness of the lamellae in relation to the diameter of the original grain is always a measure of the severity of the "rolling-out". Recrystallization affects mostly only a thickness of a few centimeters measured outward from the selvage; here the lamellae have completely disappeared. Farther away on the borders of the lamellae various stages of formation of new grains are to be observed, and finally there is the portion in which the lamellae are completely intact. In other galenas, conditions being otherwise equal, recrystallization has proceeded to development of noticeably coarser grains; here the temperature of the recrystallization must have been higher, or another factor — perhaps circulating solutions — must have caused the greater progress of recrystallization. An especially beautiful example of "rolling-out" and recrystallization in chalcopyrite is shown in figs. 40a and b.

Which ores have greater or lesser tendency to recrystallization is difficult to say; indeed, available observations are, so far, scanty. In general, medium soft to moderrately hard minerals (galena to pyrrhotite) recrystallize easily. The soft and very hard minerals recrystallize less easily or not at all. Thus, e.g., aggregates of molybdenite crystals stressed with especial severity are not recrystallized. The same is true of graphite, as far as known. On the other hand, superb recrystallization fabrics are known in *covellite* (also a layer lattice mineral!) (fig. 453), also in *orpiment*. Among especially hard minerals, I have recognized practically no recrystallization in the case of *cassiterite* and *chromite*, even where they were strongly stressed, whereas pyrite (at least in special cases), magnetite, ilmenite, and hematite show it frequently.

The *temperature* has in any case an extreme influence. From studies of sylvite and metallic lead it is known that they can recrystallize rather quickly even at ordinary temperatures; in galena recrystallization certainly happens also, at least with adequate time and in portions stressed with particular severity (fig. 37). Indeed, it was observed that during the preparation of specimens (impregnation of brittle materials at about 100 °C for twenty-four hours) distinct recrystallization occurred. This is a warning that in investigations of these problems materials should be prepared without long heating! Other ore minerals recrystallize only at very high temperature and show no trace of recrystallization in the usual occurrence. In this category belong pyrite, ilmenite, magnetite, arsenopyrite and others. Pyrite is not recrystallized in most cases but is very well recrystallized in others, e.g., in ores of Sulitelma. Fig. 44a gives a superb example of recrystallization of ilmenite in an especially stronlgy stressed portion of an ore. Arsenopyrite occurs in recrystallized aggregates only exceptionally. With exceptional ease, i.e. surely near 150 °C and below, recrystallization occurs of galena, sphalerite (figs. 41, 42, 374), chalcopyrite, bournonite, bornite, stibnite, and orpiment. In the course of recrystallization noteworthy *shifts of material* can also occur along with *changes of texture and structure*. In general, grain size becomes more uniform and also rounded* in the case of minerals that otherwise have a pronounced tendency to tabular or needlelike development. Very often, the formation of a fine granulated material (fig. 33 in stibnite) begins first along the boundaries of grains and twins of the strongly stressed grains. This material then develops more and more at the expense of the relicts. Although the parallel texture is apparently obliterated or suppressed

* An especially instructive example is given by Ödman (1933) in a recrystallized cubanite from Kaveltorp, which consists of rounded grains in chalcopyrite.

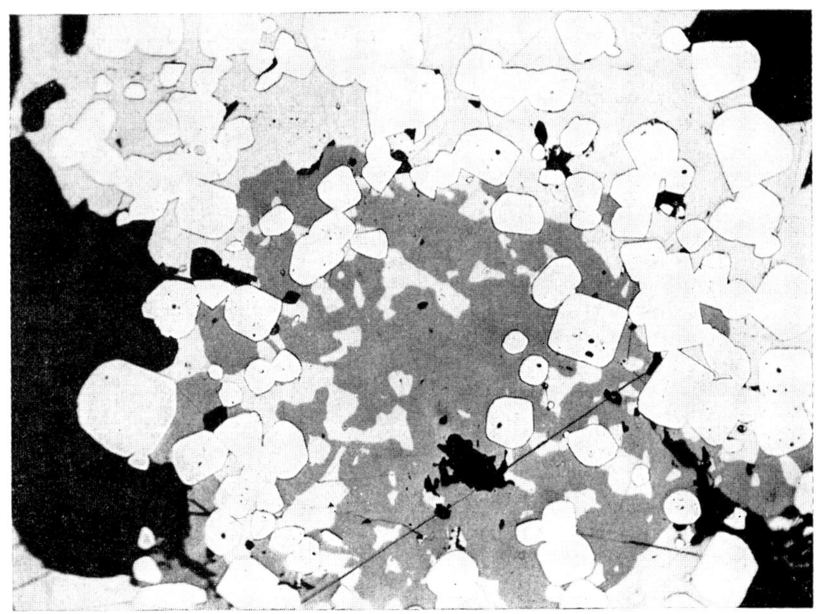

Fig. 38 250 ×, imm. RAMDOHR

Binnental, Wallis (on road Binn-Giessen), Switzerland

Splendidly recrystallized ore from sugary Triassic dolomite which was subjected to Alpine metamorphism. White, rounded idiomorphic crystals are *pyrite*; grey-white *galena*, dark grey *sphalerite*, Black is *dolomite*

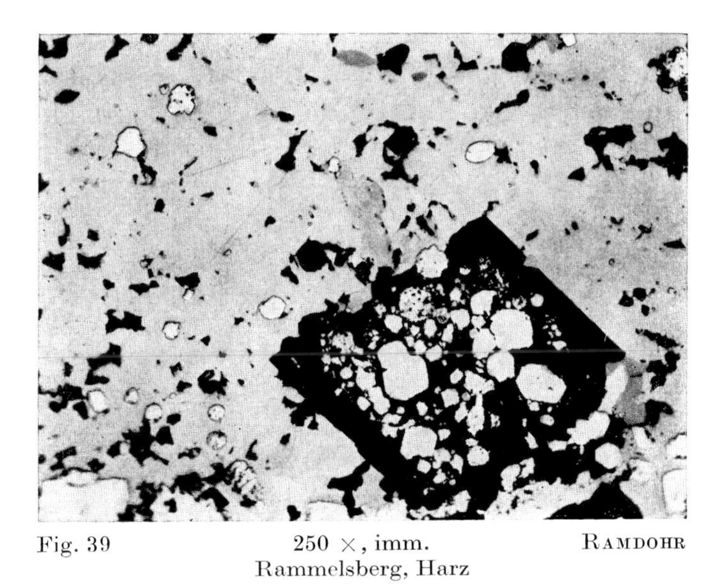

Fig. 39 250 ×, imm. RAMDOHR

Rammelsberg, Harz

Fine grained recrystallized *chalcopyrite* ore with *pyrite* grains and little *sphalerite*. A large porphyroblast of *dolomite*, with old inclusions of different kinds of *pyrite*. Such photographs show that in each metamorphism of this type migration of material is considerable

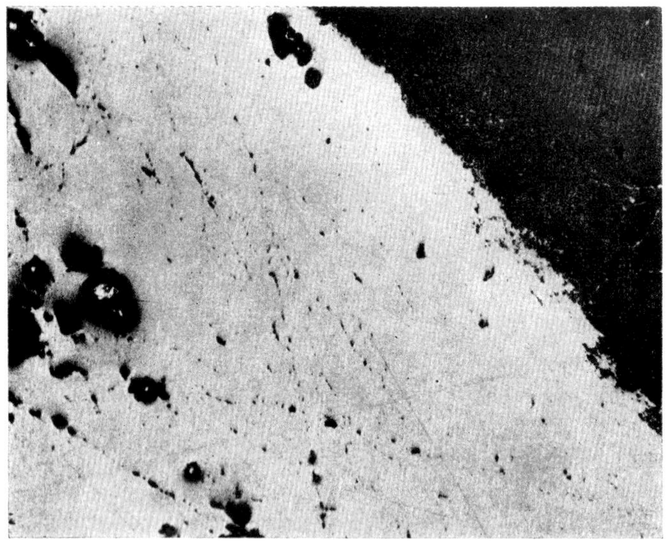

Fig. 40a 50 ×, imm. ordinary light RAMDOHR
Ems, Merkur Mine, Germany

Sheared *chalcopyrite* with a little *pyrite* (left, high relief) and *sphalerite* (dark grey, top at right). In ordinary light chalcopyrite shows scarcely any peculiarity

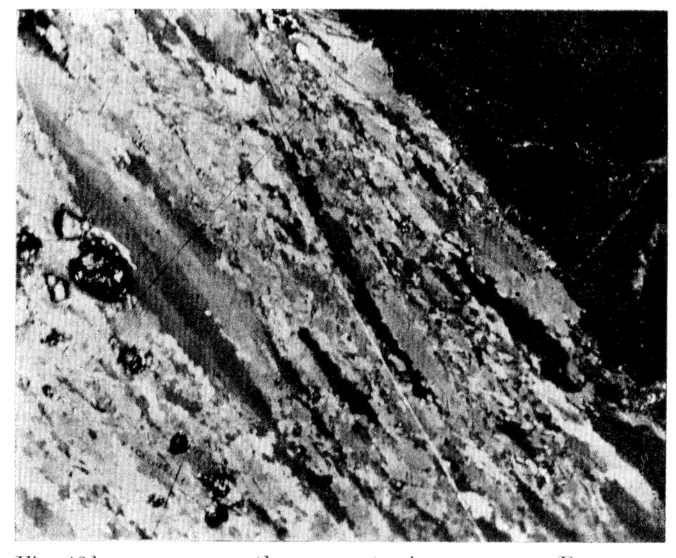

Fig. 40b the same, + nic. RAMDOHR

The streaky "rolling-out" and incipient recrystallization in chalcopyrite is excellently recognizable

through the formation of rounded grains, an order is retained and is often only then produced. It is manifested, however, more in the inner structural orientation of grains than in their form. In certain aggregates of grains this can be recognized at a glance from optical characteristics, e.g., on the basis of more or less uniform extinction (figs. 45, a and b). More detailed investigations became possible only by means of methods given by SCHACHNER-KORN in Vol. 1, first half of *Die Erzmikroskopie* on pages 268–302. An important consequence of recrystallization is the destruction of zonal structure even in minerals which otherwise regularly show these structures to an unusual extent (e.g., in smaltite–skutterudite from Gashani in Rhodesia). Also, many mixed crystals that were formed at high temperatures unmix, although they may represent "frozen disequilibria" normally durable for infinitely long times. Thus iron-rich sphalerite is unable to retain its high iron content during annealing, and exsolves it into the grain boundaries as pyrrhotite. The same is true for titanomagnetite of several occurrences; its place is taken by magnetite and ilmenite grains, often accompanied by much spinel. This can become important in mineral beneficiation. Furthermore, unmixing textures, oriented intergrowths, myrmekitic intergrowths, etc., are destroyed, so that we are often robbed of important bits of evidence for the interpretation of the genesis of the original deposits. Thus magnetite no longer shows lamellae of ilmenite or inclusions of spinel if the temperature of recrystallization was low (e.g. Otanmäki). Chalcopyrite loses the little stars of sphalerite, sphalerite the ovoid bodies of chalcopyrite, etc.

Closely linked with the phenomena of recrystallization is the *formation of porphyroblasts**, which occasionally occur in ores just as euhedral crystals of garnet and staurolite, e.g., occur in metamorphic schists. Magnetite, pyrite, and ilmenite tend especially toward formation of porphyroblasts in recrystallized ore mixtures, as is already well known from petrography. As is usual in crystalline schists the porphyroblasts often show inclusions of foreign minerals, still with the orientation and grain size they had before and during the movement ("sieve porphyroblasts" — figs. 110, 111, 113); hence they allow valuable inferences as to mechanism of movement (figs. 47, 48).

Porphyroblasts in dynamically metamorphosed ores have many times been misinterpreted. One of the best known examples is the rounded pyrite in the chalcopyrite ores of Sulitelma, which J. H. L. VOGT thought to be magmatically corroded phenocrysts in an ore magma and used as evidence for magmatic origin of this deposits. On various reasons — partly of a general geologic nature, partly of a structural nature — such an origin is surely not applicable: these crystals are without question porphyroblasts, and the rounding is a consequence of "inhibited crystal growth" in the sense of O. MÜGGE. This is the case with many porphyroblasts. The tendency towards the formation of such round crystals is again a definite property of materials. For the ores it has received comparatively little study, e.g., the danaite (cobalt-bearing arsenopyrite) that is associated with the rounded pyrite just discussed occurs in perfectly euhedral crystals although it has formed in every respect under the same conditions. Apparently, certain habits of crystals of one and the same mineral tend,

* These can have their own crystal form ("idioblasts") or, having largely uniform orientation, can be without euhedral forms ("xenoblasts"); both types of porphyroblasts occur together — e.g., in the case of magnetite in the ores of the Rammelsberg —, the form is not particularly important.

under conditions otherwise similar, more than others toward irregular rounding. Thus Sulitelma pyrite crystals with (210) as a prevailing form are very markedly rounded, those with (100) as the dominant form less rounded, those with (111) as the dominant form almost not at all.

The formation of "porphyroblasts" or "idioblasts" is in no way limited to dynamic metamorphism; it occurs in a quite similar manner as a result of contact metamorphism (e.g., the cordierites of the spotted schists [Fruchtschiefer]!) or pure load metamorphism, and porphyroblasts of the three origins are not always easily distinguished

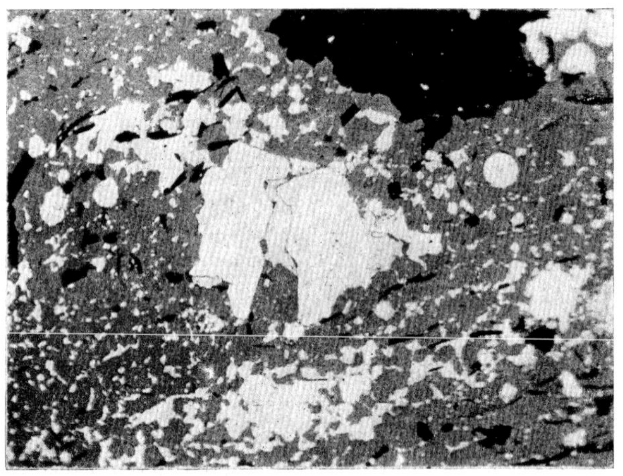

Fig. 41 350 ×, imm. RAMDOHR
Rammelsberg, Harz

Shattered former "hard body" of pyrite in sheared and recrystallized sphalerite aggregate. The crack in *pyrite* is filled with *galena* and *sphalerite* without any replacement whatsoever

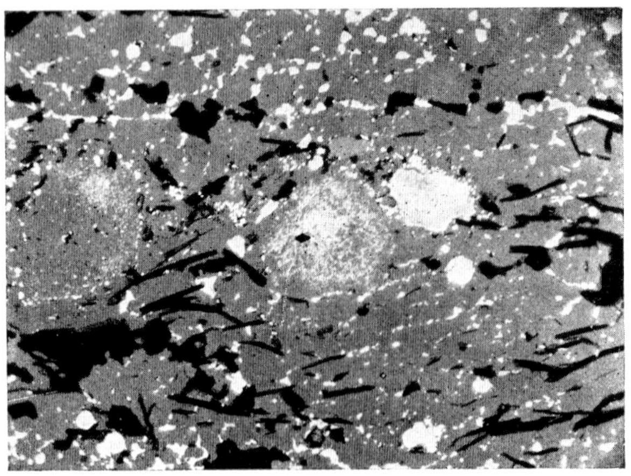

Fig. 42 350 × RAMDOHR
Rammelsberg, Harz

Sphalerite knot acts as hard body [Härtling], because of "primary", i.e. immediately after precipitation, intergrowth with a fine *pyrite* net, whereas the other sphalerite is already sheared and recrystallized

from one another. Finally, and this cannot be emphasized enough, prophyroblasts also form in the most varied mineral associations without any sign of metamorphism (figs. 111, 113). The work of SCHNEIDERHÖHN (1931) furnished good examples of load metamorphism from the ores of Postmasburg; porphyroblasts of sitaparite, e.g., have formed there. In order to determine which type of metamorphism caused porphyroblast formation one must pay attention especially to minerals formed in "*pressure shadows*" and in "*tadpole-shaped forms*" and to "*displaced relict fabrics*" in the minerals of inclusions. This is all the more true because ore minerals that form porphyroblasts — as far as known — are mostly persistent minerals: pyrite, magnetite, ilmenite, cobaltite, arsenopyrite, and spinel. They are not limited, as are staurolite and kyanite, e.g., to assemblages of a definite depth zone. The interpretation of many a deposit as yet controversial may perhaps the facilitated thereby.

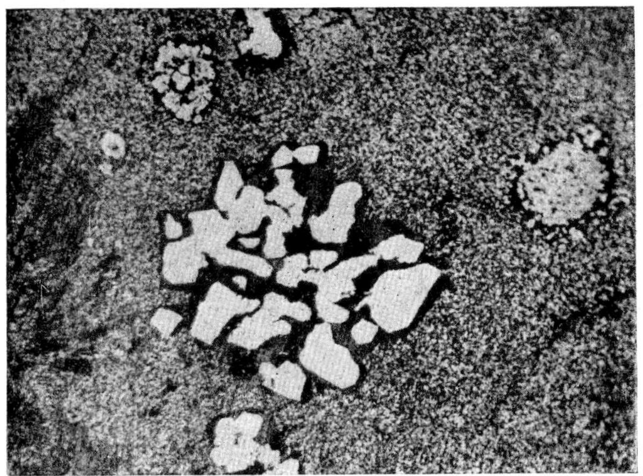

Fig. 43 350 ×, imm. RAMDOHR
Bor, Yugoslavia

Extremely fine-grained mixed aggregate of *pyrite* (white) and *covellite* (grey) shows speckled development of crystallinity in larger and smaller knots, starting spontaneously for some unrecognizable reason. In these knots the difference in reflectivity of covellite due to the difference in orientation in the section can be recognized. Note similarity in form with e.g. fig. 215, which, however, is to be interpreted quite differently with regard to genesis

With reference to the varied behavior of minerals of an assemblage formed under exactly the same type of stress, the writer was able to make systematic observations on the ores of Routivare. The primary mineral association there — titanomagnetite, with abundance of the spinel molecule, ilmenite, spinel, and locally högbomite and corundum — shows with weak stress a recrystallized product consisting of little ilmenite but abundantly preserved magnetite and spinel; with stronger stress a uniformly granular aggregate of magnetite + ilmenite is produced. Finally but rarely, the large spinels, previously showing only cataclasis, are recrystallized. The unmixed bodies have thereby mostly disappeared. Unmixed bodies newly formed in great number but in sparser amounts and smaller size show that the movement and recrystallization took place at rather high temperature (fig. 46).

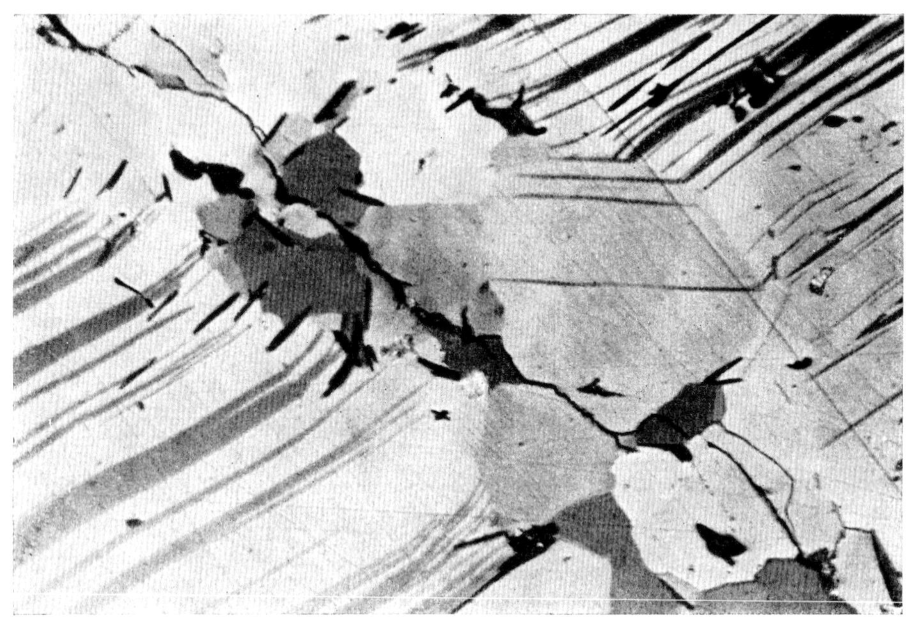

Fig. 44 a 250 ×, + nic., imm. RAMDOHR
Halsekubhammer near Sellevold, Norway

Ilmenite with zone of shearing. In the strongly stressed part recrystallized grains are almost without twin lamellae; in the bulk of the crystal is strong twinning

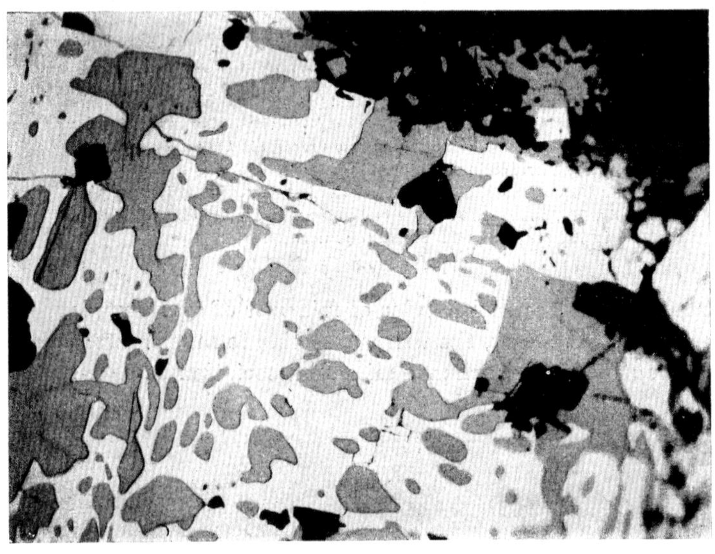

Fig. 44 b 250 ×, imm. RAMDOHR
Sierrecilla, near Rio Tinto, Spain

A porphyroblast of *pyrite* in *galena*. Typical "porphyroblast-sieve"

δ. *Changes of mineral content during the course of metamorphism.* Some evidence as to change of mineral content in the course of metamorphism, especially as a consequence of recrystallization, has been mentioned in the previous section. The phenomena discussed there, however, are for the most part not very striking. They result essentially from rapid adjustment of equilibrium in mixed crystals that have become unstable (Fe-rich sphalerite → Fe-poor sphalerite + pyrrhotite, as at Hürningskopf; titanomagnetite → magnetite + ilmenite). There is an abundance of changes that are substantially more incisive and also more difficult to interpret. These changes also repre-

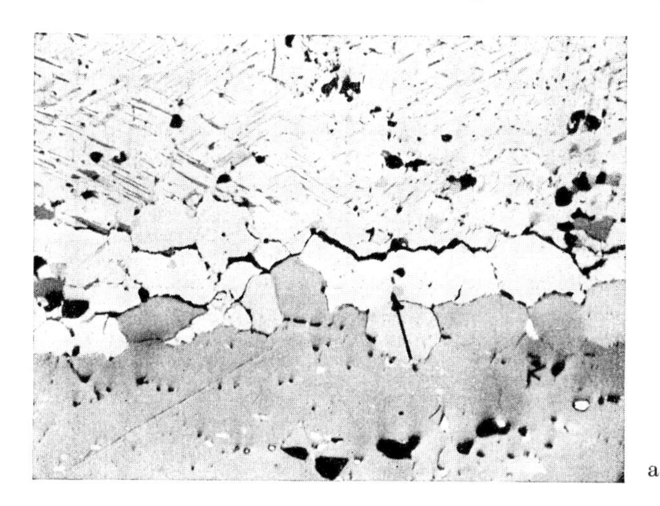

a

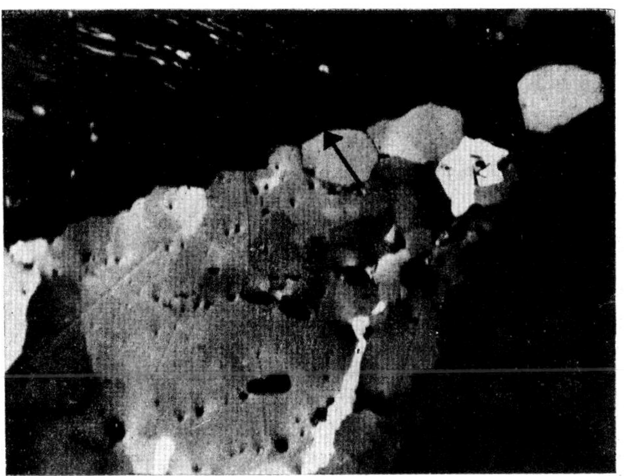

b

Figs. 45a and 45b 100 ×, imm. RAMDOHR
a, one nicol; b, nicols crossed
Routivare, North Sweden

Glide plane lying between a "large grain" ["Grosskorn"] of *magnetite* (grey white) and of *ilmenite* (grey). Along the glide plane recrystallization has begun in both minerals. The recrystallized grains are almost free of exsolved bodies and the ilmenites lie still roughly parallel with the large grain, which has acquired undulose extinction. (For orientation the same grain is designated with an arrow in each photograph)

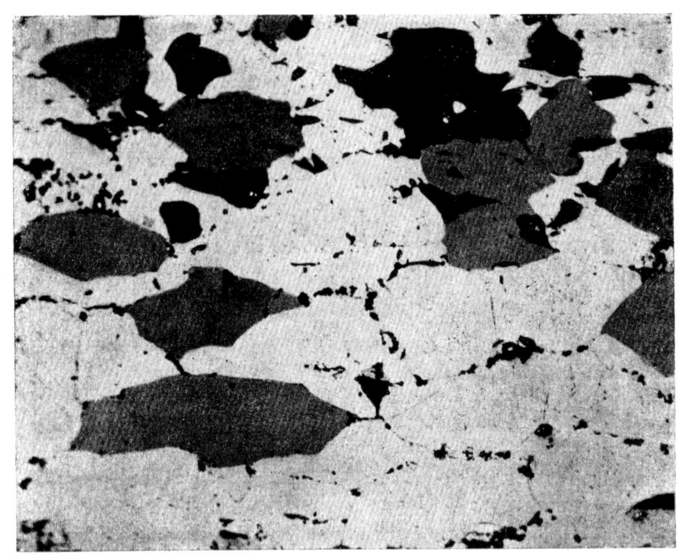

Fig. 46 325 ×, imm. RAMDOHR
Claim Pluto near Routivare, North Sweden

Recrystallized *magnetite* (light grey) and *ilmenite* (dark grey), distinctly oriented with intergranular bodies of *spinel* (small black grains), along with a few coarse grains of *spinel*

Fig. 46 a 70 × RAMDOHR
Ayoayo, Bolivia

A recrystallizate of *galena* and *sphalerite*. Grain boundaries developed by starting weathering

sent equilibria corresponding to the new pressure-temperature conditions or at least tend toward them. In these change substances (H_2O, H_2S, As) mobilized or introduced in any way through movement can be effective.

Examples introduced here are only those that appear typical and important. They are not intended to be a conclusive evaluation or an enumeration in any way complete.

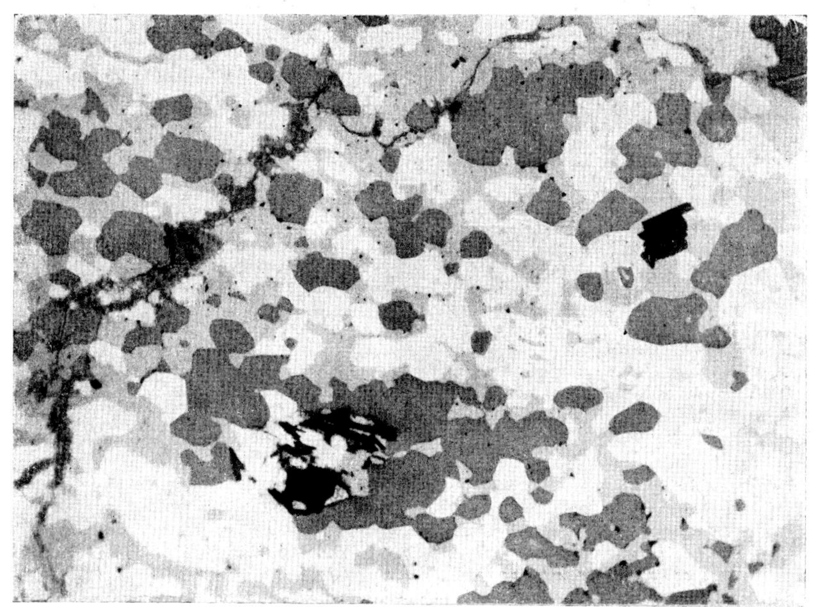

Fig. 47 170 × RAMDOHR

Gold mine Tiouit near Tinerhir, Southeast Morocco

Peculiar intergrowth, exactly the same in appearance as a recrystallized ore, of *chalcopyrite* (white), *fahlore* (grey), and *sphalerite* (dark grey). The ore shows, however, no other signs of shearing

Metamorphism at lower temperatures will mostly furnish associations which are rich in minerals compared to the original assemblage. The reverse is true for metamorphism at high temperatures because formation of mixed crystals is intensified and hydrated compounds are destroyed. These processes are reversed at low temperatures. There are, however, many exceptions; a good many minerals, e.g., those having the character of double salts, are only stable at higher temperature and appear additionally.

In *iron ores* we have especially fine examples: specularite schists that consist almost solely of hematite ± quartz arise in the course of movement at elevated temperatures (but not too high) from iron ores of the greatest variety of mineral compositions and origins. They arise, e.g., out of oolitic limonite ores, chamositic ores, and carbonate ores. They also form at temperatures certainly lower than those of original formation, out of contact metasomatic or magmatic magnetite ores. Iron-bearing contact silicates decompose to hematite and quartz just as do the silicates formed by sedimentary processes. Generally metamorphism brings about convergence, so that interpretation is often difficult (figs. 49–52).

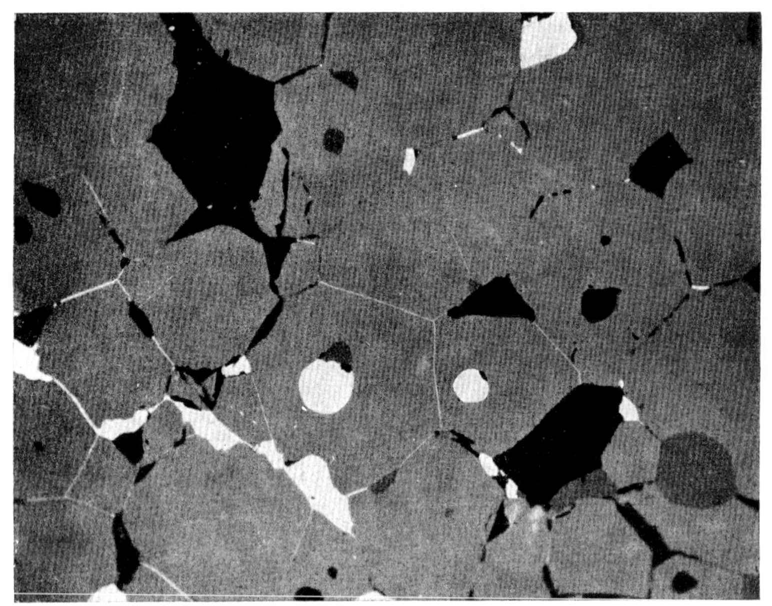

Fig. 48 170 × RAMDOHR

Broken Hill, N.S.W.

Recrystallized material composed of *manganese garnet* with round grains included in the course of recrystallization. The grains are of *pyrrhotite*, *galena* (both pure white), and *quartz*. The thin films between individual garnet grains are secondary infiltrated pyrite

Fig. 48a 10 × RAMDOHR

Zinc Corp. Mine, Broken Hill, N.S.W.

Equigranular recrystallisate of *quartz*, dark-grey, rounded, *sphalerite*, light-grey, as a filling, but of the same grain-size, little *galena*, white, in accidentally formed grains

In *manganese* ores similar relations may occur. The ores formed at low temperature, most of them definitely sedimentary ones consisting of pyrolusite and psilomelane but partly also manganite, are changed to braunite and perhaps hausmannite. HUTTENLOCHER (1936), e.g., described braunite ores of this kind which show porphyroblasts. In

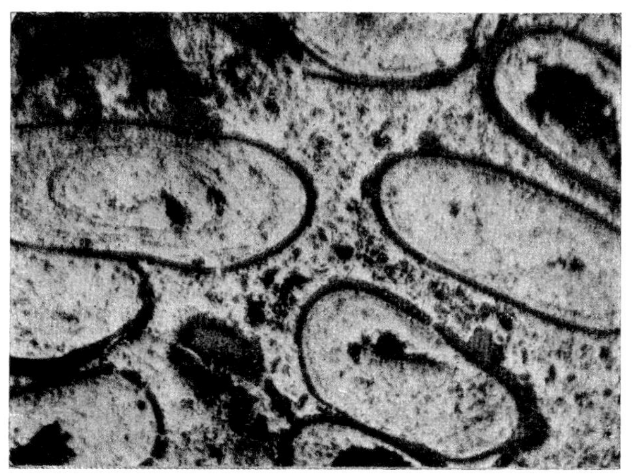

Fig. 49 170 × RAMDOHR
Giraumont, Lorraine, France

Magnetite oolites from minette ore. The formation of magnetite is here *not* a metamorphic process

Fig. 50 170 × RAMDOHR
"Golzer Berg", Switzerland

Oolite, now consisting of *hematite*; idioblasts of *magnetite*. The carbonate gangue is "dusted" with hematite. — Epizonal metamorphism

other places the original ores react with silica to form rhodonite or with CO_2 to form rhodochrosite. Little is yet known about *copper ores*, but the disappearance of cubanite with the formation of new chalcopyrite and pyrrhotite is frequent. Once cubanite is preserved as such, movement must have taken place at high temperature (cf. ÖDMAN, loc. cit.). Often simplification of the mineral association takes place as shown by the fact that most dynamically metamorphosed copper deposits are monotonous and similar in mineral content even though derived from very different types.

Regarding dynamically metamorphosed *lead ones*, hardly anything is known. In any case the association is mostly monotonous. The lead–antimony–sulphosalts, such as bournonite, are in part preserved, in part destroyed. New sulphosalts, however, are also formed.

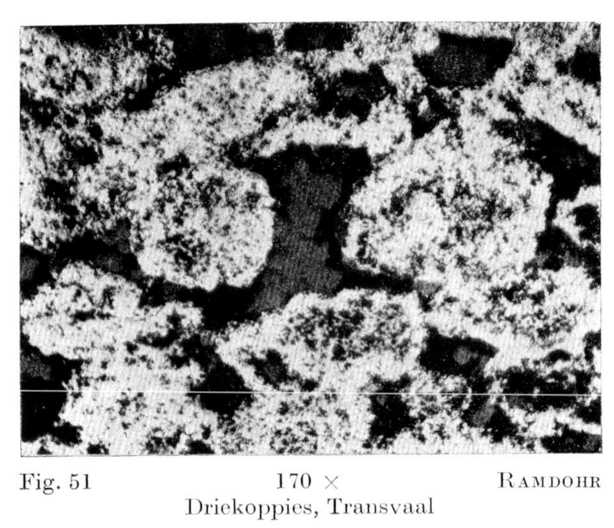

Fig. 51 170 × RAMDOHR
Driekoppies, Transvaal

Magnetite knots, apparently also former oolites. Gangue is quartz

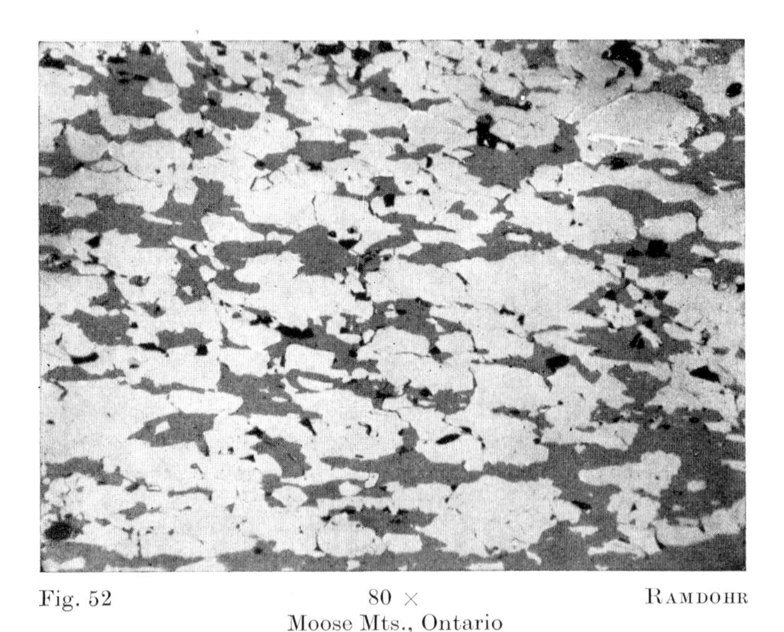

Fig. 52 80 × RAMDOHR
Moose Mts., Ontario

Magnetite-quartz ore with distinct orientation in grain form. In the light grey magnetite there is some secondary hematite

Ilmenite often decomposes into hematite + rutile or quite frequently also with a simultaneous mobilization of calcium, to sphene + hematite or magnetite.

Perhaps the best known example is the formation of graphite out of "carbonaceous substance" and hydrocarbons in many crystalline schists and the deposits within them.

A peculiar type are the replacements of hematite by pyrite in the ores of Tuscany, described by WIJKERSLOOTH. These took place along a grand-scale dislocation.

All in all, the variety of these changes of mineral associations is not comparable to that in the crystalline schists. The main reason is that there are no sulphide minerals with water of constitution, hydroxyl, fluorine ,etc. Among the oxides of the heavy metals, also, such constituents are not common.

(b) *Application of the concept of "depth zones" of rock metamorphism to ores and ore deposits*

It is well known in petrology that application of the concept of depth zones (epi-, meso-, and katazones), introduced by BECKE, GRUBENMANN, and NIGGLI and generally accepted, creates great difficulties. The definitions are clear and logical in themselves; the different minerals, however, behave quite differently with respect to temperature changes and to directed and confining pressure, so that the reactions are difficult to survey. Thus the same minerals may react very differently with changes in kind of associated minerals (or even of amounts of these minerals), with different treatment (preparatory stages of metamorphism), and with the circulation of even very slight amounts of solutions. The "depth zones" are not necessarily connected with depth. Much more do they describe reactions to a group of forces which become effective mostly with progressive subsidence to greater depth, but can develop also in other ways. It is conceivable that in one and the same mineral assemblage at the same temperatures and pressures, in the presence of traces of solution x, the mineral A may be completely disrupted or full of twin lamellae and mineral B may be completely recrystallized; whereas at still the same temperatures and pressures in the presence of only traces of solution y, A now recrystallizes and B is still cataclastic.

In addition, it may often happen that rocks pass through several depth zones successively. E.g., it is not uncommon in rocks to find that "typical" "index" minerals, e.g., of the katazone are found in rocks for which there is good evidence for excluding katazone conditions. — The difficulties become especially great if we have to deal with unusual associations and mineral compositions.

To be therefore quite careful, one is best advised (with SANDER, 1930) first of all to investigate and describe the phenomena satisfactorily and, above all, without preconceived opinions, before one thinks too much about process or place of origin.

In view of the relative monotony of the assemblages present in the *crystalline schists*, the small number of minerals involved, and, not the least, the very great number of studies already done with regard both to the possibilities involved and to the observed phenomena, it is possible to set up such depth zones at least in the form of an "average conclusion" and to designate certain minerals or mineral associations as "typomorphic". As already indicated, there are certainly cases in which minerals that otherwise fit quite freely into the scheme behave for no apparent reason in an unusual way, but this cannot lead to abandonment of our scheme or to treating it only as a "working hypothesis". Even with all its weaknesses, it is much too well founded.

The situation is quite different as soon as we try to apply the scheme to ores. For various reasons great difficulties arise. Many ore minerals that also occur in rocks especially those like hematite, magnetite, and ilmenite, are comparable to the silicates in behavior to some extent. The situation is different for most of the others. First, their behavior has as yet been very little investigated. Second, in their capacity for translation and gliding and their ease and velocity of recrystallization, they so far surpass the silicates and also differ so much among themselves that relationships quite different from those in the silicate rocks are to be expected. In the future there will be no choice but to introduce new nomenclature and new concepts.

The association of minerals of very different plasticity, important also among the silicates, introduces quite different conditions with it. At different points within the same deposit, the same mineral can react very differently to a given stress according to the kind and amount of associated minerals. As examples we may recall the variety of the textures of the individual minerals at Rammelsberg, and also of the minerals in certain alpine vein assemblages. In these, the extreme mobility of many soft ore minerals like galena, molybdenite, and graphite, which act as lubricants, has a very marked influence. Even with small stresses, phenomena result which are certainly well known in the case of the silicates but appear at quite different pressures in ores. Thus the ores of many veins and of other occurrences serve as vehicles and propagators of the process of motion and are themselves rolled out to the greatest extent and also recrystallized at a stage when the country rock is still fully intact. It is therefore no accident that "slickensides" often consist of sulphide ore minerals or graphite. Particularly excellent are cases where quite young ore minerals of the zone of enrichment react to subsidences caused by mining operations. Such cases, e.g., have been observed in the cementative chalcocite of Chuquicamata.

Everything said up to this point is therefore *quite negative* with regard to the possibility of classifying the metamorphism of ores according to depth zones!

The little that is *positive* can be said briefly. If one considers that the same depth zones hold for the ore minerals as well as the silicate minerals, then in many cases it is simplest to use the silicate minerals themselves as index minerals.

The result of this will prove that *almost all* the *softer* and *more plastic ore minerals* which still *show recognizable stress textures* (galena, chalcopyrite, sphalerite, or indeed even stibnite, orpiment, covellite, etc.) can be assigned to metamorphism of the epizone or to a range which one might in part place lower in the scale and which lies where, in silicates, only an elastic stress is present. In the mesozone and katazone these ores are already so extensively recrysallized that a distinction against an original normal paragenetic sequence has become impossible. E.g., they will always occur, regardless of their true age, as cementing or fill materials between the silicates and the "hard ore minerals". Many "younger mineralizations" in silicate rocks which under the microscope can be "proved incontestable" younger may even so have been falsely interpreted (cf., e.g., the writer's work on Broken Hill (1950a)).

The "hard ore minerals" (e.g., pyrite, arsenopyrite, magnetite, hematite, cobaltite, braunite and others, with which graphite and perhaps molybdenite are in a sense to be included on account of their inferior ease of recrystallization), *are rather to be compared with quartz and silicate minerals* of average hardness (figs. 20, 29, 41). Ilmenite also belongs here, although its decomposition into rutile and hematite, often associated with metamorphism, and its conversion into sphene in many cases make the question more difficult, so that its position can still be rather obscure. Where different sulphide and oxide stages can occur, a contrary influence is often caused by increaes in temperature and pressure with increasing depth (cf. the discussion on geologic thermometers, page 242), which without further information prevents recognition of how the equilibrium will be shifted. Thus magnetite may be found in typical epizones, hematite perhaps still in the katazone, whereas magnetite undoubtedly belongs by its nature "to greater depths". A similar and analogous relation exists between pyrite and pyrrhotite, and analogous examples are known from the study of many complete mineral assemblages. The "hard" ore minerals, likewise, always show an ease of recrystallization which is above that of many silicates. Thus, at high-temperature ore deposits or geologically old ones the structures must be considered still more carefully than it was allowed by the knowledge of convergences.

(c) *Minerals formed during serpentinization of ultrabasic rocks*

Mineralization, not easily to be included into the traditional classification, occurs where ultrabasics, i.e. the different members of the dunite family, and homologous hypabyssal rocks like pikrites, alnöites, etc. and perhaps metamorphites of similar composition are serpentinized, that is, connected with ophiolitic intrusions, orogenetic processes and perhaps other deep-seated movements as e.g. plate tectonics. These processes are at present economically unimportant, even though some platinum deposits may belong to that group and Ni-enrichment reaches 1%. Mineralogically, however, they are very important, because they are the original source of many native metals and intermetallic compounds, of oregonite, heazlewoodite, orcelite and some layer structure sulfides, and because they show peculiar — not yet really understood — alterations of chromite, partly connected with formations of chromium silicates (kaemmererite e.g.) — not regarding the trivial formation of serpentine minerals, of sepiolite, gymnite, magnetite, magnesia carbonates, etc. Of course, this not at all means that *all* platinum minerals come from this process.

Besides the very important increase of volume and the connected, partly enormous "internal" tectonics, some surprising chemical reactions are resulting. The FeO content of the olivine (Fo:Fa: e.g. 3:1) forms the well known magnetite, but besides that hydrogen i. stat. nasc., which, in its turn, can precipitate the nobler elements of the association: Pt, platinoids, Ni, Cu, Co and Fe itself in metallic form and likewise low sulphur sulfides like heazlewoodite, shandite, oregonite.

The idea that hydrogen could play a role in that process was mentioned by the author (1950) and strongly supported by BETECHTIN (pers. comm.) because he had considered similar processes for some Pt-deposits. Meanwhile, the world-wide occurrence of such deposits with a really pedantic similarity has become known, and several observers came to the same conclusion, in part apparently quite independent of older literature.

A possible reaction, to be modified with higher or lower content in Fe, is given by CHAMBERLAIN et al. (1965):

$$6\,(Mg_{1.5}Fe_{0.5}SiO_4) + 7\,H_2O = 3\,(Mg_2Si_2O_5(OH)_4) + Fe_3O_4 + H_2\,.$$

The temperature during the formation was considered by the author already in 1950 regarding the contemporaneous formation of awaruite and copper, conjectures which became approved, by the occurrence of awaruite, wairauite, and auricupride, all surely formed below 390°, or probably considerably less.

The author saw at least eight such deposits outcropping in the ophiolitic zone of the Alps, of which the most spectacular, Selva near Poschiavo, has been described very carefully by DE QUERVAIN (1945). Meanwhile, similar ones have become known and described by CHAMBERLAIN et al. (1965) from the arctic Canada, by E. H. NICKEL (1959) from Quebec, CHALLIS & LONG, New Zealand (1965), GOLDING, near Canberra (oral comm.), HAHN-WEINHEIMER & ROST, Münchberger Gneismasse (1961), KANEHIRA et al., Shikoku (1965), KRISHNA RAO, India (1964), and, exceptional and more complicated by the content of plentiful sulfides, by ANTUN, EL GORESY and RAMDOHR from Cyprus. Very many have been studied further by the author recently — actually about 50% of all serpentinized peridotites show in higher or lower degree the same association. The formerly known occurrences of awaruite or josephinite are — in other words — only members of a common association outstanding only in the size of their products. The serpentine belt of Barberton, S.E. Transvaal, shows this association very well.

The mineral content can be very complex, especially in the sulfid- and As-rich ones I give (with r and rr for rare and very rare): Awaruite, wairauite, auricupride (r) α-iron, copper, oregonite (r), orcelite (rr), heazlewoodite, shandite (rr), chalcocite (r) bornite, pentlandite, millerite, sphalerite, chalcopyrite, niccolite (r), mackinawite valleriite (r), cubanite (r), magnetite, chromite (mostly picotite), complicated alteration products of chromite, haematite, anatase (r) — and besides them a number of poorly examined, not yet identified very rare minerals.

The source of the components mostly seems to be the immediate adjacent rock with a very small migration only. In some cases high in copper perhaps primary enrichment can be assumed (Alexo Mine, NALDRETT 1965, Laxia tou Mavrou, ANTUN et al. 1966).

THE ORE TEXTURES

ORDER OF PRESENTATION

Synopsis of the bases of classification: Purely geometric — genetic — technical (economic), with references to previous attempts at classification.

A. The fabric properties considered from a purely geometric point of view

I. The properties of single grains
  a) internal nature
  b) external features

II. Intergrowths of several minerals
  a) oriented intergrowths
  b) "emulsion" intergrowths
  c) penetration textures
  d) myrmekitic intergrowths

III. The forms of aggregates
  a) arrangement
  b) grain boundary relations
  c) mineral inclusions in ore minerals

IV. SCHNEIDERHÖHN's systematic classification of the structures and textures of the ores

B_1. The fabric types considered from a genetic point of view

I. Textures of primary precipitation
  a) growth fabric (crystallization from melts and solutions)
  b) colloidal textures
  c) sedimentary textures

II. Transformation textures
  a) paramorphic replacement textures
  b) exsolution textures
  c) decomposition textures
  d) "Verdrängung" — "Replacement" — "Metasomatism"
  e) thermal transformations
  f) oxidation textures
  g) cementation textures ("secondary" enrichment, e.g.)

III. Deformation fabrics

IV. Radioactive phenomena

B_2. *Recognition of the genetic position of ore deposits*

I. Typomorphic minerals, mineral associations, (paragenesis), age relations texture types

II. Ore minerals and their associations as "geologic thermometers"

III. Relicts

IV. Further possibilities of the genetic interpretation of textural characteristics

C. Ore textures in relation to mineral dressing problems

PRINCIPLES OF THE CLASSIFICATION OF THE ORE INTERGROWTHS

The attempt to classify the textures of complex ores, and of monomineralic aggregates will be approached from several quite different viewpoints, as it should and must be. A similar variety of approach is evident in the literature, even when not clearly stated. Three lines of approach are of primary importance in this connection —

1. a purely descriptive classification, free of genetic inferences;
2. a genetic classification, based on simple assumptions on the formation of deposits;
3. a technical classification. This latter aims at revealing from the textures of the minerals the best method for an economic separation and recovery.

In investigating the textures of an ore association in detail, if one encounters textures seen previously in other ores, there is a tendency to conclude that the recurring textures are of similar origin. This is a *faulty procedure*, because we are not certain that the "characteristic" texture could not have developed by some quite different type of process. The more one studies ore textures, the more sceptical one becomes of statements concerning them in the literature, and the more one tends to revise one's earlier views and conclusions. Only by a thorough consideration of all the possible factors: geologic occurrence, country rock and its alteration, total mineral composition of the ore, the probable composition of the ore solutions, etc., one can derive from the study of ore textures to the important genetic interpretation that it can yield.

To leave the study merely in a purely descriptive stage would mean not to utilize that help; but the study must begin as purely descriptive, and in doubtful cases should not be carried further. Consider, e.g., the "myrmekitic" or "graphic" texture. At first some workers thought that this texture was proof a of eutectoid formation of the intergrown ore minerals. Later it was recognized that replacement could give rise equally well to such textures, and now many make the opposite error, of seeing in such textures proof of replacement. Today we know that such intergrowths may form in at least six or even more ways. It is better, therefore, to say: — Mineral *A* and Mineral *B* occur in a "myrmekitic" intergrowth, than in a "eutectic" or a "graphic" intergrowth, because the latter term suggests a genetic concept from its traditional usage! Too rarely is it recognized that, because of the much greater mobility of the ore minerals in the solid state, compared with the silicate minerals, ore textures are now often entirely different from what they were at the time of formation. Whereas in petrology "palimpsets" textures are relatively rare, they are common in ore textures. Thus, at high-temperature ore deposits or geologically old ones the structures must be considered still more carefully than it was allowed by the knowledge of convergences.

In this section on textures it is impossible to discuss all the innumerable descriptions that have been published, or even all the proposed schemes of classification. A number of these derive from the methods of petrology or metallography, and provide rules for the description, say, of grain shape, grain size, grain boundary shapes, and the distribution of grains relative to one another [Kornlage]. Only the major systematic or genetic classifications will be referred to.

Of these, that of GRIGORIEFF (1928), ought to be mentioned first. Unfortunately it was available only in abstract, and not in the original Russian text. GRIGORIEFF recognizes that a descriptive approach must be the basis. However, he repeatedly allows a genetic basis to creep into his systematization, so that the major groups are actually based on genetic concepts. This, however, is indeed hard to avoid, unless much is repeated. H. SCHNEIDERHÖHN (1952) too, to whom we owe the undoubtedly more consequent classification puts up with this disadvantage, and in this way attains more clarity. He also anticipates some genetic interpretations, but only to a limited extent. His classification is by no means complete, and some phenomena can be fitted into the classification only with difficulty others are missing. It may be mentioned that schemes and expressions from petrology were used, which are rather narrow and which were created, and are still used, in a different sense. A further work by SCHNEIDERHÖHN (1945) is referred to in detail on p. 137.

The excellent book of A. B. EDWARDS (1947) deals accurately with a number of problems and their solution. It is by no means exhaustive, and does not set out to be so; but it contains stimulating specific data, on which I have drawn to some extent. Equally worthy of notice is the small book by E. S. BASTIN (1950) that deals especially with certain problems of textural interpretation, and contains many references.

G. M. SCHWARTZ has twice discussed these problems in two important studies (1932, and 1951). The *earlier* and perhaps therefore somewhat outdated paper, seeks to establish criteria which will distinguish between hypogene and supergene ore textures, and thereby allow determination of the manner of formation of the orebody. Much of the detail criteria are in part quite subject to criticism. The main value of the paper is in its numerous — 138 — references. The *second* paper is the most complete and well illustrated review yet published of descriptive textural terms, and as such is extremely valuable, even though some of the terms are difficult to understand and even more difficult to translate. It is in no respect a "classification", however, although so entitled.

The third basis for classification of ore textures, the *technical view*, has not yet been carried through thoroughly anywhere, although there are several works where good foundations have been laid for individual fields. An excellent, and outstandingly well illustrated, work is that of S. R. B. COOKE (1936), which describes iron ores. G. M. SCHWARTZ deals with the general aspects in the symposium edited by E. E. FAIRBANKS (1928). The previously mentioned book by A. B. EDWARDS contains a large section describing gold-, zinc-, lead-, silver-, copper-, and tin-ores from the technical viewpoint. Two sections in the new edition of the LEITZ-Handbook (1954; RAMDOHR and REHWALD) contain many details of this type. Certainly much of the experience of large companies and concentrating plants has not been published.

In the first edition of this book it was thought to be sufficient to give a broad framework, in which single sections were more thoroughly described (myrmekitic textures, exsolutions, "replacement" textures, geologic thermometers). In the remaining sections only brief discussions and references were given, leaving much to the reader. The necessity, however, arose, to become more explicit, and to this end the writer has availed himself gratefully of the stimulations of the later work of SCHNEIDERHÖHN (1952) as well as previously of that of GRIGORIEFF. Also in this edition, the emphasis is mainly on the special part, although in the shorter sections on minerals many more illustrations and references to illustrations are included.

A. The fabric properties considered from a purely geometric point of view

I. THE PROPERTIES OF SINGLE GRAINS

(a) *Internal nature*

Proceeding, so to speak, from the smallest unit to the larger complex, our first concern after the determination of the mineral composition of an ore is with the internal features revealed by the individual grains of the minerals. Having first described them objectively, one then proceeds to an assessment of whether they are "inherent" or "acquired" features, e.g., whether twin lamellae represent "growth twinning" or "pressure twinning". It is always necessary to anticipate aspects of section B — and this is also meant to be a suggestion of work procedure.

Determination of the internal properties of grains often becomes possible only through structure etching. To omit this, at times even with good reason, may result in certain incomplete observation of features of grains, such as zoning, twinning, translation and twin gliding, subparallel aggregates and radial growths, structural flaws and mosaic textures.

Zoning

Many minerals show a very characteristic zoning, which is evident, with or without etching, perhaps by color variation or differences in hardness. If solid solutions occur, zoning is to be expected as a consequence of changing conditions of formation in a binary system. However; mixed crystals seemed to be more abundant in silicates than in sulphides, but it is now known that sulphides contain much more often solid solutions than assumed for many years. Many strongly zoned ore minerals are not solid solutions, however, but "pure substances", and owe their zoning to other processes. Common causes of zoning are interruption of growth, periodic changes in deposition of porous and dense bands, deposition of successive bands with and without inclusions of foreign minerals. On the whole, zoning is an indication of rapid growth, low temperature and impure solutions, the importance of these several factors varying from ore to ore. In many cases original zoning may be completely or partly destroyed by diffusion processes that accompany long continued heating or contact metamorphism. Zoning may also indicate repeatedly renewed growth, or rhythmic changes in the conditions attending precipitation, e.g., for the formation of zoned magnetite or garnet, so characteristic of contact metasomatic deposits.

Examples of zoning are so common that only a few extremely significant examples will be refered to here: — Copper (*p*), stibnite (*p*), pyrite (*p* and *ss*), bravoite (*ss*), skutterudite–chloanthite (*ss*), galena (*p*), cassiterite (*p* and *ss*), pyrargyrite (*p*), etc., where (*p*) means pure substance, and (*ss*) means solid solution. Most of these examples are illustrated in the special section; examples of zoning are shown here in figs. 53, 54.

Zoning parallel to grain boundaries is frequently also found in the less usual types of solid solution, the interstitial ("stuffed") solid solutions (Additions-Mischkristalle) and defect solid solutions (Subtraktions-Mischkristalle). E.g., in chloanthite, $NiAs_{2-3}$, one zone may be $NiAs_{2.5}$, another $NiAs_{2.8}$ or $NiAs_3$. This type of zoning has become

Fig. 53 70 × RAMDOHR
Mechernich, Eifel

Strongly zoned *galena* as cement of a coarse-grained quartz sandstone. The zoning is revealed by incipient alteration to *cerussite*

Fig. 53a 270 ×, imm. SCHIDLOWSKI
Loraine Au-Mine, O.F.S., S.Africa

A large crystal of porous and zoned *pyrite* is strongly rounded by the transport preceding the sedimentation in the conglomerate

increasingly recognized in recent years. It points to a variation from time to time in the chemical concentration of a component of the depositing solution.

Twinning

Twinning phenomena may be classed, according to their origin, into growth-, pressure-, and inversion-twinning; the various types of twinning may look alike, although generally they can be distinguished rather readily from one another.

Fig. 54 750 ×, imm. RAMDOHR
Horne Mine, Rouyn-Noranda, Quebec

Magnetite, with zoning revealed by small inclusions of pyrite. Set in a matrix of *pyrrhotite* and idiomorphic *pyrite*

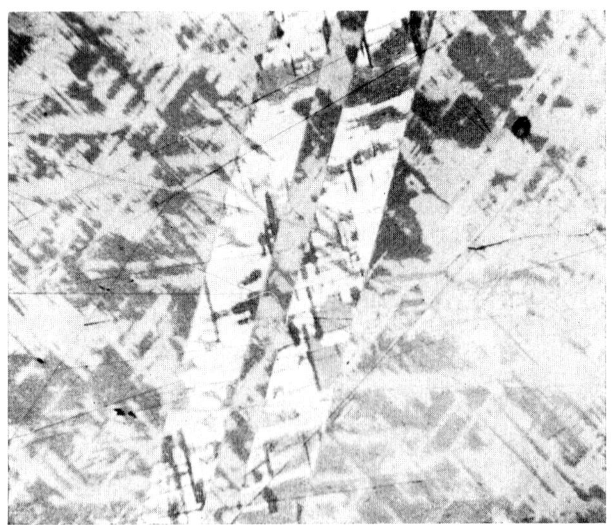

Fig. 55 90 ×, crossed nicols RAMDOHR
"Hunan", China

Stannite with inversion twin lamellae parallel (100) and two coarse growth twin lamellae along (111)

Inversion twins are commonly spindle-shaped, in other words not uniformly parallel all over the grains. They commonly form strong intergrowth networks and are hardly accompanied by strain and translation features. *Pressure twins* are also mostly lamellar, where the lamellae as far as they follow the same twin law are relatively uniform in thickness, and commonly associated with bending, cataclasis and incipient recrystallization; also, the lamellae are more or less pronounced depending on the plane of the section, and the associated minerals commonly show traces of deformation or pressure twinning. *Growth twins* can also be lamellar, and may be strongly interwoven, especially in mimetic crystals which derive from higher symmetry forms, like chalcopyrite, stannite, arsenopyrite, chalcocite, bournonite, etc. The lamellae are, however, of irregular width, and are unevenly distributed, and may often be absent from other grains of the same section. Excellent examples of these three types of twinning are shown in figs. 55–57, and also in the special section in figs. 316, 338, 390, 463, and many others.

Examples are encountered, however, from time to time, in which these criteria fail, in part through a typical development, in part because criteria occur in the "wrong" places. E.g., it has been shown that experimentally produced pressure twinning may give rise to spindle-shaped lamellae.

A special, but as yet not fully understood, twinning phenomenon is that of "*crumpled lamellae*" ["Zerknitterungslamellen"], more or less spindleshaped lamellae trending about at right angles to bent or flexed translation planes. The individuals are not altogether uniformly inclined, their margins are not always rectilinear, they may be of unequal width, and they tend to be restricted to the strongly deformed and twisted parts of a grain. Stibnite (fig. 32), franckeite, graphite, molybdenite, covellite provide typical examples.

Many minerals can exhibit lamellar twinning following one or more laws, which can be of different origins. The table that follows is far from complete. It lists the typical habits of some of the more important minerals. Rarer occurrences are referred to only when they are of special interest. Reference to p. 61 should be made for a further discussion of pressure twinning.

Table of lamellar twinning

Copper, *s*, *u*
Silver, *s*, *u*
Gold, *s*, *u*
Platinum, *i*, *u*
Iron, *i*, *s*, *P*
Arsenic, *s*, *sp*
Bismuth, *s*, *sp*
Graphite, *C*
Dyscrasite, *s*, *m*
Hessite, *s*, *sp*, *I*
Nagyagite, *s*, *m*
Chalcocite, *s*, *i*, *sp*, *G*, *P*
Umangite, *s*, *u*, *P*
Argentite, *s*, *sp*, *I*
Bornite, *s*, *u*
Sphalerite, *s*, *u*, *G*, *P*
Chalcopyrite, *s*, *u*, *G*, *P*, *sp*, *I* (rare)
Stannite, *s*, *u*, *sp*, *I*
Enargite, *s*, *m* (rare)
Luzonite, *s*, *u*
Pyrrhotite, *s*, *sp*, *P*
Niccolite, *s*, *u*, *P*
Cubanite, *s*, *sp*, *u*, *I*
Galena, *s*, *u*, *P*
Schapbachite, *s*, *u*, *I*
Teallite, *C* (rare)
Covellite, *C*
Stibnite, *C*
Wolfsbergite, *s*, *u*
Bournonite, *s*, *u*, *m*
Andorite, *s*, *u*
Sartorite, *s*, *u*, *G*
Jamesonite, *s*, *i*, *u*
Geocronite, *s*, *u*, *G*, *P*
Pyrargyrite, *s*, *u*, *G*, *P*
Marcasite, *s*, *u*, *G*
Arsenopyrite, *s*, *u*, *m*
Rammelsbergite, *s*, *u*
Orpiment, *C*
Tenorite, *s*, *u*
Hausmannite, *s*, *u*, *G*, *P*
Hematite, *s*, *u*, *G*, *P*
Ilmenite, *s*, *u*, *G*, *P*
Rutile, *s*, *u*, *G*, *P*
Cassiterite, *s*, *i*, *u*, *G*, *P*
Calcite, *s*, *u*, *G*, *P*
Siderite, *s*, *u*, *P*
Dolomite, *s*, *u*, *P* (rare)
Lievrite, *s*, *u* (rare)

Fig. 56a 100 ×, crossed nicols (unetched!) RAMDOHR
Sierra de Umango, Argentina

Umangite with prominent twin lamellae parallel to a pyramid, surrounded by *calcite* with strong internal reflections

Fig. 56b 100 ×, crossed nicols RAMDOHR
Sierra de Umango, Argentina

Umangite, showing two sets of twin lamellae probably parallel to two different pyramids; grain boundaries of lamellae excellently visible under crossed nicols

The italicized capital letters *G*, *P*, *I* and *C* indicate growth-, pressure-, inversion- and crumpled-lamellae; *s* means polysynthetic twinning, *i* refers to isolated twin lamellae, *m* refers to mimetic lamellae, *u* to uniformly wide lamellae, *sp* to spindle-shaped or oleander-leaf shaped lamellae. Frequently letters *G*, *P*, or *I* indicating the origin of twinning have been omitted.

Fig. 57 100 × GRASSELLY
Herja, Roumania

Sphalerite, rather rich in iron, with prominent lamellar twinning rendered visible by etching with sodium hypochlorite

Fig. 58 150 ×, crossed nicols RAMDOHR
"China"

Chalcopyrite showing spindle-shaped inversion twinning, from a high temperature deposit

Fig. 59 30 ×, crossed nicols RAMDOHR-AHLFELD
Ancoraimes, Bolivia

Geocronite, cataclastic and with prominent twin lamellae, cemented by *jamesonite* (white)

Fig. 60

Translation and twin gliding

These textures, owing their origin to mechanical deformation, do not truly belong to this section on the purely geometric description, but reference is made to them here for the sake of completeness. Naturally they can give very characteristic textures from a purely descriptive point of view (e.g., figs. 516a, b, of molybdenite, or fig. 59 of geocronite). These phenomena are dealt with in more detail on pp. 57—62.

Subparallel aggregates and radial growths

Free growing crystals of acicular, columnar, prismatic or leaf-like habit, very often crystallize in sheaves or radial clusters. Even some isomtetric crystals do this. Some minerals are particularly prone to form such growths, e.g., acicular stibnite or platy lepidocrocite [Rubinglimmer]. Mostly these aggregates are growths continued about early formed germinal masses probably precipitated as gels. Their central parts often reveal a transition to the botryoidal clusters, with a radial fibrous texture of the original gel. In other instances, this is not the case; the re-entrant angle of a twinned crystal commonly provides the point of attachment for a new layer of twinned crystals, and so on. This type of radial growth derives from a more or less uniform supply of material, from all directions, acting in conjunction with a preferred growth direction in the crystal to promote a spherical or globular growth. When such radial aggregates develop, as along the wall of a fracture, they eventually come into contact with each other. The crystals approaching each other become more weakly nourished

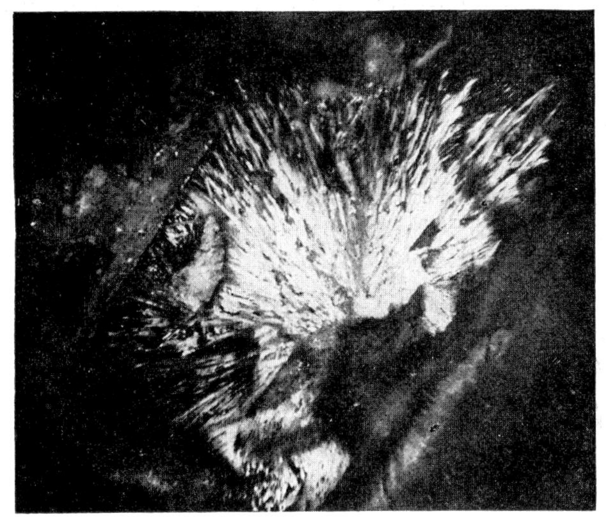

Fig. 61 40 ×, crossed nicols RAMDOHR-EHRENBERG
Andreasberg, Harz

Radial growth of *stibnite* projecting into a cavity. The anisotropy is readily apparent

Fig. 62 55 × RAMDOHR-EHRENBERG
Andreasberg, Harz

Radial aggregate of *arsenopyrite*, partly cataclastic. An outer zone of softer *löllingite* can be seen on the margins of the crystals, made evident by a slightly darker color. The slightly different color tones of the crystals are quite pronounced under almost crossed nicols. The matrix is *calcite* (dark grey)

than those growing freely outwards into the open space. As a result the successively encrusting layers tend more and more to lose their radial textures, and to become layers of parallel fibres with their long axes normal to the surface on which they grow (fig. 60).

In addition, there are crystals which form first as well developed single crystals, but which, as they grow, develop more and more incomplete faces and edges, so that finally the result is not a single crystal, but a more or less divergent radial aggregate of subparallel crystals. Such behavior commonly sets the limits to development of large grains (see mosaic texture).

Fig. 63 100 ×, crossed nicols RAMDOHR
Bisbee, Arizona

Delafossite with a typical sub-parallel *mosaic texture* most probably developed already during growth

Both these textures are very widespread, and they commonly resemble each other, despite of their very different derivation. Radial growths are favored in low temperature deposits, whereas sub-parallel aggregates of the latter of the above types are not noticeably influenced by temperature. Figs. 61 and 62 provide instructive illustrations.

Crystallographically the development of a radial aggregate involves a selection of those nuclei whose "fiber axes" — which correspond generally with their shortest cell dimension — accidentally are oriented favorably in their relation to the supplies of mineralizing solution. Minerals with a layer-lattice structure rarely form concentric banded masses in which their (0001) planes are tangentially oriented — although such an arrangement is found in artificial graphite (Acheson-graphite) and in valleriite from Kaveltorp, Sweden; mostly the tabular plates are radially arranged, as with much covellite (Mélanoa, Peko), graphite with titanomagnetite (Tahawus, Lac de la Blache, Angola), with molybdenite formed from jordisite (Bleiberg, Mečiča, Mansfeld) and with lepidocrocite [Rubinglimmer] (fig. 619).

Defect- and mosaic-structures

There is scarcely a crystal which, as soon as it has grown to a certain, though mostly very small size, does not contain inclusions of foreign matter, or individual parts with defective orientation, or patches of solid solution, which do not fit perfectly into its lattice, or some other form of structure disturbances. Even small mechanical stresses which give rise to submicroscopic translations push a lattice plane slightly out of position, introducing a distortion that at first does not hinder the growth of the crystal. These defects are commonly invisible microscopically, but become evident by irregular etching behavior, easier weathering, or anomalous optical properties. Occasionally, however, the defective areas give rise to a vague parallel banding where mosaic units of different size form, which, as a whole, still have the shape of a sound crystal, which is, to a certain extent, healed but retains scars. Such mosaics may attain considerable size, as e.g., in the galena at Joplin, but are mostly small. They can be recognized as mosaic structures only when in a near-extinction position under crossed nicols. The parts of the mosaic — with some exceptions — are not sharply defined, but pass by wavy extinction into the adjacent parts. As yet little attention has been paid to these textures in ore minerals. Fig. 63 illustrates a typical occurrence. Textures of this type due to lattice defects may closely resemble textures induced by tectonic effects.

The textures that arise from the *unmixing of solid solutions*, and from the break down of compounds that have become unstable, are in general so characteristic that it is difficult to describe them in a purely geometric way, without introducing any genetic undertones. Even so, there are examples known to the writer, where by the addition of sulphur to an iron-rich sphalerite, the sphalerite has become "dusted" throughout with a precipitate of pyrite; or by the heating of chromite under reducing conditions the precipitation of metallic iron has been induced, giving rise in each case to a texture which anyone ignorant of their mode of formation would interpret as an exsolution texture. Similar conditions may occasionally occur in nature (fig. 65).*

Inclusions

A special significance attaches to inclusions, which, by their different origins, disturb the crystal lattice. *Liquid inclusions* with gas bubbles are extraordinarily widespread. They provide means of determining temperatures of formation. Often they also act as nucleating points for exsolution. *Glass* inclusions which in principle are the same, have not yet been described in ore minerals. *Sulphide droplets* trapped as inclusions are dealt with in another connection p. 97/98.

(b) *External grain properties*

Having determined the textures and properties of the individual grain, attention must be paid to its external features. These are best divided into: grain shape, grain size, space relations, mutual grain boundary relationships, the latter of which leads us to the next section dealing with aggregates.

* Hawley & Hewitt (1948) described something of this sort: partial destruction of niccolite by loss of arsenic to form an intergrowth of niccolite and maucherite.

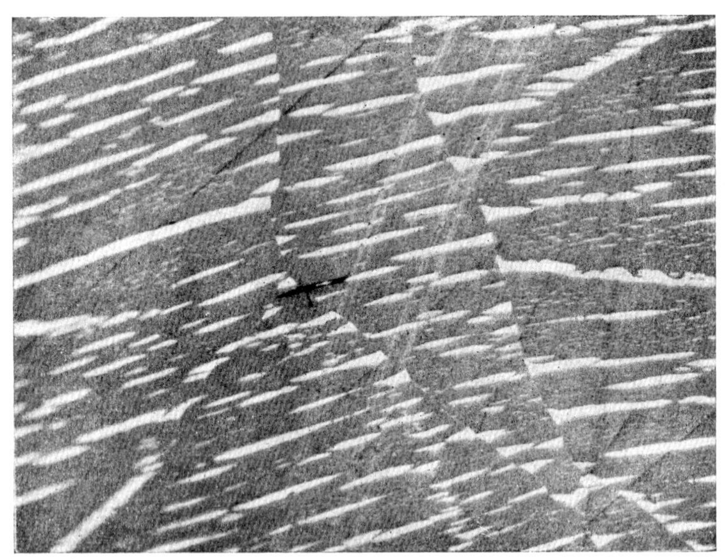

Fig. 64a 250 ×, imm. RAMDOHR
Ekersund, Norway

Mosaic texture in a coarse *ilmenite* grain. The individual parts of the mosaic crystal are aligned at several degrees from one another. Exsolution of *hematite* is independent in each individual part of the mosaic

Fig. 64b The same as Fig. 64a, but with crossed nicols RAMDOHR

The twin lamellae in the individual grains are markedly deflected, but remain roughly continuous

The observation of these properties in monomineralic aggregates is easy only if the mineral concerned is distinctly anisotropic; otherwise etching is necessary. If several methods of etching are applicable, grain boundary etching is to be preferred, whereas for the development of the internal structures grain surface etching is desirable.

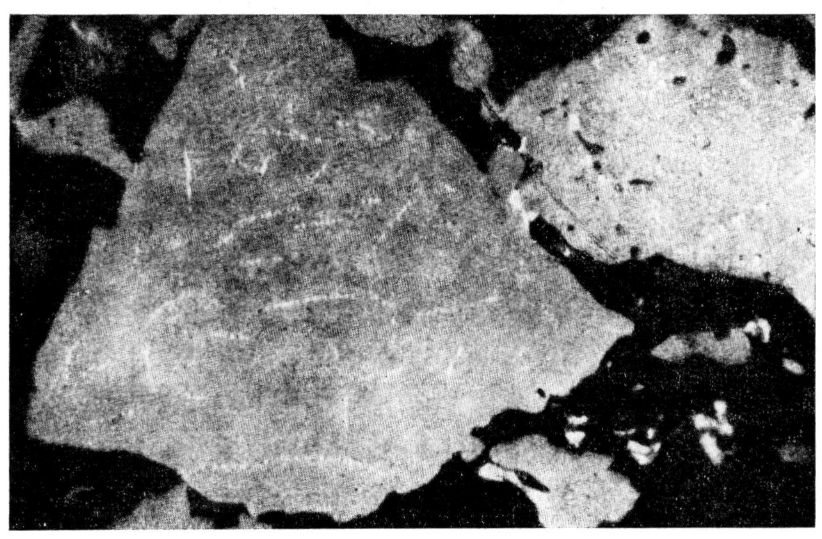

Fig. 65 600 ×, imm. RAMDOHR-LIPINSKI
Locality unknown

A precipitate of *metallic iron* in chromite parallel (100) resulting from roasting of the chromite under reducing conditions. The iron bodies resemble exsolution bodies, although not of this origin

Grain shape

The mineral grains of an aggregate can have: their own shape, partly their own shape and a foreign shape, for which the corresponding petrographic terms are idiomorphic, hypidiomorphic and xenomorphic (= allotriomorphic), or euhedral, subhedral and anhedral. "Panidiomorphism", that is, *idiomorphism* of all the grains is impossible in a monomineralic aggregate, except where an interstitial matrix fills wedge-shaped areas (figs. 79, 81;) but it is under certain circumstances an important characteristic of many minerals that they develop their own habit as completely as possible. In polymineralic aggregates one or more components may develop perfect crystals, e.g., pyrite, sperrylite, arsenopyrite, cobaltite, loellingite, jamesonite, molybdenite, graphite, magnetite at times, braunite, etc. Most of these ore minerals show this tendency to euhedral crystallization regardless of their age relations, i.e. even when late formed they will tend to develop as euhedral crystals in or upon other minerals or mineral aggregates. This is characteristic, e.g., of pyrite and of quartz, so that it is *extremely difficult* at times to determine their true age relations (figs. 112, 113, 194). Many occurrences originally regarded as indisputably "primary" idiomorphic crystallization are today doubtful, or have been recognized as being idioblastic. This applies also to rock-forming minerals, once the boundary between magmatic and metamorphic has become obscure. The development of the different possible crystal forms in an aggregate is usually not too good, yet at times individual crystals with many developed faces occur partly as a consequence of inhibited growth. All the faces developed on a crystal as a whole and its *habit*, i.e. the shape developed through the predominance of certain faces, often have genetic significance, particularly in regard to the temperature of formation — but caution is to be exercised in this regard.

Acicular shapes indicate selective rapid growth in one direction, and they are generally restricted to the non-cubic minerals. One may observe surprising and often far reaching penetrations of such acicular minerals, even in compact aggregates. As with the reticulate ("gestrickt") forms, growth commonly occurred in a viscous medium. Minerals which tend to be acicular are stibnite, many "Bleispiessglanze" (boulangerite, sartorite, etc.), millerite (fig. 428) and "goethite" (Nadeleisenerz), and under special conditions, also others.

The layer-structure minerals (graphite, covellite, tetradymite) are mostly (but not always!) *tabular*, as also, under certain conditions of formation, are others, notably hematite and ilmenite. One must be particularly careful about predictions, however, since minerals with acicular or tabular lattices can deposit as equigranular crystals, whereas others which have not preferred lattice directions can form tabular or acicular crystals as galena, which is cubic, in rolled galena (fig. 37), or cuprite in the form of chalcotrichite ("Kupferblüte").

In contrast to idiomorphism is *xenomorphism*. This also is to some extent a characteristic habit of the substance concerned, in so far as some ore minerals scarcely yield good crystals even when they are early formed, or have the opportunity to grow rapidly inside host minerals.

Only in the youngest cavities, the open druses, such minerals will develop their characteristic crystal shapes. Many minerals, especially the softer ones, behave in this way: galena, argentite, fahlore (tetrahedrite), silver, gold and many others. Grains without crystal faces are often called "crystallites" in accordance with metallographic usage. Especially important is the "panallotriomorphic" or "panxenomorphic" texture, shown by most monomineralic ore mineral aggregates. This texture reveals the internal tendency of a mineral towards a perfect or an imperfect crystallographic development, as a more or less strongly interlacing intergrowth, and as oval or almost columnar grains.

Xenoblasts are crystal grains that grow to considerably larger size than their associated minerals, but lack definite crystal faces. E.g., where a mineral replaces a chemically closely related component in a myrmekitic intergrowth, the new mineral may show uniform orientation over a considerable area. To evaluate genetically the growth of xenoblasts and idioblasts in ores is even more difficult than in petrology. The most distinctive xenocrystic texture is the "*poikiloblastic*" [Einschlußsieb] texture, particularly if it includes minerals which in their surroundings have other grain-sizes and different properties (figs. 110, 111, 113, 502).

Crystal skeletons are in all cases ultimately formed through preferred crystal growth in the locations of best supply of material. In these locations the growth mechanism which produces plane coherent surfaces cannot keep pace, during relatively fast crystal growth, with the growth rate along the corners and edges, which are more exposed to the solutions. Naturally the diffusion in the solution from which the crystals grow takes place within certain limits, which influences the rate and extent of skeletal crystal growth. Often after the first rapid skeletal growth, general growth

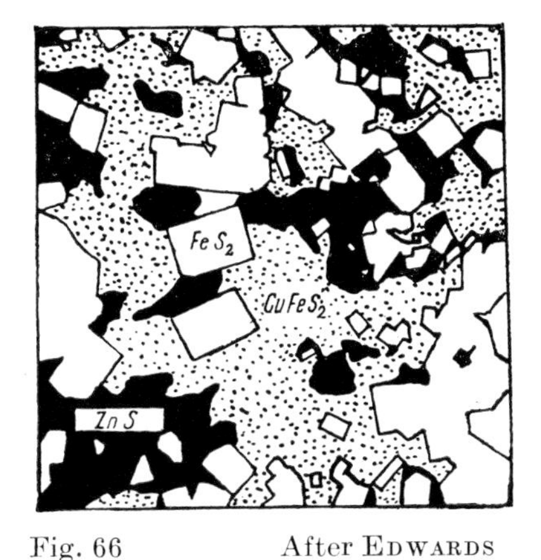

Fig. 66 After EDWARDS

Idiomorphic pyrite ("porphyritic") in chalcopyrite and sphalerite. Tasmania

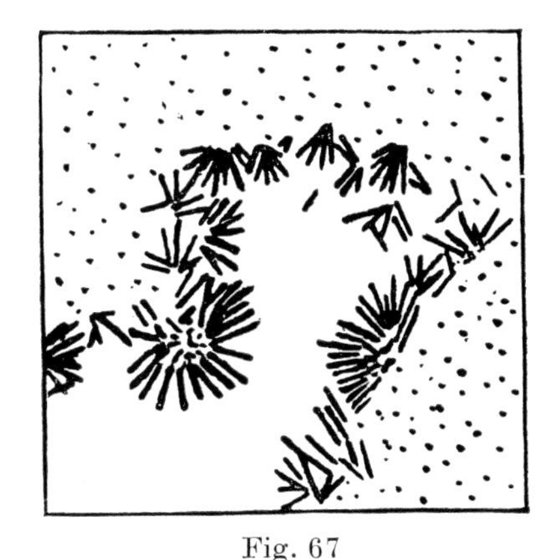

Fig. 67

Radial growth of idiomorphic crystals encrusting a cavity (chalcophanite, Sinai)

sets in that fills the re-entrant wedges gradually, as is to be expected from a kinetic consideration of crystal growth. The last spaces filled within the skeleton — e.g., in the hour glass texture of augite, or in the "black cores" of andalusite — commonly are of somewhat different composition than the earlier-formed parts, are less pure, and are less resistant to solution (and to etching!). One must not conclude, however, that a skeletal crystal has always been a skeletal crystal, for, as first shown by MAUCHER, some skeletal crystals owe their form to the leaching out of selected zones, thus leaving skeletal residues. In general, however, the fact that skeletal crystals are embedded in a matrix, which at one time was viscous and often gelatinous, shows that they are present in their original form. Examples are shown in fig. 73, of augite in a vitrophyre, fig. 70 of ilmenite in a gabbro, fig. 74 of galena, and fig. 72 of a leached out mineral of unknown composition.

Fig. 68 After NEWHOUSE

"Globules" of pyrrhotite ± chalcopyrite in the magnetite of a gabbro. They are fringed with alteration products

Of a quite different nature are the skeletal crystals left over as relicts of original exsolution bodies, whose main component has been dissolved· They should not be interpreted as skeletal crystals in the sense described here (e.g., fig·248 b). However, they are very similar to certain exsolution bodies, or to other bodies that have crystallized through "collection crystallization" [Sammelkristallisation] in their enclosing "host", e.g., "sphalerite stars" (fig. 488), or the pentlandite "flames" in pyrrhotite (fig. 71). Here, too, a medium is encountered which hinders diffusion strongly and which controls the crystal growth, which here is actually idioblastic.

Closely related to skeletal growth are the "reticulate" forms, which are described elsewhere (p. 157).

Spheroidal grains are not common, and will always attract attention. They may attain very different sizes, and may be monomineralic or complex and secondary. They are of great interest with regard to their origin, but again great care must be used in their interpretation.

Spheroidal in shape can be: trapped droplets of melt (bismuth, antimony, iron, pyrrhotite, magnetite), crystals formed under conditions of "inhibited growth", in which the surface tension is still stronger than the energy of idiomorphic development,

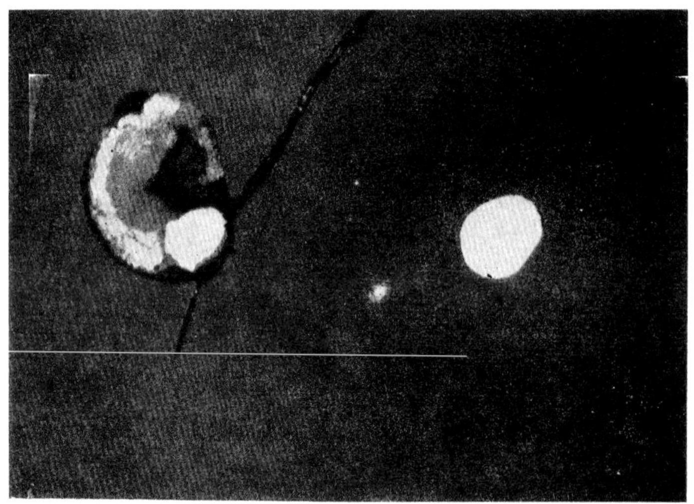

Fig. 69 500 ×, dry! Ramdohr
Unterwiesental, Erzgebirge

"Spherules" of *pyrrhotite* (in part weathered) and *chalcopyrite* in augite in a nephelinite

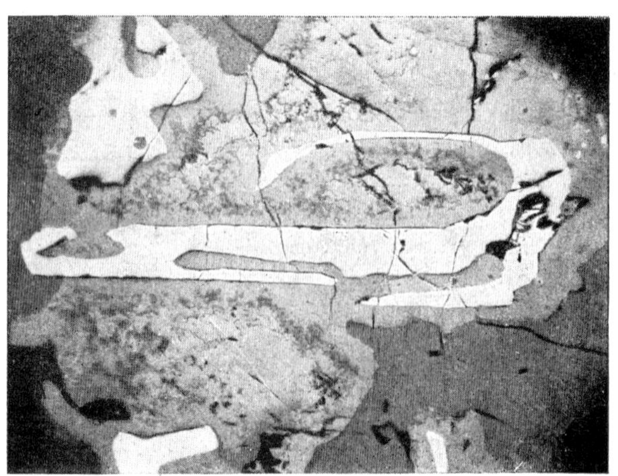

Fig. 70 50 × Ramdohr
Nubian Desert, Africa

Lobate skeleton of *ilmenite* (white) inside *augite* (light grey) which is myrmekitically intergrown with a darker component (possibly *hornblende*). The dark grey is *plagioclase*. The ilmenite is more or less contemporaneous with the augite

Fig. 71 RAMDOHR-EHRENBERG

500 ×, imm.

Sudbury, Canada

Pentlandite "flames" in *pyrrhotite*, cut exactly parallel to (0001) of the pyrrhotite, and therefore parallel (111) of the pentlandite. The trigonal symmetry of the pentlandite skeleton is evident

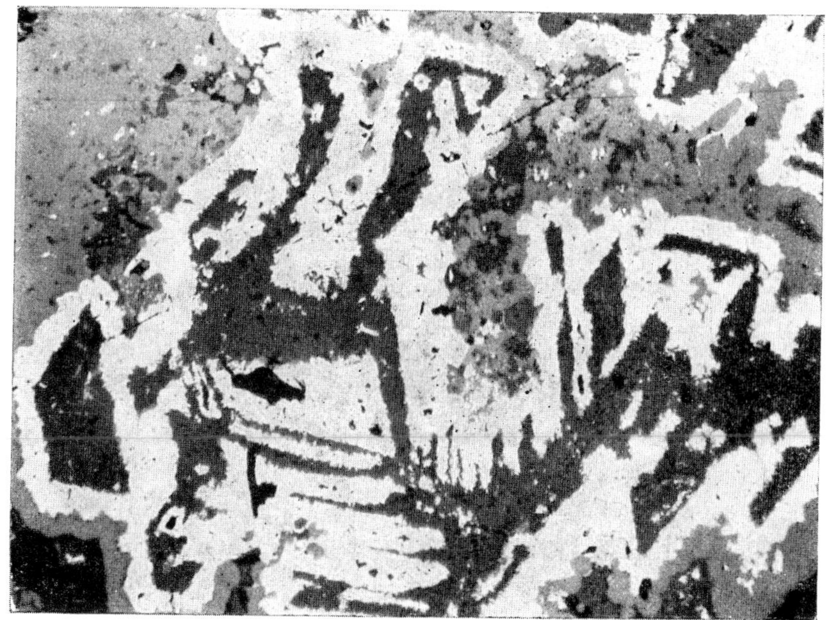

Fig. 72 60 × REHWALD

Schneeberg in Saxony, Germany

Skeletal form of a dissolved mineral, whose place has been taken up by calcite, rimmed by fine grained *safflorite*

which mostly is at a temperature slightly below that of the start of crystallization, spheroidal exsolution bodies, secondary fillings of vesicular cavities, grains occurring as inclusions in poikiloblastic intergrowths, corrosion relicts, and still others. The group first mentioned belongs to the most important geologic thermometers (figs. 68, 69, also p. 97).

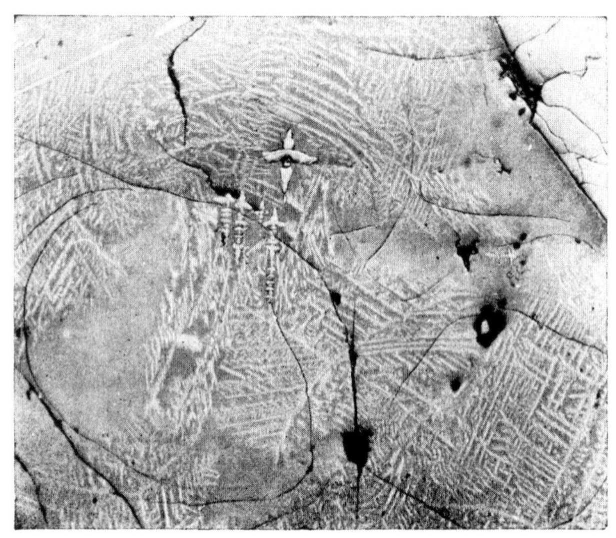

Fig. 73 100 × RAMDOHR
Vicenza, Italy

Skeletal *augite* in a vitrophyre. The skeleta occur, surprisingly, immediately adjacent to "sound" crystals of the same mineral

Fig. 74 300 ×, imm. RAMDOHR
Garpenberg, Sweden

Skeletal growth of *galena*, deposited in the zone of cementation. Notice the turbid inner axes of the skeletal crystals

Grain size

The data relating to grain shape give no indication as to the relative and absolute sizes of the component minerals. Grains with the same degree of idiomorphism with regard to crystal development may range from very coarse to very fine grained, within one particular mineral as well as in different associated minerals. Besides course of formation and therefore the physico-chemical concepts of rate of nucleation and rate of crystallization, grain size can be influenced by different factors, the most common probably being concentration. Substances in high concentrations tend to deposit large crystals, if concentrations are low the resulting crystals will be minute; this rule, however, has many exceptions, consider, for instance, the gigantic crystals of beryl, even in such pegmatites where beryl is a very rare mineral. The sparsely occurring mineral is not necessarily a late crystallization product; this is true both for rocks and for ore mineral associations. However, the terms used in petrology to express relative grain size are not without limitations applicable to ore minerals. A new, and more comprehensive terminology ought to be set up. Above all, one must be careful not to attach to ores the genetic implications which are almost unconsciously attributed in connection with magmatic or metamorphic rocks, but attempt to use these terms as purely descriptive. A "phenocryst" in an ore is by no means a product of early crystallization brought up from depth, a porphyroblast similar in form to a phenocryst may not indicate dynamic metamorphism, and neither a higher growth rate than that of the other constituents. The "phenocryst" *merely* indicates that in this instance an idiomorphic mineral grain occurs in the midst of a fine grained surrounding, mostly of different minerals, and that it is older than, or contemporaneous with, its surrounding minerals. The idioblast, in contrast, shows merely that it is younger than the surrounding minerals, and that it has grown at their expense as regards space, and probably also as regards material. The interval between the time of formation of the two can be short, and the decrease in temperature and pressure attending their formations may be quite small. Of course, the relations *can* be similar to those found in rocks, as is shown in figs. 39 and 110, picturing porphyroblasts and xenoblasts in truly kinetically metamorphosed ores. In contrast to them are similar forms (figs. 81 and 480) which are products of an apparently rapidly formed, but otherwise normal, mineral assemblage.

The *absolute grain size* cannot be described by the usual terms (gigantic-, very coarse-, coarse-, medium-, fine-grained etc.), since it is well known that especially experts will call a certain mineral grain of 5 × 5 × 5 cm giganitic, whereas another mineral of this dimension will at most be considered coarsegrained. Photographs with a scale should be used, or accurate average sizes, together with the observed upper and lower limits. Many ore-minerals are cubic or sub-cubic; for them a single diameter will suffice, so long as they have not been rolled out or otherwise distorted. For all the other crystal systems maximum and minimum diameters must be provided for the individual grain. Such measurements are not only important scientifically but they also permit estimation of the crushing size necessary to free the mineral grains if the ore is to be milled. Only then do they also permit calculation of the average grain weight. Some ore minerals show a distinct tendency to be coarse grained, notably sphalerite, less commonly galena and geocronite, as example of the rarer minerals, in some hydrothermal veins, pyrrhotite and cassiterite in pegmatites, and almost always wolframite. Fine grained even in undeformed pure aggregates, are, e.g., magnetite, hematite, bornite and chalcocite (digenite by contrast is mostly coarse grained!). For most cases, however, not even approximate rules can be given.

The extreme variability in many groups of mineral deposits within small areas of a deposit also renders indications on the absolute grain size difficult. If the same mineral occurs in very small grains and in grains as large as a cherry within a few centimeters, any indication of an average grain size is naturally misleading!

Grain bonding

The manner in which grains are "bonded" together in a monomineralic aggregate is principally similar to the bonding of complex mineral aggregates. The very different forms taken have quite different genetic explanations, and depend largely on the substance concerned and especially its bonding forces. They may be grouped as "simple grain intergrowths or fabrics", and "interlaced grain intergrowths or fabrics", although both concepts grade into one another.

Simple grain fabrics. Contemporaneous and relatively slow growth of all grains with no significantly preferred rates of growth, results in an equigranular texture with smooth contact faces. If, however, the individual grain in a monomineralic aggregate possesses considerable differences of growth velocity in different crystallographic directions, the resulting texture may be as complicated as in a polymineralic aggregate. Simple grain fabrics also develop, even more commonly through collection [Sammelkristallisation] of easily crystallizing minerals which were originally deposited as colloids, or by recrystallization following shearing, by thermal metamorphism, etc. Perfectly simple textures are not to be expected, since in mineral associations the characteristics of growth in all components are rarely homogeneous. Crystallization at one specific temperature may produce such a uniform texture, whereas at another temperature a different texture will result. The fact that a mineral is idiomorphic does, in contrast to rocks, in no way imply anything with regard to paragenetic sequence. A number of illustrations of various simple grain fabrics are given in this book, e.g., fig. 75 contemporaneous growth, fig. 607 collection crystallization of former colloids, fig. 46 recrystallization, fig. 17 contact metamorphism.

Complex grain fabrics. When the grain boundaries become more complex, fabrics develop, which, even with only one component, can be quite similar to myrmekitic intergrowth of several components. Between this extreme all transitions occur from smooth contacts to undulating contacts and interlacing contacts. It is very difficult to explain genetically the reason for these differences. They often may be due to rapid somewhat skeletal growth; or to recrystallization; or to "eutectoidal" crystallization (figs. 76–78), where water and other volatile substances form the second component; they may also be due to dissociation of compounds which formerly were an interstitial material or a replacement thereof; or to cementing of cataclastic ores, etc. Forms of this type may also sometimes result from collection crystallization of mixed gels.

Filling of available space

If the mineral grains completely fill the available space, the aggregate is spoken of as compact. Hardly any mineral aggregate is, however, entirely free of pores. Much more commonly the intergrowths are of the classes that are termed finely porous, or coarsely porous, drusy, vesicular, "ophitic", cellular, "cavernous", and all show very

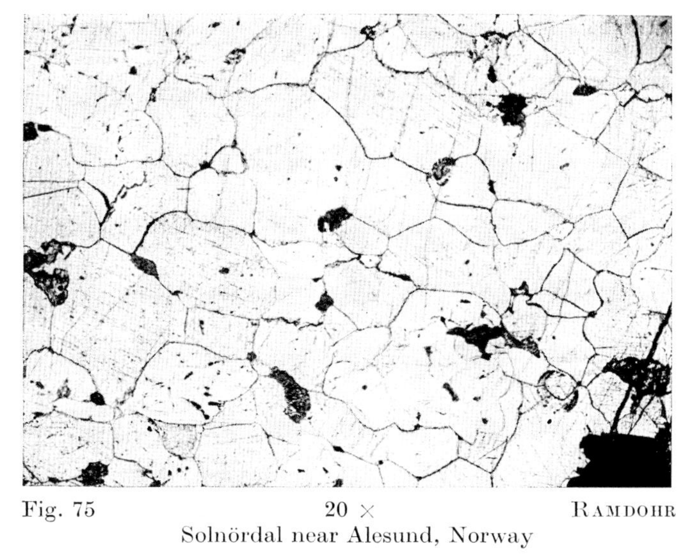

Fig. 75 20 × RAMDOHR

Solnördal near Alesund, Norway

Granular recrystallized *titanomagnetite* (grey, fine lamellae, not quite smooth), with *ilmenite* (smooth, a trace lighter) and a little *sulphide*. The irregular shape of the grains, and their weak interlocking, is typical

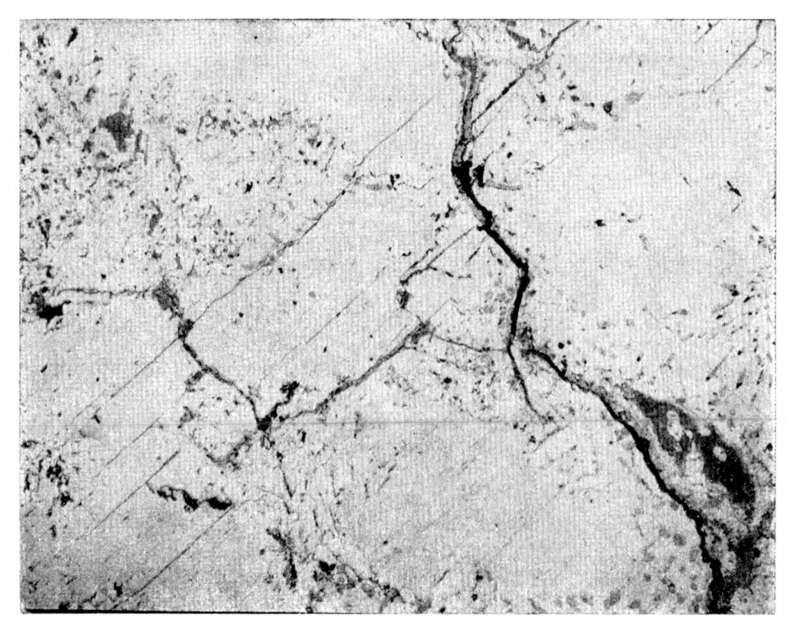

Fig. 76 350 ×, imm. RAMDOHR

Igdlokunguak, Greenland

Pyrrhotite with good cleavage on (0001) rimmed with fine-grained masses of very similar reflectivity

varying degrees of porosity. The variety of shapes is enormous, especially in near surface deposits with their many pseudomorphs, and features of leaching. Particularly difficult to deal with in polished sections are the spongy cellular oxidized ores and cementation ores. All degrees of porosity are illustrated in many of the photomicrographs in this book (e.g., fig. 79) as well as examples of compact space filling.

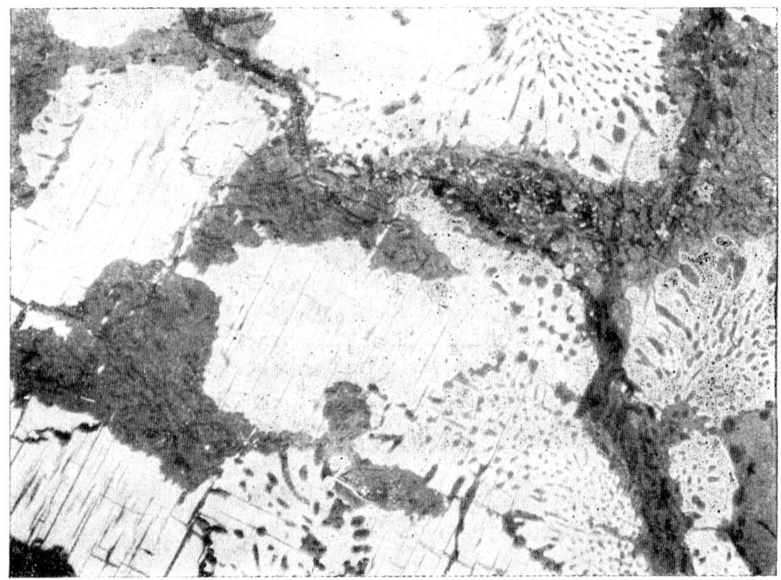

Fig. 77 350 ×, imm. + nic. RAMDOHR
Igdlokunguak, Greenland

Anisotropic *pyrrhotite*, almost homogeneously rimmed with a eutectic intergrowth of *pyrrhotite* and *magnetite*, and with an almost isotropic complex mixture of *pentlandite* and *chalcopyrite* (formed from a (Fe, Ni, Cu) solid solution). In places the "eutectic" material has undoubtedly corroded the pyrrhotite

Porosity can be "primary", in the sense, that it results from original deposition, but it can also develop later through a variety of solution processes. The determination of how and when the solution pores developed can be of considerable importance for an understanding of a particular deposit.

Crystals growing into the pores may often be especially well zoned. If open spaces transect the zones of a crystal, or especially, if they follow a zone, the cavities may for certain be attributed to solution. This statement cannot be turned around. Even solution druses can very well be filled with strongly zoned, well developed crystals of a younger generation. — Open spaces originating from the incipient deposition process should best be termed "unfilled interstices" and as such they show characteristic shapes, which can easily be recognized, even when filled by later growths (fig. 79a, 80, 81).

II. INTERGROWTHS OF SEVERAL MINERALS

It is difficult and arbitrary to divide the phenomena to be dealt with in this section from those of either the preceeding or the succeeding sections. In this edition I follow more closely SCHNEIDERHÖHN (1952), in order to facilitate comparisons. In this section, particularly, there is always a tendency to introduce prematurely a genetic interpretation into the classification — so that particular restraint must be exercised in this respect. Accordingly, the terminology chosen is as "noncommittal" as possible.

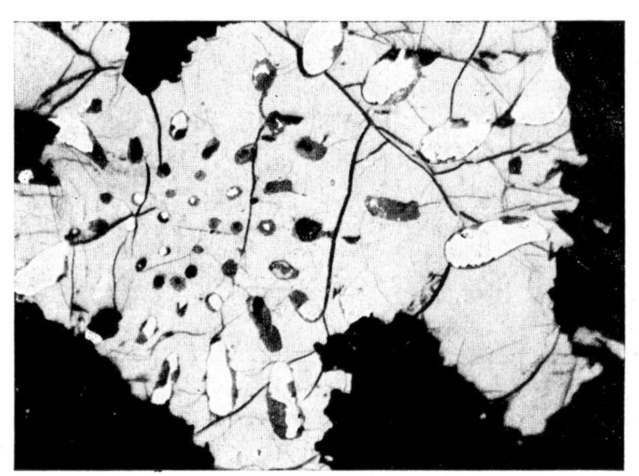

Fig. 78 350 ×, imm. RAMDOHR
Bühl, near Kassel, Germany

An emulsion, in part a true eutectic texture, of four immiscible phases: *iron* (pure white), *pyrrhotite* (grey-white predominant), *magnetite* (dark grey) and *silicate glass* (black)

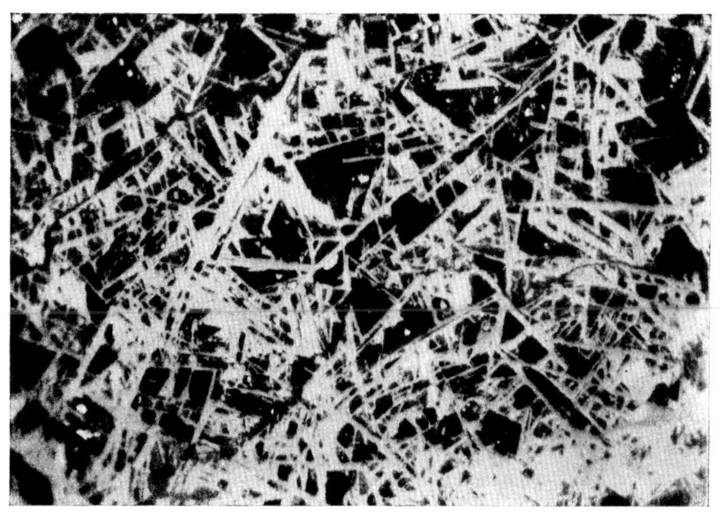

Fig. 78a 150 × RAMDOHR
Huanchaca, Bolivia

Loosely packed framework of thin tabular *wurtzite*, white-grey. The interstices are mostly open (now filled with plastics) — cavities (black), sometimes with silicate fillings, very dark-grey

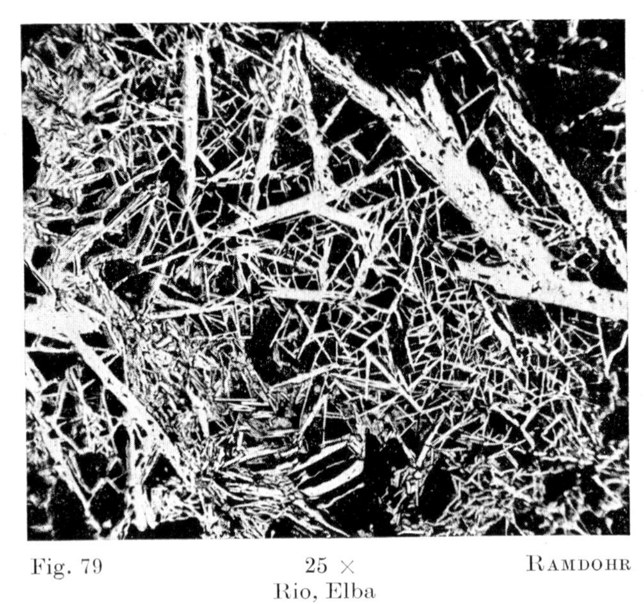

Fig. 79 25 × RAMDOHR
Rio, Elba

Hematite as an "ophitic" network, with only locally a little *limonite* as a partial filling to the cells. (Panidiomorphic cellular texture!)

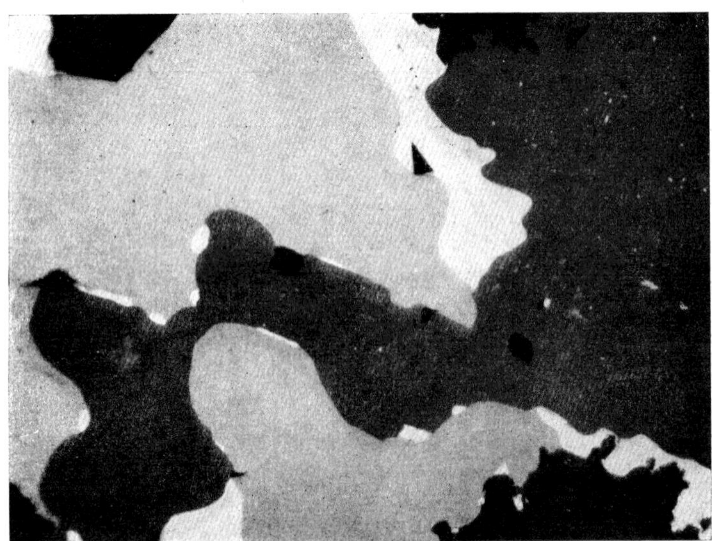

Fig. 79 a 250 ×, imm. RAMDOHR
Kapnic, C.S.S.R.

"Mutual boundaries". *Sphalerite* (dark grey), *tetrahedrite* (medium grey), *galena* (white, cleavage triangle) do not show a distinct age relation. *Chalcopyrite*, nearly as white as galena, is younger and forms thin films and tiny grains between sphalerite and tetrahedrite. The gangue, nearly black, is *rhodochrosite*

Fig. 80 60 × RAMDOHR

Saxberget Mine near Saxdalen, Sweden

Mineralization of a fine-grained matted *actinolite*, which makes the concentration very difficult. Actinolite (almost black), *sphalerite* (grey), *pyrrhotite* and *chalcopyrite* (white, not easily distinguished)

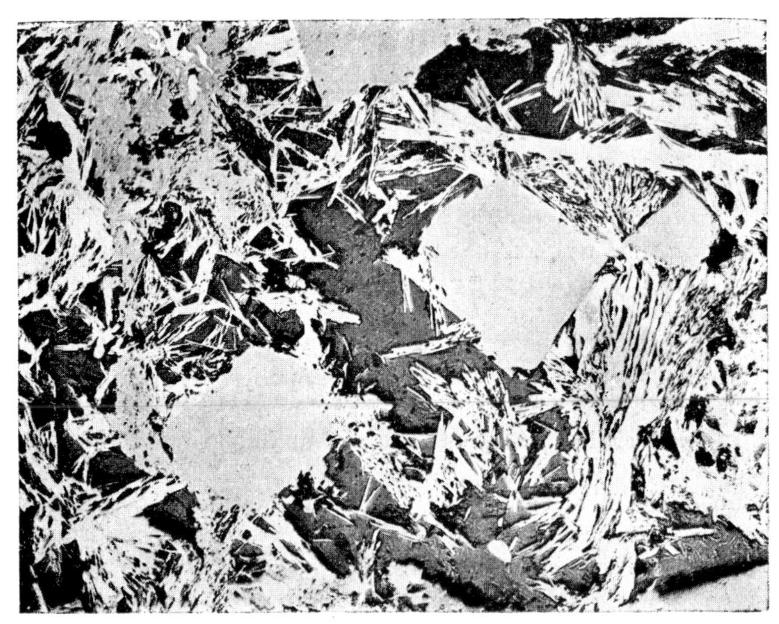

Fig. 81 150 × RAMDOHR

Slättberg Nickel Mine, Sweden

Idiomorphic magnetite (light grey) encrusted by bladed, radial *specular hematite*. In addition, gangue (very dark grey), and some sulphides (bright white)

(a) *Oriented intergrowths*

Oriented overgrowths and primary oriented intergrowths are very widespread, and are generally controlled by a simple conformity of lattice dimensions in one, two, finally all directions, or by some higher class of conformity. They have little genetic significance, since they can be formed at all different stages of mineral formation. As with solid solutions, there is greater tolerance at elevated temperatures; intergrowths of components which at low temperatures show no tendency to form, are widespread in higher temperature formations. In this respect an oriented intergrowth may give an indication, however, quite uncertain as to its temperature of formation.

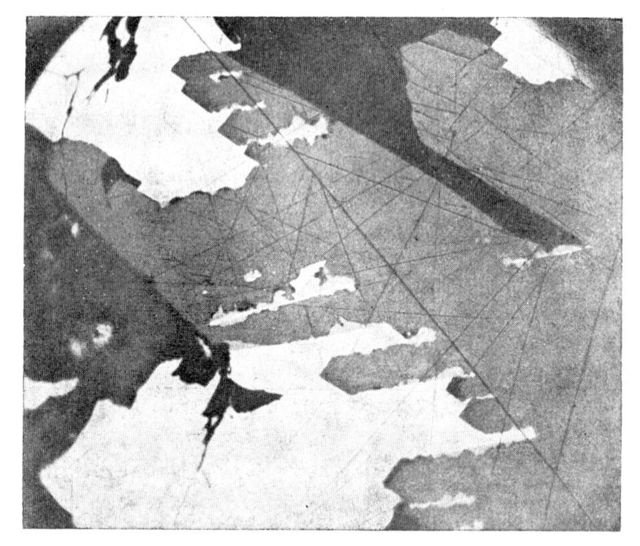

Fig. 82 85 ×, imm. AHLFELD
Veta Angeles near Tasna, Bolivia

Sphalerite (grey) with an oriented overgrowth of a little *stannite* (light grey) and also *chalcopyrite* (white)

Only very few examples can be presented here, preferably those which do not involve exsolution: graphite–muscovite (Otjimbojo); stannite on chalcopyrite, the latter on sphalerite; chalcopyrite–linnaeite; and as intergrowths of a higher corresponding order: chloanthite–bravoite, where $2a_0$ of chloanthite is very close to $3a_0$ of bravoite. Two dimensional examples are presented by: pentlandite on pyrrhotite (Miggiandone); bornite–melonite, and bornite–tetradyamite, where the sixfold axes of the hexagonal form coincide with a threefold axis of the pseudo-cubic mineral, and bornite–nagyagite, where the pseudotetragonal nagyagite conforms to the [100] of bornite, iridosmium-pyrite ($[0001]_{ir} \sim {}^1/_2 [111]_P$, $4 \cdot 6{,}0 \sim 4 \cdot 6{,}43$ Å) (figs. 82, 83, 84).

On the whole, one is often as surprised about the *absence* of overgrowths on structurally very similar minerals, as one is about the presence of overgrowths on what at first appear to be entirely dissimilar minerals.

Fig. 83 500 ×, imm. RAMDOHR
Schneeberg in Saxony, Germany

Rammelsbergite (pure white, smooth) rimmed by *chloanthite* (white, harder, partly wrinkly). In oriented overgrowth on the latter is *vaesite* (NiS_2), dark grey, despite the different lattice structures; further *bravoite*, which has still "vaesite"-like zones, and true *pyrite* (not present in this photo)

Fig. 84 30 × EHRENBERG
Miggiandone, Northern Italy

Pentlandite oriented with (111) on (0001) plane of *pyrrhotite*. A natural, unpolished crystal face!

(b) *"Emulsion" textures*

In many cases one mineral (the "guest") is found finely disseminated in a much more abundant second mineral (the "host"). The term emulsion includes a certain uniformity of distribution, equal and small grain size, and commonly rounded form, of the "guest" component. Very often these intergrowths become imperceptibly coarser grained ("coalescing" of "guest" bodies); very often also the distribution is spotty, changing regularly or irregularly, especially rhythmically shell-like botryoidal, or concentrated in alternating crystallographic zones. Many illustrations show these features (figs. 77, 78, 104, 112, 384 a. o.). — These emulsion textures can have various genetic interpretations; there may be entrapped sulphide droplets in liquid magmatic formations, undigested replacement relicts, porphyroblastic networks, dissociated precipitates from a gel, etc. By far the most important and most typical examples result, however, from "*exsolutions*", a concept which, though it could very well also be applied to exsolution phenomena in precipitates from a gel, is reserved almost exclusively for the emulsion-like precipitation, in response to decreasing temperature, of a second solid phase from crystals which at higher temperatures are homogeneous solid solutions. Not all exsolution processes give rise to emulsion textures, however. Exsolution is dealt with in a special section (p. 170) in view of its major significance. Some aspects of emulsion textures are dealt with also under "mineral inclusions" (p. 134).

(c) *Penetration textures*

Interfingering textures, like the unusually coarse intergrowth of quartz and orthoclase known as "graphic granite" are especially widespread among ore minerals, and are much more variable than the somewhat monotonous intergrowths of this type found in rocks. Beside such interfingered intergrowths of several minerals they can occur on the same mineral. The forms, too, can become sharply lamellar and can show transitions to emulsion intergrowths, or "caries-like" forms. In all cases any genetic interpretation must be postponed.

(d) *Myrmekitic intergrowths*

Among the textures shown by the ore minerals, the distinctive penetration or interfingering textures, which have been variously termed eutectic, cotectic, eutectoid, pseudo-eutectic, myrmekitic, symplectitic, granophyric, graphic, subgraphic, micrographic, play a much greater role than they do among the silicate rocks, from whose literature most of these terms are derived. Partly they belong to the "synantetic intergrowths" in the sense used by SEDERHOLM, partly they do not.

As regards *appearance* they are more monotonous; they comprise always an interpenetrating growth of large grains of two, and on rare occasions more, minerals, which are present in variable, but more or less comparable, amounts. The grain boundaries are mutually rounded, so that in a section the texture resembles finely woven fabric or the paths cut by the grubs of woodworms between bark and wood of trees.

In many cases the myrmekitic intergrowth grades into a lamellar intergrowth, but even so the characteristic rounded terminations are preserved to the laminae.*

Some intergrowths when cut in one direction appear myrmekitic, while sections cut at right angles show a lamellar intergrowth of the two components. Another characteristic feature of such intergrowths is that one (mostly both) of the components shows uniform optical orientation over considerable areas; however, this is often difficult to establish. Such intergrowths are well known in metallurgy, where some of them — by no means all! — are associated with eutectic crystallization.

Owing to the absence of crystal form, and because of the fineness of the intergrowths, it is difficult to explain the structural reasons of the myrmekitic intergrowths. Quite frequently the components have closely corresponding lattice structures at least they have existed with the original, now destroyed components. Occasionally there is absolutely no structural relationship, or the structural similarity extends to only one or two dimensions. Phenomena akin to surface tension apparently also play some role in their formation. This is evident, e.g., where on heating of a finely lamellar intergrowth, the lamellae develop a rounded "fingering" (finger-like ending) with a reduction of the free surface. — The grain size is not very characteristic. The individual "fingers" can be up to a centimeter long, and range downward to the limits of microscopic resolution.

Of the numerous terms mentioned in connection with these intergrowths, the first four suggest a definite mode of origin, or exclude another. The expression "graphic granite" is in itself purely descriptive, but has for so long had the meaning of "eutectic formation", that it must be used with caution and it is better to restrict oneself to the purely descriptive terms graphic and myrmekitic. The following will indicate, how myrmekitic intergrowths, even of the same component minerals, can originate in various ways. — It is noteworthy that some ore mineral associations favor the formation of myrmekitic intergrowths. The writer knows of examples where as many as three pairs of minerals in myrmekitic intergrowth may occur independently alongside one another (fig. 85).

Below are listed those minerals known to occur in myrmekitic intergrowth — or at least very similar textures — taken from the writer's own work (1945a), or from the literature. The predominant component is named first; and where there are sharp and distinct variations in the relative proportions of the two components, so that in some intergrowths the second component predominates, the association is named twice. Authors are only given for part of the associations; those marked R., but with no number, refer to associations observed by the present writer which are only in the work on myrmekitic intergrowth or to those not yet published.

1. Copper + silver (replacement of the less noble component). Van der Veen (1925).
2. Copper + limonite (oxidation of cementation ore). R.
3. Silver + fahlore (simultaneous ascendent replacement of a skutterudite zone). R.
4. Gold + (?) copper telluride. Stillwell (1931).

* I use the term more restrictedly than does Colony in Fairbank's "Laboratory Investigations of Ores" (1928), where the term includes several other textural variations. A sharper definition than was used there, seemed desirable to me!

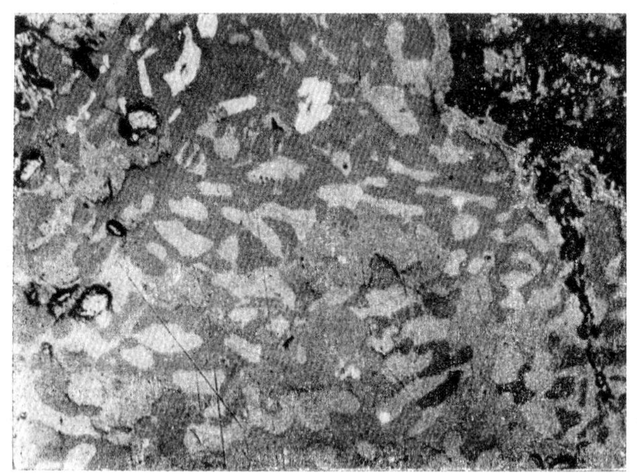

Fig. 85 150 × RAMDOHR
Yerranderie, N.S.W.

Complex myrmekitic intergrowth of *fahlore* (deep grey), *chalcocite* (light grey), *stromeyerite* (middle grey, soft), a little *galena* (white) and *pyrite* (high relief)

Fig. 85a 10 × RAMDOHR
Ruggles Mine, New Hampshire

Uraninite in a peculiar coarse form intergrown with microcline (dark grey). Some medium grey grains are zircon

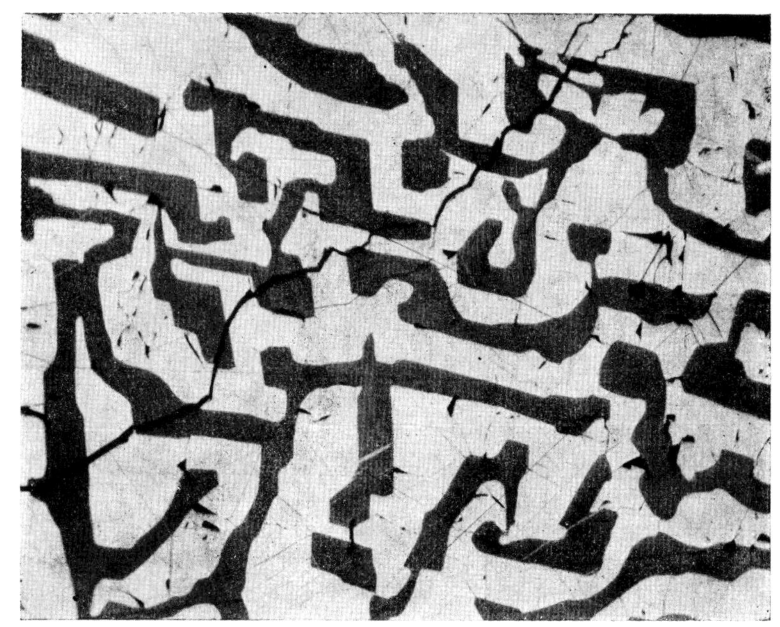

Fig. 85 E 150 × RAMDOHR
"Argentine"

A very coarse myrmekite of *pyrite* and *bornite*

5. Gold + bismuth in maldonite (breakdown of an intermetallic compound). R. (1925).
6. Platinum + sperrylite. R.
7. Iron (kamacite) + nickeliron (taenite) (eutectoid).
8. Iron + cementite (eutectoid).
9. Antimony + stibarsen + arsenic (= allemontite exsolution). KALB (1926), WRETBLAD (1941), SCHNEIDERHÖHN (1929).
10. Antimony + stibnite. R.
11. Bismuth + pyrrhotite (very beautiful, but uncertain origin). R.
12. Bismuth + galena (decomposition of cosalite). R.
13. Bismuth + bismuthinite. R.
14. Bismuth + unknown ore mineral.

14a. Dyscrasite + stibarsen. R.

14b. Tellurium + galena (SINDEEVA).

15. Hessite + gold (decomposition of a gold telluride). NEGUREI & ZEMEL (1937).
16. Hessite + sylvanite. STILLWELL (1931). R.

16a. Sylvanite + nagyagite. R.

16b. Sylvanite + krennerite. R.

17. Petzite + seligmannite. STILLWELL (1931). Reaction rim.
18. Tellurbismuth + bismuthinite. V. SZTROKAY (1946). Decomposition of tetradymite.
19. Chalcocite + bornite (various origins). Numerous references.
20. Chalcocite + galena. R.

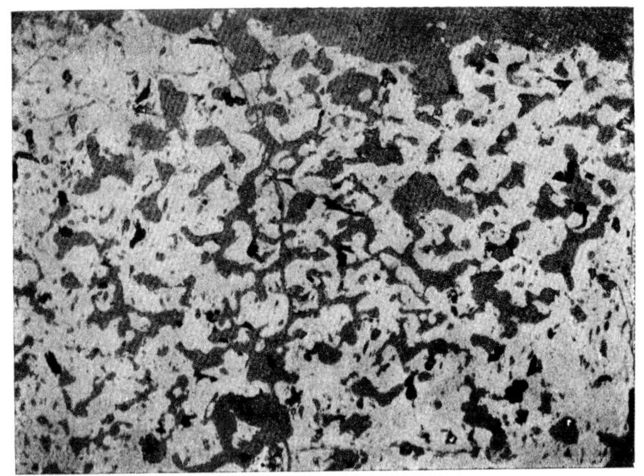

Fig. 86 170 × Ramdohr
Domokos, Greece

Chromite in a typical myrmekitic intergrowth with *sepentine* (formerly olivine). This inter-growth is very unusual, and was perhaps a true eutectic

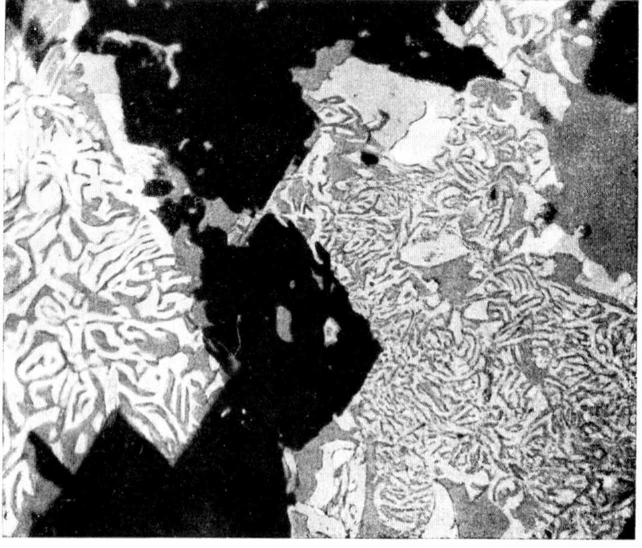

Fig. 87 500 ×, imm. Ramdohr
"Peru"

Myrmekitic intergrowths often show preference for certain ore deposits. Here adjacent to each other are a myrmekitic intergrowth of *fahlore* (dark grey) and *chalcopyrite* (grey white), and a second consisting of *fahlore* and *native silver* (pure white). The black is carbonate, probably dolomite

21. Chalcocite + clausthalite (simultaneous precipitation.) SCHERBINA (1941).
22. Chalcocite + covellite. R.
23. Chalcocite + wittichenite. RAY (1916).
24. Chalcocite + cerussite. R.
24a. Berzelianite + clausthalite. R.
24b. Berzelianite + klockmannite (decomposition of crookesite). R.
25. Stromeyerite + chalcocite. R. (1945a).
26. Stromeyerite + argentite. SCHWARTZ & PARK (1932). R.
27. Stromeyerite + fahlore. R. and others.
28. Stromeyerite + chalcopyrite (Mount Lyell Bonanza).
29. Stromeyerite + galena. SCHWARTZ & PARK, BURBANK, GRIGORIEFF (1928), GUILD. R.
30. Bornite + covellite.
31. Sphalerite + galena. GRIGORIEFF (1928), LINDGREN (1930), SCHNEIDERHÖHN.
31a. Sphalerite + cinnabar (Grafenauer 1963).
32. Sphalerite + bournonite (reaction of fahlore + galena). R.
33. Sphalerite + orthoclase (very coarse) probably pegmatitic deposition (Hagendorf). R.
34. Chalcopyrite + chalcocite (exsolution). R.
35. Chalcopyrite + bornite (devitrification of a mixed gel). R. (1945a.) (Eutectoid) VAN DER VEEN (1925).
36. Chalcopyrite + fahlore (simultaneous ascendent replacement of a smaltite component.) R.
37. Chalcopyrite + stannite (reaction rim in other cases complex origin). R.
38. Chalcopyrite + berthierite (decomposition of a complex ore). R.
39. Chalcopyrite + cassiterite (oxidation of stannite). R.
40. Fahlore + argentite (simultaneous replacement of a smaltite component). R.
41. Fahlore + bismuthinite. R. (Nueva Verdun).
41a. Tennantite + gallite (exsolution or coprecipitation). R. (Tsumeb).
42. Pyrrhotite + pentlandite (exsolution).
43. Pyrrhotite + cooperite. R.
44. Pyrrhotite + gudmundite. GAVELIN (1936), R. (1938c). (Reaction selvage of FeS with destroyed fahlore).
45. Niccolite + chalcopyrite (decomposition of a complex mineral). LAUSEN, R.
46. Niccolite + pyrrhotite. R. (decomposition of Fe-gersdorffite).
47. Niccolite + breithauptite (decomposition of solid solution "Arite"). R.
48. Niccolite + ullmannite (replacement, originating from the previous, by addition of sulphur). R.
49. Niccolite + rammelsbergite (partial replacement). R.
50. Galena + bismuth (recrystallization of tetradymite with strong interchange of material). R.
51. Galena + sylvanite (simultaneous formation). R.
52. Galena + argentite (descendent replacement of stromeyerite). GRIGORIEFF (1928), SCHWARTZ & PARK (1932), R.
53. Galena + bornite. RAY, SCHWARTZ & PARK (1932), R. (1928a), SCHNEIDERHÖHN (1920).
54. Galena + chalcopyrite ± pyrrhotite (contact metasomatic). R.

55. Galena + pyrrhotite (contact metasomatic). R.
56. Galena + fahlore (reaction selvage of galena on sphalerite, possibly also other origins, such as complex decomposition).
57. Galena + covellite (replacement of former bornite by covellite with retaining of texture), R. (1928b), WHITEHEAD (1916).
58. Galena + bournonite (from geocronite with interchange of material). R. & ÖDMAN (1939).
59. Galena + dufrenoysite (possibly supergene decomposition of gratonite to jordanite). R. (1942).
60. Galena + boulangerite (through partial replacement of bournonite). MAUCHER (1939).
61. Galena + jamesonite (decomposition of a complex compound. R., WHITEHEAD (1916).
62. Galena + cosalite (decomposition of a complex mineral). R.
63. Galena + galenobismutite (? decomposition of a complex mineral). BERRY (1940d).
64. Galena + pyrargyrite.
65. Galena + pearceite. WHITEHEAD (1916).
66. Galena + polybasite. WHITEHEAD (1916), R.
67. Galena + (?) silver mineral (possibly true eutectic). R.
67a. Galena + gallite (exsolution bodies in a sphalerite later replaced by galena). R. (Tsumeb, Kipushi).
68. Galena + quartz (decomposition of a mixed silica-sulphide gel). SCHNEIDERHÖHN.
68a. Galena + cassiterite. R. (precipitation from a mixed gel, also oxidation of teallite). R.
69. Galena + calcite.
69a. Clausthalite + klockmannite. R.
70. Altaite + tetradymite. STILLWELL (1931), THOMSON.
71. Altaite + krennerite. STILLWELL (1931).
72. Schapbachite + galena (decomposition of a high temperature solid solution). R. (1938a).
73. Bismuthinite + emplectite. R.
74. Bismuthinite + (?) mineral. R.
75. Bournonite + fahlore + galena (decomposition of wolfsbergite with addition of Pb). R. (1938c).
76. Bournonite + chalcopyrite + sphalerite (from fahlore + meneghinite) R.
77. Bournonite + galena (decomposition of different complex ore minerals). FRIEDRICH (1939b), R.
77a. Bournonite + highly reflecting carbonate. SMEJKAL & RAKIC (1957).
78. Jamesonite + pyrargyrite. ZIMMER (1936).
79. Boulangerite + galena together with jamesonite (decomposition of a complex mineral). MAUCHER (1939).
80. Meneghinite + bournonite + chalcopyrite ± sphalerite (from fahlore by addition of Pb). R.
81. Owyheeite + pyrargyrite.
82. Alaskaite, "with itself". R. (Cerro Bonete, Bolivia).

83. Polybasite + argentite. WHITEHEAD (1916).
84. Linnaeite + chalcopyrite. R.
84a. Linnaeite + bravoite + chalcopyrite. R. (decomposition product of villamaninite).
85. Pyrite + argentite (decompostion product of sternbergite). R.
86. Pyrite + chalcopyrite (contemporaneous low temperature deposition). R.
87. Pyrite + magnetite ± marcasite (oxidation of pyrrhotite). R.
88. Pyrite + linnaeite. R.
89. Arsenopyrite + loellingite (with equal orientation). R.
90. Chloanthite + pararammelbergite (ascendent replacement). R.
91. Magnetite + spinel (mostly exsolution). R. (1939).
92. Magnetite + chromite (mostly exsolution). R. (1945a).
93. Magnetite + hematite (uncertain, ?martitization). R., SCHWARTZ & PARK (1932).
94. Magnetite + ilmenite (exsolution with following collection crystallization). R. (primary intergrowth). R. (1945a).
95. Magnetite + ilmenite + spinel. R. (1945a).
96. Magnetite + rutile (from ilmenite, about Fe_2O_3:$FeTiO_3$ 1:1). R. (1945a).
97. Magnetite + garnet. R. (1945a).
98. Magnetite + augite. SCHWARTZ & PARK, NEWHOUSE (1936), R. (1940).
99. Magnetite + hornblende. SCHWARTZ & PARK, NEWHOUSE (1936), R. (1940).
100. Magnetite + biotite. SCHWARTZ & PARK, NEWHOUSE (1936), R. (1940).
101. Magnetite + plagioclase. R. (1940), NEWHOUSE (1936).
101a. Chromite + olivine (probably true eutectic). R.
102. Franklinite + hematite. R.
103. Braunite + hausmannite (excellent). R.
104. Hematite + chalcocite (oxidation). R. (1924).
105. Ilmenite + spinel. R. (1940).
106. Ilmenite + magnetite. R. (1945a)
107. Ilmenite + biotite. R. (1940).
108. Rutile + magnetite (oxidation of ilmenite). R. (1926).
109. Rutile + hematite (oxidation of ilmenite). R. (1926).
110. Rutile + limonite (weathering of ilmenite). R. (1945a).
111. Cassiterite + galena. R. (possibly decomposition of teallite).
112. Cassiterite + quartz. R. (1935a).
113. Cassiterite + fluorite. R. (1935a).
114. Cassiterite + calcite. R. (1935a).
115. Scheelite + pyrite (decomposition product of wolframite). R.
116. Rhönite + magnetite + augite (decomposition of basaltic hornblende). R. (1945a), SÖLLNER (1909).

The distribution is accordingly quite enormous, particularly if one bears in mind that naturally the table is not exhaustive.

The *classification* of the processes giving rise to graphic textures is difficult, since transitions occur in many places, and since on the other hand the processes are often less decisive than the *structural* relations between the components. I propose the following classification according closely to the work of NIGGLI, and an unpublished outline by H. SCHNEIDERHÖHN, which I have expanded considerably:

1. Simultaneous formation of the intergrowth of the components in an independent process:
   a) from a melt or contact metasomatic process (examples of certainty, or great probability: 6, 43, 67, 94 in part, and 33, 54, 55, 101 a, 112–114 resp.);
   b) hydrothermal (42, 74, 53, 19, 37, 78, 51);
   c) low temperature formation (86).

To group 1 belong also:

1A Eutectoid (with some certainty 7, 8, 35);
2. primary oriented intergrowths (certainly several cases);
3. reaction rim formation (17, 32, 37, 56, 95);
4. exsolution (9, 34, 37, 42, 72, 94 in part), and closely related thereto:
5. collection crystallization after exsolution (19 in part, 36 in part, 91, 92, 94, 95), and related types;
6. decomposition of compounds and of solid solutinons:
   a) intermetallic and similar compounds (5, 18);
   b) normal sulphides (16, 22, 46, 47, 84a);
   c) sulphosalts (12, 38, 58, 59, 60, 61, 62, 77, 78, 79, 81);
   d) oxides (101, 102, 105, 106).

To 6 belong, without the possibility of sharp distinction:

6a. Decomposition of compounds in response to (often very slight!) addition or removal of material (44, 50, 75, 80, 115, figs. 89, 90);
7. replacement (52, 57, 58 in part, 61 in part, 19 in part, 37 in part);
8. devitrification ("aging") of sulphide gels (31, 68, 69 in part);
9. oxidation of a compound (3, 87, 85, 86, 39, 103, 110).

This means then that the possibilities of formation of such intergrowths are numerous, so that extreme care must be taken when attempting a genetic explanation of *only* the texture. — The details on this topic are given in my extensive study (1945a).

The *forms* assumed by myrmekitic intergrowths are described briefly on p. 110, and further references are given in the drawings of figs. 171 and 172, and in figs. 91–95. Naturally the multiplicity of shapes cannot be exhaustively illustrated, the more so in that the very finest are difficult to photograph.

Penetrations in "myrmekitic" forms have already been dealt with in detail. "Penetrations" are not rare in other types of intergrowth either; again they can originate in various ways: as true eutectics or "cotectics", by decomposition of individual minerals or associations, especially those that undergo "collection crystallization" [Sammelkristallisation], by skeletal or reticulate, and above all by "diablastic" growth, i.e. during and after a shearing movement collection crystallization may set in, giving rise to peculiar textures which genetically and particularly in terms of crystal chemistry, are difficult to interpret. The resulting minerals occur irregularly interlaced, commonly in generally similar proportions, but at times with one predominant, such that both components in thermselves may often show largely uniform orientation. It is not surprising that this produces interconnected, even though networklike, skeletal forms. But it is astonishing when small ragged or oval crystals of one component occur with

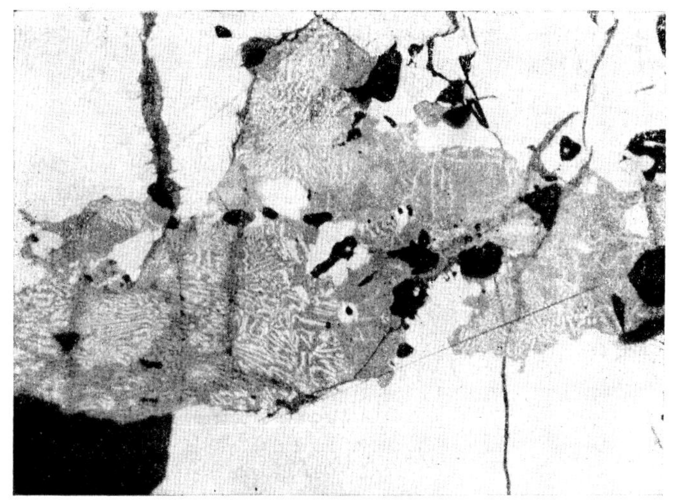

Fig. 88 250 ×, imm. RAMDOHR

Shensei Vein, Ashio Mine, Japan

Myrmekitic intergrowth of *stannite* and *chalcopyrite*, formed from *hexastannite* (dark grey, homogeneous). The large continuous patches are chalcopyrite. The secondary fracture filling consists of *chalcocite* and *covellite*

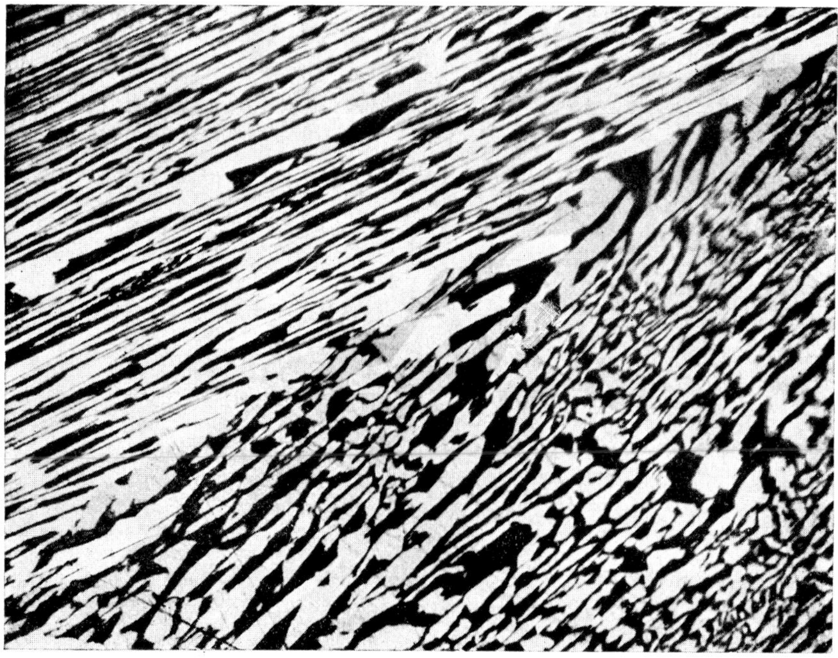

Fig. 89 150 ×, imm. crossed nicols RAMDOHR

Gladhammar, Sweden

Myrmekitic intergrowth formed through the dissociation of a radially textured complex mineral. The black (isotropic) component is *galena*, the light is *cosalite* or *galenobismutite*. The section is cut partly parallel to a radial direction, partly oblique to it

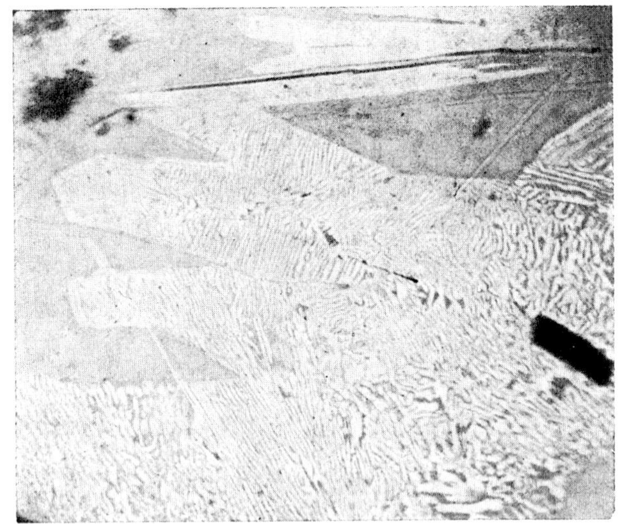

Fig. 90 450 ×, imm. RAMDOHR
Oraviţa, Banat

Tetradymite (light grey, idiomorphic plates) in *galena* (darker grey). Most of the tetradymite has been converted to a fine myrmekitic intergrowth of *native bismuth* (white) and galena

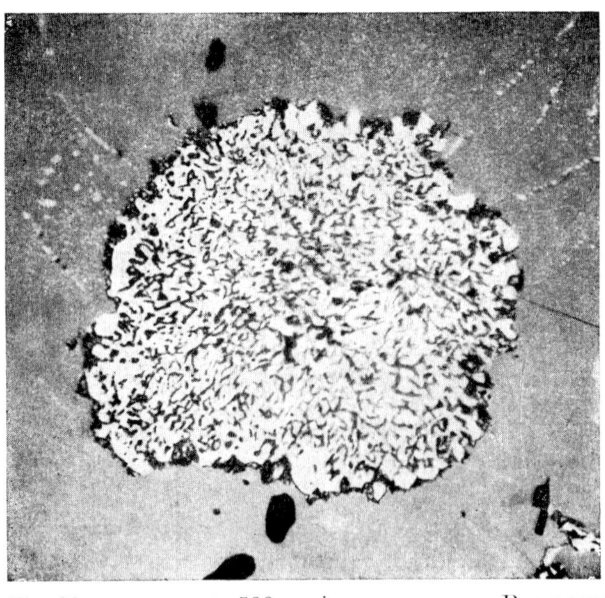

Fig. 91 500 ×, imm. RAMDOHR
"Cornwall"

Myrmekitic intergrowth of *cassiterite* (dark grey) and *pyrite* (white) formed from *stannite* (grey), perhaps through weathering. The stannite shows some exsolution bodies of *chalcopyrite*. The "myrmekitic intergrowth" because of its high relief is not entirely in focus over the whole surface

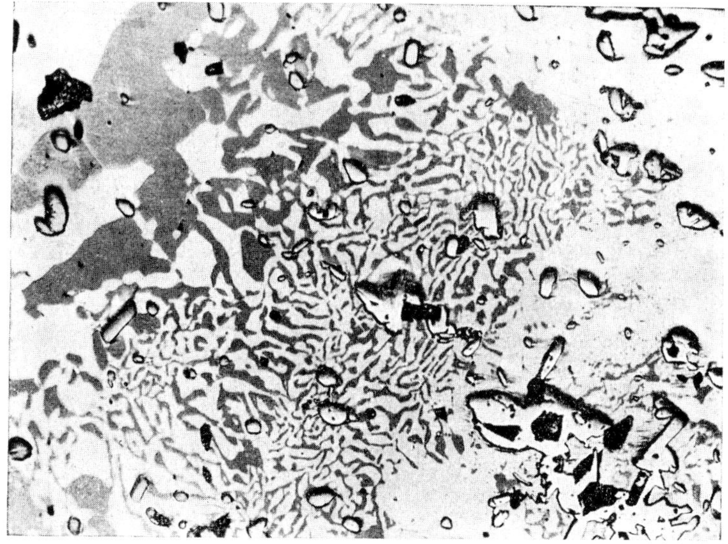

Fig. 92 210 ×, imm. Hsieh
Lou Sui Chang, Tung Chuan, Yunnan

Myrmekitic ("graphic") intergrowth of *chalcocite* (white) with *bornite* (dark grey). Nearby are idiomorphic grains of *hematite* (high relief, white)

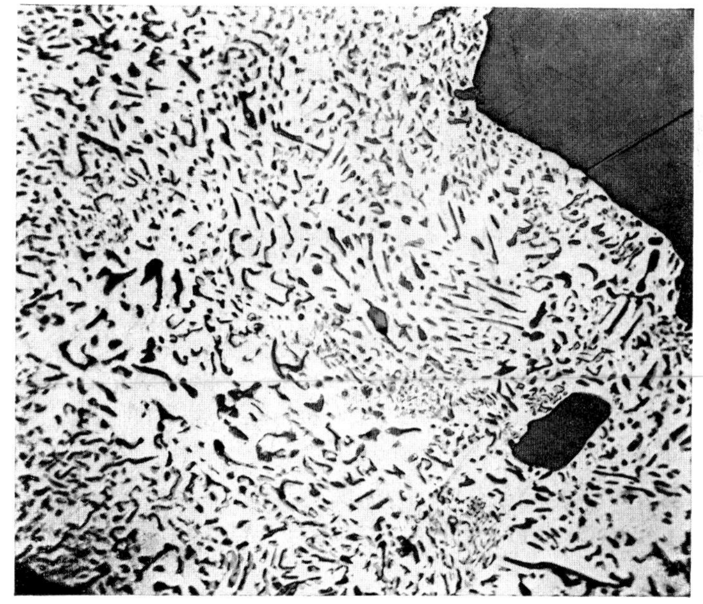

Fig. 93 215 ×, imm. Ramdohr-Rakič
Belo Brpo, Kopaonik Mountains, Yugoslavia

Myrmekitic intergrowth of a *gold telluride* (possibly krennerite or montbrayite) with *pyrrhotite* (darker)

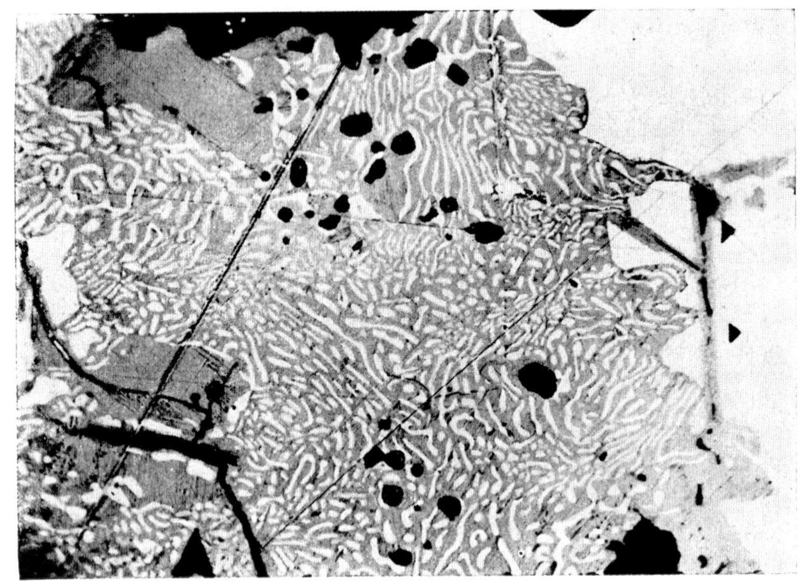

Fig. 94 250 ×, imm. RAMDOHR
Dognacea, Banate

Myrmekitic intergrowth of *galena* and cementing *chalcocite* (darker), the latter probably replacing former bornite

Fig. 95 225 ×, imm. RAMDOHR
Freiberg in Saxony, Germany

Bornite (grey) in myrmekitic intergrowth with *chalcocite* (light grey). At a short distance from the bornite, in the manner of a selvage-rim, is a medium grey, soft *mineral*. The black is *dolomite*

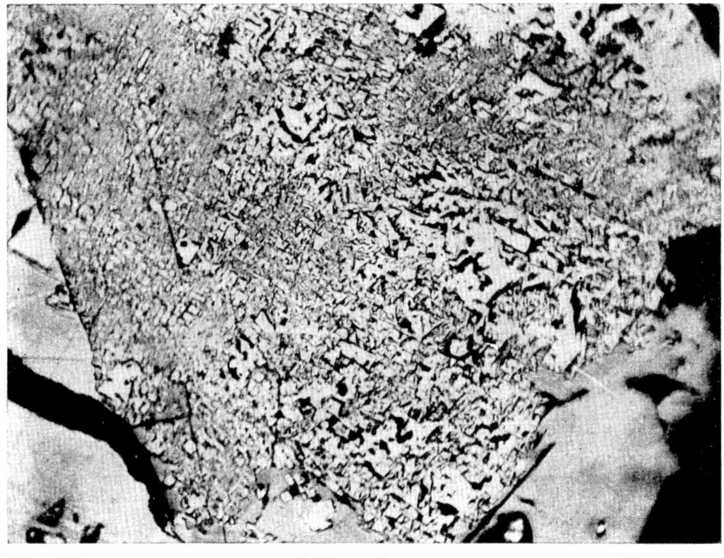

Fig. 96 500 ×, imm. RAMDOHR
Saxberget, Sweden

Pyrite (white) and *magnetite* (dark), both cut about parallel (111), as decomposition products of *pyrrhotite*, parallel to (0001). Remains of the pyrrhotite can be seen on the left and the right

uniform orientation dispersed throughout the strongly predominant component, and this without the two components bearing close crystallographic relation. Later, i.e. of course immediately after the recrystallization, the small bodies must have been connected by bridges, which were again eliminated.

Such textures are not uncommon among the ore minerals of igneous rocks. They are widespread among certain dissociation textures of complex high temperatures ores, and most widespread in metamorphic formations.

III. THE FORMS OF AGGREGATES

It has been indicated already in several places, that the nomenclature used in petrology and metallurgy has only limited application to the great variety of textures exhibited by ore minerals. On the other hand, the ore minerals commonly lack some textural characteristic of the rocks. It is not the writer's intention to introduce here a new terminology for the ore minerals; already the literature is burdened with terms, many of which are not generally used, even when — which is by no means always the case — they are well defined. Moreover, it is not necessary that the features dealt with should be established as equivalent or subordinate. What is much more important is that this chapter is a useful aid for observations and that the photographs stimulate comparisons (e.g., the genetically instructive figs. 4, 27, 78, 105, 256).

As a foremost step several petrological terms are adopted and textural relations characterized by them are shown.

First of all it is desirable to point out that while the latter could be borrowed from monomineralic aggregates, polymineralic aggregates are considerably more widespread and easier to study.

Moreover, the form of monomineralic aggregates has considerably less genetic significance, and is frequently dependent on local, often quite secondary, factors.

Of the terms introduced from petrology: eugranitic, porphyric, granoblastic, porphyroblastic, which imply a texture as well as a mode of origin, should not be used. The latter should and must be suppressed. Truly eugranitic textures, which correspond to a plutonic rock, are almost entirely absent from ores, whereas similar textures formed by secondary processes are common. Phenocrysts with good idiomorphic development are extremely widespread, but only rarely can they be interpreted as the equivalent of a phenocryst in an igneous rock. In very many ore associations "phenocrysts" develop as a result of idioblastic growth, whether with or without addition of material from outside. "Phenocryst-like" can be replacement relics, cataclastic remnants, grains of especially abundant components, as well as real porphyroblasts and xenoblasts in metamorphic ore mineral assemblages.

Attention should also be drawn to the fact that whereas in German the spatial relations of the components are referred to as "textures", in American usage such relations are referred to as "fabric".*

(a) *Arrangement in space*

Randomly oriented [richtungslos], especially equigranular ["richtungslos körnig"] texture is, in petrology, the characteristic texture of the undeformed plutonic rocks, of contact metamorphic rocks close to the intrusive igneous mass, and of the products of metamorphism at "greatest depth" (katametamorphism, which occasionally, of course, does not have to be an excessive depth!). In ore associations, this texture is also widespread, and of itself has little genetic significance, though in conjunction with other factors it may be meaningful. Fig. 97 shows an equigranular texture in a strongly anisotropic mineral; the anisotropy renders the texture visible. Some textures which are difficult to describe also show random orientation, such as the interiors of various "metasomatic veins", pseudomorphosed occurrences, and broken shredded forms ["zerhackt sperrige Formen"].

Fig. 97 100 ×, imm. crossed nicols RAMDOHR
Routivare, North Sweden

An apparently randomly oriented equigranular (actually weakly oriented) aggregate of *ilmenite*, cut about parallel to the (100) plane of the texture of a pencil-gneiss-like rock, with a "girdle" about (100)

"Oriented" or "lineated" textures are of equally diverse origin; the expression itself is not without ambiguity. The crystal fibers of a radial aggregate, the successive shells of a crystallized gel, the bladed components of a sediment, the fibrous growths filling

* But by no means exactly! Also, the words are not always used with precisely the some meaning. CROSS, IDDINGS, PIRSSON and WASHINGTON (The textures of the igneous rocks, 1906) use the term somewhat differently.

a tension gash, the individual crystals of a reaction rim, and the like, are oriented, as is also the ore occurrence that corresponds to a sheared sedimentary rock. Outwardly the texture resulting from "collection crystallization" is mostly granular, but fabric studies may reveal a distinct orientation of the grains.

In general, we have to deal with two types of oriented fabric: "orientations" resulting from the shape of the grains, as, e.g., when oval-shaped grains lie with their longer axes more or less parallel (*shape-orientation*); and structure orientations involving a parallel crystallographic arrangement of the grains (*lattice orientation*). In the first type there need to be no relation between the shapes of the grains and their optical and crystallographic orientation; in the second the orientation may not be manifested in the actual shape of grains. Commonly both factors are present, particularly in minerals with one or two predominant lattice directions, but not in minerals with an isometric or nearly isometric lattice — quartz is a good example — (fig. 98).

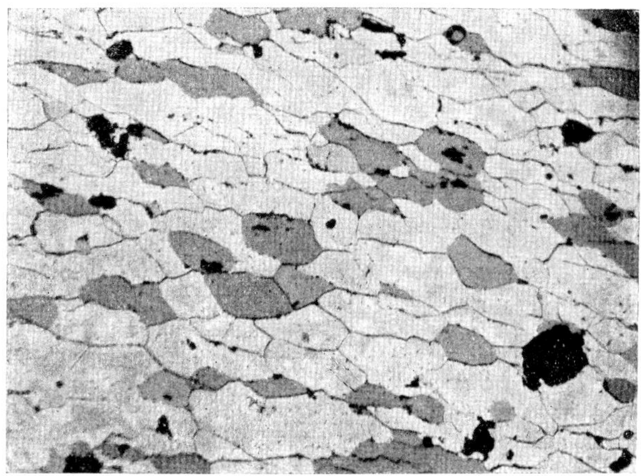

Fig. 98 100 ×, imm. RAMDOHR
Chihangi, Angola

Magnetite speckled light grey, because of partial formation of *maghemite*, *ilmenite* (dark grey) and *spinel* (black). The magnetite occurs as elongated grains with their longer axes more or less parallel, but does not show optical orientation, as is evident from the arrangement of exsolution bodies. The ilmenite shows both shape-orientation and optical orientation, with (0001) parallel to the parallel elongation of the grains. Only a few grains (which appear lighter) lie oblique to the general direction of orientation

Special reference is made to the common occurrence of *radial* arrangement and the palmetto-shaped textures of crystallized gels (examples: stibnite "suns", marcasite in schalenblende, botryoidal limonite).

A further type of oriented texture is the *fibrous texture* such as that shown by satin-spar (gypsum), which is not uncommon in ores and other minerals (fig. 101).

These fabrics are illustrated in figs. 98 to 102; fig. 99 shows parallel structure in a sediment; fig. 98 shows a parallel structure induced by shearing; fig. 100 fibrous texture in chrysotile; fig. 102 radial arrangement of prehnite; fig. 123 marcasite; fig. 128 devitrification spherulites in a gel.

Almost all the crystal grains in cockade textures, reaction selvages, annular or ring-like textures, rhythmic textures, and the like, are "oriented".

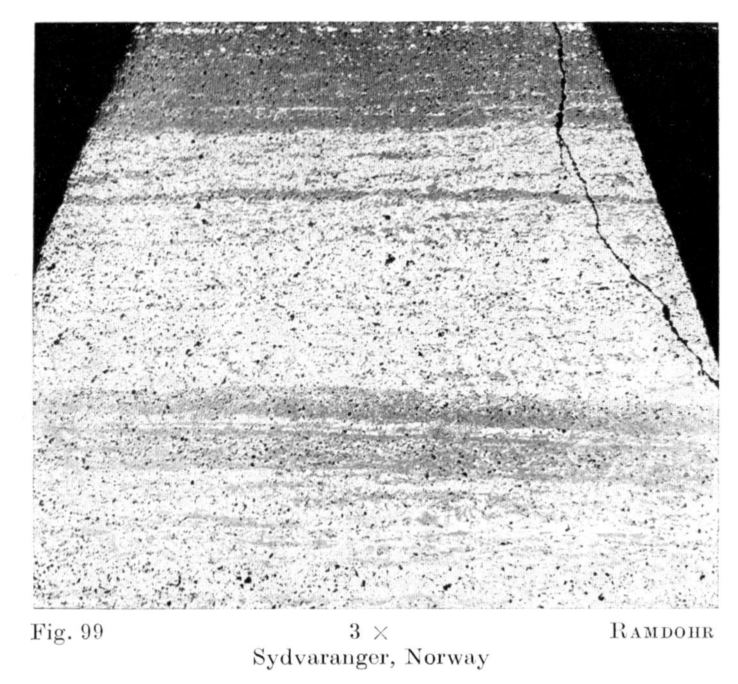

Fig. 99 3 × RAMDOHR
Sydvaranger, Norway

Finely banded, low grade ore consisting of *quartz* and *magnetite*. It is capable of beneficiation by magnetic separation

"*Random orientation*" in a strict sense is relatively uncommon; a careful examination will generally reveal at least a weak lineation. By superficial examination, however, "random orientation" is quite common. Not only the grains in magmatic and katamorphic associations, but also those occurring as suspended crystals in gels, the broken shredded aggregates, exsolution bodies in which there is no crystallographic relation evident between "host" and "guest" show "random orientation".

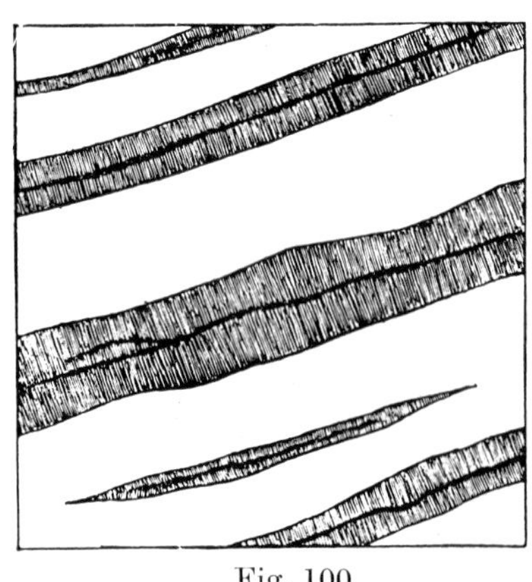

Fig. 100

Rhythmic fabrics of one or several components are widespread, but can originate in quite different ways. It is particularly difficult to distinguish between a "primary" rhythmic texture as, for instance, due to precipitation from a gel, and one due to infiltrations, which may have an almost identical appearance even without any influence by present crystal structures of older textures. The aggregate of crystal grains concerned might be thought to be younger than it is, unless the true relations are revealed by, per chance fortuitous characteristic features. The distinction is readily made, of course, in the relatively common case where secondary textures arise through weathering or cementation.

Rhythmic growth can express itself in various ways; by a change in mineral composition; by a change in the chemical composition of the individual mineral, as by

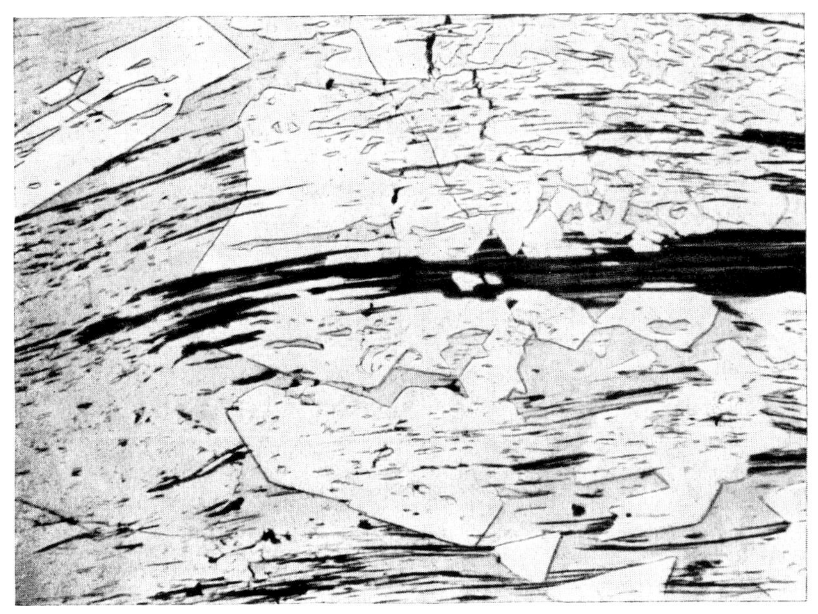

Fig. 101 35 × RAMDOHR
Ocna de Fier, Banate

Fibrous *bismuthinite* with "idioblastic" *pyrite* (white). Interleaved with the bismuthinite is a bladed gangue mineral (an *amphibole*)

Fig. 102 100 × RAMDOHR
Doros, South-West-Africa

Prehnite (grey) replaced by *tenorite* (grey-white) and *cuprite* (white), with *chalcedony* (almost black)

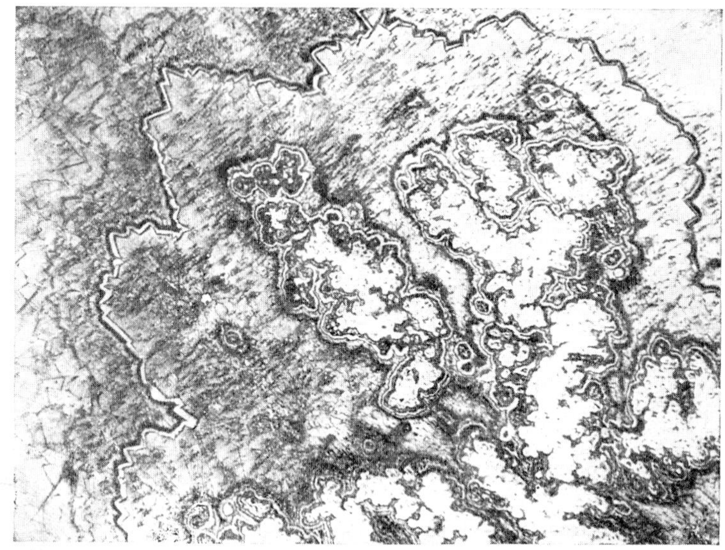

Fig. 103 50 ×, imm. RAMDOHR
Kowary (Schmiedeberg), Silesia

Native arsenic (grey) enclosing grape-like clusters of *löllingite* (white), with a rhythmic development of thin skins of löllingite encrusting successive crystalline surfaces of the arsenic

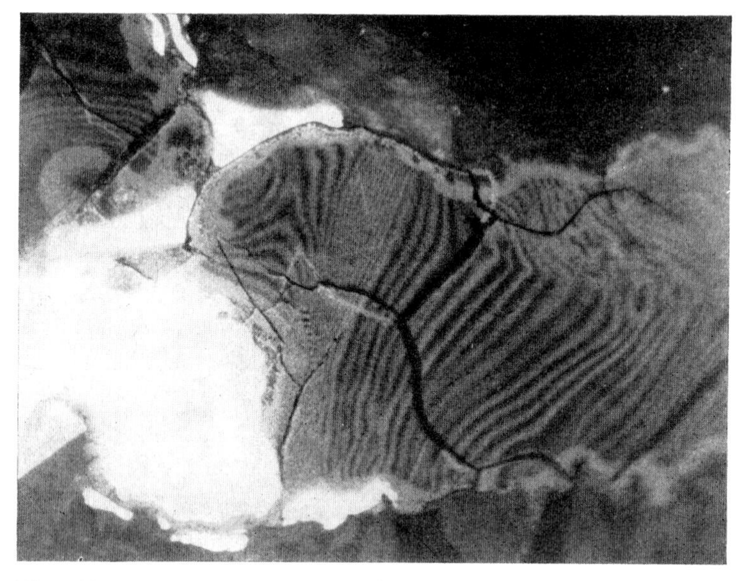

Fig. 103a 250×, imm. RAMDOHR

Rhythmical precipitation in the alteration products of native iron

formation of a solid solution; by different grain size; by change in porosity, or in the concentration of inclusions; by a change in crystallographic character, and so on. Frequently the rhythmic variation in these various properties is so slight that it is made evident only by etching, i.e. rhythmic growth shows up only through differences of lattice distortions. In most cases several different characteristic features may occur together: marcasite shells of different grain size and different intensity of twinning are at the same time separated by thin films of melnikovite-pyrite (fig. 485); botryoidal low temperature formations of chalcopyrite + bornite show not only varying proportions of bornite, but in the bornite-rich zones the chalcopyrite is particularly fine grained (fig. 385). Further examples are shown in figs. 103 and 314.

The *extent* of rhythmic textures can vary widely: individual textures can range from several meters across, down to the limits of microscopic resolution. It is thus evident that "rhythmic" growth is not limited to low temperatures and to gel formations. Magmatic resorption of sedimentary blocks can yield prominent rhythmic textures, contact metasomatic replacements give rise to such intergrowths, from very large ones down to the smallest, so do hydrothermal formations of all types, as well as also the products of weathering and even surface waters.

If in the *illustrations* gel-formations are shown with preference as rhythmic textures, it is because they provide exceptional illustrations of an appropriate scale. It must not be thought, however, that gels are only formed from low temperature, hydrous solutions. The indubitable gel masses of pitchblende in the Co–Ni–Ag–As formations, for whose origin at about 250 °C exist quite reliable "geologic thermometers", the clusters of botryoidal to rhythmic shell-like valleriite nodules in also clearly high temperature association, the cherry-sized finely rhythmic intergrowths of quartz and galena, are proof enough that, even though the physico-chemical conditions are not fully understood, most of the heavy metal sulphides can be transported in alkaline solutions mostly in a colloidal state.

Fig. 118, and those following, in the section on textures of this book, as well as in the mineral descriptions, show the similar appearance in spite of the many differences.

The tendency of different minerals to form rhythmic textures varies considerably, and depends on a variety of factors. The writer has not succeeded in finding conformity to a rule, as for instance in relation to types of lattices.

When the minerals forming a rhythmic texture, or the individual parts of a single mineral showing such a texture, are readily subject to replacement, quite unexpected associations may develop. Several examples are given in this text.

Special examples of rhythmic forms are dealt with on p. 147 etc. under layered and banded textures, annular and cockade ores, "gel textures".

(b) *Contact rims*

The term "contact rim", which is devoid of any genetic significance, will be used here for those features that SEDERHOLM recognized under the equally purely descriptive term "synanthetic". They include the numerous intergrowths which are limited in their form to "boundary fillings", and commonly also in composition are related to the adjacent mineral grains. Many of them are rightly regarded as "*reaction rims*", but some are clearly of quite different origin, and some are uncertain. The term "*intergranular film*" is a near analogous term, equally devoid of genetic meaning, but it covers only some of the phenomena — for which it is used. Some contact rims develop idioblastically as crystals penetrating the accompanying mineral (fig. 106), as complex intergrowths (fig. 105), as rows of grains (fig. 108), and so on.

They originate in various ways: as final crystallization products of solutions and melts, by exsolution and migration to the grain boundaries, as true reaction products

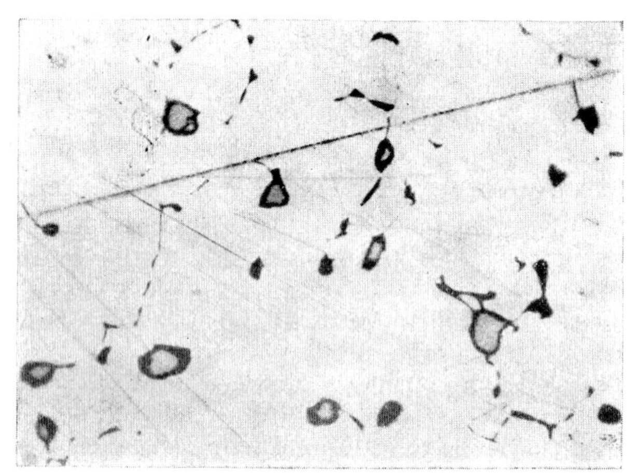

Fig. 104 600 ×, imm. RAMDOHR
Bliesenbach, Rheinland, Germany

Rounded bodies of *bournonite* (grey), with cockade-like rims of *fahlore* (dark grey) in *galena*. The fahlore also forms fine intergranular grains in the galena. The very slight differences in brightness were very difficult to reproduce photographically

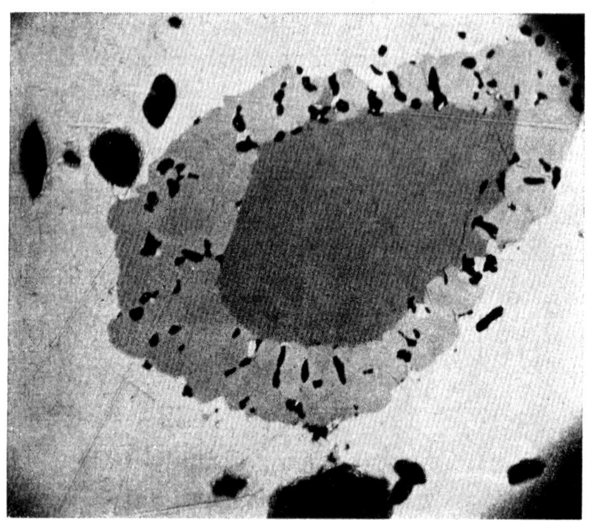

Fig. 104a 600 ×, imm. TUFAR
Myrthengraben, Austria

Tetrahedrite surrounded by a reaction rim of *bournonite* (white) and *sphalerite* (dark grey) in *galena* (white)

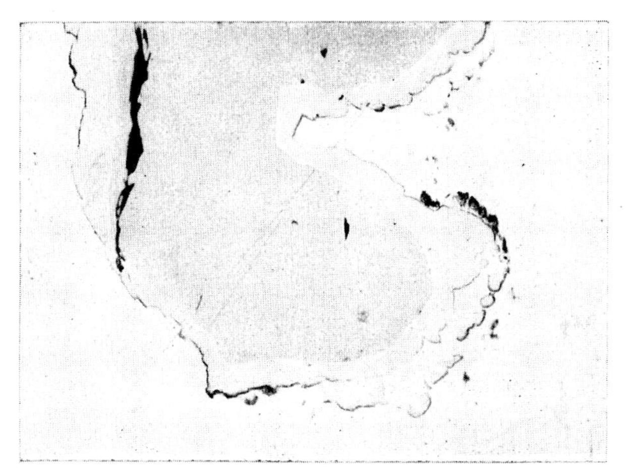

Fig. 105 350 ×, imm. RAMDOHR
Robb-Montbray Mine, Abititi Co., Quebec

Frohbergite, $FeTe_2$ (light grey) reaction rim between *melonite* and *tellurbismuth* (both pure white) and *chalcopyrite* (darker grey). Between the frohbergite and the chalcopyrite is a very thin skin which shows a film of an as yet unknown telluride or complex telluro-sulphide

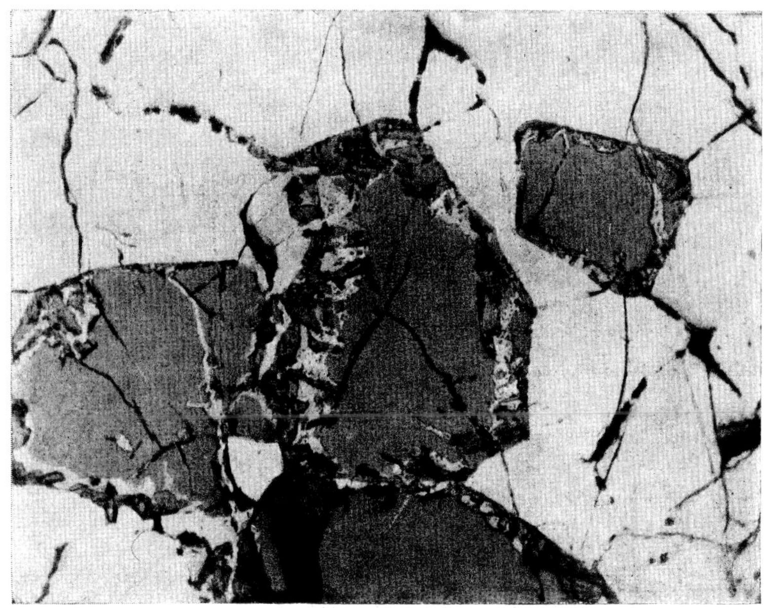

Fig. 106 170 × RAMDOHR
Chinjumba-West, Angola

Spinel in a titanomagnetite differentiate. The spinel has broken down marginally, with a precipitation of confusedly ordered *corundum* crystals. Some magnetite separated in the process

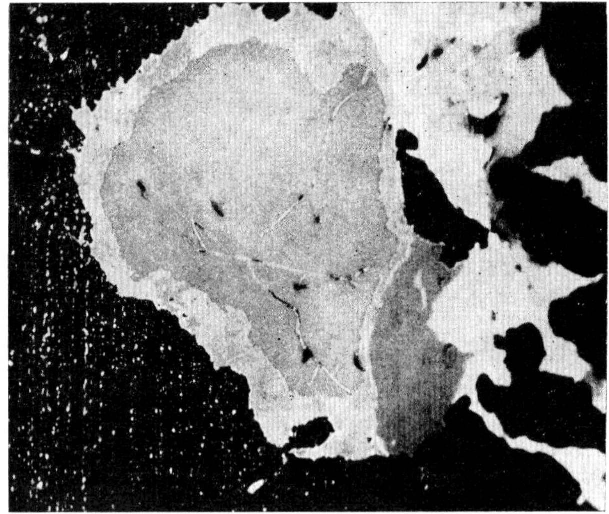

Fig. 107 250 ×, imm. RAMDOHR
Zinnwald, Erzgebirge

Reaction rim between a *fahlore mineral, isostannite* (grey) and *sphalerite* (almost black, studded with *chalcopyrite* bodies). The reaction rim consists of *hexastannite* (light grey, strong reflection pleochroism), *bismuthinite* (white) and *bornite* (dark grey)

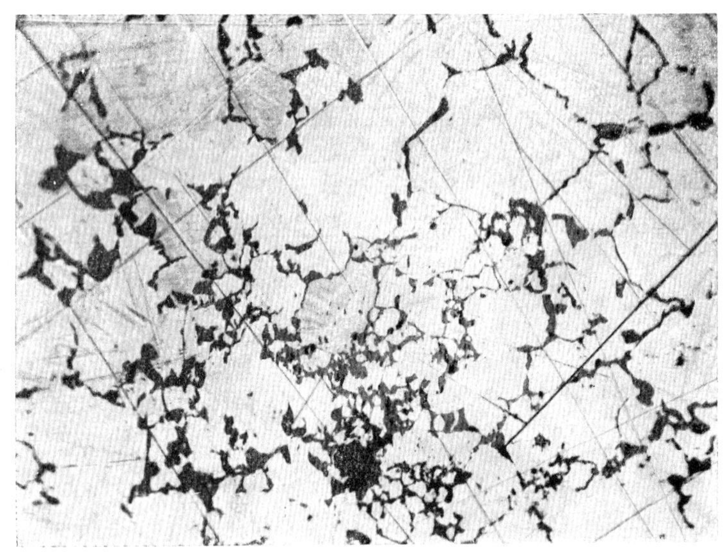

Fig. 108 50 ×, imm. RAMDOHR
Åskagen, Värmland, Sweden

Native bismuth as a granular aggregate, with intergranular films of *bismuthinite* which appear here (high contrast print) as dark grey bodies. Locally twin lamellae can be observed in the bismuth (unfortunately scratches too!)

(by solid state reaction, or in cooperation with residual solutions), by infiltrations of products dissolved somewhere else thus already to be described as replacement, as intergranular films formed by recrystallization, oxidation and cementation. Only a few examples will be considered here, but others will be found in other connections.

Reaction rims. These belong genetically, as indicated, to the great complex of replacements, and can be assigned to very different roles. Simple types of minerals may be stable in contact with each other at high temperatures, but react with each other at lower temperatures to form complex compounds, or they may be catalysed by a reaction exchange with residual solutions, or the late solutions may no longer be in equilibrium with the early-formed minerals, or there may be a supplementary addition of materials to the earlier-formed minerals, or yet other possibilities may arise. In all cases rims develop, either continuous or interrupted, sometimes of varying thickness, more often of about constant width. They consist often of two or more minerals, which as regards material are derived entirely or in part from the surroundings. Thus cases occur of the type that mineral A originally occurred in contact with mineral B, but now they are always separated by a film of a mineral A_xB_y, or are separated by an intermittent, but widespread, rim of this nature, or the rim may occur only occasionally. Various examples are illustrated in this book: wolfsbergite separated from galena by a probably later-formed, relatively thick crust of bournonite; bournonite between geocronite and fahlore (fig. 468); "hexastannite" between a stanniferous fahlore [Zinnfahlerz] and sphalerite, together with bornite (on the outside some luzonite is formed) (fig. 107); bournonite between fahlore and galena (fig. 104), besides some more unusual associations (figs. 105, 106).

A long list of similar examples could be made. They comprise the most varied combinations of elements, and include both sulphides and oxides. Reference is made to them in the descriptions of the individual minerals. The fact that some of the reaction rims are myrmekitic intergrowths of finely intergrown minerals is referred to on p. 110 etc.

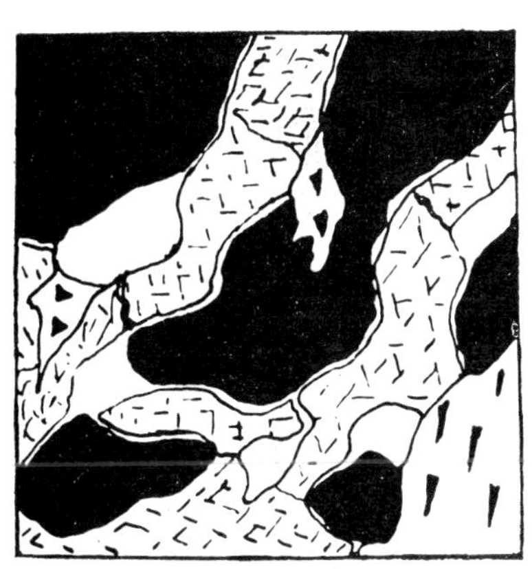

Fig. 109

A film of chalcopyrite between gangue (black), and sphalerite, crossed, a little galena, triangles

Intergranular films almost correspond in form to reaction rims — some are in fact reaction rims. They are quite a heterogeneous group of features. Quite similar or even identical textures are produced by fine films of latest liquid eutectic melts; or by residual melts, which do not react, or by a process of "self-purification" of grains, whereby the foreign components are segregated toward grain boundaries; or by the expulsion of exsolution bodies or of material till then in solid solution in response to shearing; or by deposition from solutions, which moved along grain boundaries, by incipient metamorphism, or by reaction with solutions that moved along grain boundaries, and by other means. It is not very significant, therefore, whether they take the form of complete films of uniform or variable thickness, or of "pearl-string" like rows of grains, or of more or less irregularly separated grains, or of transitions to films (fig. 108, 109).

Of the illustrated examples, which are chosen from among the most beautiful available, fig. 108 is probably a primary "magmatic" deposit of native bismuth and bismuthinite; fig. 46 represents the segregation of a former exsolution product no longer soluble as a result of recrystallization of the ore aggregate in response to shearing; fig. 109 shows a uniformly thick chalcopyrite film that has utilized the grain boundaries between sphalerite and gangue to move in. One finds similar continuous films of galena in the grain boundaries of sphalerite grains which are themselves rimmed with segregated films of exsolved chalcopyrite; and films of galena about schapbachite grains in which the galena was formerly dissolved.

(c) *Mineral inclusions in ore minerals*

Among the textures of the ore minerals, H. SCHNEIDERHÖHN has dealt in particular with the relations shown by inclusions. He presupposes, of course different from the writer, that the origin of the ore intergrowth is known. However, one can apply his classification in reverse, if in each case the clear geometric features are genetically interpreted. SCHNEIDERHÖHN has correctly recognized the inclusions for what they are by terming them "guests", a concept which the writer has adopted on several occasions: according to the presence or absence of a relation in the lattice and bonding character of the enclosing mineral or "host", he distinguishes between "related guests" [Familiengast] and "unrelated guests" [Fremdgast], respectively.

His system which is in most instances understandable without further explanations is presented below.

I. Older unrelated "guests"

1. Primary inclusions: 1a) idiomorphic inclusions, 1b) replacement remnants, 1c) included droplets
2. Incorporated inclusions: 2a) unaltered during incorporation, 2b) somewhat replaced during incorporation, 2c) completely dissolved but reprecipitated during incorporation, 2d) completely digested during incorporation

II. Related "guests"

1. Oriented intergrowth inclusions
2. Exsolution inclusions
3. Reaction inclusions
4. Devitrification inclusions

III. Transformation inclusions

1. In the thermal contact zone
2. In other heating — and cooling-zones
3. In metamorphic zones
4. In the oxidation zone

IV. Young immigrant inclusions

1. In hypogene [aszendent] domains
2. In the zone of cementation
3. In the zone of deeper weathering and in supergene [deszendent] domains.

Older unrelated "guests". The inclusions of this group existed as crystal or melt before the host crystal formed. There need be no relation of a chemical, structural or bonding nature between the host and the inclusions ("guest"). The original host mineral may remain, or it may subsequently be replaced by another mineral; in the latter case the inclusion becomes an incorporated variety, a relict texture.

1a) The examples of idiomorphic, early inclusions are so numerous that they need not be listed (see fig. 27, and many others).*

1b) Many examples of this group are known. They are mentioned especially and illustrated in the section on replacement (p. 185), so they will not be discussed here.

1c) Included droplets of native metals, sulphides and more rarely oxides, are widespread in magmatic rocks, and in some liquid-magmatic (orthomagmatic) ore deposits. They are referred to in the literature as "droplets", in part "spherules", in part "globules" (also as "blebs") (figs. 69, 70). Molten droplets tend to become enclosed by younger minerals, commonly silicates, while still molten, and crystallize only later. Metals whose compositions are known, and whose melting points are not reduced by the presence of volatile substances, serve as geologic thermometers, whereas sulphides and oxides, whose melting points are strongly influenced by such agencies, have less value in this regard.

By far the most numerous are spherules (mostly only a few microns in diameter) of *sulphides*, chiefly pyrrhotite, pyrrhotite + chalcopyrite, chalcopyrite + pyrrhotite + + pentlandite. Often these mixtures probably were originally chalcopyrrhotite. Less widespread are bornite (e.g., among others in magnetite and in "trevorite" from Barberton), chalcocite and covellite. *Oxides* are of rarer occurrence, to some extent magnetite. The "drop-like" form can be simulated by "inhibited crystal growth". *Metals* are of particular interest. Platinum occurs repeatedly in this form, e.g., in the platinum-bearing rocks of the Bushveld, yet there is here certainly a convergence of forms, since native platinum, even if strongly alloyed with iron, has a freezing point far above the highest possible crystallization temperature of silicates. Copper occurs in fresh augite in the melaphyre of Doros (Kaokoveld) South-West-Africa. Most widespread is bismuth, which solidifies at 280 °C, so that it is still quite fluid in hydrothermal ranges. Considerable care must be taken in evaluating such occurrences of bismuth, however, because various late stage dissociations can give rise to rounded bodies of bismuth.

2a) The possibility that an original host mineral may be replaced while the inclusions in it remain more or less unaffected in their old positions, has a special importance in the interpretation and the genesis of some ore mineral associations which at first sight appear quite unusual. More and more instances of such phenomena are becoming known. As an example, one finds some ilmenite — or rutile networks and lamellae derived from it — that is derived from the ilmenite content of titanomagnetites, e.g. in the pyrite of the Witwatersrand, and in the pyrite of various altered country rocks adjacent to hydrothermal veins; in the pyrrhotite and chalcopyrite of some Sudbury mines, in bornite from Ookiep South-West-Africa; or sphalerite "stars" taken over from high temperature chalcopyrite (or chalcopyrrhotite) in cubanite, in pyrrhotite, even in pyrite or gangue; tetradymite tablets deriving from probably oriented inter-

* I give here other examples to supplement SCHNEIDERHÖHN's statement, although I follow him broadly.

growths with chalcopyrite, in bornite and chalcocite. Further examples are illustrated in figs. 246, 252 and the drawing of fig. 258.

2b–2d) During the replacement of the original host the inclusions may be partly or wholly dissolved. Accordingly, the correct interpretation of events becomes increasingly difficult to follow; several statements in the literature appear unreliable. Recognition of what has happened is easier when the former host mineral remains in part, so that transitions can be seen. Examples are provided by occurrences of pyrrhotite with rounded inclusions of sphalerite, the pyrrhotite having replaced chalcopyrite enclosing sphalerite "stars". The "stars" have lost their shape to considerable extent, but remain in their original places. The final group, 2d) is even more difficult to recognize: SCHNEIDERHÖHN (1931) has described braunite with specular hematite inclusions that underwent replacement by sitaparite whereby the hematite inclusions were completely incorporated into the sitaparite structure. Similar cases induced by various types of metamorphism are as widespread in rocks as in ore minerals.

Related "guests". This type of inclusion is characterized by a close similarity of its lattice structure with that of the host mineral (or where they are precipitated as colloids, by a similarity of precipitation conditions), so that either the inclusion ("guest") grows on the host mineral, and becomes enclosed by the further growth of the host, or the affinity between the two is so great that at high temperatures they crystallize as a solid solution, from which one component is subsequently, more or less completely, precipitated as inclusions (exsolution bodies) as the temperature falls. Examples of all four categories, II1–II4 are exceedingly abundant. They are dealt with in detail elsewhere in the book (p. 162 etc.) in connection with other relationships. In discussing the "devitrification" of mixed gels, SCHNEIDERHÖHN emphasizes the fact that the primary textures of such gels are particularly liable to be eliminated through "collection crystallization" [Sammelkristallisation]. Some myrmekitic intergrowths originate in this way (p. 110), as do very many inclusions in the "rhythmic precipitates" found both in primary ore deposits and in gossans.

Transformation inclusions. This term was used by SCHNEIDERHÖHN to distinguish those inclusions, which owe their origin to some change of conditions (e.g., contact metamorphism, or change in the oxygen content of the environment) that rendered their host mineral unstable, causing it to form new minerals at its margins, or in its interior. The geometric similarities to exsolution can be quite close, and often the resulting textures are myrmekitic. The genetic aspects are dealt with in the section on metamorphism (p. 39 etc.), so that this reference will suffice here.

III1. "Transformation-trains" (progressive transformations) like pyrite–pyrrhotite–magnetite are common in inclusions in igneous rocks. Those sections of the siderite veins of the Siegerland that have suffered contact alteration by basalts are characterized by clouds of magnetite particles in the siderite.

III2. This group is a somewhat "hold-all" for phenomena which could also be attributed to processes like replacements, or the transformation that accompany normal hydrothermal processes, or changes in the H_2S concentration of the mineralizing solution. A texture should only be classed in this group if its origin can be traced. A typical example is the formation of pseudobrookite from (and in) ilmenite through the action of hot, oxygen-rich fumarolic gases.

III3. This type of inclusion is general in dynamically metamorphosed rocks (e.g., hornblende in augite, and the reverse; garnet in calc-silicate minerals, etc.), but has rarely been described in ores. Since more complex compounds, like the sulphosalts commonly become unstable at high temperatures, this type is to be expected, particularly in low temperature ore associations that have been heated subsequently. SCHNEIDERHÖHN rightly concludes that some ore deposits of the old shield areas have possessed textures of this type in the course of their history, even though the textures were obliterated subsequently. The formation of rutile in ilmenite and in titanomagnetite in the epizone belongs here, as does the development of magnetite at the intersections of lamellae in mechanically distorted ilmenite. So also do many of the inclusions resulting from recrystallization of solid solutions induced by dynamic metamorphism (or shearing) at relatively low temperatures (e.g., pyrrhotite in sphalerite at Hürnigskopf, Germany).

III4. In the oxidation zone weakly oxidized compounds change over to more strongly oxidized compounds, and sulphides are transformed to oxides, sometimes through a quite complicated course of reaction. In some cuprite one finds "free-floating" spheroids of tenorite, limonite spheroids develop in chalcopyrite, silver in cerussite; and here belong also the "bird's eye" forms of marcasite in pyrrhotite. Naturally these changes do not develop spontaneously at any one point but along grain boundaries, cracks, twin lamellae, and lead to the other geometric patterns.

Younger immigrant inclusions. This division goes somewhat too far in this instance, in that some examples that belong to III, and particularly to III4, could be classified in this category, and vice versa. As with the group III phenomena, the host mineral must be permeable in some way, so that again one encounters "free-floating" inclusions of immigrant material. The distinction between which is incoming and which is outgoing material can often be made only in terms of their relative proportions and importance.

IV1. Numerous examples occur in the hypogene ranges — many "silver-bearers" in galena belong to this group, as does idioblastic covellite in chalcocite, hutchinsonite in jordanite (fig. 469a) — a particularly beautiful example as regards the geometric pattern.

IV2 and IV3. Particularly at the boundaries between the primary ore zone and the zone of cementation, or of this latter and the zone of oxidation, the residual sulphides are commonly studded with inclusions of secondary sulphides, or of oxidized minerals.

If, therefore, SCHNEIDERHÖHN's classification goes too far in the matter of subdividing, as the writer has already indicated that he thinks it does, nevertheless it provides a clear framework within which one can organize one's observations systematically.

IV. SCHNEIDERHÖHN'S SYSTEMATIC CLASSIFICATION OF THE STRUCTURES AND TEXTURES OF THE ORES

H. SCHNEIDERHÖHN (1952) has attempted, partly in tabular form, to classify all formal appearance of the structures and textures of the ores. I consider this attempt as very successful, even though it is far from complete, and despite the fact that it adopts the nomenclature of petrology too widely, especially in the matter of interpenetrating textures. As a result, it introduces many genetic terms, partly by the meaning already attached to the terms employed, partly because an actually non genetic term in literature is often used genetically. Since many authors will make use of these tables, and since, moreover, my own efforts at classification have not yielded a geometric classification devoid of genetic terminology, I propose to avoid undue repetition by adopting them here in a somewhat shortened form, and with only few notations, adding where necessary brief references to the relevant sections and illustrations of this book.

A. Textures according to the form of the components

I. Simple grain textures

1. Xenomorphic texture (p. 140, fig. 17)
2. Hypidiomorphic texture (p. 86, fig. 54)
3. Panidiomorphic texture (p. 106, fig. 79)
4. Porphyrite texture (p. 97, fig. 66)

II. Penetrating and implicate textures

1. Ophitic or intersertal; characteristic of pneumatolytic and subvolcanic ore types (p. 14, fig. 9)
2. Myrmekitic-graphic (p. 110ff.)
3. Exsolution textures (p. 165ff.)
4. Poikilitic textures (p. 144, fig. 19)
5. Arteritic "penetration of one mineral through another" (p. 57, fig. 31)
6. Diablastic-dialytic. Somewhat unclear designations! ("Confused and mostly fibrous tangled mixture of several minerals")

B. Textures according to the size of the components

I. Relative grain size

1. Equigranular
2. Unequally granular

II. Absolute grain size

The very useful scale designed by Teuscher for petrology is recommended. (In regard to aggregates of ore minerals of the same type, the same deposit, and if possible the same ore sample, the absolute grain size is usually much less characteristic than in regard to common rocks.) I prefer therefore, to express it simply in terms of size data and typical photographs.

C. Structures according to the spatial arrangement of components

I. Randomly granular (statistically isotropic)

II. Oriented simple structures

1. Fibrous structures (p. 126 ff., figs. 100, 101, 138)
2. Foliate structures (p. 387, fig. 320)
3. Oriented granular structures (lattice-oriented, habit anisotropic, p. 125, fig. 98)
4. Radial structures (p. 127, fig. 102)
5. Concentric shell-like structures

III. Composite and rhythmic layer and ribbon structures

1. Evenly layered (fig. 99)
2. Bent layered
3. Concentric shell-like
4. Oolitic and concretionary fabric (figs. 15, 118)

D. Structures according to packing

1. Massive
2. Porous-granular
3. Saccharoidal-miarolitic (this is a combination which does not correspond to the common sense of the word!)
4. Straddled [sperrig]
5. Cellular shell-like (fig. 126)
6. Amygdaloidal structure
7. Drusy structure (figs. 10, p. 18)

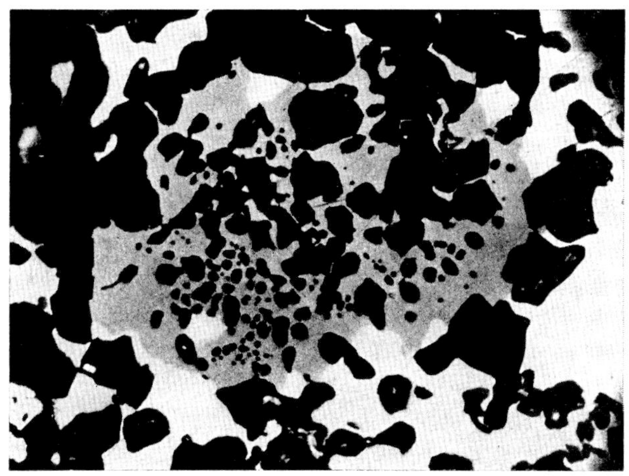

Fig. 110 350 ×, imm. RAMDOHR
Rammelsberg, Harz, Germany

Xenoblast of *fahlore* with inclusions of *sphalerite* (almost black). The sphalerite grains in fahlore are smaller than those outside, which had more opportunity to grow

B. Genetic fabric types

I. TEXTURES OF PRIMARY PRECIPITATION

General data. GRIGORIEFF's choice of "depositional textures" as a basis for classification is very precarious, particularly without further and more precise definitions in support of the divisions. It is truly impossible to follow him in classing "primary", "secondary", and "sedimentary" as subdivisions of equal status. Also, it is difficult to understand why he does not recognize metacolloidal textures, that is the textures of gels which in part became crystalline immediately after precipitation, as being primary, and similarly why "replacements" are not relegated to secondary depositional textures. Despite this we are indebted to him for many concepts and observations, while adhering to SCHNEIDERHÖHN's logically constructed scheme.

(a) *Growth fabric (crystallization from melts and solutions)*

α) Magmatic melts
β) Pegmatitic and pneumatolytic
γ) Hydrothermal
δ) Low temperature solutions

This purely genetic, elsewhere (p. 3 etc.) logically developed sequence contains in itself hardly any clear textural conception, as we have in petrology where great experience and relatively simple relations permit some acknowledged assertations. Actually almost all existing textures and structures may be found here or even be typical.

With regard *substance*, it is to be mentioned that especially sulphides exhibit quite varying tendencies of idiomorphic development in the individual mineral with differing conditions of formation, as well as in different associations. This is manifested in entirely different habits of contemporaneous components of the same mineral assemblage. Thus, genetic conclusions are less permissible here than among the much more monotonous relations of the components of igneous rocks. The relative properties of materials, which are conveyed in the expression "tendency to idiomorphic development" [Kristallisationsfreudigkeit], the nature of the solution, hot — cold, impure — pure, crystalloid — colloid, basic—neutral — acid, as well as the rate of temperature decrease and the closely related matter of crystal seeding and velocity of crystallization, all appreciably, but very obscurely influence the *development of forms*.

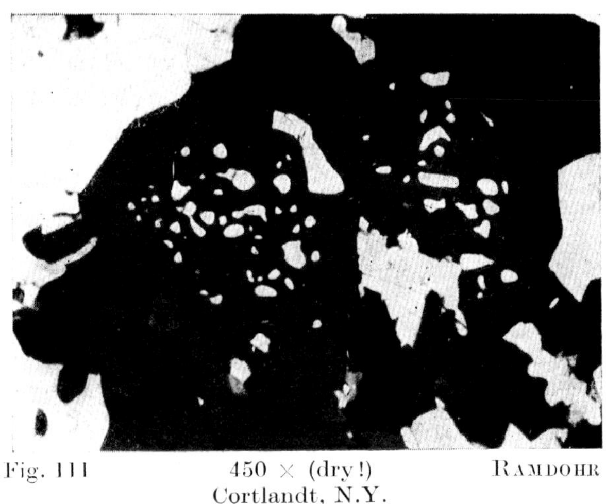

Fig. 111 450 × (dry!) RAMDOHR
Cortlandt, N.Y.

Corundum (grey) with an idioblast sieve (poikilitic) of *magnetite* inclusions. Outlying white grains are also magnetite. Observe the almost complete lack of relief in the section!

Only a very few typical examples, of course, may generally be given under each group.

"*Granular*" texture is less commonly the result of immediate crystallization in ore minerals than it is in magmatic rocks or oceanic salt deposits. This is especially true of equigranularity, which we occasionally find in titanomagnetite and nickeliferous pyrrhotite deposits. "Unequal granularity" is expectably more common in view of the differing rates of crystallization. "Allotriomorphic" (= "xenomorphic") and "hypidiomorphic" are as in magmatic rocks very widespread, but "panidiomorphic", as a texture of "ophitic" aggregates like cavity fillings, is more widespread than it is in rocks.

Spheroidal texture from melts is easily distinguished from the later described oolitic textures by the unequal sizes of the spheroids, lack of zonal structure, characteristic inclusions by other minerals, and otherwise by the nature of the material. Poikilitic ["Idioblastensiebe"] inclusions can be very similar (fig. 111). In orthomagmatic crys-

Fig. 112 150 × Duhovnik

Trepča, Yugoslavia

Idiomorphic and certainly idioblastic *quartz* (dark grey), with a fine dusting of *pyrite*, which continues in the same distribution (clouds, streaks) into the surrounding *sphalerite* (light grey) and *siderite* (medium grey). Trace of *galena* (white)

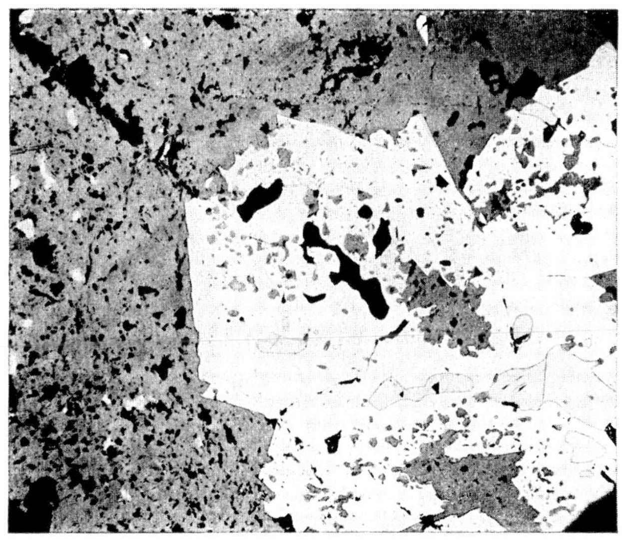

Fig. 113 35 × Ramdohr

Tsumeb, South-West-Africa

Idioblast of *pyrite* in spottily colored *germanite* (dark grey). In the pyrite as poikilitic intergrowth [Idioblastensieb] is abundant germanite, *galena* (finest pits) and gangue (black)

tallization pyrrhotite or a related $CuFeS_2(Ni, Fe)S$ solid solution crystal, can be included in droplike form in crystallizing silicates and iron oxides, and remain fluid for a relatively long time; the spheroidal shape is preserved to final solidification (fig. 69). Droplike and emulsion textures (fig. 78) may be preserved in nickeliferous pyrrhotites, and also some elements with very low melting points can be included as molten drops, even in the hydrothermal range, e.g., bismuth above 280 °, and finally, native mercury occurs as droplets at ordinary temperatures in gossan assemblages. Very rarely silicate droplets can occur in sulphides (fig. 5). The droplets of olivine in the Pallas-iron also belong here. Spheroidal forms, easily distinguished from those just mentioned as formed from melts, occur, e.g., in "inhibited" crystal growth (e.g., spinel and phlogopite, also pyrrhotite in contact rocks), in droplike exsolution bodies, very rarely in replacement growths, very abundantly — but very different in appearance — in colloidal textures, and perhaps also elsewhere.

"*Porphyritic*" textures, as a result of direct precipitation, are very rare in ore deposits. The strong tendency of idiomorphic development of pyrite, at least in relatively pure solutions, occasionally lends it the appearance of phenocrysts. Phenocryst-like pyrite (also arsenopyrite, glaucodot, cobaltite, etc.) is generally porphyroblastic.

"*Graphic textures*" are more abundant and are also widespread as eutectic formations. A remarkable controversy about these exists in the literature — dozens of examples could be given. Since ore minerals mostly do not precipitate directly from a melt, many writers maintain that "graphic textured ores cannot be the result of eutectic crystallization". This results from failure to realize that ore solutions are multicomponent solutions with water, in which very probably simultaneous "eutectic" crystallization must result along the eutectic boundaries, finally approaching the polynary eutectic of the entire system. This last, which should consist essentially of ice!, does naturally not occur. The attempt was once made to distinguish this crystallization along eutectic lines as "cotectic" in contrast to "eutectic", an attempt which cannot establish itself against the well established usage of physicochemical nomenclature and does not deserve to establish itself either! Accordingly, eutectic forms are to be expected in hydrothermal ore mineral associations; the question is only if the fabric will be preserved as "graphic" intergrowths, and if and to what extent graphic intergrowths represent eutectics. Graphic eutectics are approved in some cases, in others doubtful, and in probably very many more the "graphic" intergrowths resulted from other processes (cf. p. 110).

The same remarks hold for "*subgraphic textures*", an expression used by GRIGORIEFF for very fine intergrowths of similar nature, and one which I consider superfluous. Graphic and subgraphic are indistinguishable except by a completely arbitrary grain size distinction.

True eutectic graphic texture from magmatic melts is shown in fig. 86, and from hydrothermal solutions in figs. 92 and 93.

Poikilitic texture, resulting from about simultaneous crystallization of two components wherein one is uniformly oriented and the other is enclosed in sieve-like cells of the first, and which is so characteristic of harzburgite, occurs only very rarely in ore minerals, however, some galenas in contact deposits show it exceptionally well. Sometimes it is simulated by unusual replacement processes (fig. 223).

Zonal textures are always to be expected in solid solution crystals, especially in those where the solidus and liquidus curves are well separated. They are, however, also common among minerals that are nearly pure substances.

The cause in the latter case must be small interruptions in growth, zonal variation in growth-disturbances, unusually pore- or inclusion rich parts, and perhaps also simply faster or slower growth. The descriptive part of this book illustrates a large number of examples on both etched and unetched samples. At high temperatures

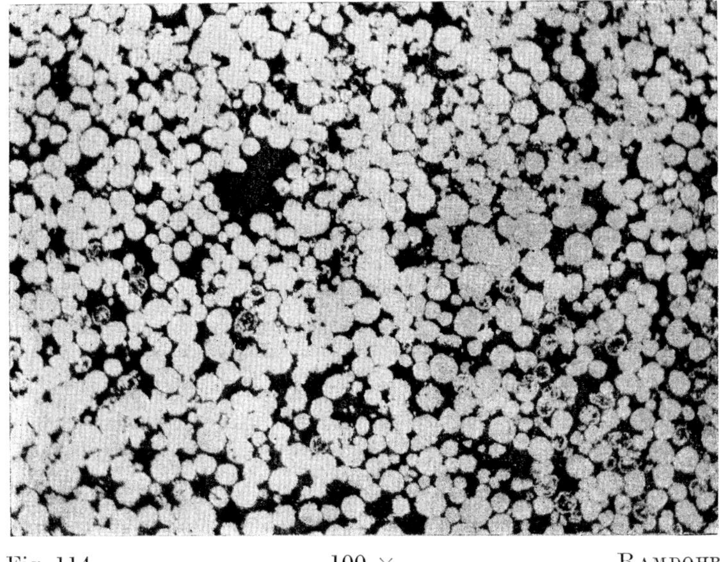

Fig. 114 100 × RAMDOHR
Meggen, "Mulde", Westfalia

"*Mineralized bacteria*" pyrite, which here comprises the bulk of the "ore". Interstitial material is mostly *sphalerite* (grey). The peculiar concentric structure is easily recognized in places

extensive exchange of material is to be expected in the solid state, so that — naturally varying with differing temperatues — zonally developed solid solution crystals homogenize. Zonal structure consisting of varying porosity, however, is likely to persist more or less. Zonal structure owing to inclusion of foreign minerals during a halt in crystallization is shown in fig. 368, and one due to varying porosity in fig. 53. Further examples are figs. 54, 314, 586, and many others.

High temperature magmatic zoning is shown by hematite–ilmenite (fig. 563), extreme pneumatolytic zoning by wolframite and cassiterite, hydrothermal by fahlore and pyrargyrite, apomagmatic by low temperature galena, etc.

Banded textures are very widespread and in part extremely fine. They can occur in single grains and are then really "textural" properties, as well as in an aggregate of one or more mineral species. They occur particularly well developed in hydrothermal veins as "banded ores" on a large scale. Finely banded, they are a characteristic property of epithermal veins (e.g., of Ag, Au, Hg, Sb — fig. 10), and also very typical for the near surface lead–zinc occurrences (type Aachen–Upper Silesia), but here not separable from the rhythmically banded colloidal textures. Of course, they are very widespread among sediments and almost the characteristic mark of dynamically metamorphosed ores. Their correct interpretation is therefore difficult, just as the diagnosis of banded plutonic rocks compared with metamorphic rocks in geologic field work. Actually, sedimentation-like processes can even take place in a magma body by interrupted rhythmic precipitation of a heavy component (e.g., chromite).

Oolitic and "bacterial" textures should be classed mainly with sedimentation fabrics, where they rightly belong, as their name implies. Yet, very similar forms grow from melts and solutions. Thus, small cockades of hydrothermal crystallization, and also at low temperatures, can be almost completely like true oolites. A part of the so-called tubercle texture (Cobalt City!) belongs here. Pitchblende of spheroidal form can be similar in size as well as appearance. The concentric structure of true oolites is mostly typical; it can, however, be diagenetically destroyed, while "pseudo-oolites" can very well at times have concentric shells.

"Mineralized bacteria" of pyrite have not always played a fortunate role in the literature of the last 50 years. Regarded by SCHNEIDERHÖHN as mineralized bacteria, these forms (fig. 115) were an object of many discussions. In view of the work of NEUHAUS (1940) and LOVE* (1957) and of the occurrence in coals and recent sapropels they could not be doubted as "fossils" for some years. But *exactly* the same forms, perhaps with more variable size (fig. 114, 116) are known from surely hydrothermal origin. And further some authors, especially KALLIOKOSKI (1969) could make them by surely a-biogenetic reactions. In any case some doubts remain. These tiny spheroids are very resistant to recrystallization and often are the most striking and sometimes the only relict in metamorphosed rocks and ore deposits.

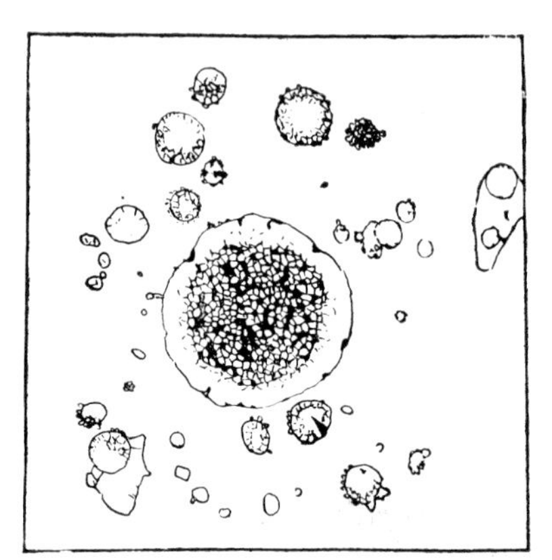

Fig. 115 RAMDOHR

"Mineralized bacteria" of several different sizes, from Rammelsberg "banded ore"

It would be amiss, without thorough discussion of accompanying textures and geologic conditions, to argue "mineralized bacteria" (here used only as a designation of shape) as being sedimentary or hydrothermal.

A very similar, dimensionally identical textural oddity is still in no way understood: peculiar pyrite spheroids in chalcopyrite, which exhibit a septarian subdivision of a cavity (fig. 149 in RAMDOHR, 1953a). Occasionally these are certain signs of incipient weathering, but elsewhere this conclusion is just as certainly excluded.

(b) *Colloidal textures*

General data. SCHNEIDERHÖHN classes colloidal fabrics, the forms of highly dispersed precipitates, as the second group of genetic fabric types. In a certain sense this con-

* The interesting studies of LOVE show that it is sometimes possible to recover the organic part by dissolution of pyrite. LOVE identified two distinctly different forms and regarded them to be most primitive fungi (not real bacteria). Later he had doubts and withdrew his opinion. The fact remains that the "pyritized bacteria" mostly are connected with coprecipitated organic material.

cept is definitely genetic, because "hydrothermal" and "room temperature" colloidal deposition, according SCHNEIDERHÖHN, involve in the first case an origin by low temperature hydrothermal action, in the second one at surface temperature — in both without formation of crystalline phases.

According to the laws of physical chemistry all colloidal precipitations are metastable with tendency to crystallize. This manifests in the loss of solvent, mostly water, of course, in becoming microcrystalline, perhaps with a later increase of grain size

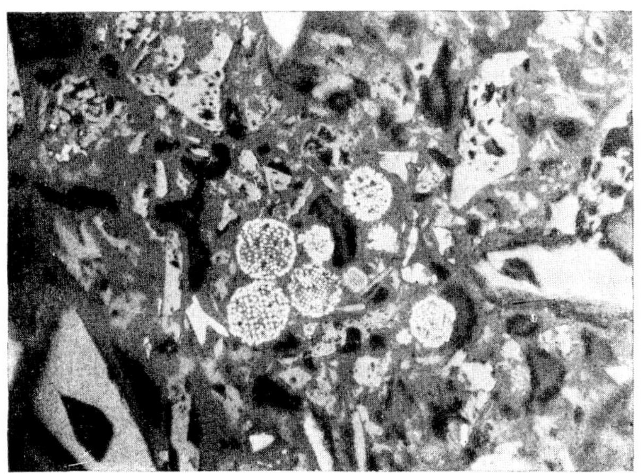

Fig. 116 400 ×, dry RAMDOHR
Tilkerode, Harz

"Mineralized bacteria", which here perhaps do not signify organic origin. They are enclosed in quartz, which cements the complex brecciated gangue. The structures in themselves are excellent and typical

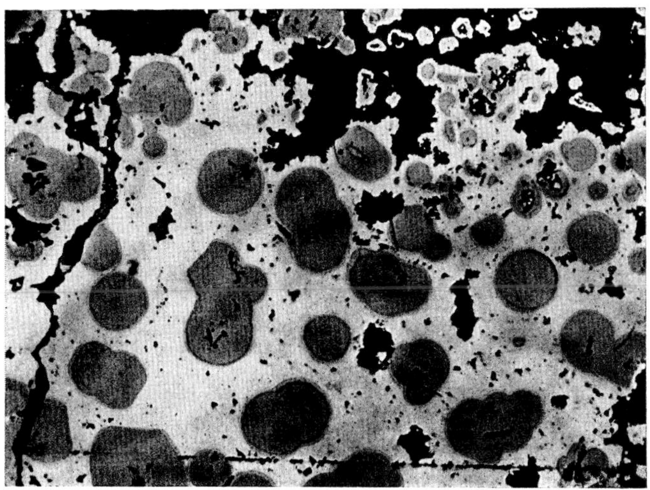

Fig. 117 250 ×, imm. RAMDOHR
Cape Cavallo near Bône, Algeria

Pyrite, colloidal texture. The spheroids are colored deep brownish in rhythmic variations the cement being normally colored pyrite. The specimen comes from an unmetamorphosed submarine exhalative Zn-Pb-pyrite deposit

and an exsolution-like precipitation of co-precipitated substances. By far in most inctances is that, what we call colloidal texture, the texture of colloids already altered to microcrystalline masses. This is so ubiquitous that occurrences of masses without powder diagram were doubted at all (e.g. for ZnS, FeOOH, etc.). That is *not* true: there is jordisite, MoS_2, metastibnite, Sb_2S_3, some limonites with a glassy fracture — all are distinctly X-ray-amorphic. First of all that is true for many low temperature sulfarsenides and antimonides of lead, briefly mentioned by the author from Wiesloch, but fully described by BAUMANN & AMSTUTZ from Cerro de Pasco.

The colloidal forms of precipitates formed directly from the gas phase (cf. fig. 322 of retort graphite) can hardly play a role in nature.

In *substance*, "colloidal" precipitation often appears to be related less to sulphides than to the accompanying gangue material, although, as is known, many sulphides typically first precipitate as colloids. The colloidal forms of gold, platinum, argentite, some Ag-sulphosalts etc. imbedded in masses of botryoidal quartz or chalcedony probably result from infiltration in an original opal mass. With zinc sulphide and melnikovite-pyrite, which mostly form at very low temperatures, the sulphide is the carrier of the colloidal properties, in still other cases it is the clayey substance. The problems of colloidal solution and precipitation in the hydrothermal region will not be explored here, where we discuss only the forms.

The mechanism of colloidal precipitation is known to be of very complex nature. The first flocculation by concentration of solution or contact with an oppositely charged colloid or an electrolyte or even organisms is almost always promptly followed in nature by conversion to the crystalline state. Strictly speaking "colloidal textures" — GRIGORIEFF appropriately named them "metacolloidal" — are actually relicts, whose original nature will always be somewhat obscured by the most variable circumstances. The nature of flocculation as well as that of the slow conversion to crystallinity is obviously different from "spontaneous" precipitation through chemical additions.

Many of the following textural forms are typical for one or the other variety of flocculation, others atypical.

As shown by the early work of LINDGREN and more recently by MEIXNER, SEELIGER, AMSTUTZ, and the writer, some sulphides can remain X-ray amorphic in some not metamorphosed and very low temperature deposits. Most striking are precipitations of lead–antimony resp. arsenic sulfo-salts, which the author briefly mentioned from Wiesloch, and in more detail BAUMANN & AMSTUTZ (1965) from Cerro de Pasco. But all these masses are so poorly durable that only by powdering can be observed that they become crystalline within very short and low warming e.g. in 2 minutes up to 100°. Regarding the circumstances and the "subjective" properties of the actual material the forms are extremely different. Surely some of them are so atypical that they cannot be recognized as such. Vice versa, other forms of precipitation can show "colloidal" forms e.g. as an extreme case, the dissociation of the gas in the coke-oven (Fig. 322).

Individual description of all forms is impossible because of their large variety. The forms can be different caused by temperature*), the time of the aging, the shrinking by dehydration, and many more or less accidental effects. All transitions are possible.

* This temperature may be appreciably different from case to case; for pitchblende it is surely often about 300 °C, mostly lower.

Especially common and typical are rhythms (fig. 118a), e. g. concentric shells, which form during the aging (and crystallisation) radiating fibrous forms often with contraction cracks. Characteristic is further the originally extreme fineness of grains, which can be veiled by quick recrystallisation.

Common is often the occurrence of "knitted" skeleton-like inclusions of easier crystallising minerals, e.g. galena in originally colloidal sphalerite or wurtzite (fig. 449). The

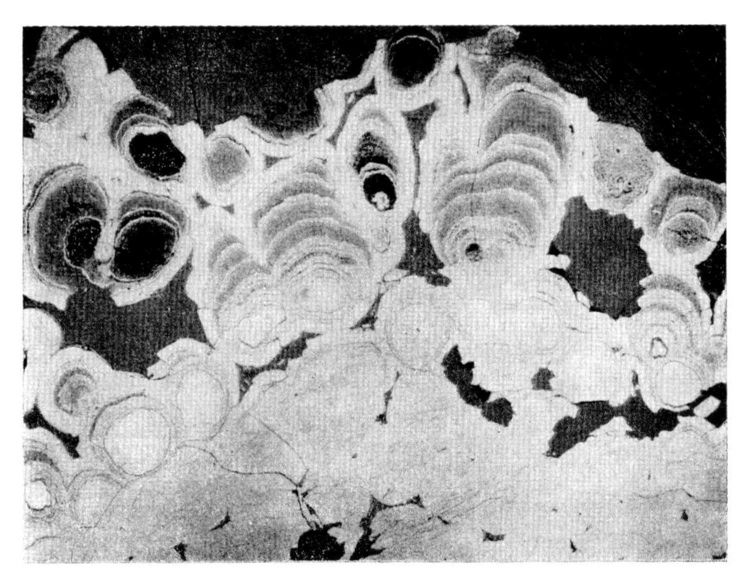

Fig. 118 8 × RAMDOHR
Bollenbach near Siegen, Germany

Colloidal textures in a complex manganese ore, principally *pyrolusite* and *cryptomelane*

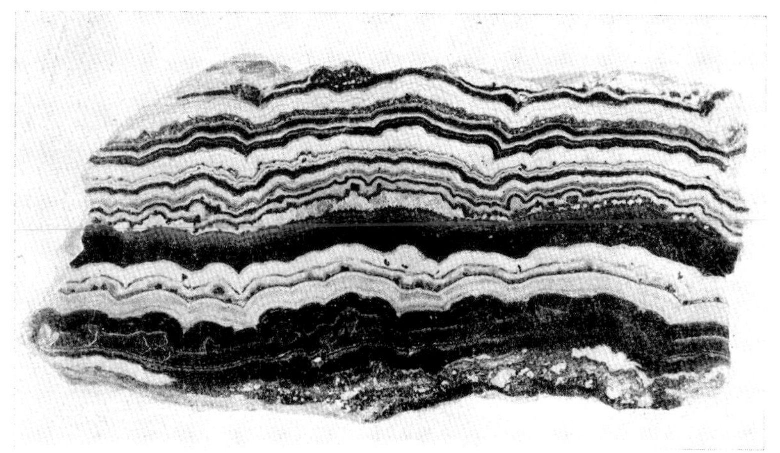

Fig. 118a 1 : 1 RAMDOHR
Wiesloch, Baden

Rhythmically layered crusts of *schalenblende*, dark-grey, with *pyrite*, white; approx. in the middle a thin band of *marcasite*

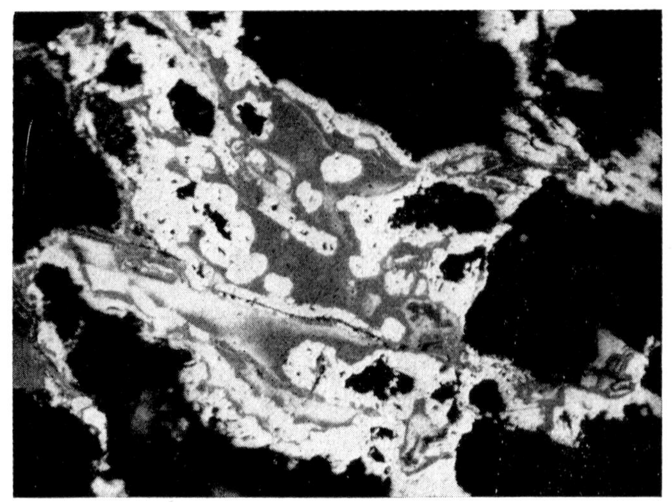

Fig. 119a 250 ×, imm. RAMDOHR

Nöten Mine, Garpenberg, Sweden

Colloidally precipitated *galena* in a recrystallized part of an old oxidation zone. Even the darker appearing parts are practically all galena

Fig. 119b $2^1/_2$ × RAMDOHR

Wiesloch near Heidelberg, Germany

Jordanite, white, overgrown by *botryoidal sphalerite* [*Schalenblende*] (light grey) with some gangue particles. Outer crust is *melnikovite pyrite*. Attached to fine earthy gangue partly intergrown with barite crystals

process is here completely analogous to skeleton crystals (e.g. of pyroxene or magnetite in volcanic glasses).

Regarding the material it seems that the colloidal precipitation depends less on the sulfidic or metallic precipitate than on other compounds in the solution in spite of the fact that some special sulfides "like" to precipitate colloidally. But colloidal forms of gold, platinum, argentite, many sulfosalts in masses of e.g. quartz go probably back to precipitation in not yet hardened opal.

The machanism of precipitation is complicated; first reversible, it becomes, sometimes in different stages irreversible by the influence of different charged electrolyts etc. All colloidal precipitation are metastable and nearly all we find in nature are "metacolloids" as GRIGORIEV stated long ago.

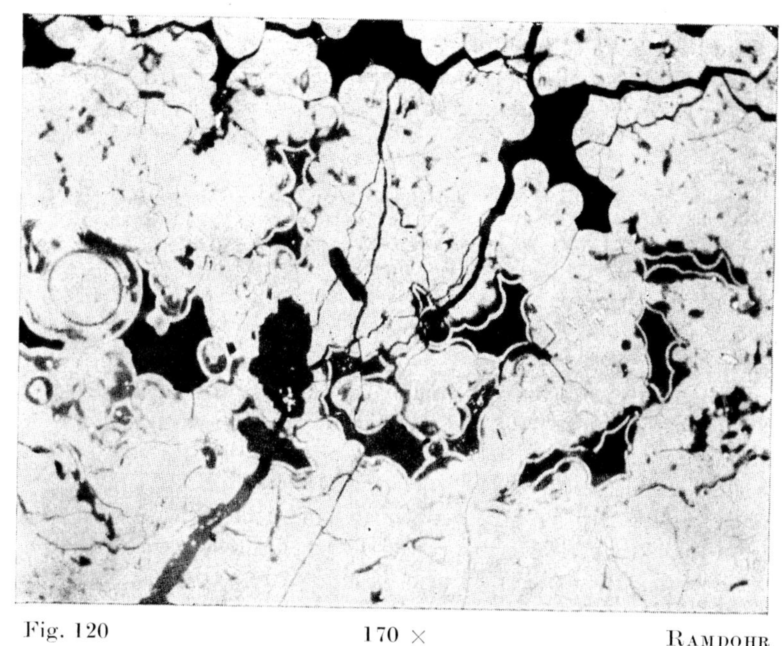

Fig. 120 170 × RAMDOHR

Kowary (Schmiedeberg), Silesia

Pitchblende, magnificent textures, rhythmic-botryoidal and spheroidal. Scattered old sulphide inclusions

I hope to have given justice to GRIGORIEFF's train of thought in the following scheme greatly expanded from his classification.

1. Botryoidal-reniform, mostly fine concentric crists ("colloform").
2. Spheroids or bubble-like masses ("vesicular") in a groundmass ("perlitic").
3. Concentrically banded, often radiating, spheroidal masses ("spherulitic" + "banded"); a subdivision of this group is useful.
4. Devitrification masses of varied form, often reminiscent of organic tissue ("gel").
5. Finely dispersed infiltrations and dispersions, often moss- and net-like.
6. Reticulate forms.
7. Feathery-flowery forms.
8. "Remainders".

GRIGORIEFF's groups, "banded" and "zoned", are included in 3. The illustrations will show especially that all sorts of transitions exist.

1. *Botryoidal-reniform.* This is by far the most common form of formerly colloidal minerals. It is so well known from the appearance of arsenic ["Scherbenkobalt"], botryoidal sphalerite [Schalenblende], pitchblende, limonite and many oxide minerals that a closer description is superfluous.

Foreign inclusions are often arranged in fine rhythmic shells; "homogeneous" aggregates, etched or unetched, also often show prominent concentric structure (figs. 124 and 487).

2. *Spheroidal or bubbly masses* ("vesicular") in a groundmass ("perlitic")*. The first germination in gels often takes the form of minute spheroidal, often radiating bodies that can form spontaneously; often, however, a clay particle, a minute air bubble, or a living or dead bacterium can serve as a nucleus. Thus, occasionally a textural transition to mineralized bacteria in its proper sense is possible. It is not known whether the vesicle cells (figs. 128 and 560) always start from a gas bubble, or whether they originate in a rhythmically concentric process; here also, textures entirely similar to organisms can occur.

The individual spheroids can be subjected to further rhythmic growth and give rise to transitions to 3. (fig. 115).

3. *Concentrically banded, often radial, spheroidal masses.*

a) Magnificent textural forms belong here as shown in figs. 121 and 122. Naturally, all transitions to 1. exist. Palmette shaped crystallization is also widespread here; many minerals, e.g., pyrite especially, are more inclined to strictly radial forms. The extent to which large pyrite nodules in clays are related to the mostly much smaller forms described here is not quite clear, since similarities may result from inhibited growth. Rhythmic precipitation of different minerals can give rise to continuous layers as well as to fine dispersions of guest mineral in its host.

b) Certainly completely similar, genetically and mechanically, but deviating in appearance, are the "tubercle" forms, especially as they occur in the rings of safflorite and smaltite described from the ores of Cobalt City, but also as they similarly occur elsewhere [Kowary (Schmiedeberg)] and also involving other minerals. Perhaps higher temperatures are responsible for the change in form.

c) *Small cockades* are closely related to the foregoing, the rings, however, being of fine-grained material and strongly cracked (figs. 126 and 605). They may be very similar to "vesicles", but also to many replacement forms, e.g., in equigranular marbles.

4. *Devitrified masses* of varied form apparently correspond to what GRIGORIEFF designated as gelatinous, since gels preserved as such ones have no "typical special form". Concentric structures appear to be again definitely predominating, often the depositional products are diffused in a matrix of gangue, e.g., opal or chalcedony, or of weakly reflecting oxidic ores. The varied forms are reminiscent of microscopic preparations of organic tissue, in part of mycelium (fig. 128).

5. *Finely dispersed infiltrations and dispersions* are genetically related and texturally similar to replacements, also with textures shown under 4. The original matrix is often dissolved or shrunken in such a fashion that the masses of ore become loose and earthy

* The otherwise appropriate expression "perlitic" is quite misleading because of its attached significance of contraction forms in glassy extrusives.

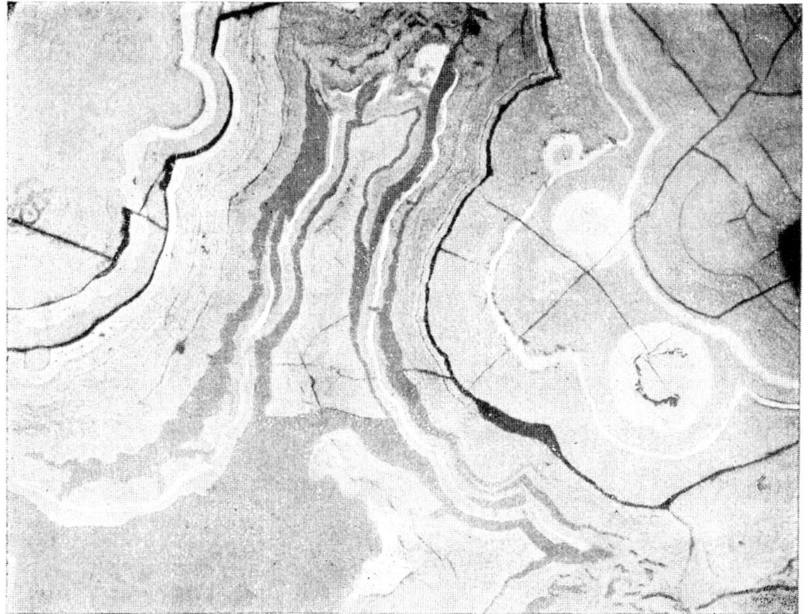

Fig. 121 170 × RAMDOHR
Saalfeld, Thuringia

Asbolane ["*Kobaltschwärze*"]. Rhythmic precipitation of various Co- and Mn-oxides, still partly isotropic, in carbonate. The cracks are nearly characteristic of former gel precipitation

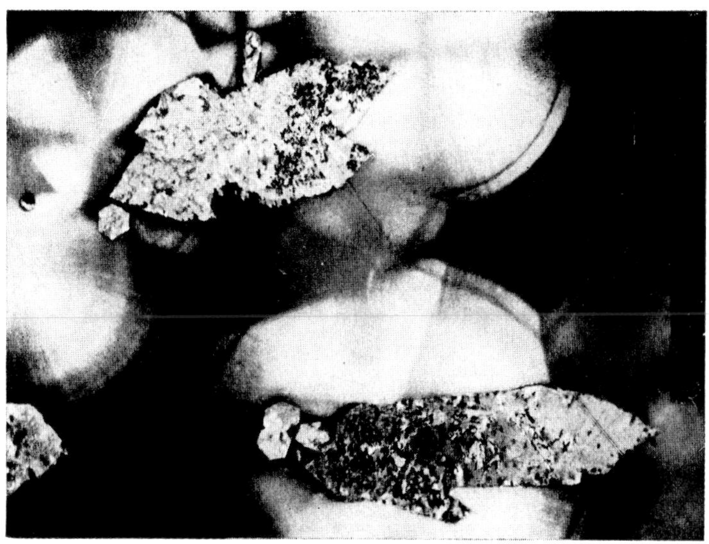

Fig. 122 ca. 400 ×, imm., + nic. REHWALD
Wiesloch in Baden, Germany

Crystals of *gratonite* (pseudomorphosed into *jordanite*) in a colloidally precipitated matrix of botryoidal sphalerite ["Schalenblende"]. Easily visible spherulitic crosses

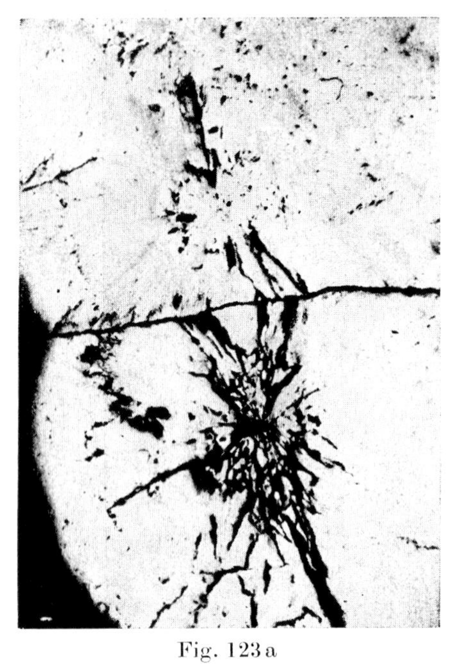

Fig. 123 a Fig. 123 b

a) one nicol, b) crossed nicols RAMDOHR-AHLFELD
Pulacayo, Bolivia

Marcasite in excellent radiating spheroidal, nearly concentric masses

Fig. 124 250 ×, imm. (not etched) AHLFELD
Pulacayo, Bolivia

Melnikovite-pyrite (unetched). Excellent rhythmic masses as a relatively very late formation. These masses are extremely reactive to weathering, even to damp air

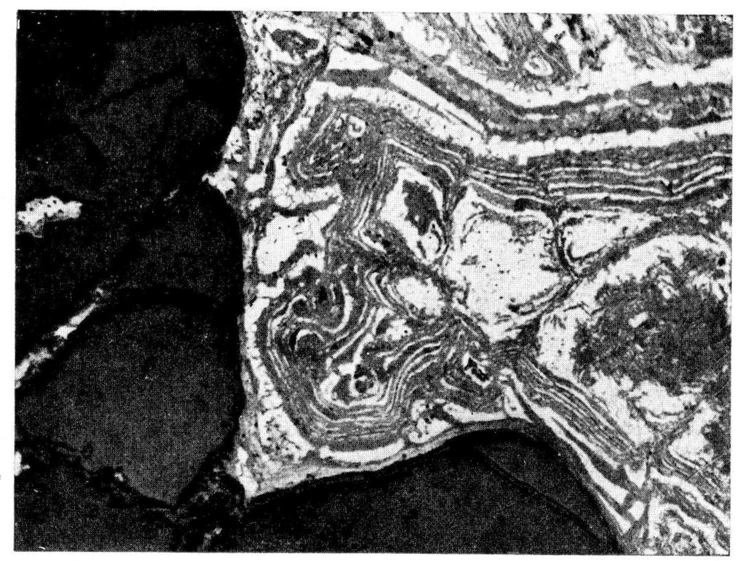

Fig. 125 100 × RAMDOHR

Kirka Mine near Dedeagatsch, East Thrace, Greece

Galena probably reconstituted from and in rhythmic growth with *cerussite* (light grey), probably first formed as a weathering product from an earlier galena. Gangue is *quartz* (dark grey)

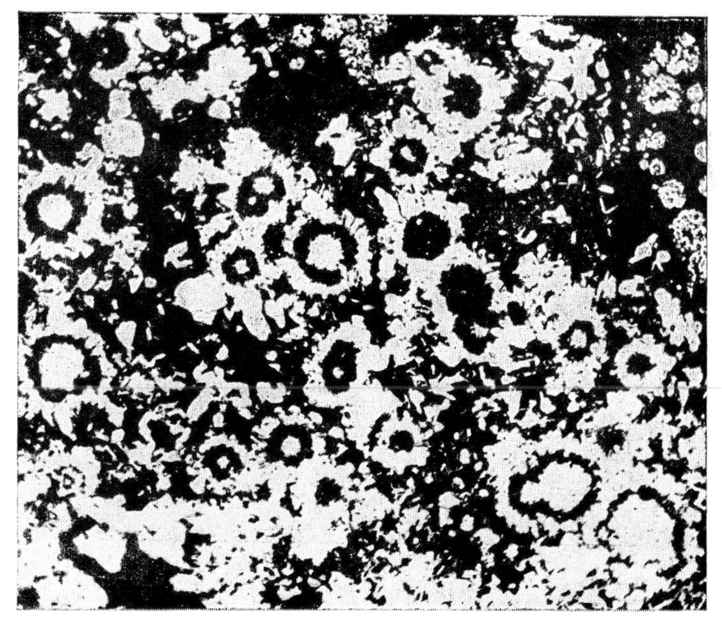

Fig. 126 200 × RAMDOHR

Meggen on Lenne, Westfalia

Concentrically shell-like *pyrite*. The central grains are about the size of the so-called mineralized bacteria

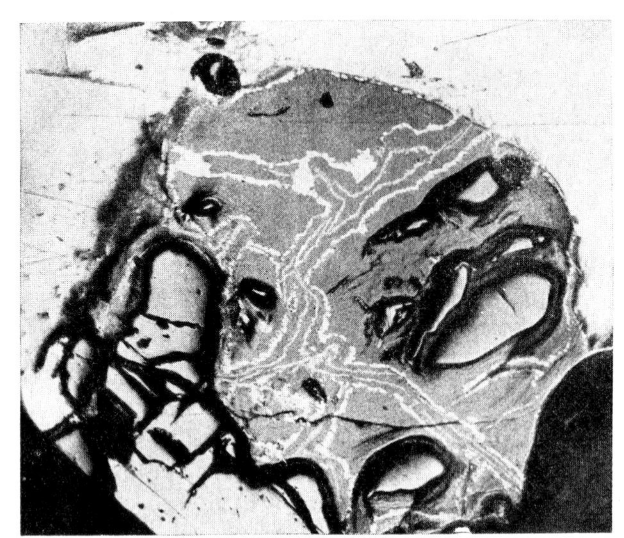

Fig. 127 40 × Ödman

Åker Mine at Kaveltorp, Sweden

Galena in characteristic garland form, in which *serpentine* (grey, smooth) originated by the disintegration of *chondrodite* (grey, strong relief)

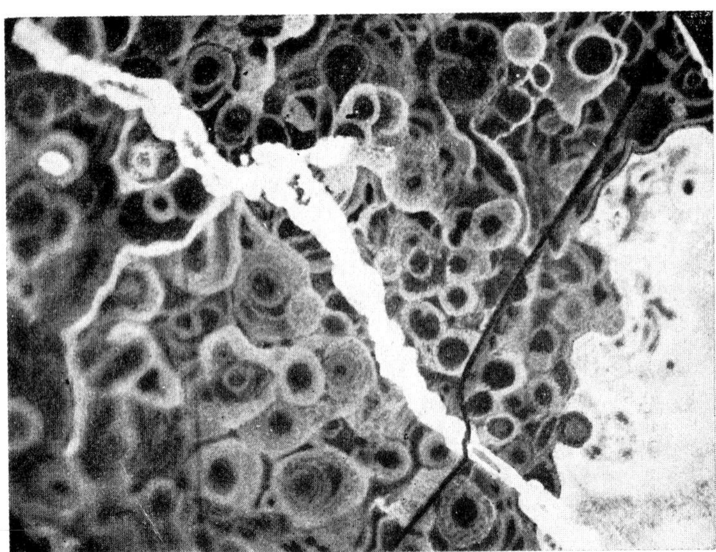

Fig. 128 500 ×, imm. Ramdohr

Doros, South-West-Africa

Gel texture of *tenorite* (lustrous white to dull grey in all shades!) in a matrix of *chrysocolla* (almost black). Transition from fine dispersions to relatively coarse crystalline and compact masses

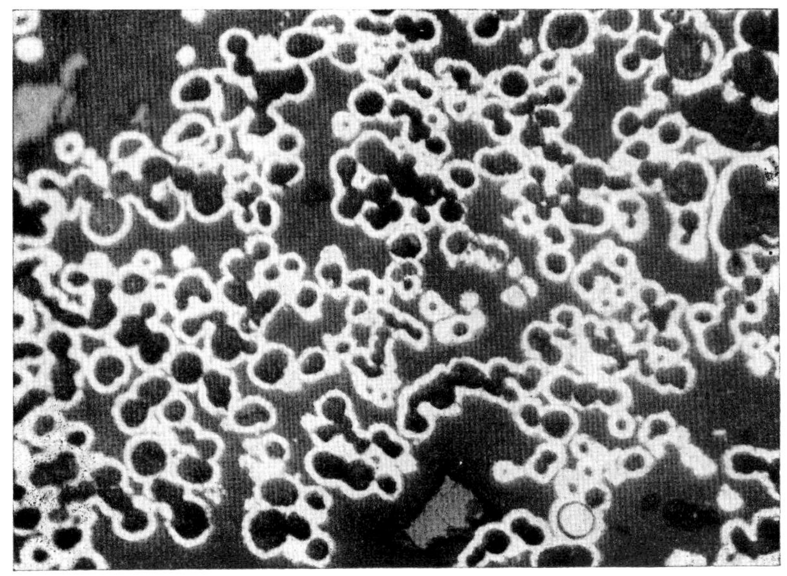

Fig. 128a 150 × RAMDOHR

Mina Capricosa, Central Peru

"Bubble cells" of *pyrite*, white, partly open, then black looking, partly filled with *quartz* or *opal*, dark-grey

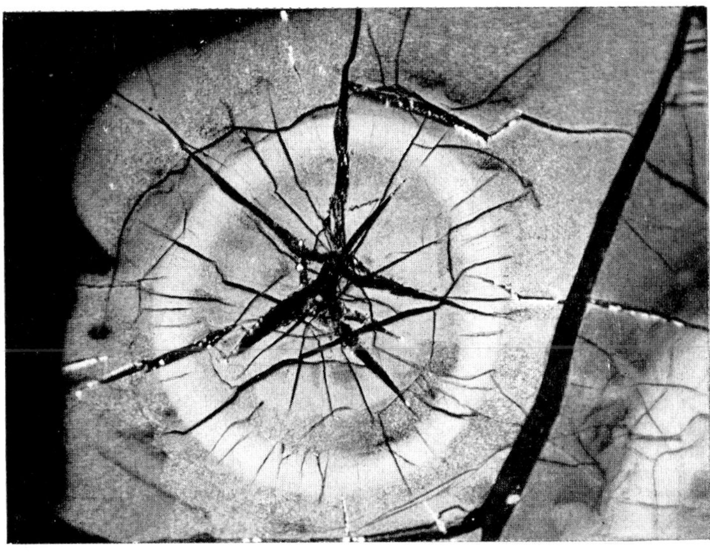

Fig. 129 250 ×, imm. RAMDOHR

Ranson property, Montreal River, Ontario

Botryoidal-reniform *pitchblende* typical of colloidal formation. Excellently developed rhythmic textures and radial contraction fissures. Very fine dusting with *galena* interpreted as radiogenic (margin of visibility)

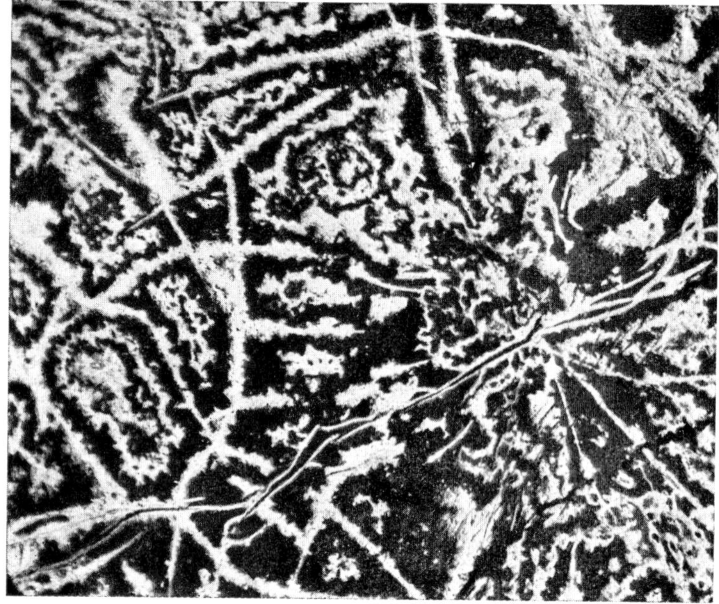

Fig. 130 100 × RAMDOHR
Kowary (Schmiedeberg), Silesia

Pitchblende (light grey), reticulate, in cracks of originally undoubtedly contracted matrix formed from a gel. The matrix here is a red colored calcite

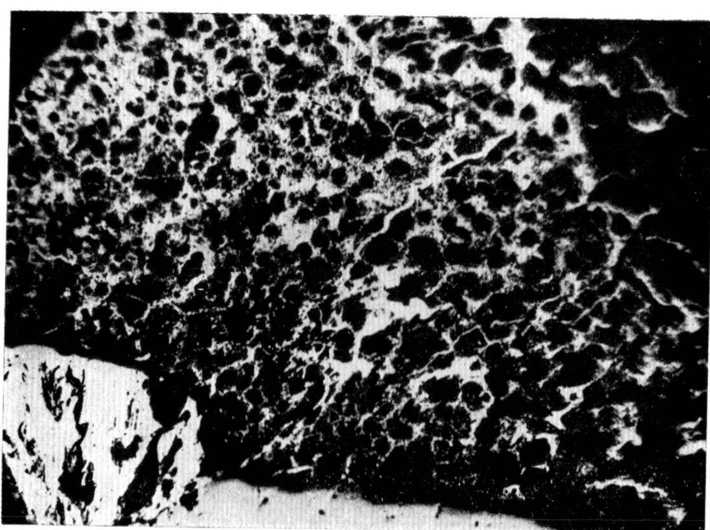

Fig. 131 ca. 60 × RAMDOHR
Grossalmerode, Hesse, Germany

Pyrite, compact masses and fine network in clay

or mossy and poor subjects for preparation of polished sections. Only where embedded in such a medium as calcite, which may be younger or older, are they easier to prepare. It is then apparent that the individual bodies are irregularly shred-like or branched and often laced together to stringers (fig. 130). Partly, later infiltrations into a shrinking gel can lead to more net-like forms. Also, they are seldom seen because the almost pyrophoric fineness of the sulphide nets causes almost instantaneous oxidation (fig. 131).

6. "*Reticulate*" *forms* have always been identified with colloidal processes and crystallization of easily growing minerals in a viscous colloidal matrix of another type. Thus these growths wholly correspond to skeletal growths in glasses. The forms are relatively monotonous, the easily crystallizing mineral forms very loose crystal nets commonly of extensive uniform orientation. These are then surrounded by botryoidal-reniform masses of the later crystallizing and very much finer grained originally colloidal matrix; the reticulate forms also remain intact. Some are of rather high temperature origin. Examples show most commonly "reticulate galena" in colloform sphalerite [Schalenblende] (fig. 449), but also many others (figs. 132 and 133)*.

7. *Feathery-flowery* forms are completely like those produced during the precipitation experiments of LIESEGANG (1913) in viscous media. A radiating pattern outward from a point is characteristic for these forms, followed by a slowly tufted spreading out and finally a somewhat sudden ending in a chrysanthemum-like form (figs. 135 and 137). According to the literature, as in my notes, these structures are rare.**

8. The numerous forms which individually have only a scant frequency of occurrence are gathered into a collective group; they cannot be classed in groups 1 to 7, but certainly are of colloidal origin. Here belong as examples long straight or bent strings, little branched, often hollow, and additionally most delicate cockades, fine earthy masses, some intergranular films in gossans and others. In other connections it has already been pointed out that convergences of textural forms with genetically entirely different groups exist; nevertheless the "colloidal textures" can, with some practice, in general be identified with relative ease and certainty.

(c) *Sedimentary textures*

General data. Sedimentary textures are not always sharply separable from hydrothermal textures, and rarely not even from magmatic ones. The characteristic layered structures are also typical for dynamometamorphism, as is shown, e.g., in figs. 40, 52, 98. If "sedimentation" in a genetic sense is to include all that belongs into the sedimentary sequence, then the great number of alteration and cementation textures should be included here, and to fix the boundaries would be nearly impossible, for both geometric and genetic classifications.

* Reticulate forms as such, which I know of, are (i = inside): Bi(i)-smaltite; Bi(i)-safflorite; Ag(i)-safflorite; Ag(i)-smaltite; Ag(i)-niccolite; Au(i)-clausthalite; galena(i)-sphalerite; galena(i)-pyrite; argentite(i)-niccolite; sphalerite(i)"gelpyrite"; pitchblende-fluorite(i); Ag(i)-pitchblende; argentite(i)-safflorite; calcite(i)-dolomite; Ag(i)-dolomite; pitchblende(i)-carbonate; (the last 5 examples are from a work by KIDD & HAYCOCK (1935)).

** They are easily overlooked and difficult to photograph, since only a single favorable plane of section shows clearly their typical structure. Platinum in platinumquartz veins and argentite in silver-gold-selenium veins occur in similar form.

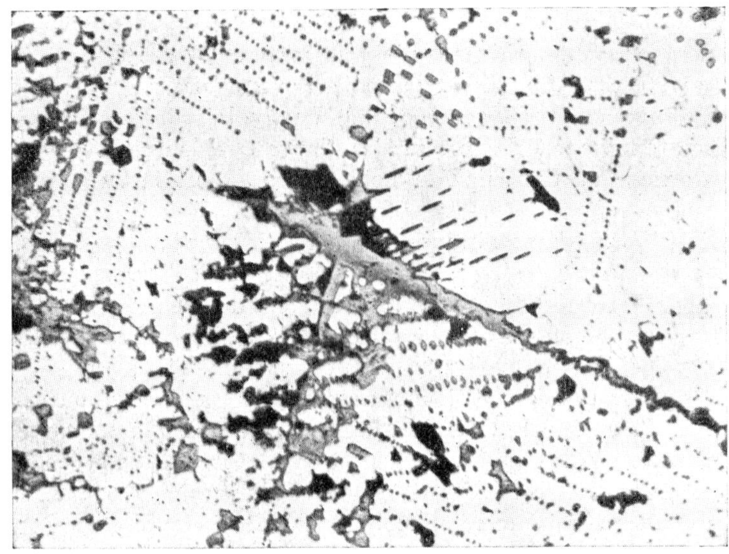

Fig. 132 250 ×, imm. RAMDOHR
Meggen on Lenne, Westfalia

"Reticulate" or "Knitted" *galena* (grey), actually in only small quantity in *pyrite*. Galena formed originally quite loose skeletons, whose interstices were then filled with pyrite

Fig. 132E 4 × RAMDOHR
Urberg, St. Blasien, Germany

Excellent skeleton of *silver* in *fluorite* and *quartz*

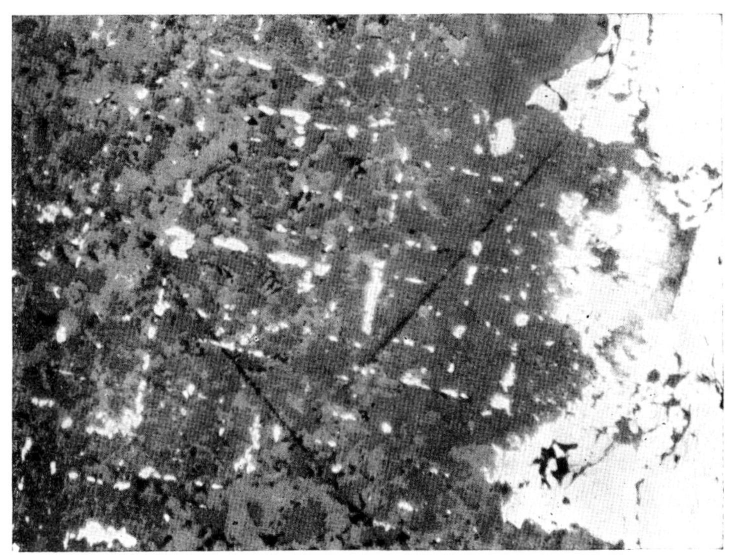

Fig. 133 150 × RAMDOHR

Zinnwald, Saxony

Galena, knitted, with "filling" of two different silicates

Fig. 133 a 1:1 RAMDOHR

Lontzen near Aachen

Excellent "knitted" *galena*, white, included in "*schalenblende*", light gray. Black are holes

The stratigraphic *layered* and *banded* structures are characterized by a laterally far extending homogeneous rhythmic alternation in material, grain size, and often also cohesion. It is just in the "ore" deposits of sedimentary origin, that the lateral homogeneity of the material applies in a strict sense only to the gangue minerals, because the ore minerals are present only as small nodules and lenses.

The true oolites (fig. 49) belong typically to the sedimentary textures. However, it is to notice that the "primary" forms often are completely copied by a later metasomatic change (e.g. calcite to siderite). In sediments with larger porosity an additional impregnation can resemble the primary ore impregnation indistinguishably.

Fig. 134 250 ×, imm. RAMDOHR

Skeleton shaped *pyrite* with wedges filled by *galena*

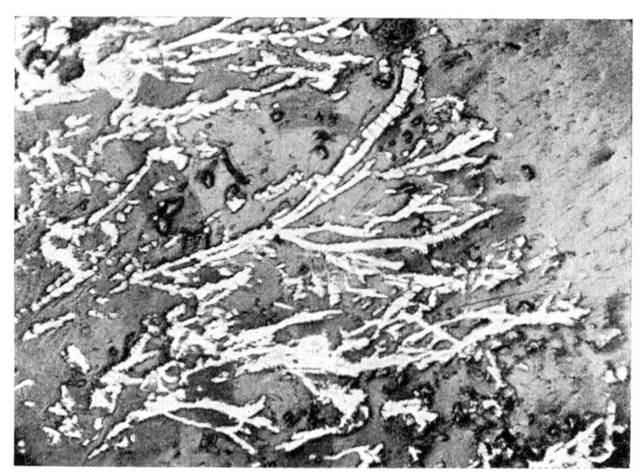

Fig. 135 25 × RAMDOHR

Kowary (Schmiedeberg), Silesia

As secretion in gel-like matrix of moss-like *arsenides* (white, mainly safflorite) in *calcite*. Little *quartz* (dark grey, strong relief)

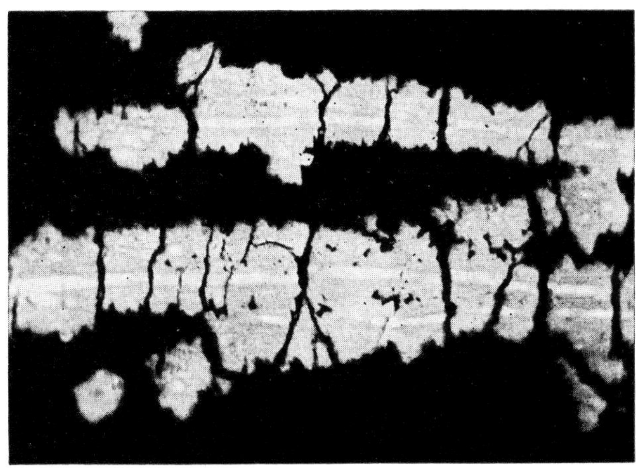

Fig. 136 500 ×, imm. RAMDOHR
Kowary (Schmiedeberg), Silesia

Tree-like and moss-like strongly articulated growth forms of *safflorite* around a "core" of a higher reflecting thread-like *mineral*, that has a hexagonal cross-section. Its nature is not certain, possibly native arsenic. Copy had to be made on high contrast paper

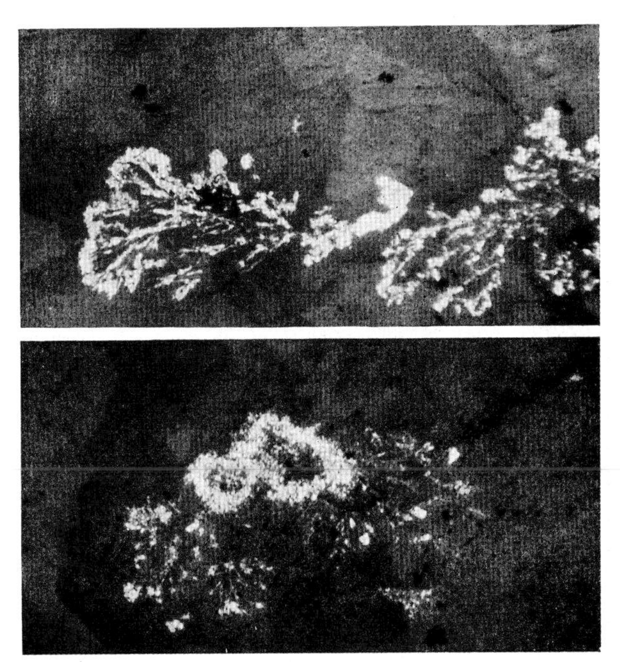

Fig. 137 85 × RAMDOHR
Eisenberg near Corbach, Waldeck, Hesse, Germany

Chrysanthemum-like forms of secretions of *gold* (white) and some *clausthalite* (also white, in the outer parts of the "blooms") in an original gel matrix, now coarsely crystalline *calcite*. — The lower figure gives a cross section

Fossils which presumably were mineralized at about the same time as sedimentation occurred will naturally be included here (figs. 139–143).

The most common type of sediments, sands and sandstones, is as in "ore", forming natural (and artificial) concentrates (placers, beach sands), often an object for ore-microscopical examination. The sizes and properties are mostly evident, the mineral content typical but often rather complicated (fig. 144a, b).

II. "TRANSFORMATION TEXTURES"

All textures which show any evidence that the "primary" mineral content resulting from deposition has been changed would be included here. In order not to have to interpret the concept "primary" too pedantically, the transformations which im-

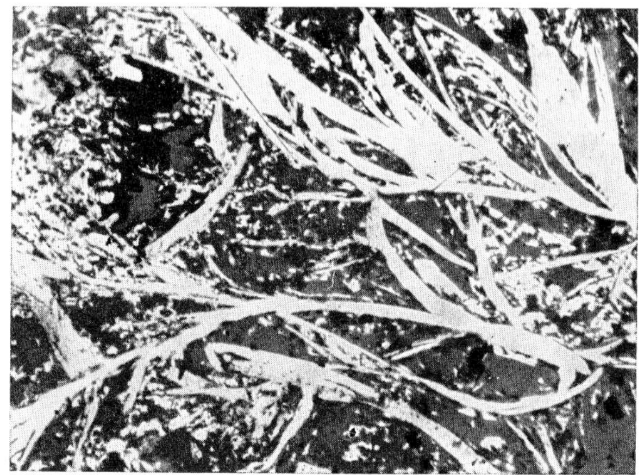

Fig. 138 150 × RAMDOHR
Veta Chica, Asunta Mine, San Vicente, Bolivia
Franckeite, feathery tufted, platy crystals in gangue

Fig. 139 50 × RAMDOHR
Gröditzer Mulde, Silesia
Foraminifera mineralized into *galena* out of "Kupferschiefer", which here is considerably sandy and limey

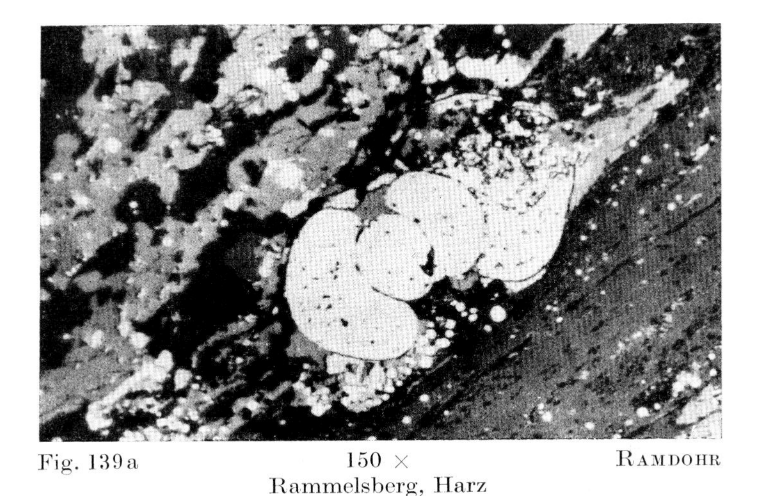

Fig. 139a 150 × RAMDOHR
Rammelsberg, Harz

A embryonic form of an ammonite (probably goniatites sp.) escaped the tectonic crushing by early pyrite-mineralization. Besides pyrite granularly recrystallized, sphalerite grey and a little gangue, dark-grey in different shades

mediately succeed deposition expressedly shall be excluded here. In this strongly confined sense, the classification of SCHNEIDERHÖHN distinguishes: a) paramorphic transformation textures, b) exsolution textures, c) decomposition textures, d) replacement textures. The last are, quite schematically, subdivided into 1. autohydration and similar ones in liquid-magmatic associations, 2. in pegmatites, 3. contact-pneumatolytic, 4. in pneumatolytic and hydrothermal associations, 5. in low temperature associations. — Further large groups would be: e) thermal transformations, 1. in contact zones, 2. outside the contacts; f) cementation textures, g) oxidation textures, h) diagenetic textures. — This last group already includes a poorly distinguishable part of the "transformations following precipitation" that are excluded above. SCHNEIDERHÖHN mentions there: "diagenesis", cementation (here in the meaning "baking together") collection crystallization, devitrification, leaching, where many writers would define the last four concepts as subgroups of diagenesis.

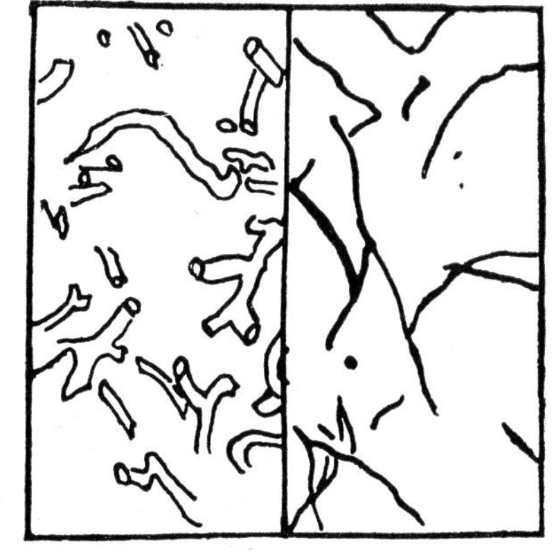

Fig. 140 accord. to GRUNER

Two types of algae in banded iron-formation of the Biwabik formation

It is impossible to mention all the resulting possibilities here. Exsolution and decomposition textures are extensively treated here, even if from another viewpoint, whereas many others are contained in preceding parts of the book.

(a) *Paramorphs*

are transformations of a compound into a stable form from a form which is becoming metastable or instable. For example marcasiste → pyrite or high quartz → low quartz are well known to everyone. They are often observable in all intermediate stages. It is most amazing that the metastable form, sometimes even the high temperature form, already partially transformed can be preserved during geologic time. Experience from metallurgy shows that small amounts of foreign components can produce such phenomena. These phenomena, however, can hardly in all cases in nature be used as an explanation. Many examples are mentioned in this book. More precisely treated and in part illustrated are, e.g., arsenolamprite → arsenic, guana-

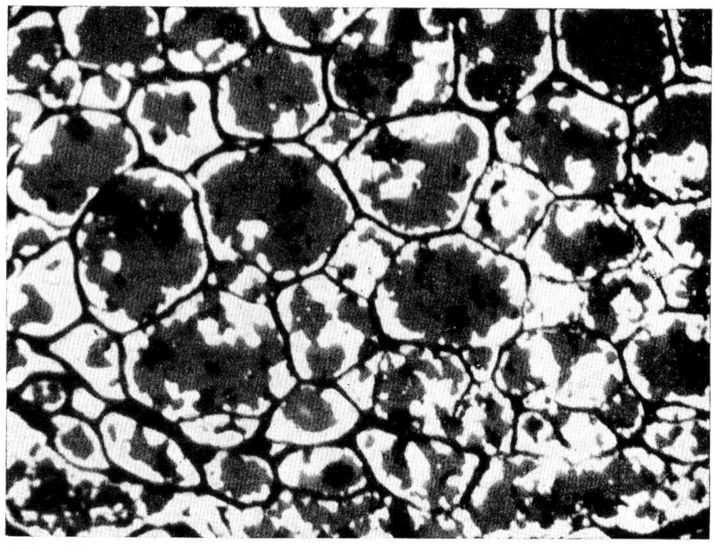

Fig. 141 350 ×, imm. RAMDOHR
Freyung, Bavaria

Wood cell texture with *galena* and *sphalerite*

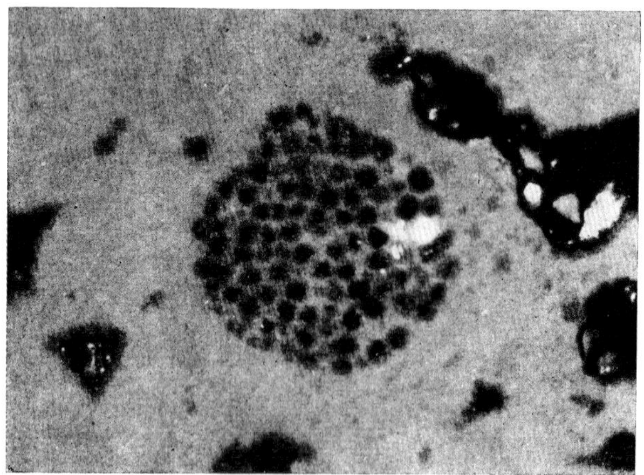

Fig. 142 1400 ×, imm. RAMDOHR
Mi Vida Mine, Big Indian Wash, Utah

Bacterial texture in *pitchblende*; clear white are some little specks of *pyrite*

juatite → paraguanajuatite, marcasite → pyrite (fig. 498), metacinnabarite → cinnabar (fig. 375), isostannite → stannite.

Completely unexplainable at first is the incontestable observation that the metastable marcasite can replace pyrite, thermodynamically a complete absurdity. Transformation by addition of radioactive energy is not meant here, although it does occur.

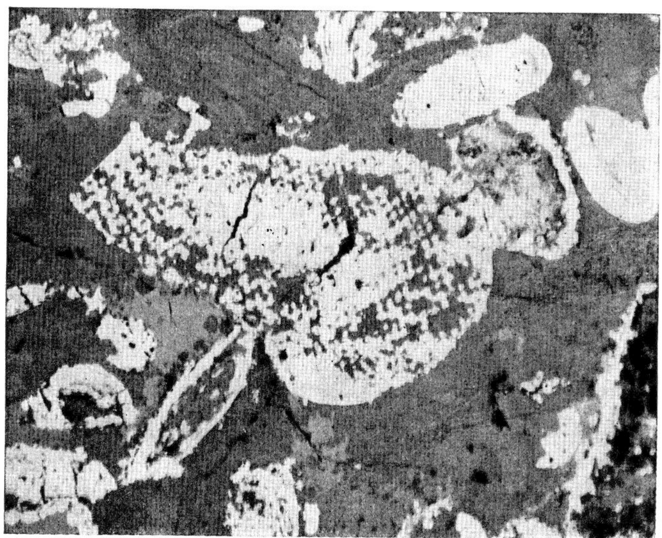

Fig. 143 70 × RAMDOHR
Eweschbour, Luxemburg

Bryozoon, which is mineralized by *magnetite*, from an oolitic iron ore

(b) *Exsolutions*

General data. It is not necessary to go into the enormous variety of formation of solid solutions in minerals. With great similarity of ionic radii, lattice structure, bonding strength and so on in the "components", the solid solution crystals remain stable almost to room temperature, and far below (N. B. not to absolute zero!). With larger differences the solid solution will be unstable with decreasing temperature; it can then "exsolve" itself. Many exsolutions occur spontaneously as soon as a certain supersaturation is reached, others freeze as "non-equilibrium" mixtures. Whether the process begins or not is dependant on the interaction of many factors and cannot be predicted. Rapid cooling, chemical affinity of the components, and lack of structural flaws in the crystals generally favor the possibility of undercooling. Addition of further "solution partners" (associated elements or minerals) can as well raise it, as also lower it (as indeed has been most accurately studied and is technically enormously important in the properties of metals). Slow cooling, anomalous mixtures, small foreign inclusions, tectonic stresses of most different type and other factors facilitate exsolution. Very stable and compact lattices, e.g., as those of the silicates and in part those of the oxides, exsolve with very much greater difficulty than do most metals and especially than do the sulphide minerals. The difference between the ionic bonding plays a role here in contrast to the metallic bonding — this is by no means the only

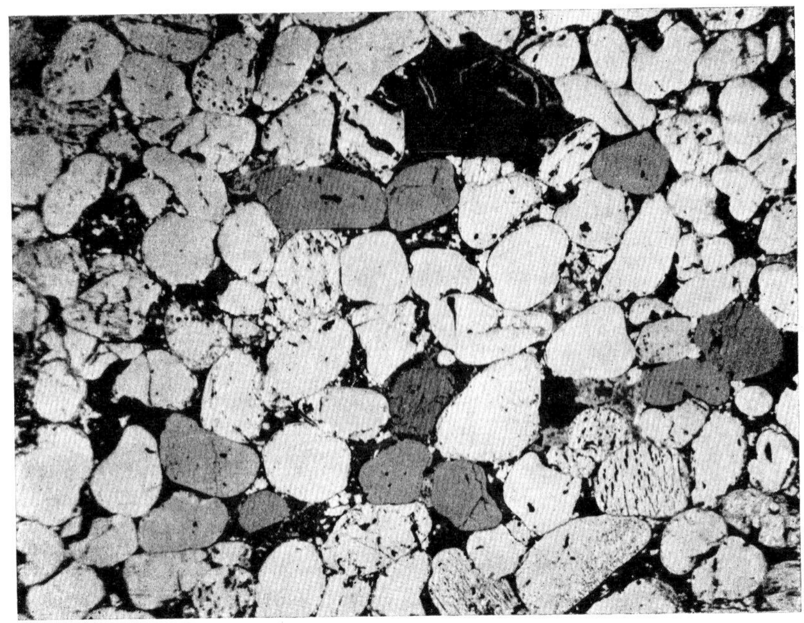

Fig. 144a Bothaville, O.F.S. RAMDOHR
35 ×

A former heavy sand consisting mainly of *ilmenite* (white). Many grains of rounded zircon (medium grey), two or three grains of *chromite* (lighter than zircon) one grain of *monazite*

Fig. 144b 40 × RAMDOHR
Monarch Reef II, West Rand Consolidated, Krügersdorp, Transvaal

Conglomeratic "sandstone" essentially consisting of small pyrite pebbles (very few chromite grains and others)

factor! — When exsolution commences, exsolution bodies are formed which at first are extremely small, and they can remain small. In most instances, however, they grow to microscopic or megascopic size by collection crystallization. If the size of the bodies remains submicroscopic, then the characteristic two phase pattern in X-ray diffractions can be recognizable, i.e., with sufficient amount of the exsolution products. Sometimes it becomes evident also, e.g., by anomalous optical phenomena.

As for *substance* it is to be noted that continuous transitions exist between the "simple exsolutions", the "complex exsolutions" (both in the sense of NIGGLI), decomposition of anomalous solid solutions, decomposition with scant addition and removal of material, and finally with complete re-mineralization.

"*Simple exsolutions*" NIGGLI called those in which the previously simple solid solution is chemically conceivable by formation of the important disintegration components. An example is a perthite of 90% orthoclase ($KAlSi_3O_3$) and 10% albite lamellae ($NaAlSi_3O_8$) that was earlier a homogeneous high temperature feldspar with 90% Or and 10% Ab composition. Hematite–ilmenite and schapbachite–galena are analogous examples.

"*Complex exsolutions*" on the other hand involve a disturbance of the stoichiometric relationships. If a feldspar in its exsolution yields hematite plates besides orthoclase, the solid solution can either not have been expressible directly by a feldspar formula, or the newly formed orthoclase must have received part of an "iron-feldspar" or finally, something must have been removed. Among those are, e.g., the pairs magnetite–ilmenite and linnaeite–millerite. Commonly the high temperature crystal will not be precisely stoichiometric and a so-called "*anomalous solid solution*" exists. Very many original exsolution materials were probably anomalous solid solutions*: e.g., ferrite–cohenite, chalcopyrite–pyrrhotite, olivine–magnetite, corundum–rutile.

Addition or removal of material is frequent in systems and events, in which larger mobility is suggested, and thus many difficultly understandable "exsolutions" may be interpreted, which in the true sense of the word are not really exsolutions any more. Since a high temperature solid solution does not need to be analogous to any of its products of decomposition in regard to crystallography or chemical composition (except for the bulk chemistry!) complete re-mineralization is not sharply separable from exsolution. Examples of re-mineralization are: decomposition of a hematite–ilmenite with $Fe_2O_3:FeTiO_3 = 1:1$ into rutile + magnetite; or decomposition of bornite Cu_5FeS_4 into chalcopyrite + 2 chalcocite.

As for *mechanism* short explanations are also necessary. The crystal lattices of the original crystal and the exsolution products are a factor in the ease of exsolution as well as in the form and arrangement of the end product. If the lattices are very similar the original textures will be quite variable and especially irregular. Moreover the textures will also, because of the ease of migration of material, become obscured and atypical, i.e. they will no longer be recognizable as exsolution textures. With greater differences, e.g., if one dimension of the lattices corresponds closely and the others vary, strong directional preferences will appear, which therefore will produce "typical" forms. Similar textures will appear, if one of the lattices involved, especially the "host", has a loosely packed lattice, as, e.g., where there are empty canals, cells, or

* The nomenclature and meaning of anomalous mixed-crystals cannot be treated here. Refer, e.g. to Lehrbuch der Mineralogie, KLOCKMANN—RAMDOHR, 14th Ed., page 217.

layer-lattice seams. These also will predetermine the shapes of the exsolution bodies.

Whether the exsolution bodies remain small or can grow to an often astounding size is dependent in an incalculable manner upon the previous history (especially the cooling rate), the lattice of the host, the "solution partners", differences of contraction with cooling, and others.

In almost all cases the exsolution bodies* in the grain of the host are not entirely uniformly distributed and besides are often unequal in size and form, so that a beginner might doubt occasionally the nature of exsolution. Apart from the fact that the original solid solutions can be strongly zoned and richer in the guest component in the core of the crystal, which necessarily causes sparser and smaller E-bodies in the outer zones, similar effects can occur by migration (of the guest) toward the grain boundaries.

Bodies aligned like a string of beads or even connected intergranular masses of the guest component can often be explained simply by exsolution. The number and size of E-bodies can be quite different in equivalent directions, consequently e.g. one individual octahedral plane can be very strongly favored. This may mainly be due to a tectonic factor. Velocity of migration and distance of migration are reduced with decreasing temperature. The large old exsolution bodies in the core will be capable of only partly "drawing in" newly forming exsolution substance, while the rest segregates between them in independent small gains. Near the border the migration possibility now falls off: small E-bodies occur also here. Most varied modifications of this scheme can result through influence of the case of exsolution by twin lamellae, cracks, inclusions and lattice disturbances of other types. — The just described associations of coarse and fine bodies can become similar to the principally different process of exsolution in two generations. This process is always then to be expected, e.g., when one of both partners suffers transformation which disturbs the continuous course and causes sudden strong supersaturation. This applies to Fe_2O_3–$FeTiO_3$, where supersaturation occurs, in all probability at that moment when the $FeTiO_3$ lattice changes from a statistical distribution for Fe and Ti to an ordered arrangement ($D_{3d} \rightarrow C_{3i}$). — Fine examples with reference to this topic are illustrated in figs. 564 to 567.

The E-bodies separated at still relatively high temperature are themselves still solid solutions which with further decrease in temperature tend to exsolve the contained portion of the guest component. Often this is not "possible"; the E-bodies remain in their properties still strongly different from the pure minerals. Often, however, a "sub-exsolution" appears, which is a diminished copy of the older, with opposite relations, or which even has its own characteristic forms (figs. 172, 386, 458).

Immediately at the grain boundaries the unmixing, i.e. the pushing out of the partner not longer fitting in the structure, is more complete than in the interior. Ilmenite, for instance, may contain in not exsolved state in average 7%, but at the grain boundary only 4%. Regarding the short diffusion path this can easily be understood. In traverses with the microprobe that can be followed very well ("Agrell-effect"), but for the oremicroscopist it is known for a long time.

Frequently the E-bodies which are first separated out are still complex, unstoichiometric products, which then produce new compounds by "subexsolution" or by ex-

* Here and later often abbreviated: E-bodies.

solution in several stages. That is the case, e.g., with E-bodies of "chalcopyrite" from sphalerite which decompose into chalcopyrite + cubanite, chalcopyrite + mackinawite, or similar ones.

Complex exsolutions of another type occur where the exsolution of one mineral species somehow induces an approximately simultaneous second exsolution. That is the case, e.g., of chalcopyrite + wittichenite in bornite, also the formation of ulvöspinel in the "magnetite" lamellae of many ilmenites.

The forms, which are discussed individually below, can be confusingly similar to those of the replacement textures, or to those of the simultaneous oriented crystallization, or of alterations of gels. Many misinterpretations may doubtlessly be ascribed to those striking similarities. Especially replacements and exsolutions have been mutually falsely explained, and accordingly attempts have been made to discover specific characteristics for each. If a solid solution, originally even in lattice dimensions homogeneous, disintegrates, and the products of disintegration merge into bodies which are easily visible under the microscope and in some cases (some titanomagnetites) grow to a few centimeters in size, considerable migration of material must have occurred. Migration must have occurred of "host substance" from the interior of the exsolution bodies, and of "guest substance" from within the host; one could even say that "Verdrängung" (replacement) occurred. DUNN has apparently called attention to this long overlooked obviousness for the first time. One should thus consider "characteristic of replacement" cautiously, and in general not exclude exsolution as a possibility. — If the expression "replacement" is to be restricted to later addtion of material as it is done in this book, then the characteristics given by G. M. SCHWARTZ may be mentioned, although they apply only to lamellar networks and even here in no way exclude cases that are doubtful in both directions: 1. When replacement has occurred along lamellae, which serve as easy ways of access of material from the outside, then the guest mineral is commonly more abundant at the intersections of lamellae. 2. Less "guest substance" is available at the intersections of lamellae in the case of exsolution, since the material is derived from the immediate surrounding. Thus the lamellae must taper or even pinch out toward the intersections (fig. 145).

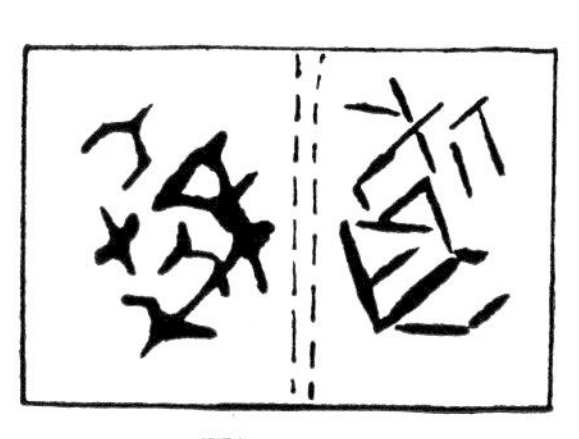

Fig. 145

In very typical exsolutions which were also experimentally substantiated, such remarkable exceptions were found, that extreme caution seems necessary. This tapering is not or hardly ever exhibited by "lamellar chalcocite", exsolved titanomagnetite, schapbachite, kamacite-plessite, and others.

Silicate and partly also oxide exsolutions may after their formation remain more or less the same, provided the cooling rate is not too slow and no subsequent rise in temperature occurs. This, however, does not hold true for many sulphides, especially soft ones. The speed of migration and the surface tension are even at moderate temperatures still so great that sharp cornered exsolution bodies are rounded off, lamellae flow out into irregular rows of dots, and the very fine bodies merge into incidentally formed partly large inclusions. With the criteria of exsolution thus destroyed, nobody would think that these textures were formed by exsolution, unless transitions occur, or an experiment reveals the true relationships. The investigations of SCHWARTZ (1931) on bornite and chalcopyrite are excellent illustrations of this. According to the nature of the process textures of an apparent replacement are also produced. Relicts of such textures not destroyed by collection crystallization are especially found in small "armored" (on all sides) inclusions, e.g., by quartz. The recrystallization of these inclusions has been prevented by lack of access of traces of solutions or gases — which apparently favor recrystallization, but are not in all cases necessary. One series of experiments by the writer on resolution of complex exsolution bodies with heating is shown in figs. 146a to 146d. One work by VINCENT et al. (1957) shows similar instructive relations.

Summary of exsolution textures

In view of the extremely variable exsolution textures which in part are always characteristic for one and the same mineral, and in part contain within the same mineral grain different types of exsolution textures, it is not surprising that attempts have been made to work out a summary of exsolution textures and to present a scheme of classification. The most comprehensive attempt is that by GRIGROIEFF, and another is that by SCHWARTZ. GRIGORIEFF distinguishes cellular, lattice-shaped, emulsion-like, graphic and spotted exsolution textures. He lumps with the last-named textural type all those which do not fit into the other groups (but neither do they fit into this last one!), including oleander leaf, lilac, doughnut and other textures.

I would like to introduce a new classification which utilizes the two previous attempts and in part leans on that of SCHNEIDERHÖHN, but is more extensive, and where possible more sharply demarcated; like the others it does not include all possibilities. Individual examples will be cited and documented by illustrations in this book.

Exsolution (unmixing, "Entmischung") is surely a well defined conception, but it does not say anything on the forms. In the same process and in immediate neighborhood objects can be formed of very different appearance which may suggest different origin. Obviously it is here very difficult to separate the formal point of view from the genetical one, but we try first to give only the forms with data on their structural explanation.

A separation also according to even or uneven distribution, as SCHNEIDERHÖHN and I myself have attempted, seems to me no longer feasible.

Exsolution segregations in the interior of grains

1. *Very irregular forms.* They are for the most part products which are the result of collection crystallization from all other forms, but they can also originate spontaneously (fig. 148). They occur preferentially with cubic lattices which are closely related in size and bonding and hence are subject to easy change in position (fig. 381a). They can hardly be distinguished or not distinguished at all as for shape from some replacement forms (fig. 197) or recrystallization products of a former gel precipitate (fig. 132).

2. *Dispersions (emulsions) of rounded or oval drops of varying regularity.* Dispersions (emulsions) should be called those textures in which finely-divided guest particles are dusted through the host which is present for the most part in considerable excess. In many cases the guest particles may in one immediate neighborhood be once irregularly distributed, once more or less prominently aligned in strings parallel to certain directions in the host, or may form peculiar garlands (figs. 389, 366, also

152, 299). Transitions exist to 1, 3, 8 etc. — Such textures can typically appear with larger deviations of the lattice relations, but also form other textures in the early stage of "running together".

3. *Dispersions (emulsions) of more or less regular-shaped particles.* In cubic minerals these are frequently octahedra, stars, crosses or lines, and with non-cubic host or guest they may be small plates, flat stars, "commas", "flagellates", "oleander leaves"

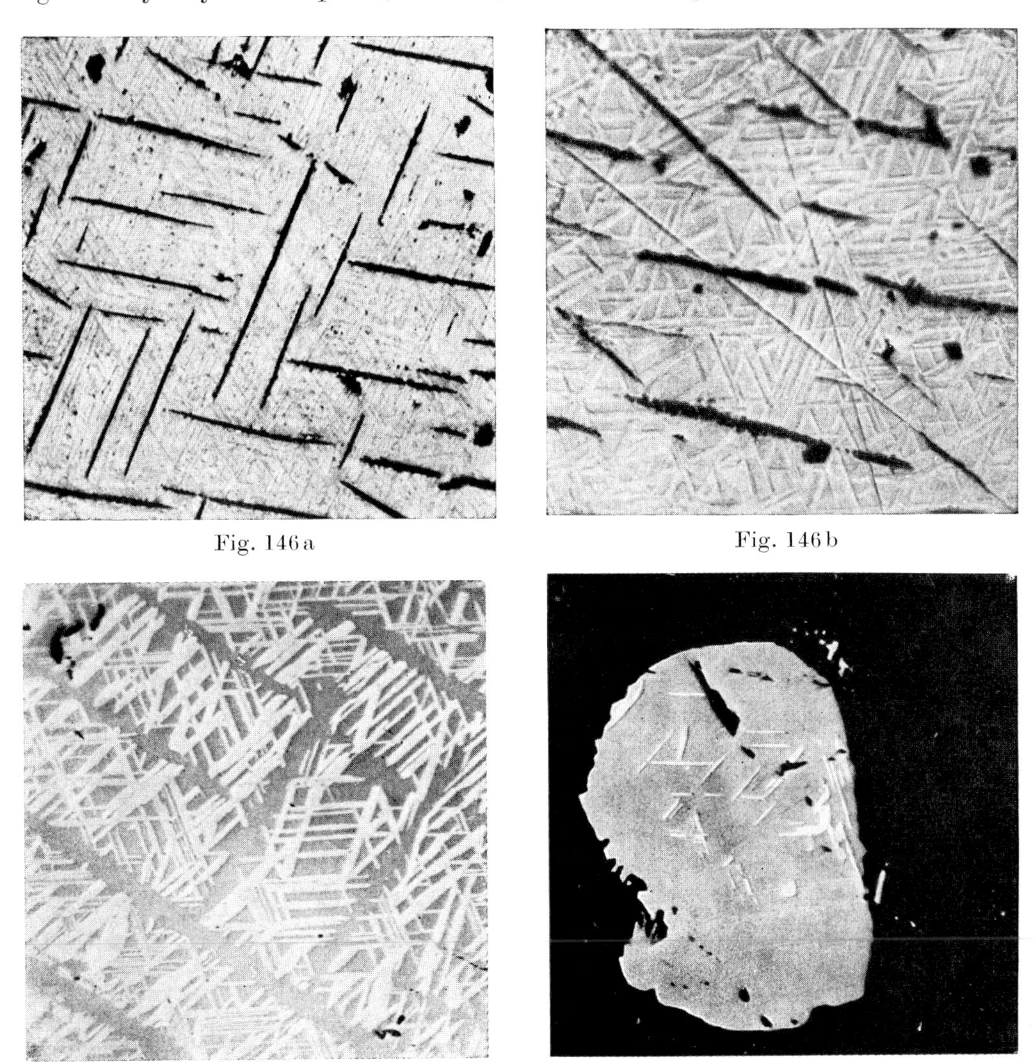

Fig. 146a Fig. 146b

Fig. 146c Fig. 146d

Figs. 146a–146d 480 ×, imm. Ramdohr
Smålands, Taberg, Sweden

A series of heating experiments on *titanomagnetite* which had unmixed with the formation of *spinel* // (100) and *ilmenite* // (111). 146a, unheated. 146b, 12 hours 800 °C, the ilmenite lamellae have been considerably enlarged and with oxygen addition have taken up material (Fe_2O_3) from magnetite; thereby they have become lighter. 146c; 12 hours at 1000 °C. A more advanced stage in the process. Spinel has been resorbed, and the ilmenite lamellae have further become increasingly lighter. 146d, 12 hours at 1200 °C. Homogenisation has been essentially completed; some ilmenite relicts remain

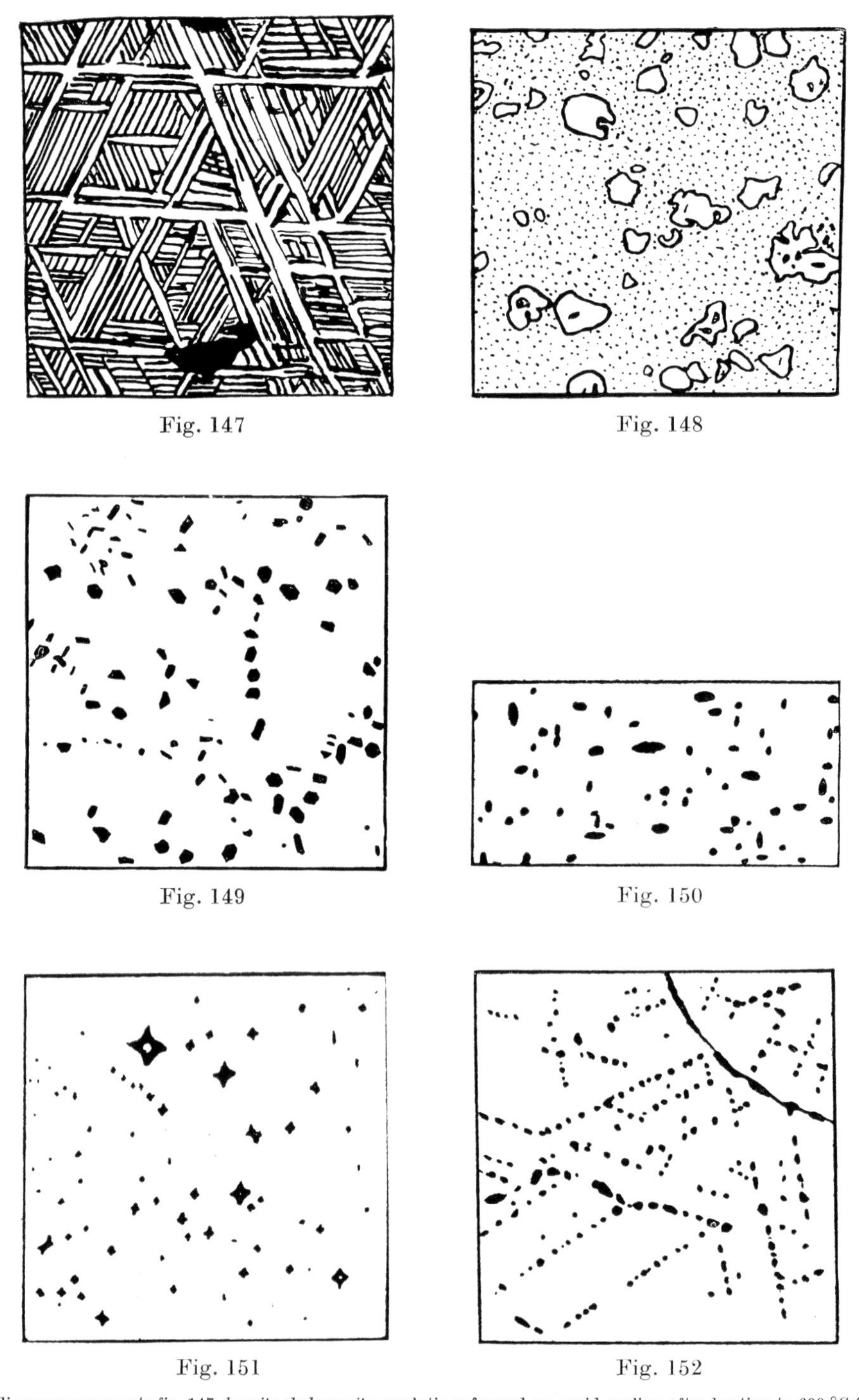

Fig. 147

Fig. 148

Fig. 149

Fig. 150

Fig. 151

Fig. 152

The diagrams represent: fig. 147, *bornite-chalcopyrite* exsolutions formed on rapid cooling after heating to 600 °C (acc. to SCHWARTZ); fig. 148, same after slow cooling; structures entirely obliterated; fig. 149, pyrrhotite in alabandite; fig. 150, chalcopyrite in stannite (incipient uniformity); fig. 151, chalcopyrite in sphalerite; fig. 152, sphalerite in chalcopyrite

and rarely small columns or needles. It is characteristic that even minerals which, otherwise are typically acicular, often become tabular as E-particles (millerite, cubanite — fig. 435). — It is not uncommon that a pore or foreign particle is apparently the point of origin of these particles (figs. 149, 151, 153, 160, 166, 183). Transitions exist to 2, 4, 8. — It has happened that these particles have been mistaken for replacement relicts.

4. *Lamellar textures* with longer and more continuous lamellae of the guest. The textures differ somewhat with cubic host and cubic guest, and cubic host and hexagonal guest, or with hexagonal host and cubic guest. They also differ as a result of especially high temperatures of formation (which for all minerals are, however, "subjectively" different!) or especial similarity of the lattice bonding, and are in both cases more lancet-like (similar to the oleander leaves in 3).

a) Host and guest cubic. For the most part cubic nets of lamellae are formed, i.e. a maximum of three swarms of lamellae (figs. 146, 147, 155, 543 E). The lamellae can instead also be platy after the octahedron (in four directions! e.g., meteoritic iron, fig. 454). Of course, this is not the result of the lattice dimensions, but is somehow conditioned by the bonding.

b) Host cubic, guest hexagonal with lattice dimensions $\perp$ (111) $\sim$ $\perp$ (0001). In this especially frequent case octahedral nets are formed (4 directions!). Even here "oleander leaves" occur. — An especially frequent case as illustrated in figs. 156, 157, 158, 159, 160, 169, 175, 342, 532 and many others.

c) Host hexagonal, guest hexagonal (or even cubic). With great similarity of the lattice dimensions very irregular shredded disks, and also skeletons as in 3, are formed. With greater differences, smooth and ultimately thinly plated lamellae are formed (figs. 162, 163, 164).

What has been said under 2, 3 and 4a, b, and c applies also in the same sense to hypocubic and hypohexagonal lattices.

d) In addition there are of course still other lattice dependent lamellar textures, where neither host nor guest are hypocubic or hypohexagonal. They are relatively rare (figs. 179, 181) and often lead to 6 and 8.

5. *Netlike textures.* These can be closely linked to 4a and may be closely lattice dependent, but can also be dependent on old cataclasis, latent tensions, cleavage or old lattice block structures. They are not common (figs. 165, 341, 538). Transitions exist to lamellar textures, and under certain circumstances to others. Some of these nets have been incorrectly interpreted as cementation textures and vice versa.

6. *Skeletal, feathery, and filigree textures.* These textures bear close relation to some cases noted under 3, 4, and 5, and distinction between them can be quite arbitrary. They depend first of all on the guest, and are dependent on its tendency for idioblastic development. Frequently all of the "filigrees" are crystallographically of uniform orientation. Even convergences may occur with idioblasts of entirely different origin (figs. 174, 178 b, 180).

7. *Snake and garland shapes.* Especially apparent as a case of very close lattice relationship is an unmixing form which shows no preference whatever to any direction. Within *uniform* and as far as can be determined *undisturbed* grains, this unmixing product occurs in the form of thinnest, snakelike winding films, which resemble the grain-boundaries in a myrmekitic intergrowth (figs. 181, 305).

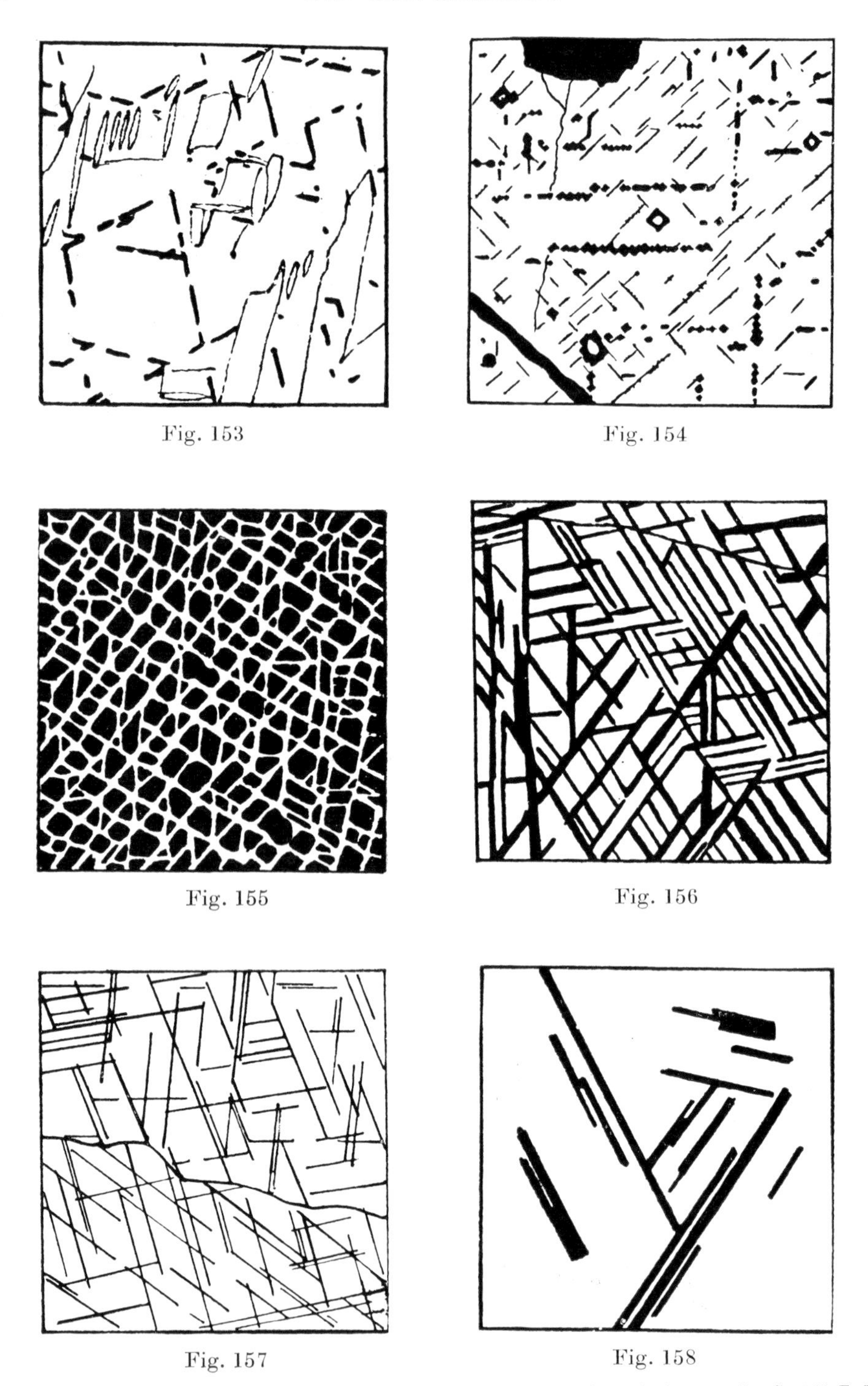

Fig. 153 Fig. 154 Fig. 155 Fig. 156 Fig. 157 Fig. 158

The diagrams represent: fig. 153, mackinawite in chalcopyrite; fig. 154, spinel and ilmenite in magnetite; fig. 155, Fe_2TiO_4 in magnetite ("cloth structure"); fig. 156, ilmenite in magnetite; fig. 157, β-Cu_2S in neodigenite (chalcocite); fig. 158, cubanite in chalcopyrite

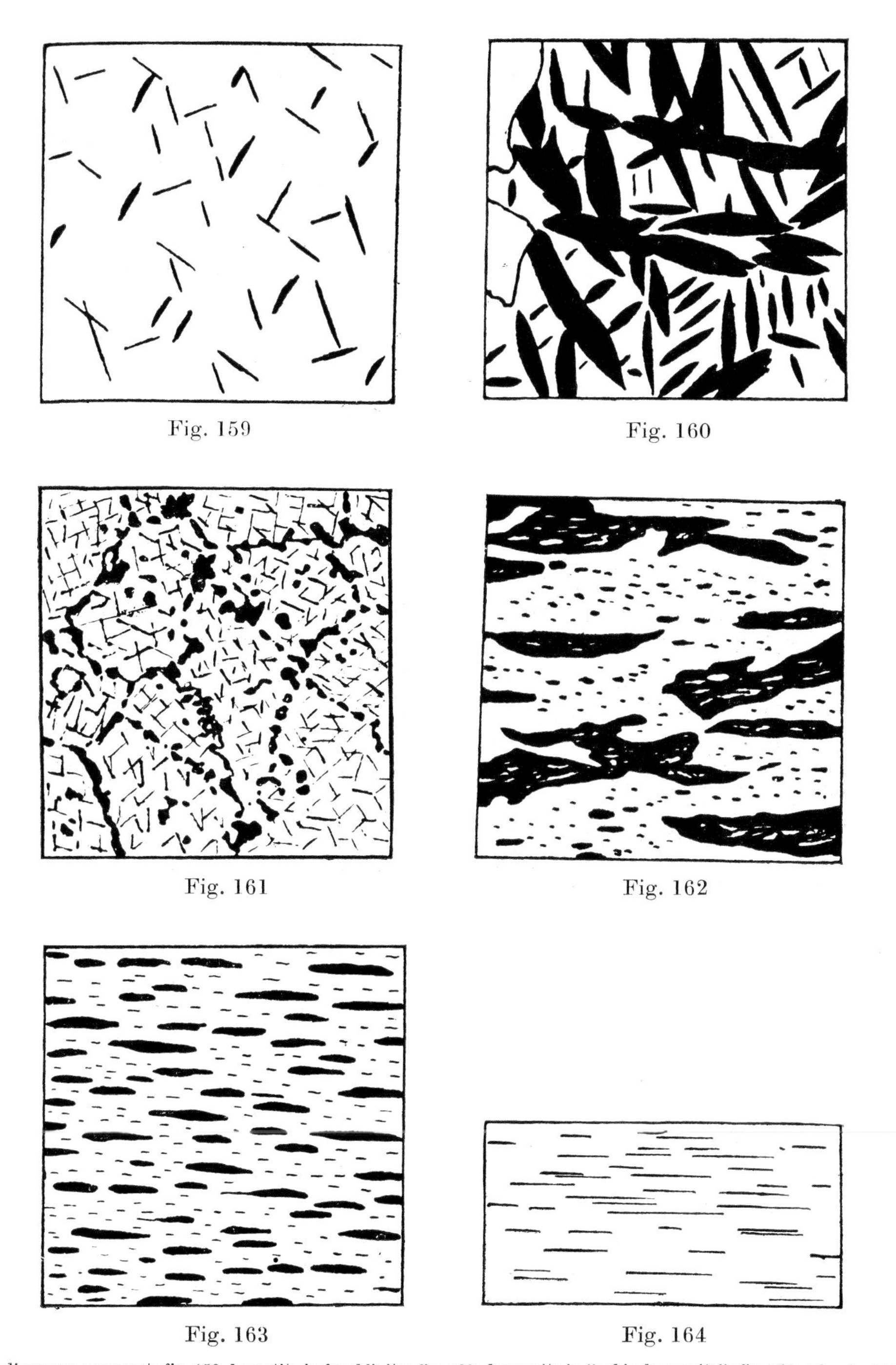

Fig. 159 Fig. 160 Fig. 161 Fig. 162 Fig. 163 Fig. 164

The diagrams represent: fig. 159, hematite in franklinite; fig. 160, dyscrasite in "cubic dyscrasite"; fig. 161, schapbachite-galena-intergranular material (accord. to ANDERSON*); fig. 162, hematite in ilmenite; fig. 163, ilmenite in hematite; fig. 164, magnetite in ilmenite

* Accord. to ANDERSON (personal communication) the identification of brongniardite-cosalite is incorrect.

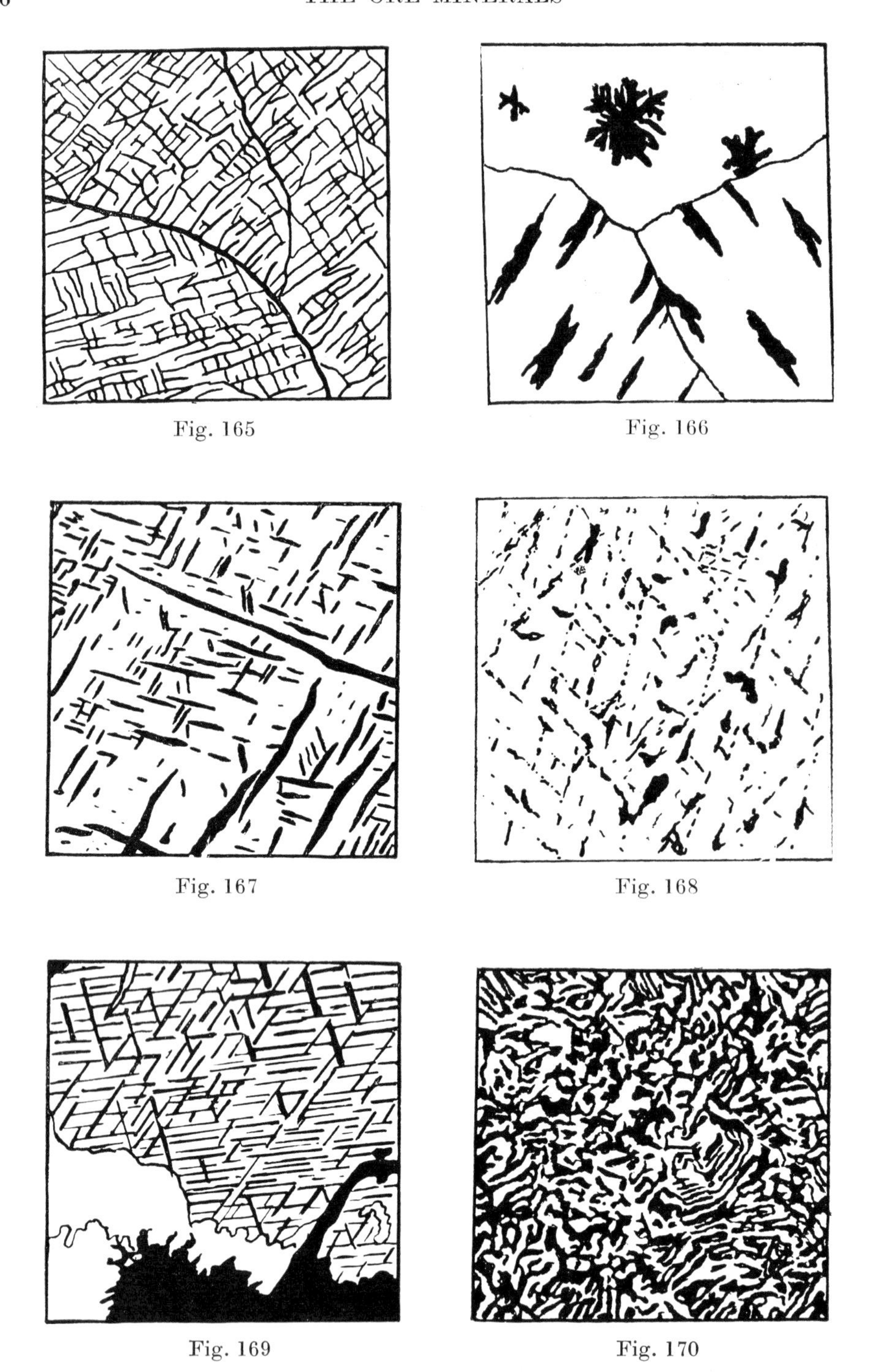

Fig. 165 Fig. 166

Fig. 167 Fig. 168

Fig. 169 Fig. 170

The diagrams represent: fig. 165, mackinawite in pentlandite; fig. 166, pentlandite in pyrrhotite; fig. 167, chalcopyrite in bornite; fig. 168, ilmenite in tantalite (accord. to EDWARDS); fig. 169, covellite in neodigenite, Khan; fig. 170, stibarsen arsenic

8. *Myrmekitic or "graphic" textures* are not very common, but for individual cases extremely characteristic (fig. 315). They can occur as exsolutions, especially where the lattice structure of the components is very similar. If this agreement holds true in only two dimensions, then several sections cut in different directions may provide different textural configurations (fig. 171). Numerous transitions exist; especially important to note is that lamellar textures may yield, as a result of collection crystallization, completely analogous textural configurations.

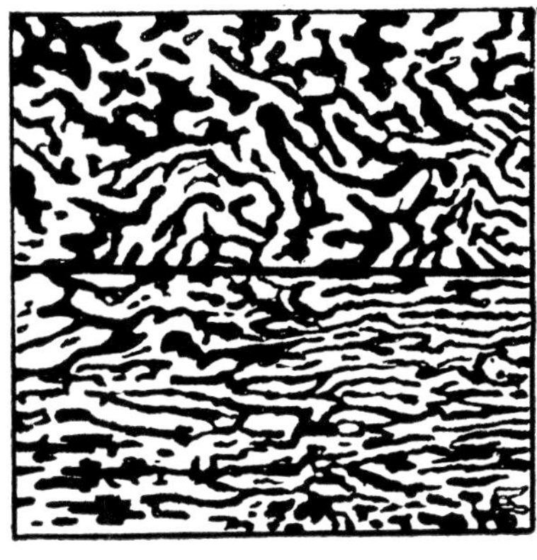

Fig. 171

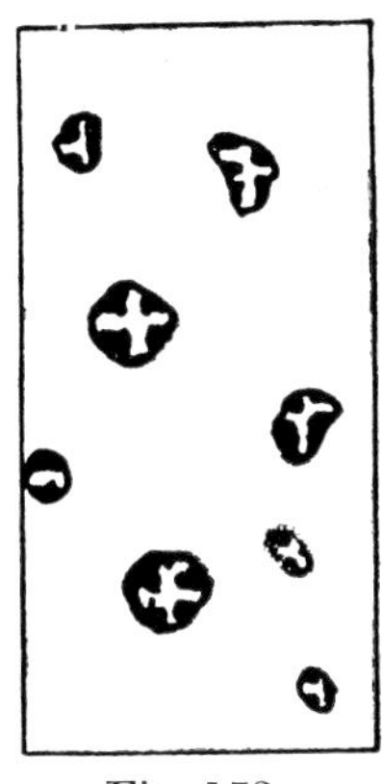

Fig. 172

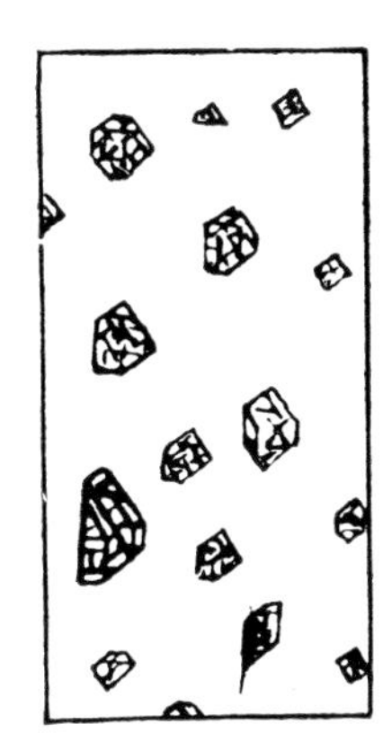

Fig. 173

The diagrams represent: fig. 171, galena-bournonite for falkmanite ⊥ and // to its columnar texture; fig. 172, chalcopyrite with mackinawite in sphalerite; fig. 173, pyrrhotite in pentlandite

All of these texture or at least very similar ones may result from entirely different processes.

Of course, complications of many kinds exist: where two guest components exsolve (e.g., sphalerite + chalcopyrite in stannite, fig. 389), the first exsolved guest may subsequently become transformed (the cubic exsolution product which at first forms in pentlandite breaks down with resultant pyrrhotite formation), or may disintegrate later into several components ("chalcopyrrhotite" in sphalerite disintegrates into chalcopyrite + mackinawite).

Exsolutions can be simulated by several other processes which yield similar textures. A detail in a sample, even an entire sample, may leave open the question of its true origin, whereas other samples or even other parts of the same section may clarify the subject at once. Sometimes, however, it takes several years until a mistake can be corrected. Considered especially susceptible to incorrect interpretation are poikilitic textures ["Idioblastensiebe"], certain replacement textures, rarely true eutectics, eutectoids (which, however, can hardly be distinguished with certainty from exsolutions), and primary oriented inclusions. Many stages of development of replacement textures can, in fact, always be observed next to each other, whereas true eutectics have constant ratios in the amounts of the different components which make up the eutectics and never show zones of different composition. In contrast, primary oriented inclusions are commonly restricted to sharply separated especially rhythmically-repeating zones. Intergranular recrystallization products are found especially in typically metamorphic ores.

Exsolution segregations at grain boundaries

On p. 170 ff. reference has been made to the mechanism of unmixing and the segregations of "intergranular films" and chainlike intergranular grains resulting from unmixing. In very fine-grained aggregates and with great velocity of migration and very slow cooling all unmixed substances can migrate outwards and therefore the diagnostic unmixing texture can disappear entirely.

A very frequently observed texture is sketched in fig. 161; figs. 180b, 329 and 370 illustrate similar textures. Some sphalerites in which for various reasons abundant

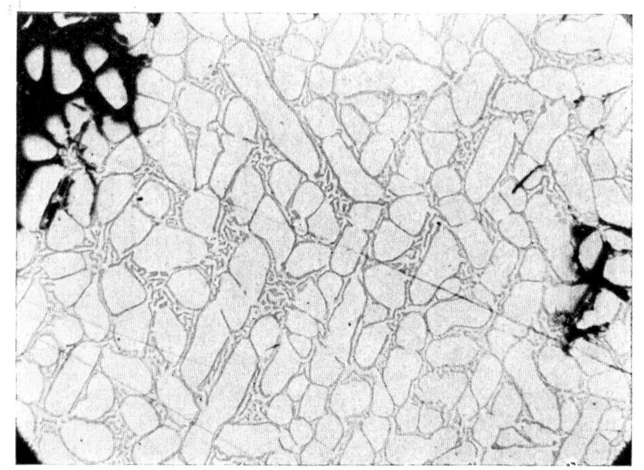

Fig. 174 40 ×, imm. RAMDOHR
Varuträsk, Sweden, from pegmatite

Stibarsen with exsolution net of *arsenic* (grey), section approx. // to base

Fig. 175 450 ×, air-etched RAMDOHR
Harmsarvet near Grycksbo, Sweden

Dyscrasite (white); spindlelike lamellae occurring in cubic Ag and Sb-mixed crystals as unmixing product. The surrounding area consists of *galena* (grey, cleavage cracks). The etching took place entirely by exposure to the air in approximately ten days

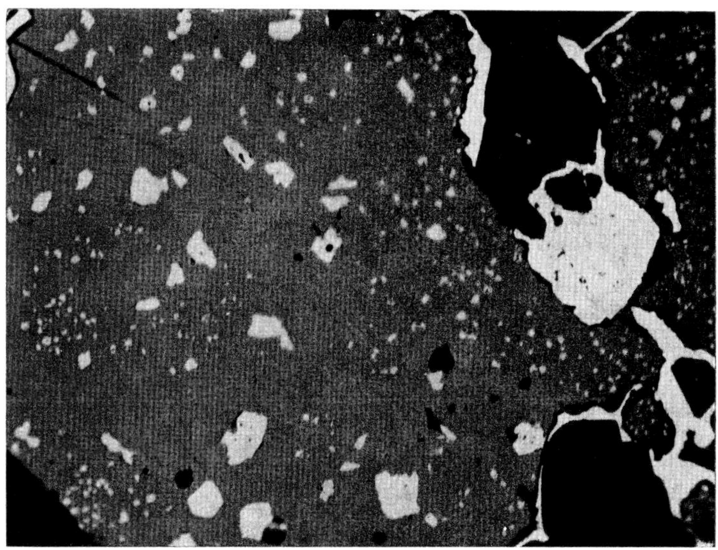

Fig. 176 250 ×, imm. RAMDOHR
Noda Tamagava Mine, Japan

Alabandite (dark grey) with small (marginal) and large (central) exsolution bodies of *pyrrhotite* (light grey). Moreover *pyrite* (pure white) occurs as a newly-formed mineral in zones where alabandite has been dissolved

Fig. 177 250 ×, imm. RAMDOHR
Tsumeb, South-West Africa

Exsolution of a spindlelike lamellar *grey* mineral (gallite) in *sphalerite* [probably according to (100)]. It occurs also in more compact patches between sphalerite grains. In places where *galena* has replaced sphalerite, these grey lamellae were subjected to replacement only later (i.e. to a smaller extent)

Fig. 178a 70 ×, imm. RAMDOHR
Waterfall Gorge, Insizwa, S.Africa

Mackinawite as exsolution in *pentlandite*. The lamellae are of different sizes and are // (100) of pentlandite (3 groups of lamellae); in this section two of these groups appear to be almost parallel At ordinary illumination, i.e. with one nicol, only one group of lamellae stands out well

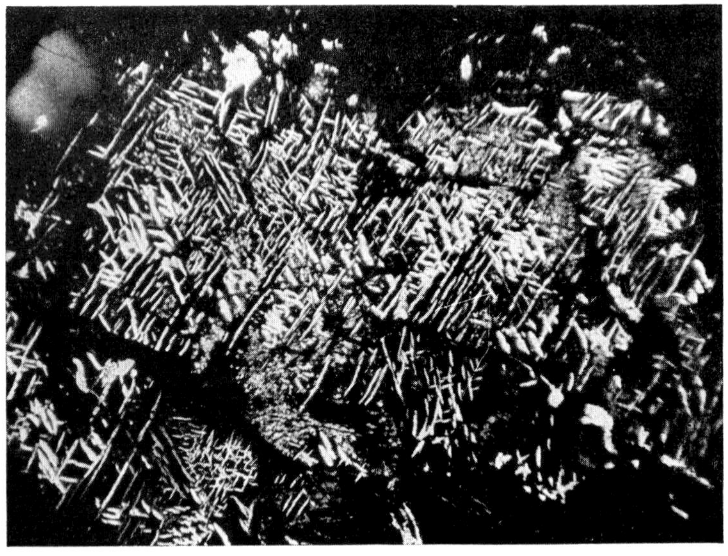

Fig. 178b Same; crossed nicols RAMDOHR

Fig. 179 800 × (approx.), imm. RAMDOHR
Mortsbäkken, Norra Storfjället, Västerbotten

Exsolution of *rutile* in *hematite*. Section approximately // (0001); the rutile crystals are perhaps embedded // (22$\bar{4}$3). Twin lamellae in hematite, one large rutile grain

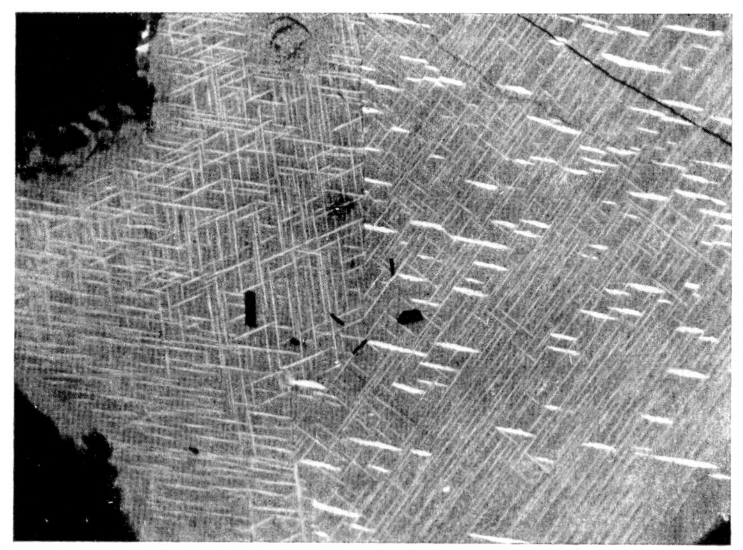

Fig. 180 a 250 ×, imm. RAMDOHR
Ida Mine, Namib, South-West-Africa

Exsolution of *chalcopyrite* in *bornite*. The lamellae are in three directions following the (pseudo-) cube of bornite. A contact between grains of two different orientations passes through the approximate center of this figure

Fig. 180b 250 × RAMDOHR
Lauterberg, Harz, Germany

Exsolution of *chalcopyrite* in *bornite*; the exsolution occurs in part in the form of coarse lamellae, and in part as finest dust. In approximately one half of the section the lamellae have become patchy as a result of "collection crystallization"

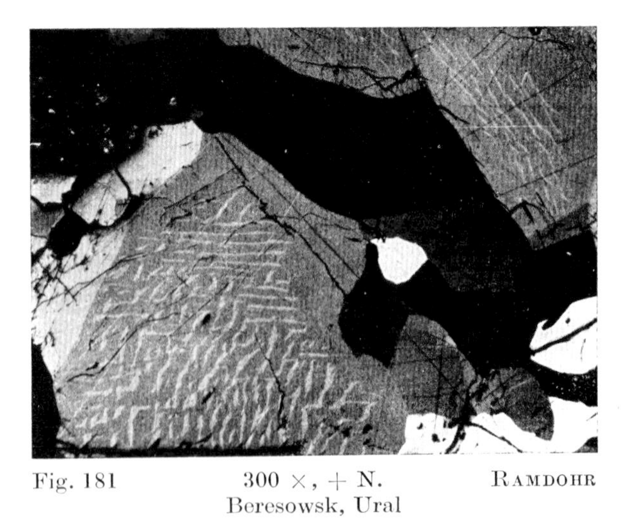

Fig. 181 300 ×, + N. RAMDOHR
Beresowsk, Ural

Patrinite with exsolution network (zonal texture, only in center of grain) of probably *emplectite*

FeS exsolutions would have been expected, appear to lack these. It was found that the exsolutions had "run together" into crescent-shaped marginal areas of the sphalerite joining galena and other ore minerals. Analogous examples can be cited for sphalerite-chalcopyrite, chalcopyrite-stannite, and others.

Examples of exsolutions

Without claiming to be complete, presumably certain examples of unmixing from the literature and from the writer's experience will be presented. Cases in which the supporting data are too doubtful have been purposely omitted; the literature is burdened with a number of obvious misinterpretations.

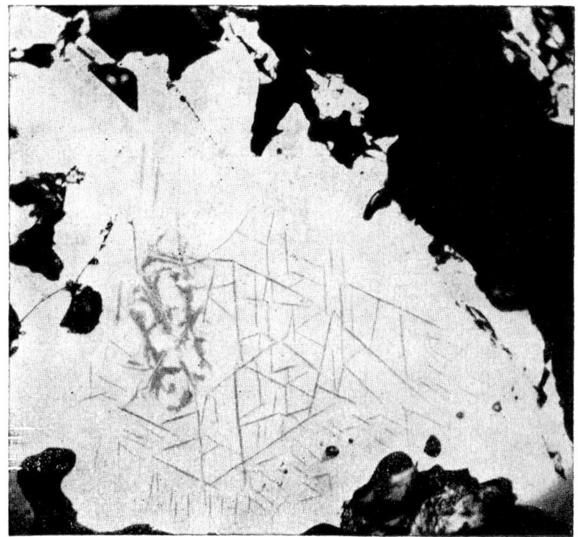

Fig. 182 500 ×, imm. STUMPFL
From a black-sand concentrate

Exsolution of *rutile* in *hematite* probably according to (22$\bar{4}$3). Part of the lamellae has been converted into a peculiar filigree texture, a result of collection crystallization

In the list of examples, for which more details are available in the descriptive part of the book, the predominant component of the decomposition product is given first and arranged according to the sequence adopted in this book. A great number of examples are known where the ratios of the amounts of the components alternate; they are then listed twice and an appropriate reference is made. If such reference is lacking, it means that in the unmixing association for instance predominant sphalerite occurs with minor chalcopyrite and also chalcopyrite with minor sphalerite, *but that to this date* ratios of approximately 1:1 are not known, even though they may be expected from synthetic work at high temperature. — In many minerals the structure of the exsolution is variable, depending on origin, rate of growth, solution partners and especially relative amounts; nevertheless the nature of the intergrowth can be characteristic. This is indicated in column 3. — Only for rare occurrences, or in the literature unknown or barely known examples, are writer (sometimes with appropriate reference year) and locality given. The column "relative amount, guest:host" presents, of course, only approximate relationships.

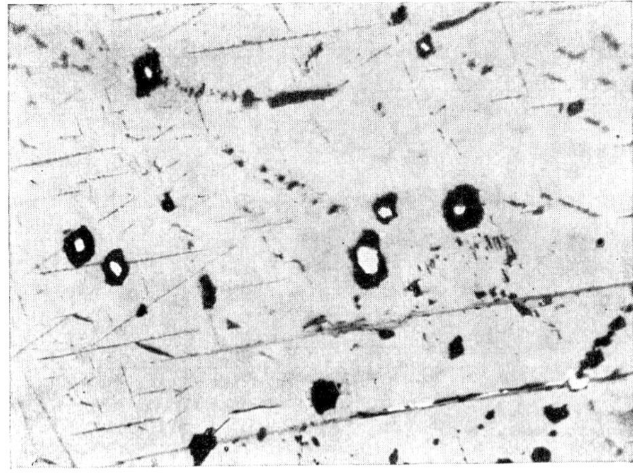

Fig. 183 a 500 ×, imm. RAMDOHR
Solnördal, Norway

Magnetite with exsolution of *ilmenite* in the form of lamellae and equant grains of *spinel*. The coarser spinels commonly begin their exsolution on "spheres" of sulphides

Fig. 183 b 250 × RAMDOHR
Chongwe River, Rhodesia

Chalcopyrite, main mineral, contains stars of *sphalerite* which are often formed on inclusions, e.g., on a bent lamella of *molybdenite*

No.	Predominant ("Host")	Subordinate ("Guest")	Characteristic type of intergrowth	Rel. amount Guest: Host	Writer and locality	Remarks
1	Silver	Dyscrasite	oleander leaf	all		Actually more complicated
2	Gold	Copper	lattice shaped	1:3	SCHOLTZ, Insizwa RAMDOHR, Mooihoek	decomposition of intermediate compounds
2a	Gold	Bismuth	myrmekitic	1:2	RAMDOHR, Maldon	
3	Ferrite	Cohenite	variable, mostly myrmekitic	all	uncommon, of course, LÖFQUIST and BENEDICKS	as "exsolution" only with small amounts of Fe_3C
4	Kamacite	"Plessite"	lattice shaped	all	generally known in meteorites	
5	Platinum	Iridium	"emulsion" (dispersion) "snakes"	1:4	RAMDOHR, BETECHTIN	
6	Platinum	Iridosmium	(0001)-platelets (// (111)	small	RAMDOHR	
6a	Iridium	Platinum	"sub-exsolution"	1:5	RAMDOHR	
7	Arsenic	Stibarsen	myrmekitic	all }		
8	Stibarsen	Arsenic	myrmekitic	all }	KALB, RAMDOHR,	
9	Stibarsen	Antimony	myrmekitic	all }	QUENSEL, WRETBLAD	
10	Antimony	Stibarsen	myrmekitic	all }		
11	Domeykite	?	oleander leaves	1:2		
11a	Domeykite	brown lamellae	thinly-lamellar	1:10	RAMDOHR, Talmessi	
12	Algodonite	?	—	—	—	
13	Tellurobismuthite	Tetradymite	finely lenticular	all }	RAMDOHR and	frequently also with
14	Tetradymite	Tellurobismuthite	finely lenticular	all }	WATANABE	bismuthinite (ÖDMAN)
15	Argentite	Hessite (?)	emulsion (dispersion)	1:5	RAMDOHR, Chihuahua	
16	Chalcocite	(Neo)digenite	cubic network in which chalcocite lamellae form fillings in digenite	all }	e.g. SCHNEIDERHÖHN, BATEMAN, RAMDOHR	

No.	Predominant ("Host")	Subordinate ("Guest")	Characteristic type of intergrowth	Rel. amount Guest: Host	Writer and locality	Remarks
17	Chalcocite	Bornite	network of variable shape	1:2		to 17: in synthetic products all ratios possible
17a	Chalcocite	Chalcopyrite	—	—	Ramdohr Mt. Lyell-Bonanza	
18	Chalcocite	?Lamellae	discs (001)	1:3	Ramdohr, Scherbina	probably derived from α-Cu_2S
19	(Neo)digenite	Chalcocite	network in which chalcocite lamellae form fillings in neo-digenite	all	—	
20	(Neo)digenite	Covellite	network of (0001)-platelets // (111)	1:2	Ramdohr, Khan	
21	(Neo)digenite	Bornite	cubic network with bornite, plessite	1:2	Schneiderhöhn	forms mostly only on decomposition in β-Cu_2S
22	Eucairite	Klockmannite	—	—	Ramdohr, Umango	
23	Bornite	Chalcopyrite	individual lamellae // (100) or (111), also network	1:3	—	unlimited under artificial conditions
24	Bornite	? pure white mineral	—	—	Ramdohr	
24a	Bornite	Linnaeite	—	—	Edwards	uncommon
24b	Bornite	Fahlore	—	1:10	Edwards Pine Valley	rare
24c	Bornite	Chalcopyrite + Idaite	grains	1:5	Ramdohr, Natas	exsolution of similar age
25	Pentlandite	Pyrrhotite	emulsion (dispersion)	1:4	Ramdohr, Ehrenberg	
26	Pentlandite	Chalcopyrite	cellular-netshaped	1:10	Ramdohr, Pauly Igdlokunguak	uncommon as exsolution; similar and more common as a replacement

27	Pentlandite	Linnaeite	lamellar // (100)	1:3	—	addition of material (?)
28	Pentlandite	Mackinawite	net-shaped-lamellar	1:4	SCHNEIDERHÖHN, meanwhile repeatedly reported, especially in S. African Pt-ores	
29	Sphalerite	Pyrrhotite	—	—	STILLWELL, RAMDOHR	
29a	Sphalerite	Mineral similar to sphalerite	emulsion (dispersion)		RAMDOHR, Zinnwald	
30	Sphalerite	Chalcopyrite	emulsion (dispersion)	1:4	—	
31	Sphalerite	Chalcopyrite + Pyrrhotite + Cubanite $\pm$ Mackinawite	emulsion (dispersion)	1:4	—	
31a	Sphalerite	Chalcopyrrhotite	emulsion	$\sim$1:5	RAMDOHR, many places	
32	Sphalerite	Fahlore	lamellar	1:5	—	very rare!
33	Sphalerite	Stannite	emulsion (dispersion)	1:10	—	
34	Sphalerite	Isostannite	platelets // (100)	1:5	RAMDOHR, Tsumeb	apparently closely-related to No. 33
34a	Sphalerite in Germanite	Gallite	platelets // (100)	1:4	RAMDOHR, GEIER and STRUNZ	
35	Chalcopyrite	Bornite	—	—	LATSKY, Ookiep	
36	Chalcopyrite	Pyrrhotite	—	—	STEVENSON	confusion with chalcopyrrhotite ?!
37	Chalcopyrite	Millerite	rare!	1:6	RAMDOHR	
38	Chalcopyrite	Fahlore	finest emulsion (dispersion)	1:10	RAMDOHR, Bliesenbach and Mt. Lyell, EDWARDS, Thologolong	very rare
39	Chalcopyrite	Mackinawite	irregular "whips", also skeletons	$<$1:10 1:1	SCHNEIDERHÖHN	extremely common only in chalcopyrite grains in sphalerite
40	Chalcopyrite	Chalcopyrrhotite	irregular spots	1:2	RAMDOHR, Querbach, Insizwa etc.	

No.	Predominant ("Host")	Subordinate ("Guest")	Characteristic type of intergrowth	Rel. amount Guest: Host	Writer and locality	Remarks
41	Chalcopyrite	Cubanite	thin platelets // (111)	all	Borchert	very frequent
42	Stannite	Sphalerite	emulsion (dispersion)	1:5	—	
43	Stannite	Chalcopyrite ± Sphalerite	emulsion (dispersion)	1:5	Reinheimer, Schneiderhöhn, etc.	
44	Stannite	Chalcopyrite	emulsion (dispersion), rarely network	1:5 1:2	— Ramdohr, Schwarzenberg	apparently high-temperature
45	Stannite	Hexastannite	complex lamellae // pseudocubes	1:4	Ramdohr	
46	Stannite	Isostannite	lamellae (100), also quite irregular	1:2	Ramdohr	
46a	Stannite	Cubanite	small E-particles	—	Ramdohr	
47	"Brown stannite"	Fahlore	network // (100)	all	Ramdohr, Zinnwald	
48	Fahlore	Chalcopyrite	lamellar network	1:5	Ramdohr, Urique	rare, but with unusual fahlore varieties quite typical
49	"Fahlore" (colored) similar to bornite)	Fahlore + ? red-brown ore + sphalerite	network // (100) (?)	1:3	Ramdohr	
50	Fahlore	? Mineral	emulsion (dispersion) extremely fine	1:10	Ramdohr, Gardsjö	
51	Fahlore	Cinnabar	emulsion (dispersion)	1:5		perhaps first induced by weathering
52	Colusite	Tetrahedrite	—	—	Landon, Butte	
52a	Germanite	Renierite	—	1:4	Ramdohr, Tsumeb	
52b	Germanite	"grey mineral"	—	quite variable	Ramdohr, Tsumeb, variable in different zones, Landon, Butte	
53	Enargite	Chalcopyrite	—	1:10	Ramdohr	rare

54	Pyrrhotite hexag.	Pyrrhotite monocl.	lenses // (001)	all	VAN DER VEEN, RAMDOHR, KISKYRAS, SCHOLTZ, etc.	
55	Pyrrhotite	Pentlandite	myrmekitic // (0001) ("flames")	1:5	SCHNEIDERHÖHN, EHRENBERG, HEWITT	see p. 600
56	Pyrrhotite	Chalcopyrite	—	1:4		only high-temperature magmatic rocks which cooled fast!
57	Pyrrhotite	Valleriite	platelets // (0001)	1:10		very rare
58	Pyrrhotite	Magnetite	platelets	1:10	in meteorites, RAMDOHR	very striking as exsolution; another interpretation does not seem possible!
59	Niccolite	Breithauptite	myrmekitic	all	RAMDOHR, Les Chalanches	uncommon, mixed crystals often preserved
60	Cubanite	Chalcopyrite	irregular and pseudocubic, tabular	all	—	
62	Argentopyrite	Pyrrhotite	in part orientated and in part unorientated but uniform network	1:10	RAMDOHR	Pyrrhotite, almost always converted to pyrite
63	Galena	Tetradymite	(0001)-platelets // (111)	1:5	ÖDMAN, RAMDOHR	primary intergrowths are similar but much coarser
64	Galena	Argentite	emulsion (dispersion)	1:10	SCHNEIDERHÖHN	
65	Galena	Schapbachite	lamellae // (100)	1:10	RAMDOHR, Schapbach, Zinnwald, Bustarviejo	
65a	Galena	Galenobismutite	—	—	BERRY	
66	Galena	Polybasite	platelets // (100)	1:5	RAMDOHR	
67	Altaite	Aguilarite	—	—	STILLWELL, Kalgoorlie	
68	Alabandite	Pyrrhotite	(0001)-platelets // (111)	1:10	RAMDOHR	
69	Schapbachite	Galena	network // (111) with filling of plessite	1:2	RAMDOHR Schapbach, etc.	
70	Bismuthinite	Emplectite	lattice shaped	1:2	RAMDOHR	

No.	Predominant ("Host")	Subordinate ("Guest")	Characteristic type of intergrowth	Rel. amount Guest: Host	Writer and locality	Remarks
70a	Bismuthinite	Argentite	—	—	SCHWARTZ	
71	Patrinite	?Emplectite	—	—	Beresovsk	in part formed upon heating
72	Geocronite	?Boulangerite	thin plates (010)	1:5	Waldsassen	
73	Polybasite	?Mineral	—	<1:10	RAMDOHR, Schemnitz	
74	Linnaeite	Chalcopyrite	network	1:2	—	
75	Linnaeite	Millerite	network // (100) (or (111)	1:3	—	
76	Pyrite	?Mineral	—	1:5	—	
77	Villamaninite	Chalcopyrite + Linnaeite + others	—	—	—	no true exsolution, complete disintegration
78	Blockite ("Penroseite")	?Mineral	emulsion (dispersion)	1:3	RAMDOHR, Pacajake	
79	Smaltite	Anisotropic mineral	zonal	—	—	
79a	Smaltite	probably Rammelsbergite	finest leaves, probably // (100) and (111)	1:3	RAMDOHR, Allemont	
80	Zincite	Hausmannite	complex platelets (0001)	1:5	FRONDEL, Långban	
81	Periclase	Manganosite	ordered emulsion	1:10	—	
82	Manganosite	Zincite	(0001) platelets // (111)	1:10	RAMDOHR, FRONDEL	
83	Spinel	Magnetite	cellular network, mostly emulsion	all 1:10	RAMDOHR, many writers	very high temperature, moderate temperatures uncommon
84	Spinel	Ilmenite	(0001)-platelets (111)	—	RAMDOHR, Ronsberg	
85	Magnetite	Spinel (± ilmenite)	(100)-nets (ilmenite as usual)	1:5 rarely more	RAMDOHR and others	common
86	Magnetite	Ulvöspinel	network // (100)	1:3, or even more	MOGENSEN, RAMDOHR	widely distributed

87	Magnetite	Franklinite	nets // (100)	1:5 to 1:1	Ramdohr	
88	Magnetite	Ilmenite	(0001)-platelets // (111)		—	
89	Magnetite	Geikielite	(0001)-platelets // (111)	1:5	Ramdohr, Nittis	
90	Magnetite	Pyrophanite	(0001)-platelets // (111)	1:10	Ramdohr, many times	
91	Chromite	Hematite	(0001)-platelets, in part in closed networks	1:5	Ramdohr, Mooihoek	as a rule unommon in rocks
92	Chromite	Ilmenite	(0001)-platelets // (111)	1:10	Ramdohr	
93	Franklinite	Zincite	(0001)-platelets // (111), also irregular	1:10	Ramdohr	
94	Franklinite	Spinel	network // (100)	1:2	Ramdohr	
95	Franklinite	Magnetite	network // (100)	1:4	Ramdohr	
96	Franklinite	Hetairolite	nets, lamellar	—	Mason, Ramdohr	"Zn-Vredenburgite"
97	Franklinite	Hausmannite	nets, lamellar	all (?)	Mason	"Franklinite-Vredenburgite"
98	Franklinite	Pyrophanite	(0001)-plates // (111)	1:10	Ramdohr	
99	Jakobsite	Hausmannite	nets, lamellar	—	many writers	"Vredenburgite"
99 a	Hausmannite	Jakobsite	network (100)	all	many writers, especially Mason	
99 b	Davidite	Rutile	irregular grains	< 1:4	Ramdohr	
100	Braunite	Hausmannite	network	1:5	—	
101	Braunite	Hematite	finely-lamellar network	—	Ramdohr, "Argentine"	
102	Braunite	?Mineral	finely-lamellar network	—	Ramdohr, "Argentine"	
103	Corundum	Hematite	(0001) lamellae	< 1:10	J. M. Bray	
104	Hematite	?Hematite	discs // (0001)	1:3	Ramdohr and others	unexplained, since it occurs even in extremely pure Fe_2O_3
105	Hematite	Ilmenite	discs // (0001)	all	—	commonly two generations
105 a	Hematite	Braunite (?)	—	—	Ramdohr, Otjosondu	

No.	Predominant ("Host")	Subordinate ("Guest")	Characteristic type of intergrowth	Rel. amount Guest: Host	Writer and locality	Remarks
106	Ilmenite	Magnetite	thin plates (111) // (0001)	1:5	—	apparently all ratios; high magnetite contents depend on replacement of hematite
106a	Ilmenite	Magnetite + Ulvö-spinel	—	—	RAMDOHR	
107	Ilmenite	Corundum	thin plates (0001)	1:5	RAMDOHR, Kussa, Routivare, Solsta	
108	Ilmenite	Hematite	discs (0001)	all	—	often 2 generations
109	Ilmenite	Rutile	plates probably // ($22\bar{4}3$)	1:10	EDWARDS, RAMDOHR, SCHERBINA	
110	Ilmenite	Gangue mineral	thin plates	1:5		
110a	Manganite	?Mineral	emulsion (extremely fine)	1:10	RAMDOHR, Saarbrücken	
111	Rutile	Hematite	—	—	RAMDOHR, SCHERBINA	
112	Rutile	Ilmenite	lamellar (cf. p. 1008)	1:10	RAMDOHR, Bodenmais, St. Yrieux, Ohlapian	
113	Ilmeno-rutile	Columbite (?)	emulsion, irregular	—	RAMDOHR, Iveland	
113a	Ilmeno-rutile	Rutile	fine lamellae	1:4	RAMDOHR, several localities	
114	Cassiterite	Tapiolite	quite irregular but parallel axes	< 1:10	RAMDOHR	
114a	Cassiterite	Tantalite	rounded particles	1:10	EDWARDS	
114b	Cassiterite	Columbite	zonal, rounded	1:2	RAMDOHR, Nigeria	
115	Cassiterite	?Mineral	needles // (902)	1:10	MÜGGE	
116	Tantalite	Ilmenite	irregular spots cryst. directions	1:3	EDWARDS	
116a	Tantalite	Uraninite	myrmekite like or fine octahedra	1:3	RAMDOHR	

117	Olivine	Magnetite	finest lamellae, common pseudohexagonal or trigonal axes	$< 1:10$	Ramdohr	
117a	Olivine	Chromite	mostly rounded dots	1:10	Ramdohr, common in meteorites	
118	Olivine	Ilmenite	finest lamellae, common pseudohexagonal and trigonal axes	—	—	
119	Garnet	Magnetite	emulsion	$< 1:10$	—	as exsolution not approved
120	Bronzite and Clinopyroxene	Ilmenite (± Magnetite)	emulsion of commashaped particles	1:10	known for a long time	

In my notes and literature excerpts I find the following examples in which replacements have resulted through material introduced by solution, e.g., in example 3, and in some cases even weathering must have contributed. They have in part already been listed under exsolution.

1. Galena + bismuth from cosalite.
2. Galena + bismuth + bismuthinite + cosalite from "chiviatite".
3. Galena with arsenic inclusions from fahlore.
4. Galena + boulangerite + arsenic from geocronite.
5. Galena + realgar from gratonite.
6. Chalcopyrite + gudmundite + pyrrhotite ± arsenopyrite from fahlore.
7. Jordanite from gratonite.
8. Pyrrhotite + gudmundite from fahlore.
9. Linnaeite + bravoite ± chalcopyrite from villamaninite.
10. Klaprothite ± emplectite + bismuth + bismuthinite + chalcopyrite from wittichenite.
11. Cosalite + galena + chalcopyrite + bismuth ± bismuthinite from klaprothite (?).
12. Pararammelsbergite + niccolite from chloanthite.
13. Enargite + arsenopyrite from tennantite.
14. Magnetite + anatase from ilmenite.
15. Hematite + braunite + pyrolusite from bixbyite.
16. Hematite + rutile from pseudobrookite (Havredal).

Figs. 184, 185 and 200 present appropriate examples.

(c) *Decomposition structures*

Decomposition structures which cannot be sharply separated from exsolution structures, as has already been noted, are therefore discussed immediately after exsolution. If we suppose that unmixing in its strictest sense (with the justified reservations made by NIGGLI) requires that host and guest form from the same original mixed crystal, with at least one of the two in crystallographic relationship with it and moreover that the overall chemistry remains unchanged, this supposition strictly is frequently not found to be true. The product can be crystallographically quite unrelated to the original material, the chemistry can be changed to varying degree by introduction or removal, and moreover reactive products can form on decomposition which cause reciprocal attack on the associated minerals. Hence transitions to replacements are also found here, and are especially in the assemblages of lead-, bismuth-, antimony-, and arsenic sulphosalts not separable from them. In appearance they completely resemble exsolutions in the ordinary sense.

(d) "*Verdrängung*" — "*Replacement*" — "*Metasomatism*"

General data. The three terms in the above heading, derived respectively from German, Latin and Greek roots, have essentially the same meaning. "Verdrängung", taken precisely, has the most active meaning: an object (*a*) with application of a certain amount of force places itself in the position of another (*b*) immediately following this application. The two other terms have reference only to the fact that (*b*) is now in the position originally occupied by (*a*) without saying anything about the process and any intermediate stages. In the American geologic literature, and taken from it in the Dutch literature, the term "metasomatic" has taken on, quite unjustified, in the literal sense the meaning of hydrothermal, and i.e. as an active substitution, in the sense of hydrothermal "Verdrängung", whereas the English word "replacement" has retained for the most part, but not invariably, its original colorless meaning.

Some writers, more or less consciously, refer to the original meaning of the word, others use the expressions completely synonymously, while a third group thinks of all three terms as describing a hydrothermal (or even magmatic), definitely epigenetic, and therefore distinctly later process. For this reason discussions in the literature arise where almost no differences of opinion exist. Conversely, because of convergence of the terms, several different observations or concepts may frequently have received an identical label; but this has not often been brought out in the open!

To prevent erroneous conclusions, the term "Verdrängung" will be used *throughout* the ensuing discussion irrespective of whether substance (*b*) substitutes for (*a*) immediately, or perhaps at a later geologic time indirectly, or whether the processes are hydrothermal, magmatic "autometamorphic", or metamorphic, or the result of weathering; moreover irrespective of whether the process is characterized by complete or partial exchange of material or absence of such an exchange. The observation that even exsolution processes occasionally presuppose "replacement" has been stated in the discussion of exsolution.

It is already apparent from this introduction that both according to the process as well as according to the structural characteristics "Verdrängung" can be extremely variable. At first the most frequent "Verdrängung" structures will be described and illustrated, without attempt at exhaustive treatment. The genetic significance of certain structures and conversely their interpretation for explaining the formation of the ore deposit will be deferred till later. *It is most emphatically pointed out that many textures can form in a variety of ways and that even where at present only one mode of formation is known, the possibility of several modes of formation exists.* "Multa fiunt eodem, sed non semper in eodem modo!" ("Many things are formed to look the same, but not always in the same way!").

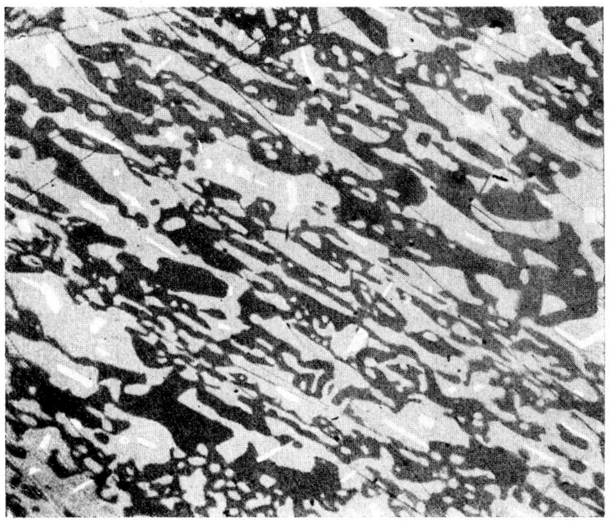

Fig. 184 250 ×, imm. RAMDOHR
Falun, Sweden

Intergrowth of *bournonite* (grey), *galena* (white), and *native arsenic* (light white) formed from geocronite

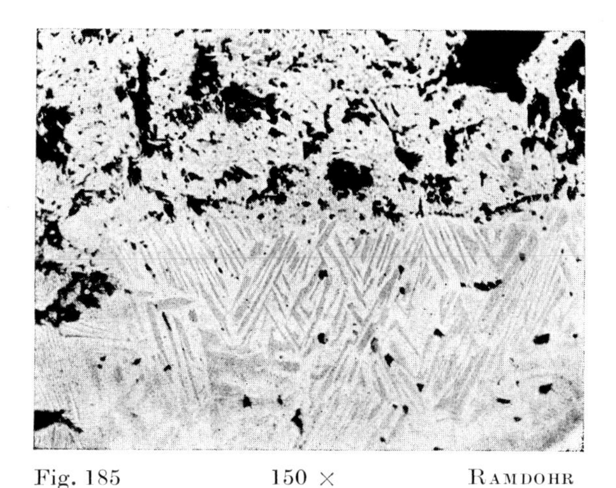

Fig. 185 150 × RAMDOHR
Havredal, Norway

Former *pseudobrookite* decomposed into two sharply-separated forms of myrmekitic intergrowth in *hematite* (white) and *rutile* (grey)

GRIGORIEFF (1928) has essentially recognized this and has worked out a classification in which the textural forms recur in part several times as subgroups within the genetic master groups. He states emphatically that metasomatic processes (in the sense commonly used in the U.S.A.) do not necessarily require replacement textures.

His geometric sub-groups of the genetically conceived master group "replacement" are:

Filiform	"Verdrängung" (replacement) in the form of a network of finest veinlets.
Cellular	where only small "Verdrängung" (replacement) relicts have been preserved.
Shredded	"Verdrängung" (replacement) relicts in angular shreds, often with concave sides.
Skeleton-shaped	where marginal parts are in part preserved
Graphic	allegedly very abundant*
Lattice-shaped	where "Verdrängung" (replacement) prevailingly follows crystallographic orientations.
Zonal	(allegedly very rare as a replacement structure).
Dendritic	where calcite has been replaced along cleavage planes.
Cement shaped	where, e.g., the intergranular cement of a sandstone has been selectively replaced.

This classification is not exhaustive and is burdened by a preconceived genetic opinion; nevertheless it offers a useful framework to which the following will refer to a limited extent. As always, it has to be understood that all transitions between the different geometric patterns exist, and that each attempt at classification is therefore somewhat arbitrary. Reference has also been made to the difficulty, in the discussion of the different groups of structures which is to follow, that each structure can form by processes other than by exchange of material. It should be noted at the outset that cataclasis, dynamometamorphism and recrystallization, unmixing, "inhibited growth", formation of porphyroblasts, skeletal growth, infiltration of void spaces and healing of fractures, and a pronounced tendency to idiomorphic development in general, perhaps also other processes, can simulate replacement textures.

Even the course of crystallization of the simplest binary system from a melt, such as the solid solution series of plagioclase, or the system with an incongruently melting binary compound such as olivine–quartz, requires that the crystals which have been precipitated first are attacked by the rest melt and are replaced by solid solution crystals of different composition or by another type of crystal possibly with excellent replacement structures. In ternary systems, particularly in those with volatile constituents, and especially in multi-component systems, such replacements can occur with recurrent formation of the crystal which was at first resorbed. Phases of different composition can form successively, such as salts in different stages of hydration, or sulphosalts with increasing antimony or lead content, with complete or partial resorption of phases which were initially precipitated.

In all these cases there is neither an indication of a hiatus in the supply of material (such as a change in the solutions bringing in the ore), nor of an especially rapid

* The allegedly means the observations of the author differ distinctly

temperature decrease, nor of rejuvenation, and certainly not of "epigenetic" formation. This is overlooked frequently, and continuous processes are considered as discontinuous in the most diverse ways. Much divergency of opinion on the origin of the Sudbury ore deposits, especially that of the "offsets", can be settled with the same explanation as that of hornblende formation around augite or the replacement of the latter in gabbros, or more generally the corona formation in many magmatic rocks. Of course, subsequent introduction of material, sudden temperature drops, changes in the ore solution, or renewed raising of the temperature are not excluded, but surely often take place. Nowhere is it at first clear whether the patterns are ascendent or descendent ("hypogene" or "supergene"), genetically related or independent, or whether they have formed with or without the addition or removal of material. Likewise low-temperature associations which are normally characteristic for the zone of cementation may be in direct sequence hydrothermal, even high-temperature hydrothermal, in origin, and conversely low-temperature structures, such as those found in the zones of cementation and oxidation, may contain, although only under unusual circumstances, minerals and textures which are normally considered unhesitatingly as "hydrothermal".

The genetic interpretation of replacements is a task which requires much experience and often intuition, even where process and paragenetic sequence are established without doubt!

Very much has been written about the exchange of *material* in replacements, but the literature is also burdened with many errors. The rule that the "more noble" element dissolves out the "less noble" from its compounds, e.g., that argentite normally replaces chalcocite, suffers from so many actual or apparent exceptions that caution is nescessary in its evaluation. A great excess of the less noble cation can explain even at lower temperatures a reversal of this sequence. It has been established beyond doubt — although it is very uncommon — that even in the "zone of cementation" (fig. 237) chalcocite has been replaced by bornite and bornite by chalcopyrite. At more elevated temperatures, in addition to the concentration of the solutions, reference must be made to the temperature dependence of the concept "noble" which means that even at constant concentration the compound of one element may replace another or be replaced by it. The very strong change of the pH in the normal sequence, e.g., from pneumatolytic to high-temperature hydrothermal and then to low-temperature hydrothermal, results likewise in surprises. Also in the relationship of ore minerals to gangue, such as in the frequent and most surprisingly selective and reciprocal replacements of quartz and silicates by sulphides and especially by oxides at both high and low temperatures, one has to be alert against preconceived opinions, otherwise it is easy to make mistakes in the interpretation of the paragenetic sequence and other relationships!

It is beyond the scope of our task to consider more closely the significance of replacement in ore deposits and to go further into the chemical relationships; this would require hundreds of pages without being even remotely exhaustive. We are limiting ourselves therefore in both description and in the figures to the presentation of examples which include some of the most well known occurrences, but also purposely those with which the beginner is not familiar and even exceptions and rarities.

The mechanism is often very easily understandable: solutions, in very rare cases even melts, circulate along zones of weakness, such as: fissures, faults, thin hair cracks between the individual mineral grains or within a single grain; they search for access into the crystal lattice of the individual grains which often but not necessarily coincides with the cleavage, and seek for latent tension, and finally even proceed by "frontal" movement, i.e., without such zones of weakness.

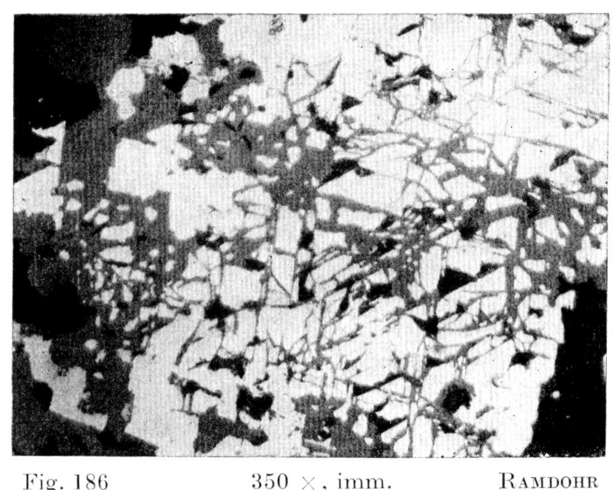

Fig. 186 350 ×, imm. RAMDOHR

Pyrite (white) replaced by *sphalerite* (dark grey) along (100) cleavage

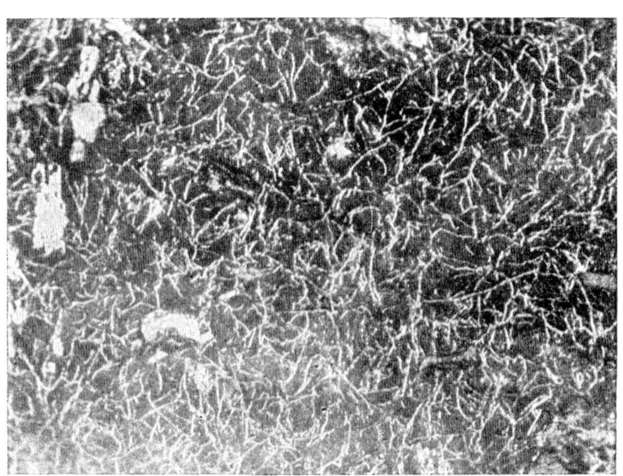

Fig. 187 350 ×, imm. RAMDOHR
Namib Mine near Swakopmund

Filigree of manganese ores in carbonate gangue

Unexpected complications may set in which are difficult to interpret. Minerals which are known to be easily replaceable may appear to prove themselves replaceable only with difficulty. The explanation may be that the solutions themselves contain abundant ions of the mineral concerned. More difficult to understand are cases, where minerals are attacked which are known to be chemically particularly resistant to replacement (chromite, quartz, ilmenite), especially if they are replaced in preference to other ore minerals. This presumes solutions of unusual composition, and often also enormously long duration of exposure.

The observation that certain silicate minerals are replaced, while at the same time others of very similar composition remain intact, may, in the course of future investigations, provide information on the nature of the ore solutions.

The *forms of the replacement structures* are the result of many circumstances, such as cleavages, planes of translation, hardness, plasticity, chemical stability in the lattice structure and lattice bonding, and temperature, duration of reaction, tectonic stresses in the geologic history preceding the reaction, and *p*H (acid or basic character), combination of elements, and relative concentration which lie in the nature of the reacting residual or foreign solutions. Only in a few cases can these factors be differentiated by their effects, even though some rough rules exist. Instead of detailed descriptions, reference is made to the diagrams and pictures.

It is most emphatically stated again that the designations applied here to the types and forms are by no means exhaustive, and frequently not very fortunate, and that the writer only singles out examples, without on his part, increasing the jumble of the nomenclature.

1. *Filigreeform networks* are those forms which occur especially with ore minerals lacking definite cleavage. They are favored by cataclasis or latent tensions. Of course, hidden cleavages can lead to preference in one directions (figs. 187, 188).

2. *Cellular-island shaped* is probably the most pronounced and most frequent replacement form. The part made up by replacement relicts is variable, in general it is not high, since this group results partly from groups 4, 6, 7, 8 and also in part from 9. Originally well-pronounced preferential directions may have been present; later these disappear to a large extent. This is illustrated in diagrams (figs. 211–217) as also in part in fig. 189–206.

The somewhat unfortunate designation of "exploded bomb" texture for cementative pyrite replacement represents a special case.

In the American and English literature the very appropriate term "carieslike" has come to stay for incipient replacement which ultimately leads to type 2. Of course cleavages and fractures which guide the replacements modify the textures (cf. figs. 191, 195 and others).

3. *Shredded textures* in their final form can resemble closely the islandform structures, but in their typical stage they leave concave replacement relicts (shreds), in contrast to the convex ones for the islandform structures. The original mineral is often characterized by a lack of distinct cleavage, however, latent tensions could have existed. The forms and their characteristic development are shown in fig. 349, and in fig. 213 which is a diagram made from fig. 349, as well as in fig. 189.

4. *Skeleton-shaped textures* appear especially when originally the crystal had developed a skeletonlike framework by accelerating the growth of corners and edges, and the wedges of the framework accreted somewhat later and for the most part fast. Especially galena and pyrite of low-temperature deposits tend toward this growth and therefore to such replacement textures. Moreover cases belong here where preferentially impure interior portions of crystals which are rich in inclusions of mother liquor or which have grown rapidly are more susceptible to replacement than the pure marginal parts of the crystals. Transitions to this zonal type also exist (figs. 223, 205, 218).

5. *Graphic textures* which according to my experience form considerably more infrequently in replacements than is commonly assumed, are expected especially

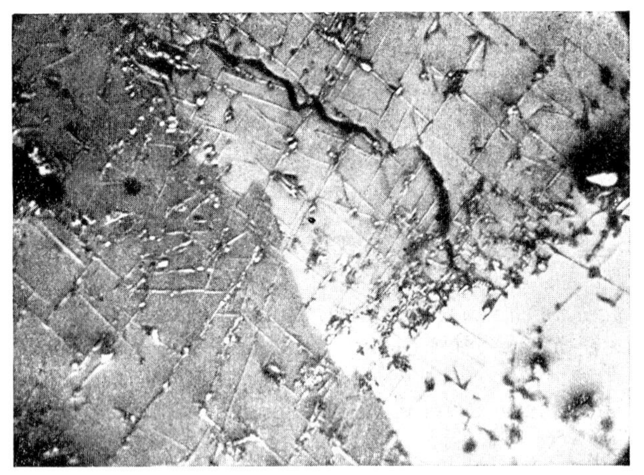

Fig. 188 500 ×, one nicol, imm. RAMDOHR
Freiberg in Saxony, Germany

Argentopyrite with *pyrite* network as forerunner of decomposition in early weathering. "Filiform" texture. Note strong bireflection

Fig. 188E Sunshine Mine, Idaho RAMDOHR
Magn. 225 ×, imm.

Tetrahedrite infiltrates and replaces *pyrite*

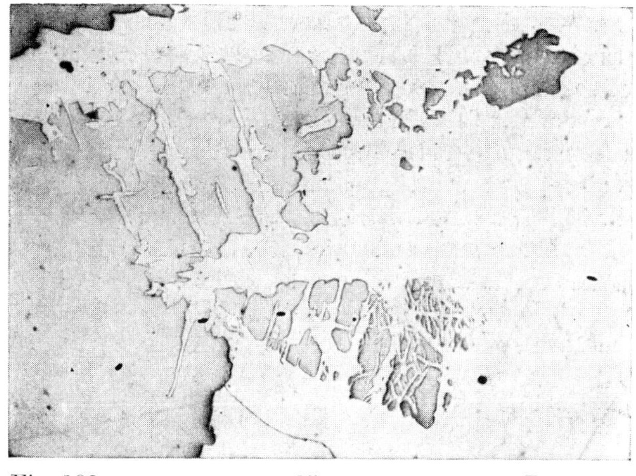

Fig. 189 65 × RAMDOHR
Lauterberg, Harz

Sphalerite (grey, strong relief) replaced by *bornite* (white). "Shredded" to "island shape" texture. A few grains of *chalcocite* (light white)

Fig. 190 250 ×, imm. RAMDOHR
Temagami Copper Mine, Ontario

Sphene with a little *rutile* formed from ilmenite skeleton of titanomagnetite. This is an example where "leucoxene" does really consist of sphene

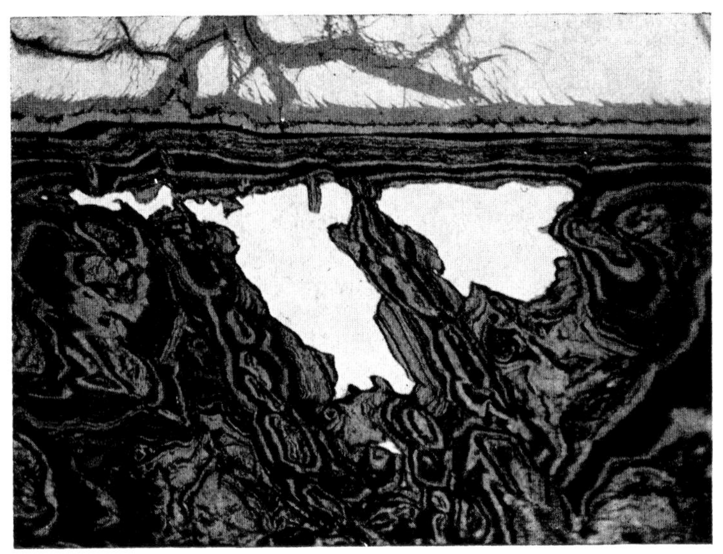

Fig. 191 250 ×, imm. RAMDOHR
Cligga Head, Cornwall

Arsenopyrite (white) entirely surrounded by *chalcopyrite* (at the top) is much quicker oxidized than the latter. *Goethite* (Nadeleisenerz) and *lepidocrocite* are formed in variable ratios, in addition perhaps scorodite and others Only few very strongly reflecting relicts of arsenopyrite remain

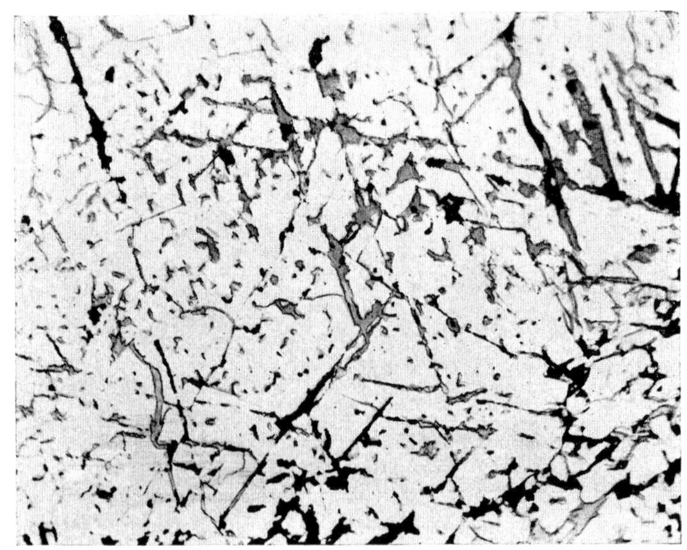

Fig. 192 250 ×, imm. RAMDOHR
Georg Mine near Horhausen, Nassau, Germany

Cubic fractures in *pyrite* are filled with *bournonite*, some *chalcopyrite* and gangue (black)

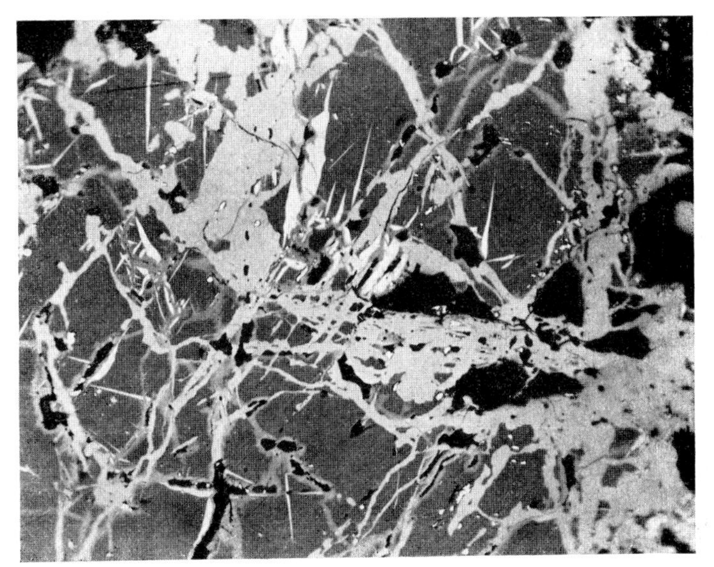

Fig. 193 350 ×, imm. RAMDOHR
Saalfeld in Thuringia

Complex replacement: *bornite* (dark grey) is veined through and replaced by *fahlore* (light grey); later *chalcopyrite* formed (white spindles). All three are veined by narrow stringers of *chalcocite* (medium grey). Gangue and holes are black

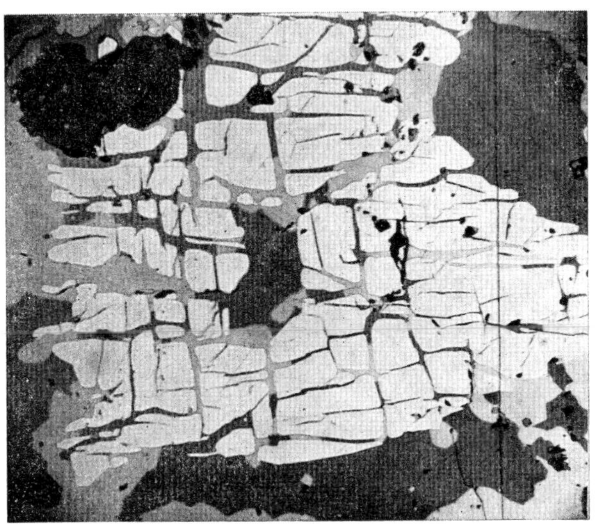

Fig. 193a 225 × RAMDOHR
Mt. Lyell, Tasmania

Linnaeite (white) replaced, following an otherwise nearly not visible cleavage, by *chalcopyrite* (greyish white) and *bornite* (dark grey). Black is *quartz*

Fig. 193E 20 × RAMDOHR

Umkanto, S. Rhodesia

Former crystals of *pyrite* are first entirely replaced by *chalcopyrite* (white) which, in its turn, is replaced by *bornite* and surrounded by thin rims of *goethite*

Fig. 194 150 × RAMDOHR

Carrick Dhu-Quarry, Cornwall

Very peculiar replacement! *Chalcopyrite* with well-developed small exsolution bodies of *stannite* is replaced by perfectly idiomorphic *quartz*. The observation that the stannite bodies are completely intact and in their old orientation within quartz provides the only clue, but a conclusive one, of the replacement nature of the quartz. Both the chemistry and mechanism of this sparing and selective replacement are enigmatic

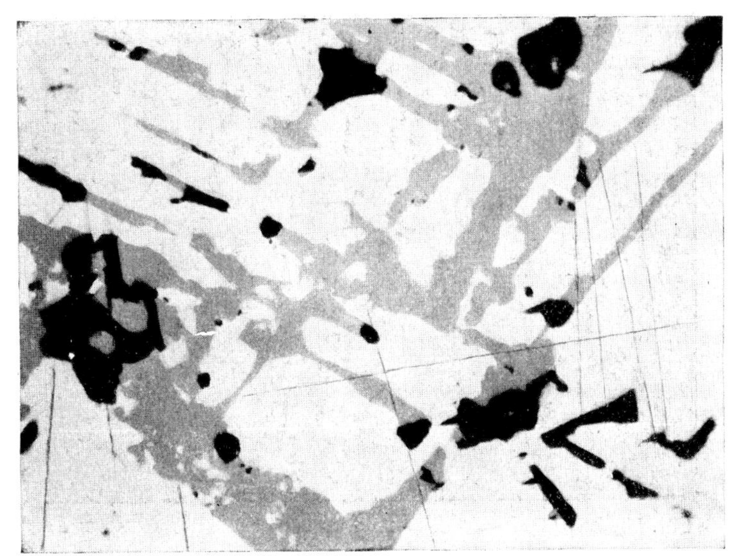

Fig. 195 250 ×, imm. RAMDOHR
Zinnwald, Erzgebirge

Fahlore (grey) replaces *galena* (white) (undoubtedly ascendent) along (100) (cleavage cracks are black)

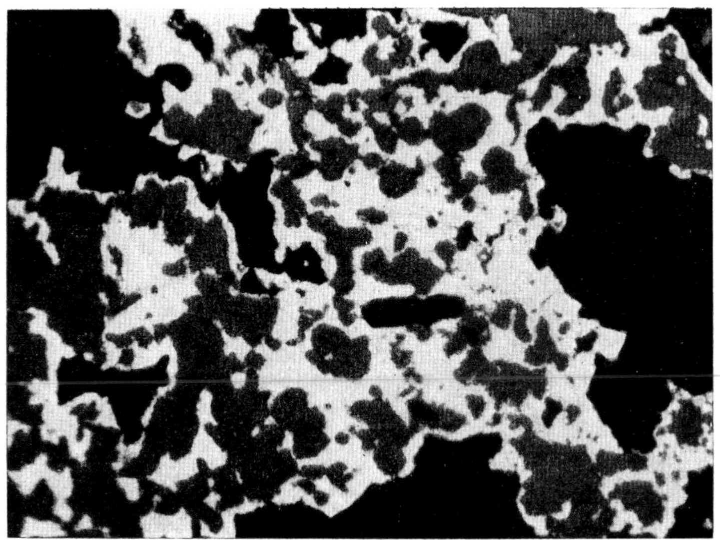

Fig. 196a 250 ×, imm. RAMDOHR
Shensei Gang, Ashio Mine, Japan

Stannite (light grey) replaces *cassiterite* (dark grey). The solutions responsible for replacement have apparently advanced along the grain contacts between cassiterite and *quartz* (black). *Chalcopyrite* (white) is found locally

Fig. 196 b 250 ×, imm. RAMDOHR
Carn Brea, Cornwall

Bornite and *hematite* as decomposition products of *chalcopyrite*. Hematite lamellae make up a network parallel to (111). This relationship between these minerals must be considered an ascendent oxidation reaction, but may in other analogous cases perhaps be descendent:

$$\sim 5\,CuFeS_2 + 18\,O = Cu_5FeS_4 + 2\,Fe_2O_3 + 6\,SO_2$$

Fig. 197 150 × RAMDOHR
Outukumpu, Finland

Chalcopyrite (dull white), some *sphalerite* (very dark grey) and *pyrrhotite* (light grey) replace *pyrite* (pure white)

Fig. 197a 315 ×, imm. SCHIDLOWSKI
Loraine Gold-Mine, Oranje Free State

A *chromite* grain with unusual unmixings is broken between several grains of *pyrite*

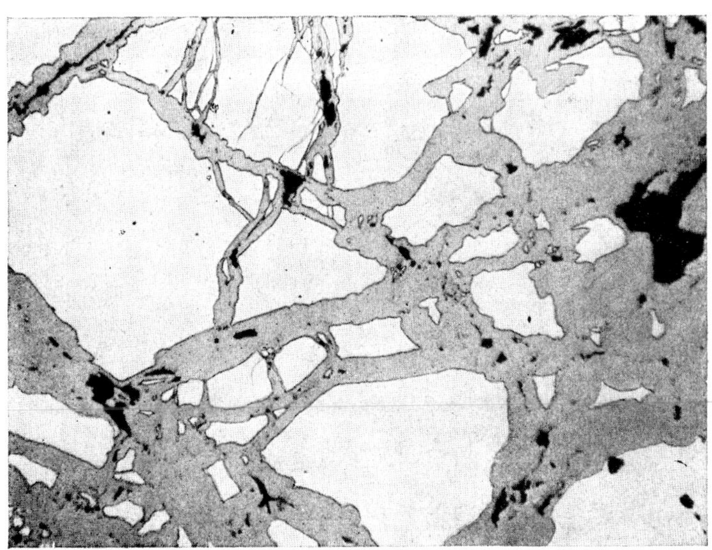

Fig. 198 70 × RAMDOHR
Morenci, Arizona

Pyrite is replaced along fractures in all stages by descendent Cu_2S

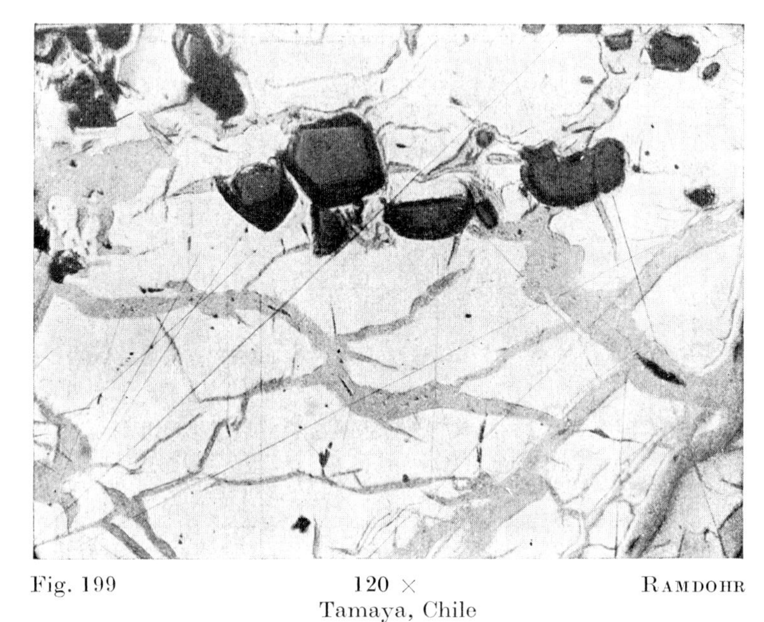

Fig. 199 120 × RAMDOHR
Tamaya, Chile

Typical cementation structure. Descendent replacement of *chalcopyrite* (pure white) by *chalcocite*. "Cellular". Grey idiomorphic grains of *quartz*

Fig. 200 250 ×, imm. RAMDOHR
Silver Bar Mine, Cobalt, Ontario

Chloanthite (grey), large strongly zoned crystal is zonally replaced by *pararammelsbergite* (white) which occurs in part in disordered idioblasts and in part in lamellarlike (for the most part much smaller) crystals embedded // (100) of the chloanthite

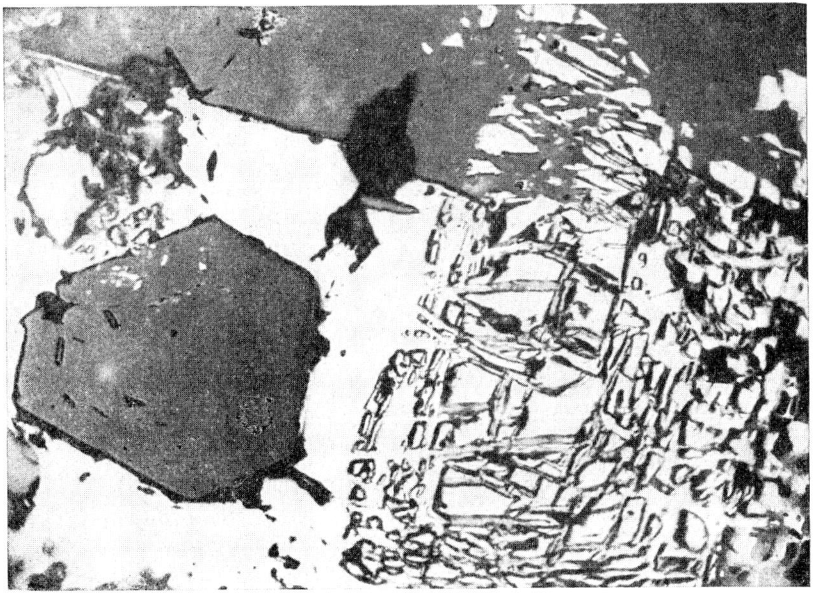

Fig. 201 60 × Ramdohr
Hunan, China

Pyrite (white, strong relief) is replaced along cleavage // (100) by *quartz* (grey) and *chalcopyrite* (soft white). The structure betrays that pyrite was originally embedded in another medium. Note cross-section of quartz. "Lattice shaped"

Fig. 201 a 250 ×, imm. Ramdohr
Sohland, Spree

Unmixed titanomagnetite, partly replaced by sulfide (pyrrhotite and pentlandite)

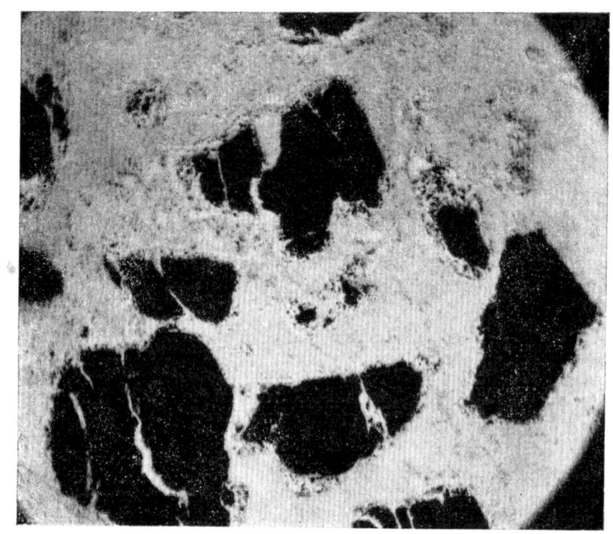

Fig. 202 40 × RAMDOHR

Well at Nasb, Sinai

Hematite (white) replaces intergranular cement, and also quartz grains of a sandstone. "Cement shaped"

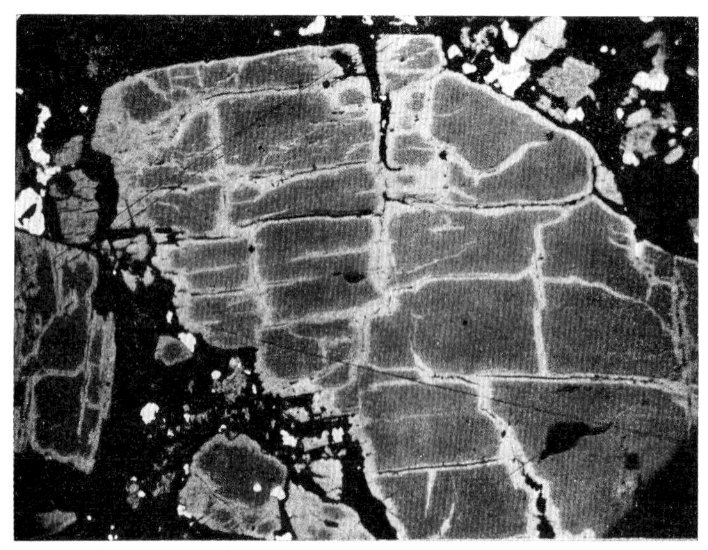

Fig. 203 100 × RAMDOHR

Kirka Mine near Dedeagatsch, Thrace

Cementing *chalcocite* replaces *wurtzite* along (0001) and (11$\bar{2}$0) cleavage. "Lattice shaped"

Fig. 204 100 × RAMDOHR

Doros, Kaokofeld, South-West-Africa

Plagioclase (dark grey) is replaced by *cuprite* (white) and *tenorite* (grey white), whereas *augite* remains fresh (unaltered). The progress of the replacement // to the cleavage is easily recognizable. "Lattice shaped"

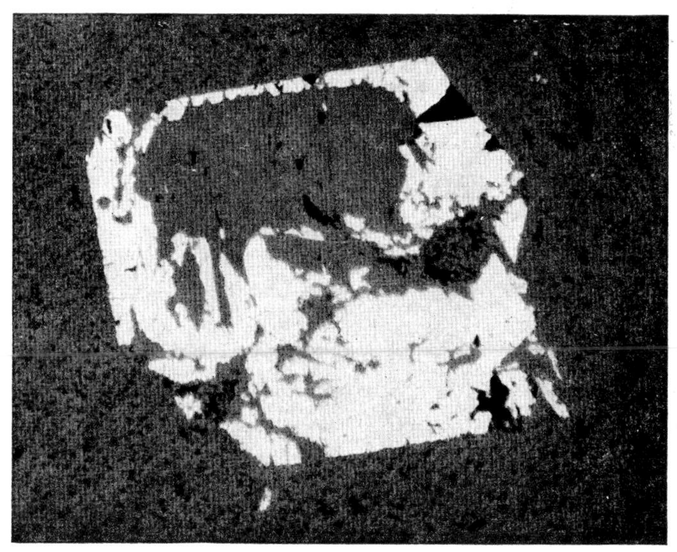

Fig. 205 35 × RAMDOHR

Imsbach, Palatinate, Germany

Orthoclase phenocrysts in quartz porphyry are selectively replaced by *chalcocite* (white). Some parts of the core remain unreplaced where the replacing and the replaced mineral have great structure similarity and approximately the same lattice dimensions (cf. p. 113).

6. *Network-like forms* are likewise often linked to certain structural coincidences and are especially expected with well-developed cleavage and thereby paths in the object that is being replaced (fig. 186). The similarities as well as the possibilities for distinguishing textures which belong here from those which form by exsolution are discussed under that subject. Of course, lamellar unmixed minerals can also be especially amenable to replacement both in the host as in the guest mineral, and

Fig. 206 280 ×, imm. RAMDOHR
Gorob Mine, South-West-Africa

Covellite (different shades of grey) replaces *chalcopyrite* (white) in lamellae // (111). In addition some newly-formed *limonite* (grey, some irregular garlands in the fractures)

Fig. 207 280 ×, imm. RAMDOHR
Tocopilla, Chile

Similar to fig. 206, but process has progressed much further. Only sparse relicts of *chalcopyrite* have survived

render faithful pictures of exsolution textures with what at first would seem surprising partners (fig. 258). Thus there are octahedral ilmenite nets in pentlandite and pyrrhotite, chalcopyrite with pyrite lamellae (in place of cubanite), quartz with pyrite relicts, bornite with sphene skeletons, etc. Very beautifully latticed forms can also appear when pressure twin lamellae are selectively replaced (fig. 221).

7. *Zonal replacements* should be rare according to GRIGORIEFF. I have become acquainted with so many examples of this type of texture that I cannot concur with this opinion. Very characteristic is, e.g., the zonal martitization of very pronouncedly zoned magnetites of contact deposits which is well revealed by etching (figs. 530, 542). Smaltite is subject to replacement by many minerals, and associations of these show up especially well (fig. 200). Fig. 218, a drawing, presents a rather unusual case (Los Jarales, Spain): here in addition to silicates chromite is replaced by niccolite along fractures, and preferentially along its marginal zones (its outer parts). In cementation replacements similar occurrences are most abundant. Transitions to 6, to 4, and to still other cases are common, since the outer parts of crystals as a result of slower growth are purer and therefore more difficult to replace (fig. 200). Since zonal development is especially pronounced in minerals which have formed at low temperatures, selective replacement is also appropriately displayed by these minerals.

8. *Dendritic textures* (the term by GRIGORIEFF is here very misleading) form, e.g., when calcite is replaced along its three cleavage directions and here preferentially along lines of intersection of these (i.e., edges). Similarly, and perhaps even more pronounced, this type of structure develops as a result of the easy invasion of the solutions along twin lamellae (fig. 219, 221).

9. *Cementation structures*, designated cement-shaped by GRIGORIEFF, are those where the bonding agent of sandstones, but also, e.g., that of fault breccias, has been replaced. The replacement then also often extends to the cemented grains themselves (figs. 202).

Extreme caution in interpretation must be exercised in many respects: sometimes very unusual cementing minerals are "primary", but then conversely some cements are unexpectedly so completely replaced that the cement has the appearance of being "primary". Moreover, sandstones are known which have such widely distributed and abundant cement that one thinks of an original rock which had a very large amount of cement, whereas such was almost entirely lacking; rather the quartz grains themselves have been extensively replaced from the grain boundaries toward the inner parts of the grains so that wide cement areas are simulated. The fact that quartz itself can be replaced extensively even in sedimentary association is surprising, but cannot be denied (fig. 202).

10. *Frontal replacements* have not been mentioned by GRIGORIEFF, but are probably very common. If the replaced mineral has no cleavage, fractures, grain boundaries, and foreign inclusions, and no latent zones of weakness exist, then "Verdrängung" can set in on a broad front in irregular, but really fortuitous forms, often in uniformly rounded and smooth boundaries. "Mutual boundaries" (i.e., grains of different minerals with uniform and fortuitously alternating jutting contacts) are often considered as undeniable "evidence" for deposition of the same age and against replacement (however, usually with the secondary meaning of "replacement by cementation"), but this is by no means valid here. Relicts, e.g., exsolution bodies from the original mineral, in the replacing mineral can, without doubt, prove "Verdrängung" (replace-

Fig. 208 70 × RAMDOHR

Magnet Heights, Bushveld, Transvaal
from the immediate footwall of the ore bands)

Ilmenite skeleton of *titanomagnetite* is preserved; the filling of magnetite has been entirely removed (and has been utilized in building up sulphides); the cavities have been filled with silicates

Fig. 208a 500 ×, imm. RAMDOHR

Rheingold, Lußheim near Karlsruhe, Germany

Tiny pebbles formed by a copper-gold alloy, very probably AuCu, of which remnants, white, are preserved, and a nearly pure brownish-yellow very porous *gold*. The contrast is made visible by special filters

ment) even with smooth boundaries (fig. 201). Of course without such relicts it is often barely possible or entirely impossible to prove or even suspect replacement (figs. 248, 393).

11. *Other replacement forms* are surely also numerous; the textures reflecting the reciprocal exchange of material can be so manifold that no single classification can be entirely exhaustive. Moreover many convergences and transitions have, necessarily, led to setting up this group.

"Verdrängungen", examples.

Replacements in the before described forms among ore minerals, between gangue and ore minerals, and also among gangue minerals are so extremely manifold and numerous, and so generally well known that it appears superfluous to cite examples. The "rule" that hard minerals are replaced by softer ones, such as in the sequence pyrite → sphalerite → chalcopyrite → galena → pyrargyrite → argentite, has been documented by hundreds of instances, but even so, it is only a rule.

The few examples which are cited, here, have reference either to unusual cases which have barely been considered in the literature, or to multiple sequences with reversals and the like. The common occurrences are well known or can be found in that part of the book which deals with descriptions of the individual minerals.

Antimony → pyrargyrite
pyrrhotite → galena (especially well-developed as graphic form!)
niccolite → rammelsbergite → gersdorffite
pentlandite → chalcopyrite
stibiopalladinite → sperrylite
löllingite → arsenopyrite → pyrrhotite and reverse, in part in the same deposit
gudmundite → galena
calaverite → gold (WILLEMSE)
luzonite → enargite (a very rare exception; usually the reverse takes place!)
fahlore → luzonite (ditto!)
fahlore → bournonite → jamesonite → boulangerite → galena (accord. to O. FRIEDRICH) an excellent example of addition of Pb and removal of Cu and Sb
cassiterite → sphalerite
cassiterite → chalcopyrite
cassiterite → stannite → cassiterite
cassiterite → herzenbergite
titanomagnetite → quartz + bornite; the ilmenite skeleton is preserved so, that the appearance is given as though it had exsolved from quartz or bornite!
ilmenite → rutile → ilmenite
quartz → hematite
mica → bornite
mica → graphite!!
"silicates" → galena
All cited examples are surely ascendent (hypogene)!
The schematic illustrations 209–224 show additional replacement examples.

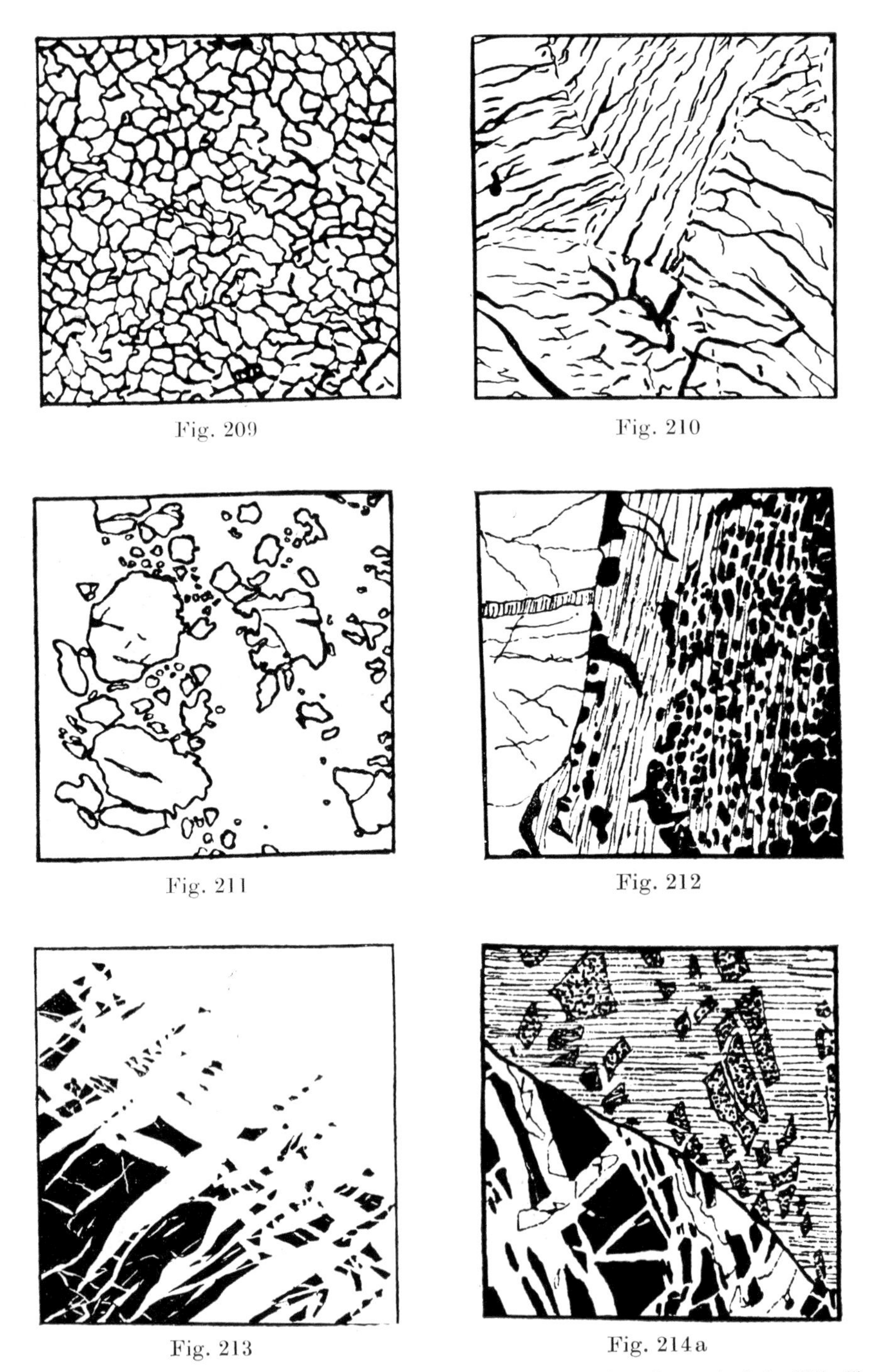

Fig. 209

Fig. 210

Fig. 211

Fig. 212

Fig. 213

Fig. 214a

Figs. 209 to 214a: fig. 209, cuprite, netlike (filiform) in chalcocite. Fig. 210, silver along cleavage in chalcostibite. Fig. 211, pyrite relicts in chalcocite, islandlike. Fig. 212, pyrite replaced by copper ores, limonite appears along the margin of the gangue. Fig. 213, neodigenite "shreds" bornite. Fig. 214a, pyrite (black) replaced by chalcopyrite and bornite, in upper part of diagram limonite and blue descendent chalcocite, "shredded" or "cellular"

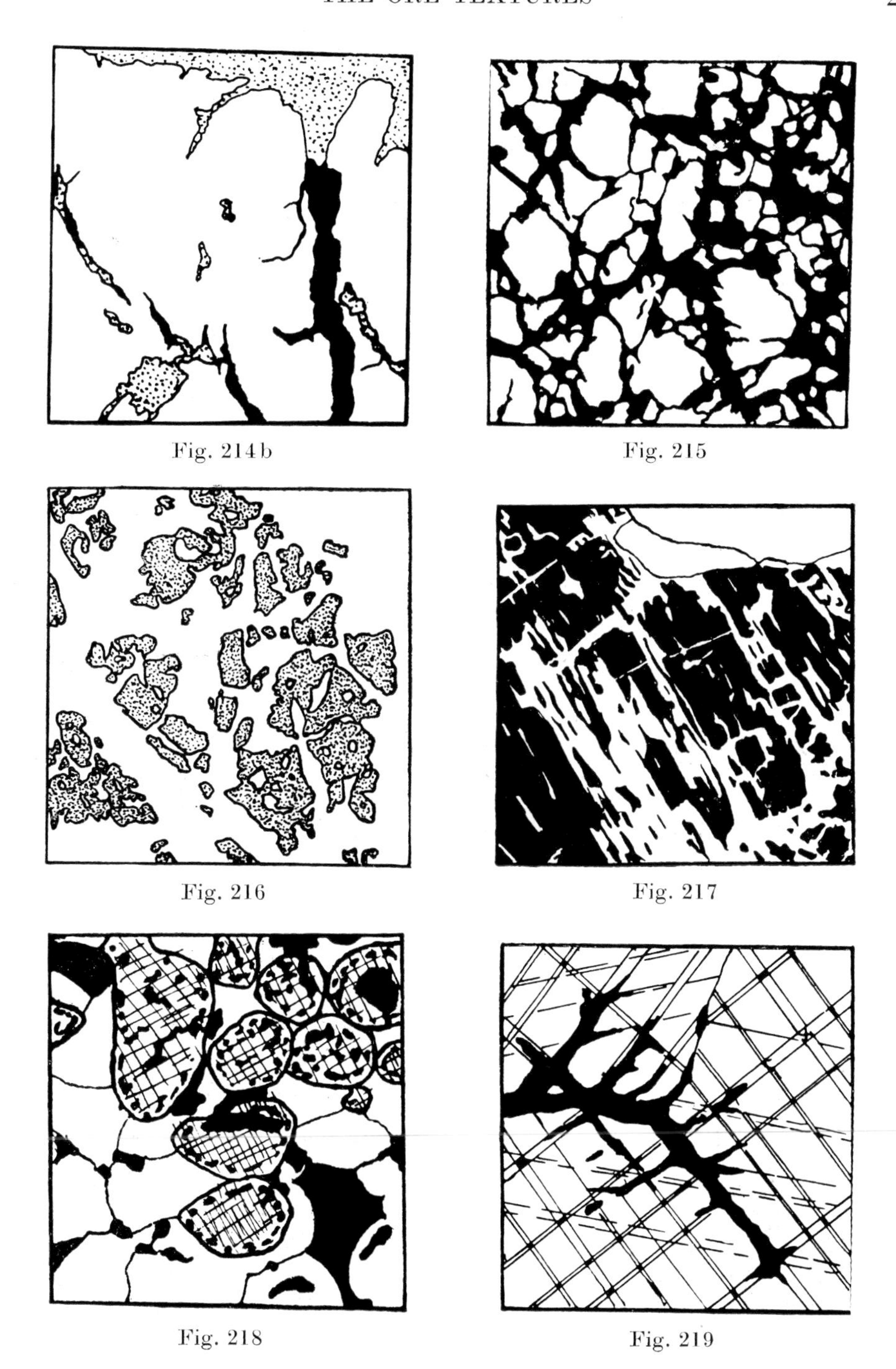

Fig. 214b

Fig. 215

Fig. 216

Fig. 217

Fig. 218

Fig. 219

Figs. 214b to 219: Fig. 214b, silver and calcite penetrating into arsenic minerals, Cobalt Ontario, accord. to BASTIN. Fig. 215, pyrite, "netlike", replaced by chalcocite, accord. to EDWARDS. Fig. 216, chromite replaced by serpentine! Fig. 217, plagioclase replaced by tenorite. Fig. 218, chromite peripherally and zonally replaced by niccolite (black). Fig. 219, calcite replaced by galena along twin lamellae, "dendritic"

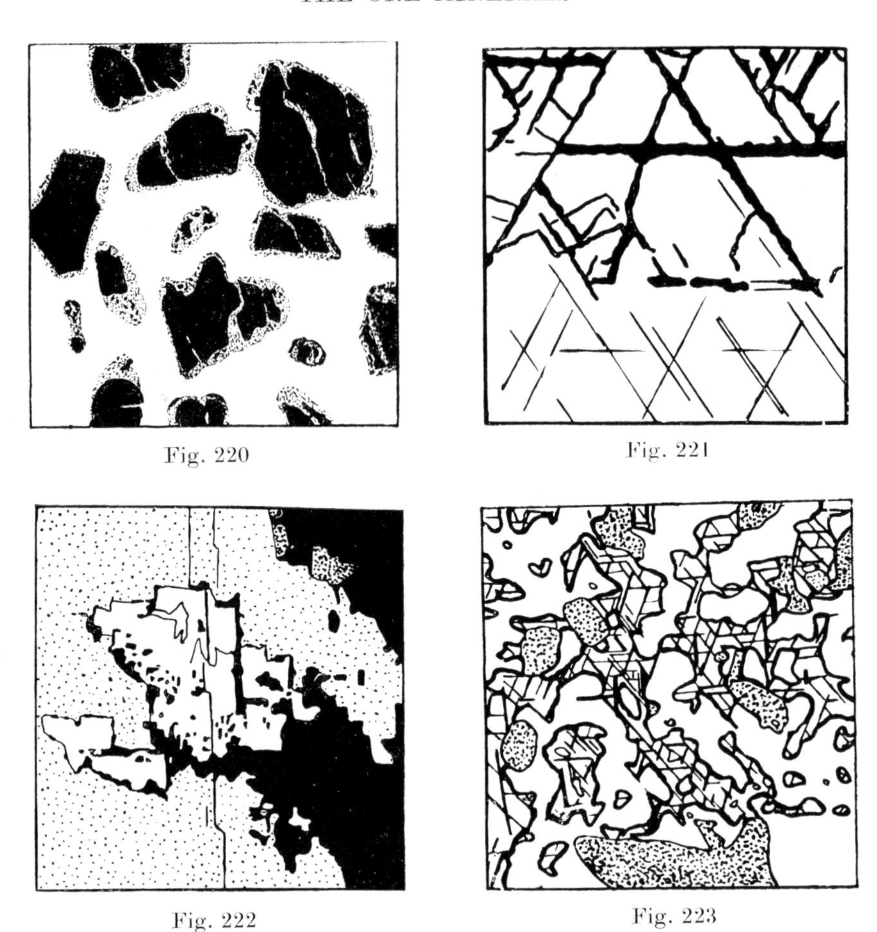

Figs. 220 to 224: Fig. 220, hematite replaces quartz in sandstone. Fig. 221, meteorite of Tombigbee. Oxidation of homogeneous iron follows twin lamellae (NEUMANN lines), accord. to PERRY. Fig. 222, chalcopyrite, calcite, and sphalerite. Fig. 223, coarse-grained spinel with tabular exsolution bodies of ilmenite replaced by an unknown mineral, pseudo-poikilitic texture

(e) *Thermal transformations*

Structural changes in contact aureoles or those outside aureoles which are brought about by hot solutions are briefly treated elsewhere in this book, e.g., on page 39 etc.

(f) *Oxidation textures, and* (g) *Cementation textures*

Textures of the zones of oxidation and cementation. Introduction.

Topics which should be discussed in this section have already been dealt with elsewhere, and in part in some detail. Because of the great diversity and significance of these textures, they will be reviewed here.

From a geologic and economic point of view a knowledge of the characteristic mineral associations and textures of the oxidation and cementation zones is of greatest importance. It permits decisions on whether a valuable assemblage of minerals can be expected more or less unchanged at greater depths or whether enrichment is of secondary origin and thus restricted to a narrow range of depth. The importance of such information has been recognized for surely one hundred years, and both megascopic and microscopic diagnostic criteria were therefore studied very early in some details. Some of the oldest ore-microscopic studies deal with problems of this kind. There are also a number of reviews and textbook-like treatises available on this subject: among others are those of SCHNEIDERHÖHN (1924) who gives a complete summary of the older literature and LINDGREN (1934), FAIRBANKS (1928), LOCKE,

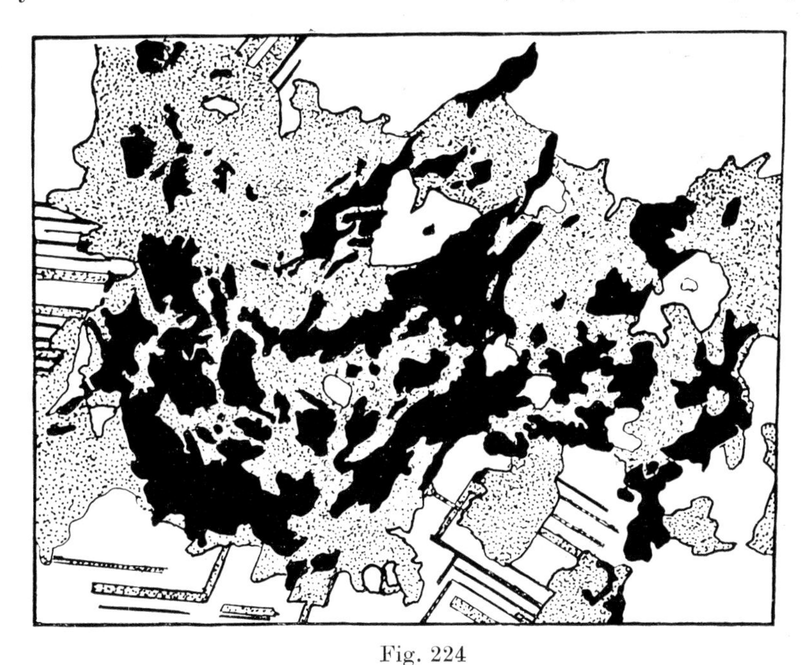

Fig. 224

Fig. 224, pyrrhotite (black) replaced by cubanite (dotted), the latter by chalcopyrite (white) (accord. to ÖDMAN)

HALL & SHORT (1924); many additional papers deal with some aspects of this subject — It is assumed that the reader is familiar with the general relationships and these will not be discussed; only a few examples will be given to illustrate textures and mineral assemblages which have to be determined microscopically. However, a preferred role will be attached to the possibility (and also impossibility) that "descendent" (supergene) fabrics can be distinguished from "ascendent" (hypogene), often misleadingly called "primary", fabrics. Ore-microscopy has shown that some supposedly cementation assemblages are not really such, and that "index minerals", such as argentite, chalcocite, stromeyerite, not to speak of bornite, are frequently, even almost entirely, of ascendent (hypogene) origin. Moreover textures which by definition appear to be in "the middle", e.g. those which have resulted from the alternate effects of ascending and descending waters, appear to have greater significance. Many an unusual mineral association in epithermal deposits which has been interpreted as "telescoped" may have to be explained in this manner. Here apparently the temperature and pressure conditions as well as the pH fluctuate strongly. The pseudomorphic replacement of pyrrhotite by pyrite with or without marcasite and magnetite in the same structural relationship as in the initial stages of undoubtedly weathering may serve as example (figs. 417 and 419a resp.).

Mineral dressing significance. The oxidized ores of *copper* despite their complexity can be leached as such or after solution in sulphuric acid; primary ores and pure cementation ores are amenable to flotation. Mixed ores require double treatment, if one is not willing to take appreciable losses. By no means will laboratory experiments always indicate the necessary treatment. This is shown only by careful microscopic examination which frequently will reveal and explain quite unexpected complications. *Lead* ores, if grains of galena and lead sulphosalts are rimmed by cerussite, anglesite and pyromorphite, are no longer amenable to flotation without prior treatment. Here also will analysis or other chemical tests not always give an indication as to the mineral dressing behavior to be expected. Thus it frequently happens that cerussite is intergrown in small amounts with fine covellite, so that in contrast to what has been said previously it can be floated off without prior treatment. Encrustations of even the smallest thickness ($\sim 1\mu$) can impart to the various minerals the "wrong" behavior on flotation, e.g., PbS crusts on sphalerite, etc. Thin crusts of limonite, manganese oxide, or cerargyrite in placer minerals can lead to great losses in gold for both, amalgamation or cyanide leaching procedures. Similar observations can be given for hundreds of examples (figs. 261, 262, 287, 340).

Textures of the zone of oxidation. Oxidation can take place with or without significant volume changes. The volume increase which results from addition of oxygen is often, perhaps mostly, compensated by removal or even by appreciable volume decrease. This brings about characteristic fabric types, of which examples are illustrated in figs. 225, 226, 227, and 617. Very commonly, for instance, the cavities that are formed by weathering of chalcopyrite can be filled with copper carbonates between the masses of different kinds of limonite, chalcocite, and covelliite. If arsenic, phosphorus or vanadium are present in the original ore, then complex arsenic minerals, etc. will exhibit such textures. With *lead ores* dissolution and transportation are generally minor; cerussite and the somewhat scarcer anglesite form thick crusts around galena remnants which are thus protected from further decomposition. Typically, in very many cases finely powdered secondary galena is precipitated again in these crusts, sometimes in botryoidal-rhythmic forms or parallel to older galena zones (fig. 125). This "dusting" of galena in cerussite greatly resembles dissemination with Ag_2S and may be mistaken for it (fig. 353). Pyrite, arsenopyrite and some other minerals always give rise to cavities of varying dimensions, unless carbonates immediately neutralize the free acid which forms and slow down the process of disintegration by forming limonite. *Sphalerite* leaves particularly extensive cavities since its oxidation product $ZnSO_4$ is an extremely soluble compound which is carried away entirely. Even the presence of carbonates inhibits this effect only partially, since the sulphates $CaSO_4$ and $MgSO_4$, which are produced when smithsonite forms, are dissolved and sphalerite is always laid bare again. These different behaviors are distinctive of the characteristic variable appearances which in complex ores may change from millimeter to millimeter.

There is not adequate room to discuss the manifold oxidation textures for each mineral, for each pH, temperature, and mineral association; some of the textures are documented by figures and are briefly explained in the captions.

In general, oxidation destroys all sulphides, selenides, tellurides, also arsenic and antimony compounds of metals and, of course, sulphosalts. The ease of oxidation is

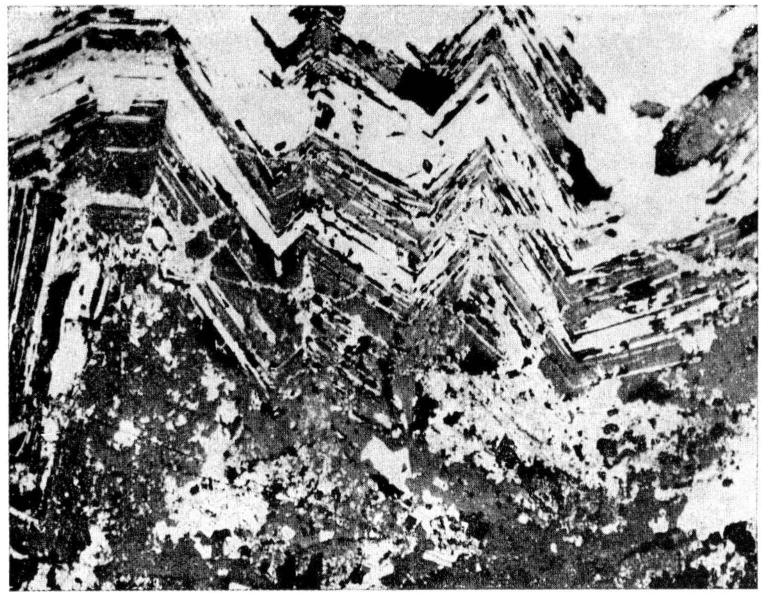

Fig. 225 170 × RAMDOHR

Wiesloch near Heidelberg, Germany

Zonal weathering of low-temperature *galena*. Galena is white, *cerussite* is dark grey, in part with internal reflections. In lower part of figure finely-divided, frequently rearranged galena

Fig. 225a 7 × RAMDOHR

Conrad Mine, Qld., Australia

Galena, white, with cleavage cracks // (100) of oxidation products, mostly cassiterite, medium grey, immediately beside galena sometimes a light grey sulfide. Dark grey, idiomorphic is quartz

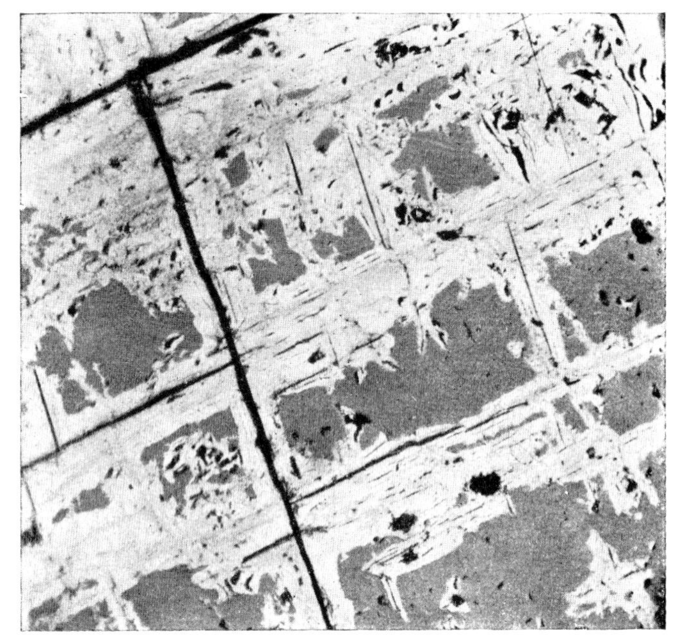

Fig. 226 170 × RAMDOHR

From unknown locality in Australia

Sample from vein with *galena* and *sphalerite*. *Rhodochrosite* (grey) replaced near fractures by "*psilomelane*" (white), which in itself is quite complex, and *limonite* (grey white). It is remarkable that these extensively oxidized minerals are in immediate contact with completely unaltered sphalerite

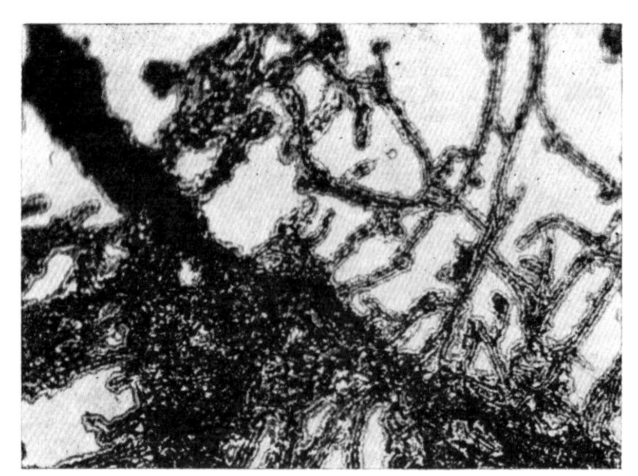

Fig. 227 170 × RAMDOHR

Djebel Ouénza, Algeria

Tetrahedrite with quite extensive weathering along irregular fractures. The series of lines are composed of *chalcocite*

roughly in proportion to the position of the metal in the electromotive series and to its lattice packing (and thereby to its hardness and tenacity). For some minerals (e.g., sperrylite and laurite), however, the extent of weathering is extremely small. After CaS which is present only in meteorites (and slags), MnS is the easiest compound to disintegrate, followed closely by FeS. The lower oxides of Fe, Mn, Co, and Ti are likewise very amenable to oxidation; those formed by iron and cobalt yield for the most part hydrated compounds and those of manganese form waterfree or at least water-poor compounds. In some cases the direction in which the changes take place is not clear, e.g., on oxidation both CuO and Cu_2O appear to form from copper. For titanium, ilmenite and many other titanium compounds probably always yield anatase and rutile.

Fig. 228 250 ×, imm. RAMDOHR
Reichenbach in Odenwald, Germany

Galena (pure white) zonally replaced and encrusted by *cerussite* (different degrees of very dark grey), and locally replaced by cementing *chalcocite* (light grey, spotted)

Not in all cases does *oxidation* of sulphides proceed without complication; e.g., pyrrhotite frequently yields at first pyrite with or without marcasite, or chalcocite gives rise to covellite. In both examples the minerals which have formed are richer in sulphur which presupposes rather complex steps.

Pentlandite, like pyrrhotite, is also very sensitive to oxidation; it also yields initially in the form of bravoite and violarite compounds with a higher sulphur content which have extremely typical forms (figs. 360, 361). Only later form the soluble and under certain circumstances very mobile sulphates of nickel. — *Pyrite* is very stable to weathering, if carbonate is present in the gangue. "Limonite" formation is often conspicuously rhythmic (fig. 480). Contrary to the rule that α-$Fe_2O_3 \cdot H_2O$ ("Nadeleisenerz") is more common than γ-$Fe_2O_3 \cdot H_2O$ (lepidocrocite), the latter predominates here for the most part (fig. 617). *Magnetite* passes over into α-$Fe_2O_3 \cdot H_2O$, at least in temperate climatic zones, for the most part directly but slowly; in tropical zones an intermediate martite commonly forms (fig. 541). In that case it is often impossible to decide whether the formation of martite is "ascendent" (hypogene), or the result of straight weathering. The formation of maghemite (fig. 580) formerly considered to be rather rare and apparently also almost restricted to the tropics is much more common. Titanomagnetite and *ilmenite* frequently yield "leucoxene", the thin lamellae much faster than the massive grains. "Leucoxene" is mainly complex; in ores anatase surely contributes more to its formation than does titanite (fig. 246). Ulvöspinel which is present in many titanomagnetites forms ilmenite before it actually gets weathered.

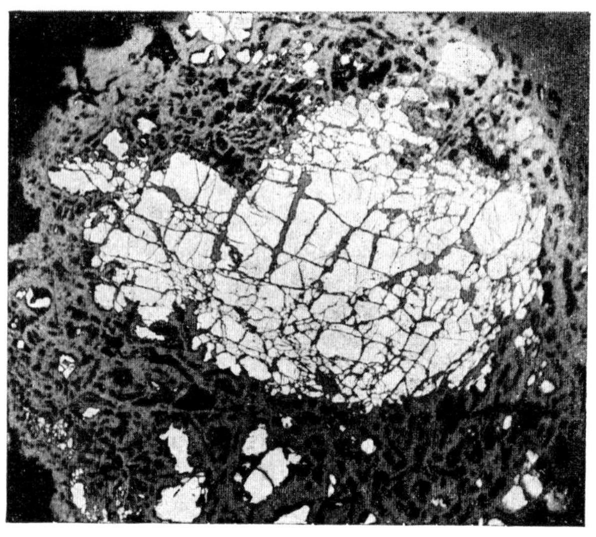

Fig. 229 350 ×, imm. RAMDOHR
Chuquicamata, Chile

Pyrite grain, squashed tectonically, is replaced on weathering by cellular *limonite* (grey). An unusual observation made here is that pyrite does not form lepidocrocite

All *manganese ores* tend to change towards pyrolusite–polianite and "psilomelane", whereby with a higher content of Ba or Pb ions the last-mentioned should more correctly be designated hollandite or coronadite respectively. They commonly form beautifully developed rhythmic textures. *Sphalerite* produces $ZnSO_4$, which upon the disintegration of calcium carbonates and silicates forms smithsonite, willemite, hemimorphite, hydrozincite, etc., processes which could be included with some justification under cementation.

The "*tin gossan*" ("tin hat") which played an important role in the literature at the turn of the century, and was said to include cassiterite deposits, especially those in Bolivia, which result from the weathering of stannite and tin-lead-sulphides, does not exist as such. True weathering of these minerals yields extremely fine-grained masses which resemble light yellow powder, often even colloidal souxite (varlamoffite); they evade all concentration techniques. What has been mistaken for weathering products are fine-grained cassiterite aggregates, often made up of acicular crystals, which are undoubtedly of late hydrothermal origin. Careful investigators have established this by megascopic studies, and microscopic investigations have confirmed it unanimously.

Textures of the zone of cementation. "Cementation" is only typically to be expected where the solution of a more "noble" element is precipitated on a compound of a less "noble" element, such as $CuSO_4 + ZnS = CuS + ZnSO_4$. As a general rule "cementation" restricts itself to the effect of sulphate solutions on sulphides and occurs for the most part below ground-water level, in the so-called zone of cementation. Both conditions, however, are by no means essential. Cementation structures are especially important for Cu and Ag; they are analogous but less common for Au, Pt, and Hg. Quite unusual, but surely still belonging here, is the occurrence of FeS_2 on alabandite, and some others. Lead as a sulphide is in part precipitated on ZnS, even though $PbSO_4$ dissolves only with great difficulty and, as a rule, barely gets transported. However, there are known examples where PbS is destroyed considerably faster than ZnS. Very remarkable are clear deviations from the above rule, where again certainly descendent (supergene) the less noble element replaces the more noble element. This "reverse cementation" is shown for instance by $Cu_2S \rightarrow$ bornite $\rightarrow CuFeS_2$ (figs. 237, 193). — Whether one includes under "cementation" also cases where native metals are directly precipitated and even in a completely oxidized environment (Au, Ag, Hg) is a matter of definition or opinion. The resulting textures are the same. Even precipitation caused by carbonaceous matter, bituminous material, etc., yields similar textures; for these textures the application of the term "cementation" is likewise fluctuating.

Precipitation occurs for the most part in crusts, commonly rhythmic, with special preference of cracks, open or latent tension fissures, porous or chemically deviating zones, and very often at grain broundaries, of the same mineral species or differently composed ones. In the initial stages surface films and intergranular materials, develop which are very annoying in concentration processes (p. 270). With further progress the relicts of the mineral causing precipitation become smaller and smaller despite the thick crust, which, one should think, would protect against further attack, until only the shapes of the textures betray the events that have taken place.

The pH, governed especially by the content of weathering pyrite and its rate of decomposition, but also by the capability of reaction of the associated minerals (such as carbonates!), and by the distance to the ground-water level, etc., is decisive in the development of the different textures which are present.— A more exact knowledge of the conditions still requires further study.

Photographs of cementation textures are shown abundantly so that only reference is made to the captions explaining them (figs. 230–237, 283, 257, 350, 407, 480). They predominantly refer to Cu and Ag ores on their most varied associated minerals, which corresponds to their true significance. Peculiar is the role of millerite which in nickel-rich ores occurs almost like a cementation mineral. It is certain that millerite can form hydrothermally (figs. 429, 481), but more frequently it is of descendent (supergene) origin; the latter has hardly ever contributed to minable enrichments. In the border field between oxidation and cementation, the crusts of greenockite on Cd-rich sphalerites should be mentioned, which locally lead to appreciable Cd-enrichment, but which for the most part result only in thin yellow films.

The *recognition of cementation textures* as opposed to other replacement textures, such as ascendent or lateral secretion textures, or others, is from the economic point of view, as has been said initially, of the greatest importance. It is much more complex, however, than has sometimes been assumed since also in ascendent (hypogene) formations the replacement of the less noble by the more noble element generally pre-

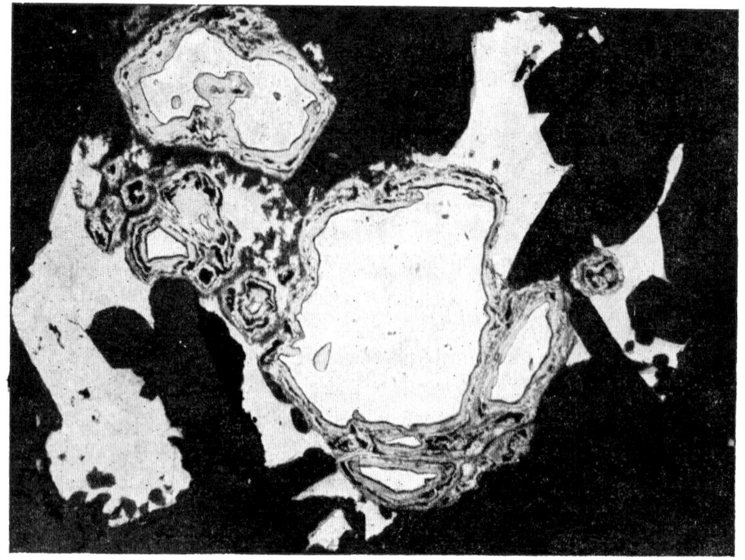

Fig. 230 35 × RAMDOHR
Chuquicamata, Chile

Pyrite (white) rhythmically replaced by *goethite* (Nadeleisenerz) and *lepidocrocite* (medium grey); beyond these rhythmic replacement textures are large amounts of precipitated *chalcocite* (light grey). The very dark grey mineral is *quartz*. A few inclusions of primary enargite are found in pyrite

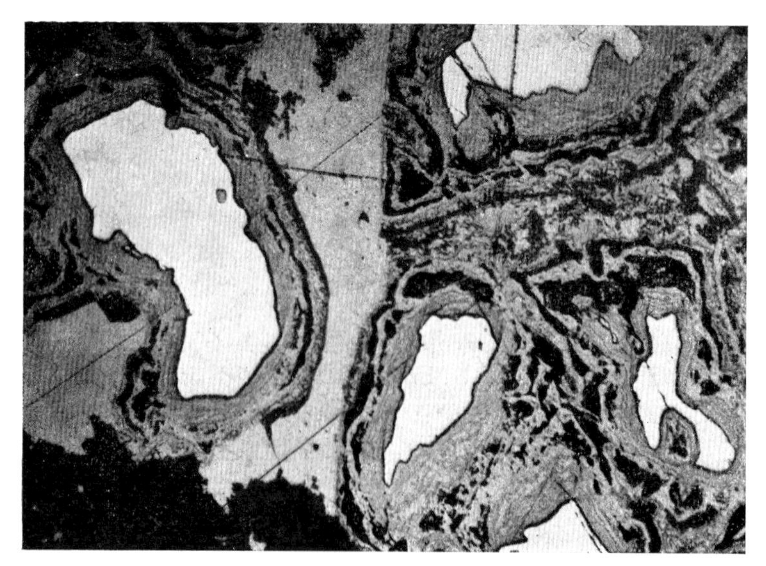

Fig. 231 150 × RAMDOHR
Chuquicamata, Chile

Detail of a texture similar to fig. 230. The rhythmic transformation of pyrite is much more clearly defined

dominates. Good, reliable indicators, but not without exception are: loose, commonly cellular masses; the appearance of "sooty" covellite, chalcocite, argentite; almost always the presence of native silver with chalcocite but without compact argentite or other silver-rich ores; thin rhythmic encrustations of chalcocite, limonite, and sometimes copper carbonates, often disseminated with finest covellite, especially its blue remaining variety; often the presence of pyrite in association with chalcopyrite, which replaces the pyrite and is being replaced by chalcocite in the absence of major amount of bornite. In the other hand, the occurrence of low-temperature chalcocite is by no means any longer a decisive factor since the relationships in the Cu_2S–CuS system have become known; the same applies to native silver and some silver-rich

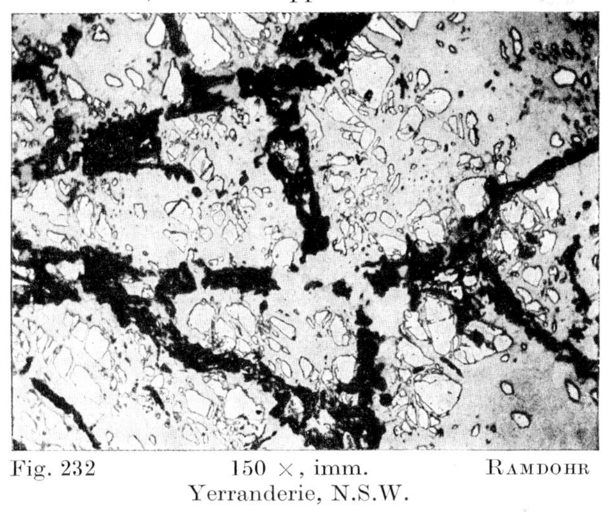

Fig. 232 150 ×, imm. RAMDOHR
Yerranderie, N.S.W.

Pyrite (pure white) replaced by *chalcopyrite* (light grey). In addition some *galena* (lighter than chalcopyrite) and gangue minerals (black)

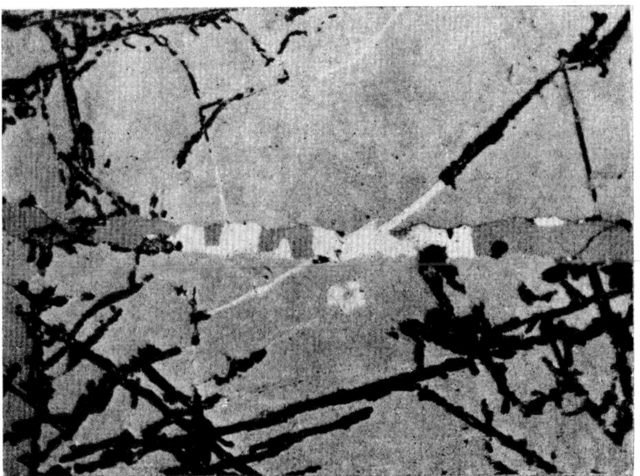

Fig. 233 350 ×, imm. RAMDOHR
Pacajake, Bolivia

Blockite ("penroseite") (grey) with very delicate exsolutions (?) // (100) is replaced near irregular fractures by *clausthalite* and *naumannite* (dark grey). The black fractures are filled with low reflecting weathering minerals

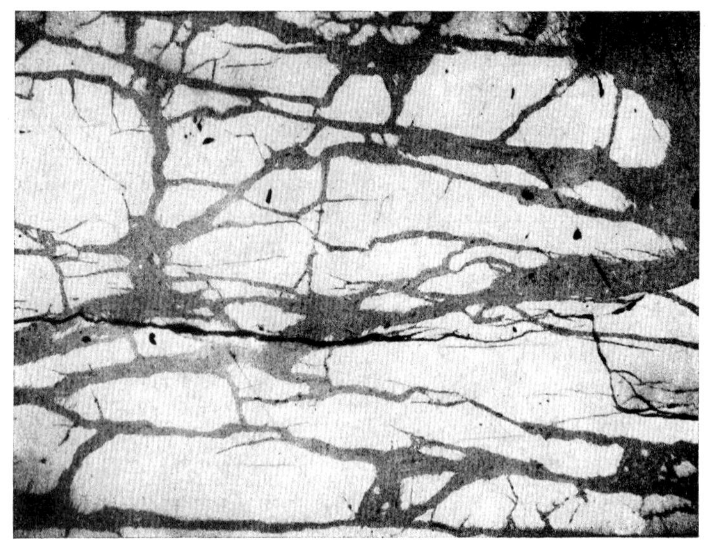

Fig. 234 40 × RAMDOHR

Reichenbach in Odenwald, Germany

Chalcocite, cementlike, replaces *chalcopyrite* along irregular fractures, but essentially to the same direction

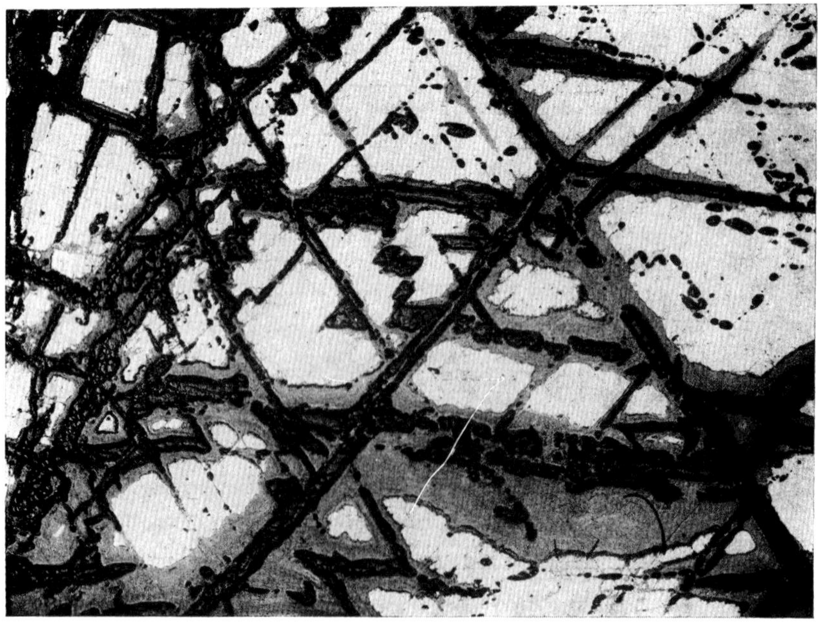

Fig. 234a 150 × RAMDOHR

Jawlennoj near Nertschinsk, Altai

Galena with old cleavage, the fractures filled are with anglesite, gray, and quartz, black

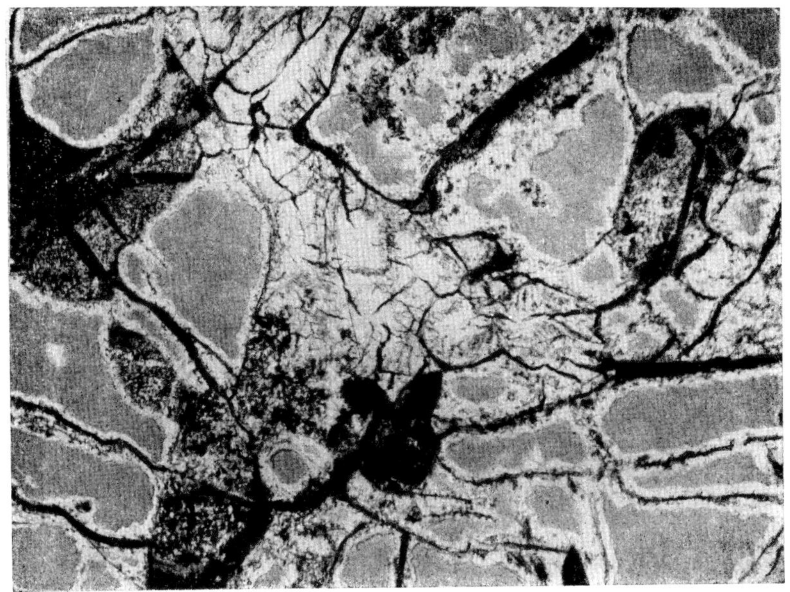

Fig. 235 170 × RAMDOHR

Belussow, Ural

Sphalerite (grey, smooth) replaced by cementing chalcocite (white) which is intergrown with *covellite* (in part white and indinstinguishable from *chalcocite*, and in part light grey) and oxidation ores (almost black). Fabric and association are typical of certain cementative replacement

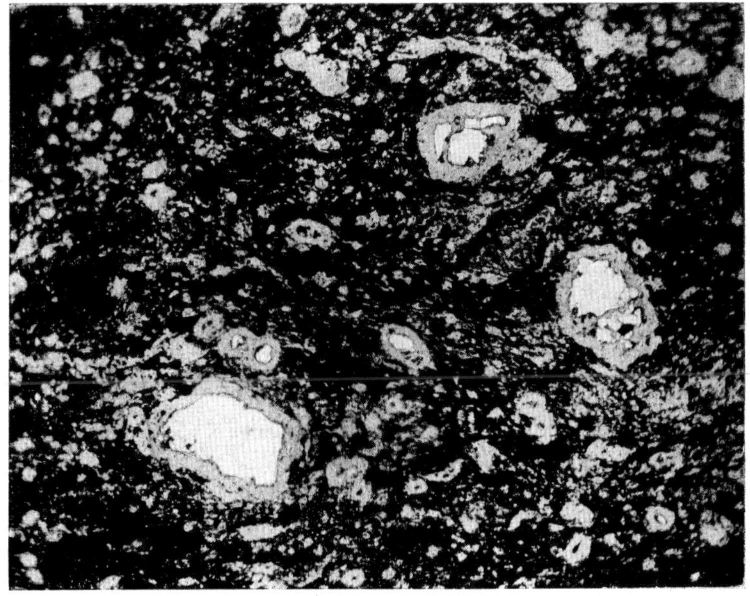

Fig. 236 350 ×, imm. RAMDOHR

Chuquicamata, Chile

Deformed rock along fault zone, in which the embedded *pyrite* grains (white) are pulverized and extensively, often completely, replaced by *chalcocite* (grey). The gangue minerals (chlorite, quartz, etc.) appear, of course, black on immersion

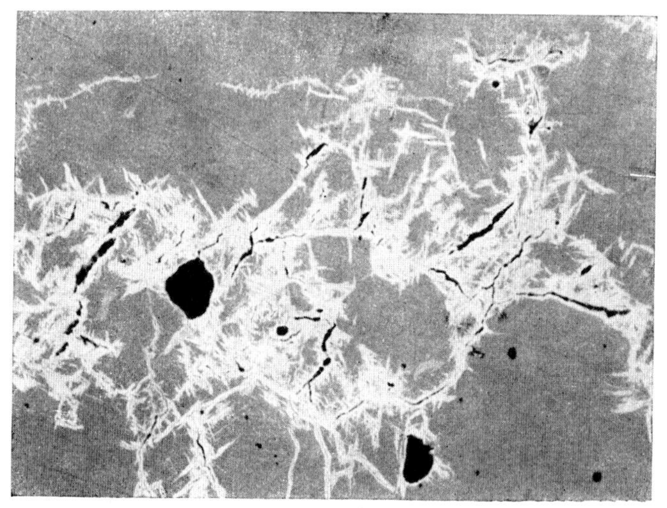

Fig. 237 170 × RAMDOHR
Rheinbreitbach, Germany

Descendent *chalcopyrite* (white) replaces *bornite* which represents 'reverse cementation!" Two holes (black)

ore minerals. Many details on this subject are presented in the descriptive part of this book.

III. Deformation fabrics. SCHNEIDERHÖHN's deformation fabrics correspond apparently to GRIGORIEFF's "pressure structures", which the latter, however, has not treated exhaustively. SCHNEIDERHÖHN distinguishes 1. cataclastic, breccia, and mylonite fabrics, 2. shearing fabric, 3. recrystallization, and 4. complex and polymetamorphic fabrics. Even he is far from including all possible fabrics, perhaps because it has so far not been feasible. The writer has described and documented with illustrations many aspects pertaining to deformation fabrics in his study of the Rammelsberg (1953a); reference is made to this publication.

Since the proceess of metamorphism and its resultant textures has been treated and partially explained with illustrations in the special section on ore metamorphism (p. 39), further explanations are not required here. Reference is made in tabular form to illustrations which are relevant to certain concepts, without being anywhere near exhaustive:

1. Bending and translation: fig. 34
2. Pressure lamellae and "corrugations": figs. 32, 33, 445, 420
3. Cataclasit: figs. 26, 28, 29, 364, 374, 461
4. Breccia: fig. 59
5. Broken "hard bodies": fig. 41
6. Boudinage and augengneiss structure: fig. 42

6b. Pencil gneisses

7. Recrystallization: figs. 46, 441, 442
8. Xenoblasts and idioblasts: fig. 110
9. Banded forms: figs. 37, 40
10. Structures of blastesis: figs. 45, 46, 383, 453.

III. RADIOACTIVE HALOES, LATTICE DESTRUCTIONS, BLASTING

Small discoloration haloes around inclusions of strongly radioactive minerals are known in approximately 35 translucent to transparent minerals, mostly silicates, but also phosphates, sulphates, carbonates, fluorides, and oxides. The dimensions of these haloes have been correlated with the range of α-particles for the individual decomposition products of U and Th. Of the minerals discussed here, the haloes have been

Fig. 237a 250 ×, imm. RAMDOHR
(reflected light, but focused a little below the surface)
Stiepelmann Mine, near Arandis, S.W.Africa

A radioactive halo (uranium halo) in cassiterite. The outer halo is formed by RaC′, the inner one by RaA. The internal reflections show, that the cassiterite is zonally and patchy colored

more accurately investigated in cassiterite. — In opaque minerals radioactive haloes had not been previously observed and have only recently been described by the writer (1957a; 1958a) in ilmenite, hematite, columbite, pyrrhotite, löllingite, safflorite, arsenopyrite, pyrite, carbonaceous matter, and graphite. STACH (1955) has briefly drawn attention to an occurrence in anthracite coal.

The interpretation in the individual case is not very simple and rests surely on the lattice disturbances which have been designated "Frenkel defects", which are caused in the paths of α-particles, and in part by the extreme heat which produces local melting at the end of the path of such an α-particle. Haloes are most easily observed in strongly anisotropic ore minerals, in which local changes in anisotropy are relatively easy to recognize. Weakly anisotropic minerals require extremely careful observation. Why some ore minerals show these effects, whereas numerous others apparently do not, is not clear as yet (figs. 238–240).

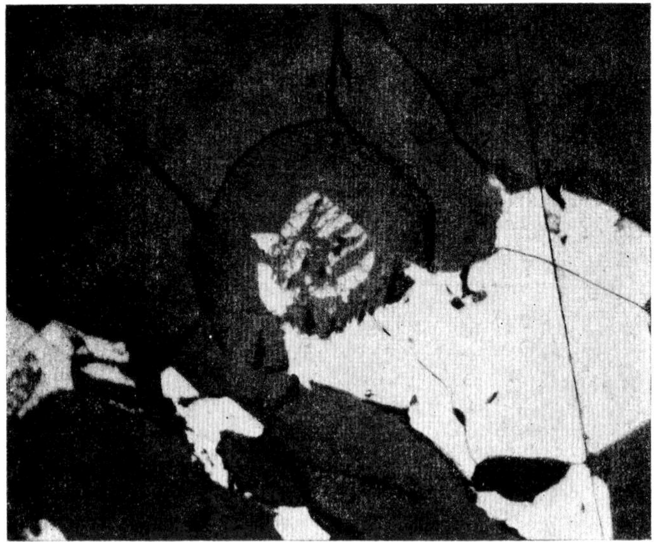

Fig. 238 280 ×, dry RAMDOHR
Tybble, Östergötland

Radioactive halo around broken-up crystal of *uraninite* in *pyroxene* (dark grey) and for a small distance in *magnetic* (white)

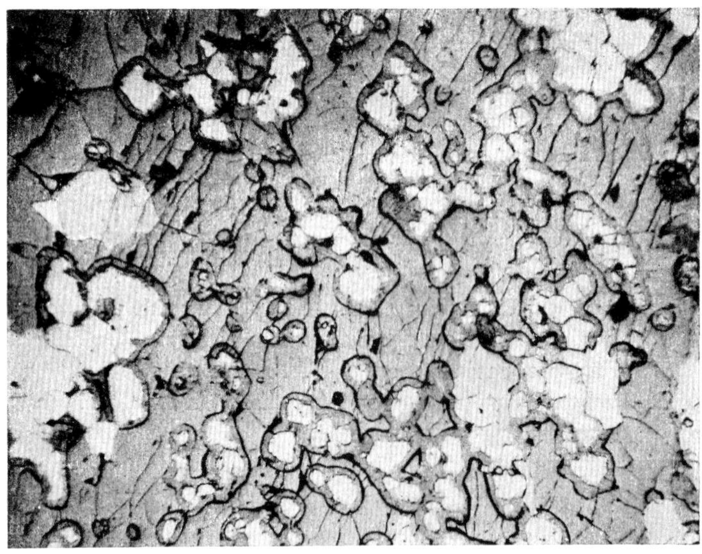

Fig. 238 a 70 × RAMDOHR
Itomba Hill, Nyassa

Radioactive haloes surrounding *pyrochlor*, white, in *pyroxene*, grey. In the neighbourhood a silicate (?titanite) grey-white, practically not altered by the irradiation

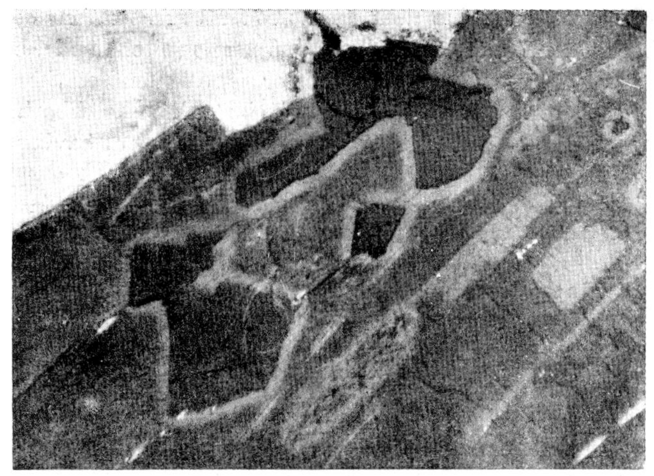

Fig. 239 70 ×, imm., crossed nicols RAMDOHR
Hagendorf, Bavaria, Germany

Inclusions of *uraninite* in *columbite* produce in the latter radioactive discoloration haloes. (The photographic documentation of this texture is very difficult, since the light intensity is extremely low for these haloes which can only be recognized under crossed nicols.
A highly sensitive photographic plate was used with a xenon high-intensity lamp and 30 minutes exposure!)

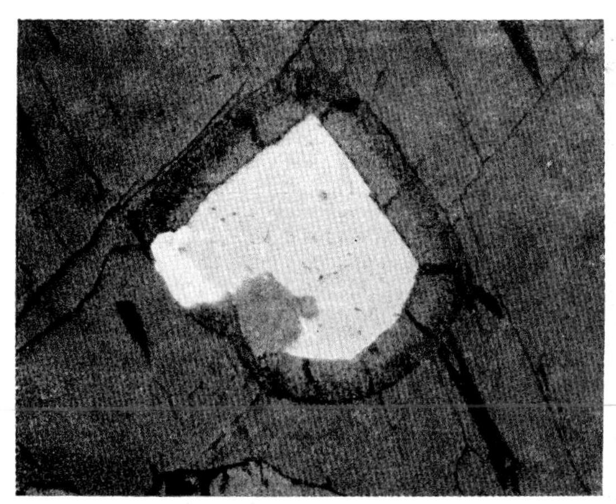

Fig. 240 150 × RAMDOHR
Faraday Mine, Bancroft, Ontario, Canada

Well-developed *uraninite* crystal (light grey) produces a destruction halo in *orthoclase*. The material which has become isotropic is likely to take up less volume than the unaltered ("fresh") orthoclase which has a very loose lattice, and therefore no break-up results.
A small inclusion of cyrtolite in uraninite, however, forms a break-up in the latter. Younger sulphide is white in color

Apart from radioactive discolorations extensive *lattice destructions* occur in the immediate neighborhood of strongly radioactive minerals (pitchblende, brannerite, thorianite) which pave the way for dissolution or replacement of the affected areas. Grains of pyrite and arsenopyrite which are in the immediate neighborhood or enclosed in the above-mentioned minerals may have the appearance of having been "nibbled on" (figs. 238–243).

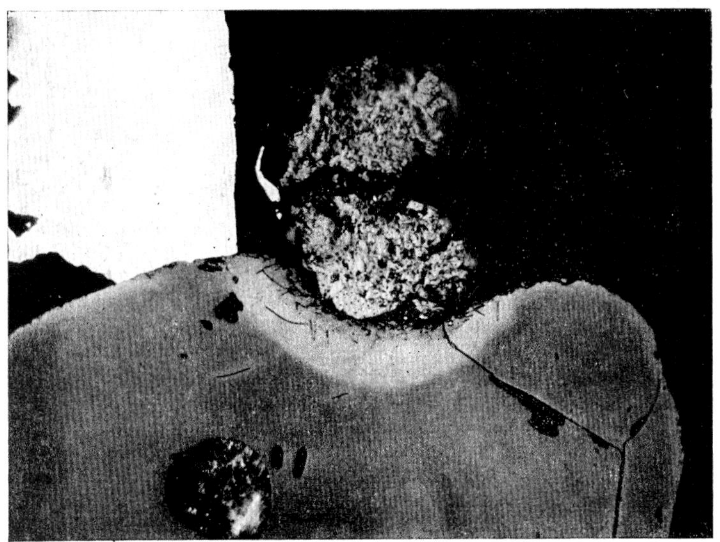

Fig. 240 a 250 ×, imm. SCHIDLOWSKI
Loraine Au-Mine, O.F.S., S.Africa

Radioactive halo in *chromite* around a grain of "*thucholite*" with plentiful remnants of *uraninite*. The width of the halo is about 32 μ, roughly corresponding with that radius of RaC′ in chromite. The coarse alteration of the chromite by irradiation (see) e.g. the tiny cracks!) allowed the thucholite to sink about 40 μ into the chromite. — White is pyrite

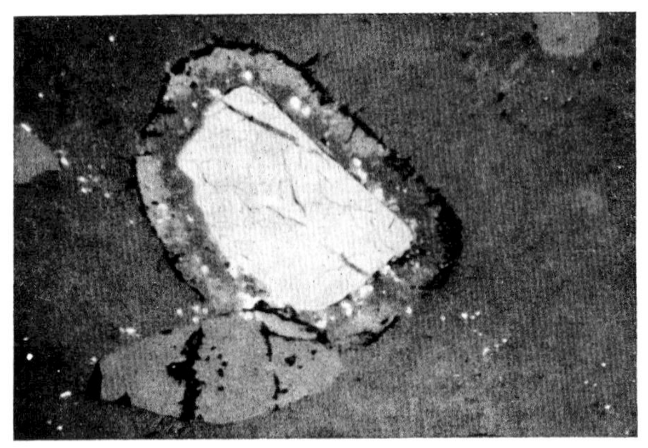

Fig. 240 b 150 × RAMDOHR
Little Man Mine, Washington State

Radioactive halo around uraninite, white, enclosed in pyroxene, dark grey (main part) and sphene, medium grey and idiomorphic, Easily visible is the fact that the halo in sphene is distinctly thinner than in pyroxene

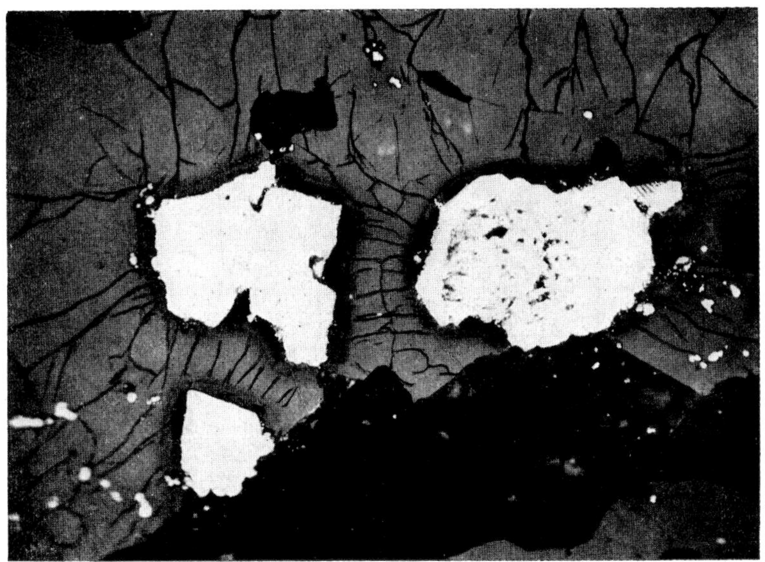

Fig. 241 250 ×, imm. RAMDOHR

Mary Kathleen Mine, Queensland

Uraninite (here pure white) causes haloes with lattice destruction in *allanite* (grey). In these haloes the volume has increased so that the intact allanite is broken up. The volume of uraninite has not changed. Gangue minerals are black (internal reflections) (Induced blasting)

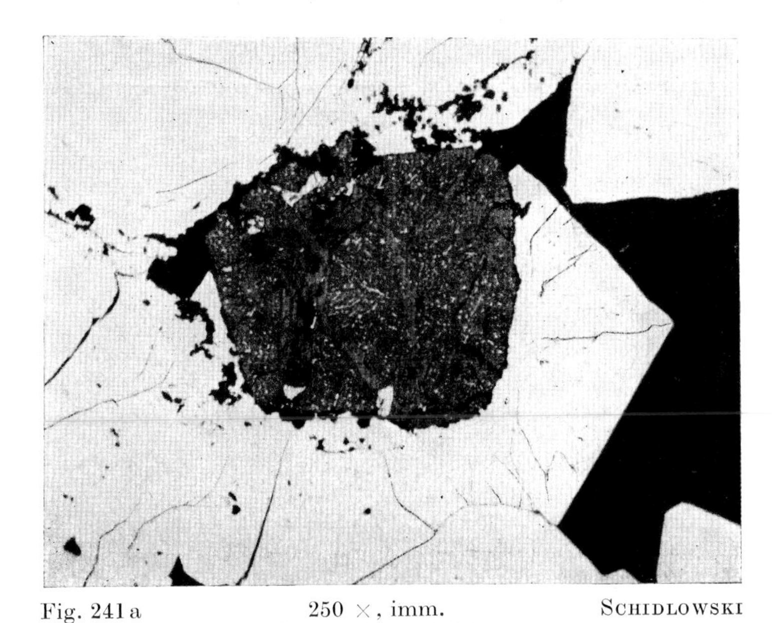

Fig. 241a 250 ×, imm. SCHIDLOWSKI

Loraine Mine, O.F.S.

A crystal of uraninite, dark grey, a bit rounded with a dust of radiogenetic galena, is overgrown, by pyrite, white. This is isotropized in the immediate neighborhood and became more voluminous. By this increase of volume the intact outer parts of the pyrite became blasted

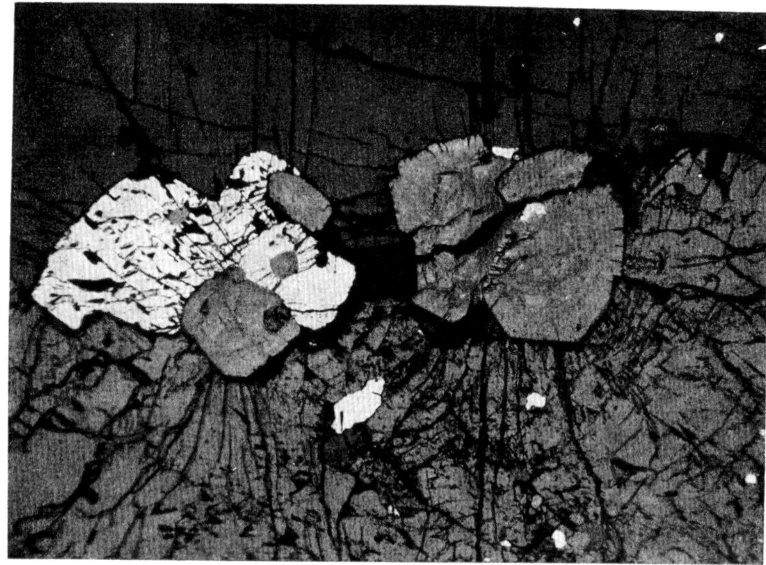

Fig. 242 30 × RAMDOHR

Faraday Mine, Bancroft, Ontario, Canada

Zircon ("*cyrtolite*"), rich in uranium and/or thorium has become isotropic in spots to varying extent and has thereby gained volume and broken up *plagioclase* (dark grey), *pyroxene* (somewhat lighter grey), *magnetite* (white, and *uraninite* (grey white). It is quite apparent here and elsewhere that the break-up seems to occur suddenly and all at once

Fig. 242a 80 ×, imm. RAMDOHR

Faraday-Mine, near Bancroft, Ont.

Crystal of *uraninite*, cut about // (111), causes a blasting halo in titanite. Following the cracks the latter is decomposedin a low reflecting silicate + *anatase* (tiny grains, nearly white). Enclosed in uraninite a small grain of *pyrite*

"*Radioactive cracking*" is also the result of radioactivity. Where "metamict" minerals or minerals which have been made isotropic and whose lattices have been destroyed by their own content of U or Th are embedded in other, preferably brittle, minerals, these host minerals are commonly broken up. The conversion to isotropy is mostly connected with an appreciable increase in volume, and this swelling can exert substantial forces. Examples that have been observed include break-up of pitchblende and magnetite by zircon, ilmenite by monazite davidite, pyrite by samarskite, and others. — Additional information about radioactive manifestations is presented in two papers by the writer (1957a; 1958a).

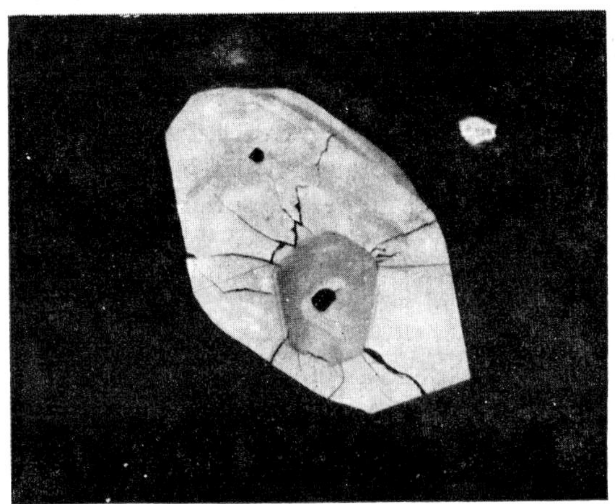

Fig. 243 250 ×, imm. RAMDOHR
in Rapakiwi, Finland

Zircon with a interior part which has become isotropic and thereby increased in volume, breaks up the normal exterior zone

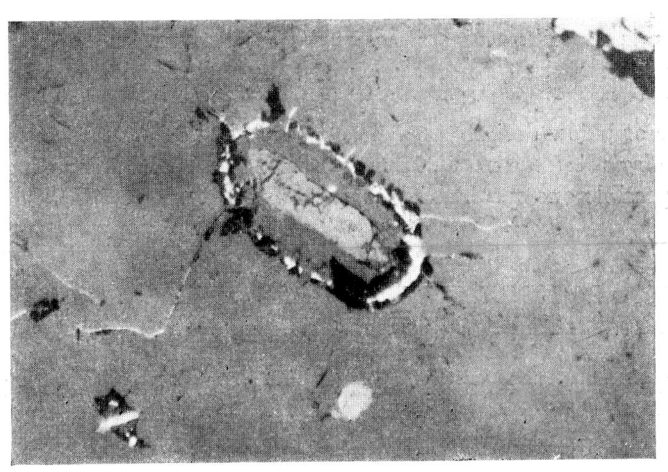

Fig. 243a Wadi el Gemal, 180 × RAMDOHR

A crystal of a coffinite-like mineral or thorite is surrounded by a rim of pyrite + galena. In the surrounding the gangue mineral shows some blasting cracks. The small and lighter blebs in the gangue are probably pyroxene en tirely

Induced blastings, i.e. such ones where the inclusion, e.g., uraninite itself, has not increased its volume but by isotropisation the host-mineral, is shown e.g. in allanite (fig. 241) and similarly columbite, pyrite and others.

RECOGNITION OF THE GENETIC POSITION OF ORE DEPOSITS

The knowledge of the genetic position of ore deposits is of extreme scientific and practical importance. Ways of solving questions related to it have in part long been known, in part they have only recently developed, and in part have not been found as yet.

The purely geologic methods have often been very helpful and will always retain their significance. Yet, they fail easily, e.g., with deposits in highly metamorphosed rocks and with mineral associations where identification of the components alone and recognition of the fine textures especially presented previously has been an insoluble problem.

A short summary is presented here on the possibilities — mentioned in other places in this book (parts I and II) — to answer problems of metallic mineral deposits by means of ore-microscopy. Only in a few cases will information not discussed as yet be added, more often only references will be given.

In our system of classification of mineral deposits the main concern is characterized by questions of temperature, pressure, place of formation, and history after formation. The mineral content as such is in many instances of secondary importance, but we shall need it where certain minerals are characteristic — which at first will have to be determined empirically — for certain ranges of temperature, pressures, and concentrations, or, where applicable, for mineral associations, paragenetic sequences or fabric types. Such then are described as "*typomorphic*" minerals, typomorphic assemblages, etc.

Many minerals as well as mineral assemblages, textures and so forth, are not only characteristic for certain types of ore deposit, and thus for temperature ranges and geochemical associations, but also owing to their physico-chemical position were formed only above or below definite or at least more or less definite temperatures. They are called "*geologic thermometers*". In rare instances, where similar conclusions on the pressure are possible one speaks of *geologic barometers*.

I. TYPOMORPHIC MINERALS, MINERAL ASSEMBLAGES, PARAGENETIC SEQUENCES AND FABRIC TYPES

Originally it was purely an empirical rule, that, e.g., quartz appears in quartz porpyhries as hexagonal bipyramids (hexagonal trapezohedral class) and in Alpine veins as prismatic crystals (trigonal trapezohedral class). In the case of the rock-forming minerals we know in many instances now the reasons for these characteristics, but in the case of the ore minerals we are often still in the first stage of empirism. Thus it is still unknown to us, why covellite in cementation zones in one case exhibits in oil a color change toward red while in another case it remains blue, etc. Likewise we do not know yet, why in some zinc ore deposits of low temperature of formation sphalerite is found in one instance and in other cases wurtzite. In this latter instance we know empirically, however, that wurtzite (often secondarily paramorphosed) is restricted to very few types of ore deposits, specifically to the lower temperature metasomatic lead-zinc occurrences. Wurtzite is typomorphic in these occurrences!

With the application of typomorphic minerals to the determination of the genetic position of ore deposits one must, however, be very cautious, because through microscopic studies many ore mnierals have been found to be much more widespread, than was thought earlier. Part II of this book gives very many examples. On the other hand microscopic work determined a very large number of new typomorphic minerals and likewise it has shown that also in persistent minerals certain typomorphic criteria can be present. For example little star-shaped sphalerites in chalcopyrite of high temperature deposits, zonal structure in magnetite in contact metasomatic deposits, etc. Naturally one will be able to evaluate correctly as typomorphic minerals only those that are not especially rare. The tables present in a certain sense a reversal of the abundance of individual minerals (p. 33 etc.). An exact review of the abundance shows also the generalizations made *here*. (See tables p. 240—243.)

Typomorphic mineral assemblages are occasionally enumerated in many types of deposits; a large number of the old "formation"-names is nothing but a description of such assemblages. But unfortunately economically important minerals were too strongly considered in that connection, and minerals of little economic importance which perhaps are more typical, were given too little consideration. Thus in many medium- to low-temperature veins, e.g., the normal hydrothermal lead-zinc veins, the short prismatic to pyramidal zoned quartz crystals are typical, even more typical than the galena-sphalerite assemblage itself, which is also contact-pneumatolytic, or can be found as well in many gold-quartz veins, or which forms also at quite low temperature, and can occur even in sedimentary deposits in different genetic positions.

Here too, an abundance of surprises have been encountered recently in regard to the typomorphic assemblages, which show that many are yet to be clarified. "Nickeliferous pyrrhotite occurrences" have, e.g., pronounced pegmatitic portions containing molybdenite, which at the same time contain chromite, sperrylite, native gold, pyrrhotite, cubanite, and pentlandite, possibly with yet unexpected age relationships.

The *age relationships* are also frequently typomorphic. From petrography one knows, e.g., that in granites magnetite and ilmenite are often idiomorphic, whereas in gabbros they are generally xenomorphic, but in the corresponding effusive rocks they are again idiomorphic. Naturally, as in the case of all conclusions made with the aid of age relationships, the relative quantities of the components are essential.

Typomorphic structures are to be evaluated with the most critical precaution, although some are clear. Thus, e.g., strongly zoned magnetites (moreoever even garnets and some other silicates) are very characteristic of contact metasomatic ore deposits. Magnetite occurs, however, on the one hand in undoubted contact metasomatic environments but shows no zoning (perhaps homogenized through very slow cooling!). On the other hand, e.g., orthomagnetic or hydrothermal deposits contain magnetite, that is very beautifully zoned. In the higher degrees of metamorphism, which are reached much more quickly by ore minerals than by silicates (see pp. 77—79), the coarser structure is often one that is quite similar to those formed at high magmatic temperatures (e.g., with pure magnetite ores). Certain relicts or the absence of exsolu-

Typomorphic

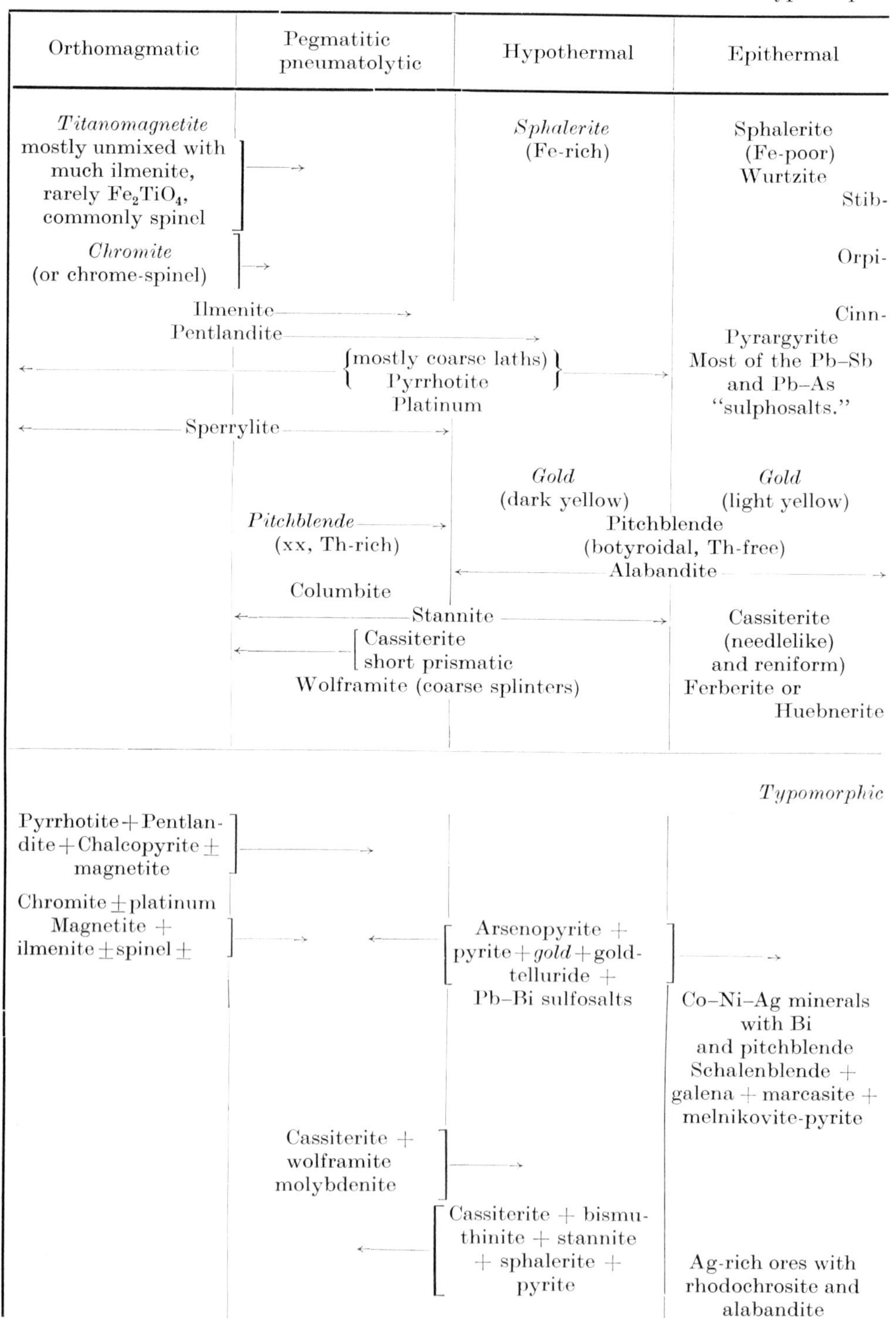

Orthomagmatic	Pegmatitic pneumatolytic	Hypothermal	Epithermal
Titanomagnetite mostly unmixed with much ilmenite, rarely Fe_2TiO_4, commonly spinel	→	*Sphalerite* (Fe-rich)	Sphalerite (Fe-poor) Wurtzite
			Stib-
Chromite (or chrome-spinel)	→		Orpi-
Ilmenite →	Ilmenite →		Cinn-
Pentlandite →	Pentlandite →	Pentlandite →	Pyrargyrite
←	(mostly coarse laths) Pyrrhotite	(mostly coarse laths) Pyrrhotite →	Most of the Pb–Sb and Pb–As "sulphosalts."
	Platinum	Platinum	
← Sperrylite	Sperrylite →		
		Gold (dark yellow)	*Gold* (light yellow)
	Pitchblende → (xx, Th-rich)	Pitchblende (botyroidal, Th-free)	Pitchblende (botyroidal, Th-free)
		← Alabandite	Alabandite →
	Columbite		
	← Stannite	Stannite →	
	← Cassiterite short prismatic		Cassiterite (needlelike and reniform)
	Wolframite (coarse splinters)	Wolframite (coarse splinters)	Ferberite or Huebnerite

Typomorphic

Orthomagmatic	Pegmatitic pneumatolytic	Hypothermal	Epithermal
Pyrrhotite + Pentlandite + Chalcopyrite ± magnetite	→		
Chromite ± platinum			
Magnetite + ilmenite ± spinel ±	→	← Arsenopyrite + pyrite + *gold* + gold-telluride + Pb–Bi sulfosalts	→
			Co–Ni–Ag minerals with Bi and pitchblende
			Schalenblende + galena + marcasite + melnikovite-pyrite
	Cassiterite + wolframite molybdenite	→	
	←	Cassiterite + bismuthinite + stannite + sphalerite + pyrite	
			Ag-rich ores with rhodochrosite and alabandite

minerals

Hot Springs and Exhalations	Cementation zone	Oxidation zone	Zone of dynamic metamorphism (dislocation-metamorphic)
	Silver (finely porous and mosslike)		
nite	Chalcocite and covellite (both sooty)		
ment	Stromeyerite		
abar		Limonite	
Specularite Magnetite *Tenorite (platy)*	native copper, cuprite, tenorite (botryoidal, powdery) *Gold* (nearly) brownish-yellow		Specularite ("micalike") *Jakobsite* Sitaparite, hollandite
Pseudobrookite		"Souxite"	
Mineral Assemblages			
Martite formed by heating + specularite + pseudobrookite	Chalcopyrite + bornite ± covellite on chalcopyrite	Limonite + cerussite ± covellite ± many Cu, Pb, Zn oxidized ores	Danaite + cubanite + pyrrhotite ± pyrite
			Specularite + magnetite in porphyroblasts
Stibnite + Sb_2S_3 gel + cinnabar		Limonite ± Mn oxides ± Cu-carbonates	

Typomorphic

Orthomagmatic	Pegmatitic pneumatolytic	Highthermal	Lowthermal
Generally high temperature: *Exsolutions*			
Zonal structures on magnetite (and garnet) →			Zonal structures in galena, pyrargyrite, fahlores
Chalcopyrite with spindle shaped lamellae →			Chalcopyrite in equidimensional lamellae
Globules of FeS, $CuFeS_2$, bornite in silicates			*Gel textures*, of sulphides in general "chrysanthemums" of Au, Pt, and others, "reticulate" forms, three-dimensional dendrites

tion textures, however, can reveal the origin. The discovery of such relicts will be attained often through time-consuming and patient examination of many polished sections. Besides those structures only recently recognized and evaluated by the microscope, textures that have been long known, and that are in part very coarse, and can be recognized with the naked eye, fully retain their significance. Examples of such megascopic textures are crustifications, breccias, pseudomorphs and many others.

II. ORE MINERALS AND ORE ASSOCIATIONS AS "GEOLOGIC THERMOMETERS"

For the scientific interpretation of each mineral assemblage, particularly of each rock and each ore deposit, information about temperature and pressure of formation, if possible with reliable quantitative data is of the greatest importance. The literature of the last 70 years contains an immense number of statements concerning the "geologic thermometers" with widely varying value. They pertained originally to magmatic rocks and to the specific conditions (of origin) of the oceanic saline deposits; since 40 years they refer more and more also to ore minerals. — The most direct method, i.e., measurement in natural occurrences themselves, is unfortunately possible only in very rare cases.

The silicates of the magmatites are insofar quite appropriate for such investigations, because as crystallization conditions are hardly disturbed through the appearance of a gas phase, the reactions are quite sluggish, and in the original conditions the participation of water, especially of watery solutions, is quite unimportant. On the other hand they are methodically less suited, as, e.g., the establishment and determination of the very fine exsolution and decay textures in transmitted light is much more difficult than in reflected light. The introduction of polished surfaces and the possi-

structures

Hot springs and exhalations	Cementation zone	Oxidation zone	Metamorphic
			"*Porphyroblasts*" of pyrite, magnetite, danaite, cobaltite
	Instructions about primary sulphides		"*Granoblastic*" shape of very many other ore minerals
Rhythmical textures Thinly laminated, botryoidal sinter	"Cracked" pyrite	Limonite in different, with the original assemblage varying, relict structures	
Slender individual crystal on fractures and cavities			

bility of high magnification has likewise made possible the recognition of minute exsolution bodies in bronzite, hypersthene, diallage, some olivines, aventurine, etc. and has given substantial help to the study of rocks.

Whereas the study of the ore minerals is methodically rather simple, in reality a series of large difficulties arise: The formation of most ore minerals, especially of sulphides, results at lower temperatures and under a very substantial influence of water, in conjunction with more moderate temperature decline, and consequently with less possibility of supercooling. Strong recrystallization is common in spite of lower temperature. The rate of reaction in the solid state is often so large that one cannot expect to find any kind of "primary textures". It has been shown indeed rather early by SCHWARTZ (1931a), e.g., that the beautiful intergrowths of minute lamellae of bornite and chalcopyrite which were in case of quenching excellently preserved, disappeared entirely, and a granular fabric remained after a cooling period of 60-hour duration. — The effect of water and the hitherto little known pH-concentrations of the ore-forming solutions furthermore make the investigations of ore formations in the laboratory especially difficult although certainly the foundation of all works concerning the geologic thermometer must be experimental. Some of those methods which are possible with silicates are here impractical but others are excellent as shown by the works of KULLERUD and his coworkers.

Especially disturbing is the additional fact, that there are many closely related ore minerals which are different only by their states of sulphurization, in contrast to the silicates where consideration hardly needs to be given to different oxidation states. Whether the one or the other state will be reached, depends naturally upon the temperature and pressure, but it depends also upon whether or not sulphur is present in excess, whether sulphur releasing or sulphur combining associations are present and particularly upon whether the chemistry of the country rock permits sulphur transfer into the country rock. Indeed very simply appearing relationships can produce great difficulties, e.g., the reaction $FeS_2 \rightleftharpoons FeS + S$ at a known pressure could be evaluated as a geologic thermometer in the case of a close association of pyrite and pyrrhotite. Because nearly all these deposits also contain sphalerite, which can dissolve quite large amounts of FeS, the reaction will be influenced in a manner which is hard to conceive. The *appearance of different sulphides-* and selenide-associations and as well as the *oxidation state* of the iron in iron ores *can consequently* be used as a geologic thermometer *only exceptionally* and with *greatest caution*!

One must consequently restrict oneself to reactions that are dependent neither on the accompanying minerals nor on the partial pressure of sulphur or of oxygen. Those reactions are especially melting points, allotropic changes, exsolutions, and the occurrence of phases *together*, which would be miscible or could combine at high temperatures, and which according to the observation could not have been the result of mixings and combinations. Here the simplest relations indeed furnish the clearest conclusions.

1. *Melting points* actually are useful only quite rarely, because they are mostly treated within complex multiphase systems of highly volatile components. In treating such complex systems one cannot use the beginning of solidifications or crystallizations of the individual components as geologic thermometers because these temperatures do not correspond to the experimentally established melting point of the components in a simple one-phase system. Melting points thus are useful only where single phase systems exist which will practically never be the case, or where immiscible liquids occur, where one or the other can be regarded as a single-phase system. A further limitation is that quite few minerals in the case of the most common ore-forming processes are molten at the temperatures of importance in the study of deposits. A different situation exists in the case of ore minerals in orthomagmatic rocks. Thus drops of native iron crystallized from the melt occur in some basalts. The iron (e.g., Ovifak or Bühl near Kassel) contains only very little Co and Ni, but abundant carbon (figs. 310, 311). The basalt melt must have been above 1145°, the eutectic temperature in the system Fe–C. In addition these same basalts show spongy iron masses, which did not coagulate into drops. If the drops and the spongy masses have equal C contents, which is not easy to establish, the possible temperature of formation would be narrowed down considerably. At a somewhat equivalent temperature — 1112° — smelting of galena begins in rare inclusions of this mineral in basalt (RAMDOHR, 1931 b). The high-temperature solid solution $CuFeS_2$ (resp., $Cu_xFe_yS_{x+y-n}$)–FeS ("chalcopyrrhotite") becomes separated into drops in many silicate minerals of magmatic rocks and thus shows that the separation temperature according to the diagram of MERWIN & LOMBARD (1937) should have been situated somewhere near 950° (actually this is, of course, a complex three-component system and pressure of sulphur vapor is undetermined!). — At lower temperatures, *bismuth* is molten (at atmospheric pressure, about 280°, at increased pressures somewhat lower than 280°). There are many occurrences, where it is certain that bismuth was separated already in molten drops, e.g., in feldspar masses of some pegmatites, in some galena crystals. In scheelite even "frost-blastings" ["Frostsprengung"] could be observed by crystallizing of such bismuth drops. In other minerals assemblages crystalline bismuth was precipitated undoubtedly directly from solution, therefore the temperature was below 280°. Where bismuth is found associated with silver, and without forming a solid solution with it, it must have been formed below the eutectic point in the system Bi–Ag which

is at about 250°. — The changes in the melting point with pressure naturally are to be considered, but are not large if the volume change during melting is small.

2. *Transition points.* For our purposes they can be divided into monotropic and enantiotropic or reversible, according to the suggestion of H. SEIFERT. The latter again can be divided into those in which a complete rebuilding of the lattice takes place, and into those in which only a small lattice-modification occurs; these we name "α–β transitions".

a) The *monotropic* transitions can, according to their nature, give indications only, since their "transition"-temperature actually marks but the temperature when the conversion begins in the laboratory at a measurable rate. Nevertheless, a transition which has *not* occurred is a clear indication that this temperature was not attained. Examples would be *metacinnabarite* $\xrightarrow{400^\circ}$ *cinnabar*, and *marcasite* $\xrightarrow{\text{ca. } 400^\circ}$ *pyrite*. If, e.g., in the Rammelsberg deposit the marcasite remained preserved from Devonian until now and if it has endured the shearing movements, it is strongly suggestive that 400° has *not* been reached after the formation of the deposit. In this group belong further:

Maghemite, which changes rapidly to *specularite* at $< 550^\circ$; and *aragonite–calcite* (at about 460°), as abundant as gangue minerals, as also a few others.

b) The *reversilbe* transitions are much more important and are numerically very widespread. The two groups of reversible transitions cannot be sharply separated. Nevertheless complete lattice reconstruction requires considerable changes of the internal energy. Such transitions can easily become retarded or passed by rapid cooling or by heating whereas the α–β transitions set in rapidly. The former commonly are also strongly pressure-dependent, but the latter only rarely so. So-called isometric chalcocite previously classed in this group does not belong to the reversible transitions because it represents, accord. to the investigations of BUERGER (1941), a special phase between Cu_2S and CuS, whereas the transition of *hexagonal high-chalcocite* into the common *rhombic* variety at 103° is a typical α–β transition, that apparently always occurs with decreasing temperature (fig. 338). The conversion of *cubic argentite* formed at higher temperatures into monoclinic acanthite occurs apparently exactly at 179°. A close association of both, the paramorphic and the primary monoclinic forms that occurs without strong paragenetic hiatus in Freiberg, e.g., means that the part of the deposit in question was certainly formed very near to 179°. Similar indications can be derived from *naumannite* (133°) and *hessite* (150°), as well as *stromeyerite*; the last, however, is almost always formed at low temperatures; stromeyerite converts already during preparation through slight heating, and shows relict textures of the cubic form after reconversion. The relations between guanajuatite and paraguanajuatite have not yet been investigated. Also *bornite* still presents many problems. It often shows textures that, clearly approved, indicate an origin from a cubic high temperature form. Since at higher temperature a very remarkable miscibility with FeS, NiS, and ZnS exists, a sharp transition temperature cannot be given. In any case it is high, approximately at 500°. Likewise *cubanite*, which is orthorhombic at normal temperatures, appears to have a hexagonal high temperature form. The rare upgrown crystals are rhombic in form and habit, those which occur as exsolution bodies in chalcopyrite are made up of a mimetic hexagonal lamellar pattern (fig. 431). *Stannite* (cf. fig. 390) is cubic at higher temperatures and becomes at lower temperatures in most cases tetragonal. Because stannite is a complex solid solution of varying composition, the exact determination of the transition temperature has little significance. In case of need the transition temperature must be determined for each instance. Further examples of allotropic transitions are: *schapbachite* (see below), probably *canfieldite* (in part), *cobaltite,* this only with transition delays that are difficult to explain during artificial heating (thus *not* useful as a geologic thermometer!), *iron.* because of strong dependence on inpurities (and its rarity!) is not useful here, and *sphalerite–wurtzite,* which are treated elsewhere.

3. *Mixed crystals and exsolution, decomposition of compounds.* Many substances with similar lattice structures are hardly miscible or miscible in traces only at common temperatures, but they rapidly become miscible at higher temperature. With decreasing temperature these high temperature solid solutions then become unstable again, but can remain preserved any length of time and can then be used as a geologic thermometer only through the fact of the miscibility or immiscibility; or the unstable compounds decompose and can show then through the "exsolution textures", that at one time a solid solution existed, and that the temperature required for its formation had been reached.

One experimental difficulty must be discussed: The beginning of miscibility is determined by measurement of the homogenization temperature of an exsolved solid solution. If I bring one such crystal through heating perhaps to about 500° for 24 hours just to the beginning of homogenization, it could be conceivable, that I should attain the same effect at 400° if heated for 10 times 24 hours, and perhaps at 300° with 100 times 24 hours. In this way of reasoning lies naturally the possibility for the largest mistakes. The sparse experimental data seem to show, however, that for sulphide and oxide ore minerals, the miscibility at a certain temperature rises so rapidly, that even with very long exposure *below* this certain temperature *no* homogenization occurs, above that certain temperature, however, *complete* homogenization occurs in a very short time.

In contrast to the above said one could think that the absence of miscibility or of exsolution textures were evidence that the miscibility temperature in question has never been reached. In many cases this may be true; in others, however, recrystallization processes make exsolution textures disappear completely even in a very short time (e.g., the work of G. M. SCHWARTZ 1931a). Besides, oriented primary intergrowths or oriented replacements must not be confused with exsolution, since texturally the oriented intergrowths and the solid solution formation are very closely related phenomena (similarity of the lattice dimensions!). Exsolution textures are generally, if not always, oriented textures.

To the first group, which is long known and evaluated accordingly, belong e.g., the assemblages of *native copper* and *native silver* from Lake Superior, from Corocoro and elsewhere. Through lack of miscibility it can be recognized that these assemblages must have been formed at low temperatures. — That the ruby silver ores (*pyrargyrite* and *proustite*) practically never form solid solutions in natural occurrences, although they often occur associated with one another, shows that the temperatures of formation are very low, since from a melt they form a solid solution in all proportions of the two endmembers. This agrees with old empirical knowledge. — *Sphalerite* can dissolve very much FeS at high temperature; if iron-poor sphalerite occurs isogenetically with pyrrhotite, this is an indication of a low temperature of formation. In individual cases the ZnS–FeS-solubility can quantitatively be determined quite exactly. If during mechanical movement the pyrrhotite exsolves from iron-rich sphalerite, this means surely that this movement took place at low temperature. — *Fahlores* of different compositions associated with each other indicate low temperatures of formation, those with complex compositions, especially with unusual components, formation at high temperatures.

Ilmenite–hematite solid solutions change into pseudobrookite by strong heating (around 1000 °); pseudomorphs of this type — they are known from kimberlite and basalts — therefore indicate very strong heating. Because of the special influences which can lead to different iron contents and oxidation states, the absence of pseudobrookite, however, is not to be interpreted to be definite proof of temperatures lower than 1000°.

At Cobalt City abundant breithauptite is found with niccolite in textures which exclude that a former solid solution of both existed. In the deposits of Les Chalanches, France, which has a mineral assemblage quite similar to that of Cobalt City, both minerals, however, occur in myrmekitic intergrowths in varying proportions. These intergrowths correspond clearly to former "arite"-crystals, the optical orientation of which is preserved in the intergrowth aggregate, and whose coating by other arsenides is still present. Les Chalanches must therefore have had a higher temperature of formation during the period in question.

The presence of a zonal structure alone can occasionally suffice to establish that no especially high temperatures have been attained, because at high temperature here too homogenization is rapidly attained. Exact data are as yet not available but one knows, e.g., that galena possesses strong zonal structure only in deposits of relatively low temperatures of formation. Also the different possibilities of double salt formation or their absence could be evaluated, e.g., of PbS with Sb_2S_3, Bi_2S_3, As_2S_3 or Ag_2S and Sb_2S_3, however, almost no preliminary work has been done.

The complicated conditions in the system Ag–Sb cf. part II under "dyscrasite".

Exsolutions of solid solutions or analogous *decomposition of compounds* are by far the most important. They can give in very many cases a good, and in part of the cases a quite safe indication of the temperature of formation. Of course, the exsolution character of the aggregate must be clearly ascertained. Finger-thick cubanite lamellae and coarse-lobed irregular intergrowths can be exsolution features, but uniform, very fine-grained, scattered disseminations need not be exsolution textures. If an exsolution texture has been ascertained, then this means that a homogeneous high-temperature solid solution previously has existed and later has become unstable. Thus exsolution textures, e.g., are known from the systems: Au–Cu, Pt–Ir, Sb–As, Bi_2STe_3–Bi_2Te_3, FeS–(Fe, Ni)S, ZnS–FeS, Cu_2FeSnS_4–$CuFeS_2$–ZnS, $CuFeS_2$–$CuFe_2S_3$, PbS–$AgBiS_2$, PbS–Bi_2STe_2, Ni_3S_4–NiS, Cu_2S–AgCuS, Geocronite–Jamesonite, ZnO–Mn_3O_4, $FeTiO_3$–Fe_2O_3, Fe_3O_4–$FeTiO_3$, Fe_3O_4–$FeAl_2O_4$, Fe_3O_4–Fe_2TiO_4, $FeCr_2O_4$–$FeTiO_3$, Mn_3O_4–$MnFe_2O_4$ and so forth. Almost all are hardly pressure-dependant, since the volume effect in solution and exsolution is commonly very small; unfortunately most are still scarcely studied experimentally, and some are abundantly complex in miscibilities. More acccurate data about temperatures exist for Au–Cu, Sb–As, PbS–$AgBiS_2$, $FeTiO_3$–Fe_2O_3, Fe_3O_4–Fe_2TiO_4, "vredenburgite", and others.

Some experimental data also exist on the relations of cubanite and chalcopyrite, Unfortunately the results of SCHWARTZ (1935a) and BORCHERT (1934) contradict each other somewhat on this important subject. The former evidently has identified intergrowths as cubanite which should better have been called "chalcopyrrhotite" (or i.s.s. in the sense of *Cabri* (cf. p. 545). Therefore the common form of cubanite indicates a temperature of 250° accord. to BORCHERT, not 400° to 450°, as SCHWARTZ thought (certainly false!). The absence of cubanite although for very different reason, does not demonstrate a formation temperature below 250°. Recent work done by KULLERUD will lead to more exact data.

From the beautiful bodies of hematite in ilmenite and ilmenite in hematite, which commonly separate out as two generations, one knows that the small bodies of the younger generation are exsolved between 500° and 600°, and the large bodies of the older generation at about 700°. Likewise similar relations exist for titanomagnetite but experimental investigation is made difficult by accompanying reactions. The preliminary investigations on the relation of bornite–chalcocite of G. M. SCHWARTZ (1939) show that at around 225° unlimited miscibility exists, a temperature which is doubted.

4. *Other applicable observations* have been discussed in great number in the literature. One of the oldest methods of general geologic thermometer measurements, already traced back to SORBY, was investigation of the mother liquid inclusions in relation to the included libellae. This method has also obtained importance for ore minerals at the present. Because of its nature this method is practical only on nonopaque minerals, such as sphalerite, ruby silver, cinnabar, since undisturbed liquid inclusions have to be studied. The fundamental idea is this: The mineral grew in the mother liquid and at the instant of formation enclosed a small quantity of it *without* inclusion of air. With temperature decrease the solution contracts much more rapidly than the crystal, therefore an "air" bubble must appear in the liquid inclusions, which in reality is filled instead with water vapor. If then the mineral is heated again to the starting temperature (at which the inclusion occurred), all bubbles will disappear at once. This temperature can be easily measured. To be precise a correction factor should be applied to the pressure that existed during formation, taking into account the compressibility of the crystal. This is, however, so small that in most instances it can be neglected. NEWHOUSE (1932) thus has quite accurately established the temperature of formation for many hydrothermal sphalerites. For beryls of gem quality the writer succeeded in doing the same.

More recently the SORBY method was extended through the decrepitation method. A little above the temperature where the bubble disappears the fluid exerts a pressure on the enclosing mineral so strong that it cracks with a slight "pop". With coarsely powdered material the "popping" can be made audible by means of an amplifying arrangement. This naturally is possible also on opaque minerals. One must, however, be aware of many sources of error.

In some more accurately investigated metals after mechanical stress and heating, *recrystallization* occurs in such a regular manner that one can conclude from it the temperature of heating. The attempt to apply this to ore minerals, is in itself not hopeless. Unfortunately in the only case, which appeared in the literature, it seems that it failed, as apparently the time factor is to be accounted for in a manner presently not yet known. In any case recrystallization is so easy for many ore minerals (PbS, Sb_2S_3, CuS, $CuFeS_2$, ZnS) that they certainly cannot be used in this method. The method appers promising, however, for magnetite and ilmenite, and especially for the only exceptionally recrystallizing ore minerals pyrite, arsenopyrite, cassiterite, chromite. *Tables of mineral assemblages* which, because of superior experimental data, especially for the saline deposits, permit excellent determinations, in part accurate to a few degrees, have, with respect to the ore minerals not yet evolved beyond roughly qualitative indications. Very many associations hitherto almost axiomatically regarded as rules without exceptions, have been found to be unreliable. Molybdenite in Kupferschiefer, magnetite in unmetamorphosed iron ores of the minette type in the Liassic, tourmaline in epithermal veins in Bolivia; on the other hand tetrahedrite in pegmatites, covellite in contact limestones, specularite in basalt inclusions, native Ni–Fe along with native Cu in serpentines, all show that surprises and accordingly false inferences are easily possible here.

The *literature* about geologic thermometers and about investigations which can be evaluated for geologic thermometry is quite dispersed. Worthy of mention are works of BORCHERT (1945), BOWEN (1928), BUERGER (1941), EDWARDS (1947), GAUDIN (1938), GUILD (1928), MASON (1943a), MERWIN & LOMBARD (1937), NEWHOUSE (1932), PEACOCK (1940), SCHNEIDERHÖHN (1922), SCHÜLLER (1952), SCHWARZ (1931a), SEIFERT (1930), and the writer (1931a), more recent ones by KULLERUD (1953), BUDDINGTON & LINDSLEY (1964), MOH (1961).

Geologic barometers are much less important. Disregarding the very roughly qualitative and sometimes purely speculative data, scarcely anything is found about it in the literature. The difficulties are of varied nature:

1. In most conceivable cases one has to work with the sulphur or oxygen contents of the mineral assemblages. Individual cases with Se, As and others, are obviously also to be evaluated; these are on their part, however, very temperature dependent, naturally always in the sense that the higher S or O states will be stable with decreasing temperature.

2. We are dealing here surely not with a system in which only one pressure as total pressure or partial pressure was active, as, e.g., the H_2O partial pressure at one (or few) atmospheres total pressure in regard to the water content of zeolites, but on the contrary with the partical pressure of many substances, in particular S, As, O (or their combinations as H_2S and so forth) combined.

3. The effect of the pressure is very strongly influenced by the relative and absolute affinities of the different metals, for sulphur and oxygen, as examples.

4. As in another connection already stated, the partial pressure of sulphur is dependent on the receptivity of minerals in the country rock outside the deposit.

Difficulty 1 can be removed through knowledge of the temperature from geologic thermometers, just as now and then 2 can be eliminated through observations, which, e.g., exclude substantial amounts of S from the reaction. Difficulty 3 in addition can be removed somewhat similarly through the establishment of the fact that only one metal was present, and finally 4 can be eliminated through an impervious and completely inert country rock. However, it can hardly be expected that all difficulties could be removed *simultaneously*.

Further hints: a) Rather rare cases are known, where chalcopyrite is changed into covellite + pyrite ($CuFeS_2 + S = CuS + FeS_2$). In regard to the high stability of chalcopyrite this is naturally possible only at very high partial pressure of sulphur. b) Assemblages such as that of the Pinge Riekensglück, where pyrite + pyrrhotite + + magnetite + fayalite occur together, could become useful geologic barometers connected with reliable geologic thermometry, since the oxygen partial pressure obviously was very low (the FeO-compound fayalite is not decomposed into Fe_3O_4 + + SiO_2) and metals other than Fe are lacking. The association pyrite + pyrrhotite is therefore dependent only on pressure and temperature, the pressure thus perhaps being determinable. c) The association of exsolutions of Fe_2TiO_4 and $FeTiO_3$ in many titanomagnetites requires a very low, but with cooling apparently rapidly increasing partial pressure of O (RAMDOHR, 1953b). These existing relationships could permit experiments. d) Frequently occurring magnetite and chalcocite, in the siderite veins of the Siegerland, which were altered by a basalt contact, presuppose already such complicated reactions that in spite of the simple and clear relationships an evaluation of such occurrences appears hopeless — and so will it mostly be!

III. RELICTS

Relict structures from preceding stages of development are of special significance for the recognition of the history of a deposit. Let us remember, that already through the course of formation of even the simplest hydrothermal deposit, only with respect to the temperature, at least three stages must be passed through: 1. The entering of hot solutions into a cooler environment, whereby precipitations must occur, 2. the *climax* of the influence for supply of material and heating where under certain circumstances the early formed mineral associations again can be changed by additions of other elements, or simply into high temperature associations, and finally 3. the time of *decrease* of the thermal activity, where again, due to a different chemical composition of the solution, the minerals that were precipitated in both of the first stages are corroded; and perhaps elements that were already previously in the solution, but did not precipitate are now precipitated. The numerous changes in mineralization in the late phases thus by no means need to represent additions of new material. I refer especially to very frequently occurring late mineralizations of certain types of deposits (Sb-minerals in lead-zinc ore veins, Te-minerals in gold quartzes, Ag-minerals in Co-, Ni-, As-deposits). Fundamentally everything is a relict which originated in earlier stages of formation and which is no longer in equilibrium (materially, texturally, or paragenetically) with the later phases. We will mention only the cases, where a thorough transformation occurred, and where minerals and textures in their present environment appear to be "foreign" in regard to matter and forms. E.g., if in a limonite mass of a gossan instead of botryoidal textures characteristic for this mineral cells in rhombohedral forms occur, it becomes evident that once a siderite mass had occurred here, or if magnetite remnants are found in a dynamically metamorphosed mass of specularite-schists of the epizone, and so forth. Naturally the number of these relicts, to which belongs each remnant of replacement, each exsolution, each pseudomorph, is legion and it is not possible to discuss here more than a tiny number of the observed materials. Therefore I present only a quite small selection of examples which appear remarkable to me; for part of them, also, instructive photographs are available. Classification is made only quite roughly into relicts of matter and into relicts of form, however, it should be noted that they often cannot be sharply divided. A textural relict, e.g., can only be attributed in regard to its origin with certainty, if there is still some of the original material left, possibly thoroughly altered, and vice versa.

Relicts of matter

Replacements can actually only be recognized with certainty if they contain material or textural relicts. Of importance can be that during the replacement processes sluggishly reacting minerals are formed as intermediate stages, which then prevent or delay the complete destruction of the original material. One speaks of these minerals in petrography as "armored relicts", a term that one can well introduce into ore-microscopy.

Fig. 244 40 × RAMDOHR
Otjosongati, South-West-Africa

Mesh texture of *limonite* with some *chalcopyrite* (clear white). A typical relict fabric of originally everywhere present chalcopyrite parallel to (111). Gossan material

An almost trivial example is given by the chalcopyrite remnants in limonite masses. Frequently they are protected against fast destruction by chalcocite shells, which advance slowly towards the center of the chalcopyrite (fig. 244). In other cases a relict is formed, apparently through presence of quite acid and highly concentrated ground waters, parallel to (111) of the chalcopyrite with (0001) covellite plates embedd-ed. These plates in part protect the chalcopyrite, in part may be recognized (fig. 207), after the chalcopyrite has been dissolved as a framework which allows recognition of its former distribution and even of grain outlines.

Equally very common is the case, wherein hypogene tetrahedrite is replaced by galena with the formation of bournonite. The reaction selvage of bournonite, that develops on the fahlore protects for some time the fahlore from further destruction and only thus permits identification of the latter.

Coatings without replacements occur similarly. Occasionally magnetite crystals containing fragments of crushed chromite are found in some highly metamorphosed deposits of talc schists or compact talc rock ["Topfstein"], which are not immediately recognized as former peridotites. No indication exists that these chromites were possibly replaced by magnetite; chromite only forms crystallization nuclei for magnetite. Without these chromite relicts such rocks that result from conversion of dolomite to talc would be very similar, and could scarcely be distinguished with certainty from those resulting from peridotites.

The *original material* can be completely *destroyed*; a *remainder* of its *material composition*, however, can be preserved. In diabases, green schists and similar rocks we find spots of titanite or anatase, earlier named "leucoxene" in petrography, in roundish clusters in place of former titanomagnetite crystals, oblong clusters in place of former ilmenite. The interpretation is ascertained, since all transitional phases from titanomagnetite to the skeletal ilmenite pseudomorphs are present (fig. 246). Fig. 247 shows the octahedral arrangement of a former magnetite in josephinite, fig. 245 an example of one such relict of unknown explanation perhaps andradite.

In this connection the investigation of the ores of the Witwatersrand is of great genetic importance. They carry abundant rutile as the most common mineral except for quartz, pyrite and perhaps sericite. This rutile is to be explained as a relict of "black sand", titanomagnetite — ilmenite — haematite sands, which originally accompanied the placer of the Witwatersrand conglomerates just as any other gold placer. The iron content of these sands in part has supplied the material for the pyrite. — Since occasionally, but quite rarely, very typical texture relicts occur just as those mentioned, no doubt exists as to interpretation (fig. 252).

Texture relicts exist in an enormous variety. Some examples are mentioned in roughly the same order of presentation as in the section on page 254 etc.

Zoning can still be recognized and allows inferences to be made about the original mineral, where this has completely vanished. Thus I have observed the complex intergrowths of niccolite, chalcopyrite, fahlore, and silver minerals, which reveal, even though they are very fine grained and finely interlaced intergrowths, through the zonal arrangement of these components the former zoning in very coarse smaltite crystals, which have completely disappeared. This interpretation is verified in another way, since also former coatings of these smaltite crystals by safflorite are preserved in their typical form (fig. 249).

Exsolutions can produce good relict textures, if the exsolution bodies are resistant toward replacement of any sort. The best known and really universally distributed examples are the ilmenite skeletons of titanomagnetites, which are mostly preserved especially in sulphide replacement of magnetite. Thus ilmenite skeletons were found in chalcopyrite, bornite, pyrrhotite, pyrite, arsenopyrite, pentlandite, several gangue minerals and in intergrowths of these. The form was often such that it could be recognized that aggregates of fine grained sulphides or sulphide mixtures exist now in place of very coarse grained titanomagnetite, or also vice versa (figs. 248 a, b).

Titanite skeletons, resulting from the ilmenite skeleton of exsolved ilmenite-hematites, are also not rare, and, e.g., often present in the country rock of hydrothermal veins.

Oriented inclusions can be of equal significance. Fig. 258 (sketch) shows krennerite (pseudomorphous after nagyagite), which is included in a former digenite (?). The latter is completely replaced by chalcocite and bornite.

Reticulate forms can permit the original net mineral to be recognized in the case of its complete solution by their form and the mineral assemblage. That very many occurrences of "reticulate" silver, argentite, and others in the Co–Ni–Ag-assemblage originate from previously existing reticulate bismuth was first recognized and emphasized by KEIL.

Characteristic *cataclastic textures,* characteristic *exfoliation* of minerals with layered lattice structure, *cleavages, translations,* and *formation of pressure twins,* all can be preserved excellently. They can aid in the determination of the original material and a part of its history. Many smaltites are traversed by fractures that are lined with crusts of safflorite. It happens that the smaltite crystals are completely dissolved, the

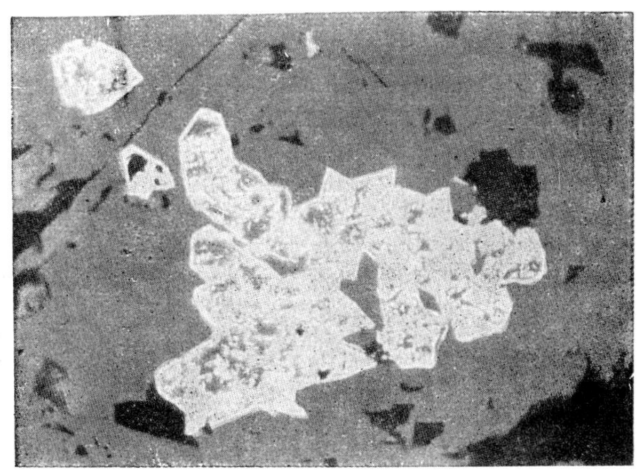

Fig. 245 150 × RAMDOHR
San Domingo, La Huerta, Mexico

Pseudomorphs of unknown nature, perhaps from whorl-like trillings of *enargite,* now consisting of *chalcocite*

Fig. 245a 1000 ×, imm. RAMDOHR
North Mine, Broken Hill

Cubanite, section // the pseudohex. (001) breaking down to chalcopyrite

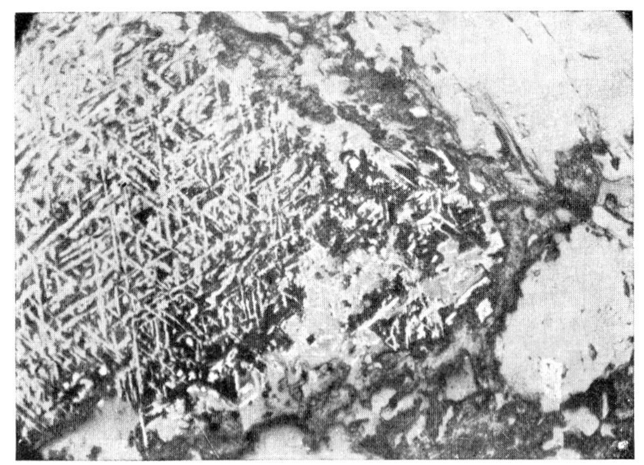

Fig. 246 150 × Ramdohr

Melteig, Fen Region, Norway

Former ilmenite skeleton of titanomagnetite in "Damkjermite". The somewhat lighter reflecting part of the lamellae (top) consists of *rutile*, the main mass of *titanite*

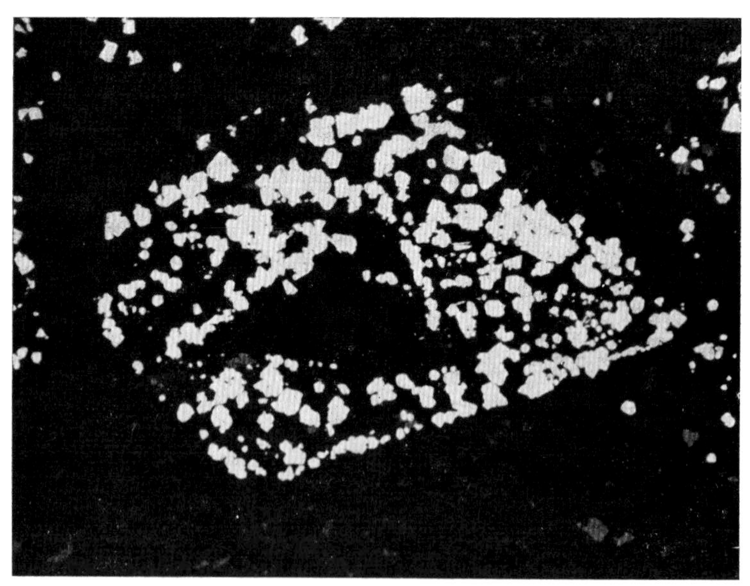

Fig. 246 a 55 × Ramdohr

Loraine Mine, O.S.F.

A large crystal of *amphibole* is destroyed forming white rounded grains of *pyrite*. The surrounding silicates are nearly black

fracture fillings, however, are preserved (e.g., specimens from Niederramstadt, which the writer earlier illustrated; fig. 871).

Intermediate weathering textures can reveal the original materials after the mineral assemblage has been completely changed to limonite, even if the compact limonite mass is in itself not very typical. Thus the gossan of the Matooster Mine (Farm Vlakfontein 902) contains coarse limonite in which the polished surface still shows pronouncedly the "bird's eye" texture that is so characteristic of the change of pyrrhotite to pyrite, yet there is no material relict whatsoever of a sulphide (fig. 250).

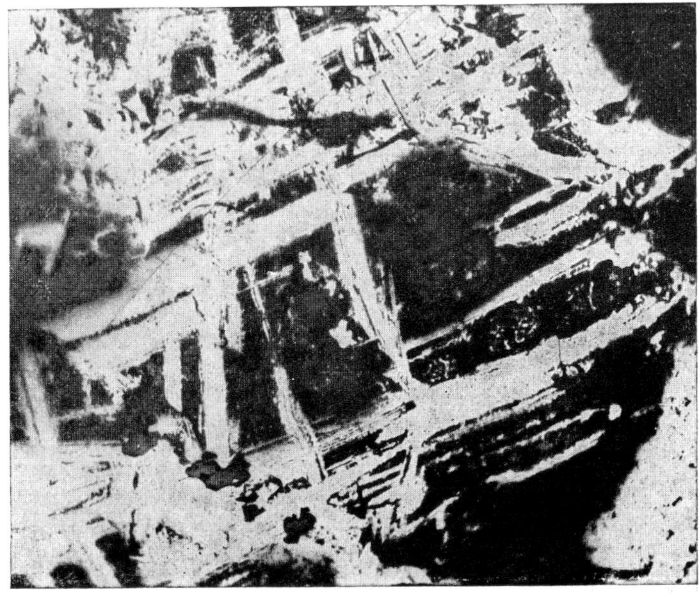

Fig. 247 60 × RAMDOHR
Josephine Co., Oregon, U.S.A.

Josephinite (Ni, Fe) pseudomorphous after magnetite, whose exsolution net is well preserved. The josephinite portion is in itself extremely fine grained*

* Recently it became probable that this texture is a remnant of dissolved andradite, not magnetite

Myrmekitic intergrowths have occasionally lost one component through weathering or some other type of replacement. In some cases one can interpret the nature of the vanished mineral solely from the form. Regarding the amazingly large distribution of most different myrmekites, however, only conclusions that are very cautiously drawn are permitted.

Gel structures can be recognized in regard to composition as well as to structure through the globular-botryoidal form, e.g., of the arrangement of foreign inclusions, even if the main mineral has undergone rather coarsegrained "collection" crystallization [Sammelkristallisation], or even if it is completely replaced by another mineral.

This number of different possibilities could be extended considerably, and each new possibility could be illustrated through very many examples. It may be sufficient to refer to figs. 253—258.

Fig. 248 a 170 × RAMDOHR
Ookiep, Cape Province, S.Africa

Characteristic lamellae of *ilmenite* in four directions in *bornite* and *gangue*. These lamellae are a relict of exsolved titanomagnetite. Here the advance of the softer bornite against the photographically also light *magnetite* is clearly visible. The ilmenite skeleton remains intact

Fig. 248 b 250 ×, imm. RAMDOHR
Kaulatunturi near Petshenga (Petsamo)
White Sea coast

A peculiar relict structure! Unmixed titanium-magnetite is replaced by *pentlandite* (white) and *pyrrhotite* (greyish white in different tints). The unmixed ilmenite-lamellae remain nearly unaltered

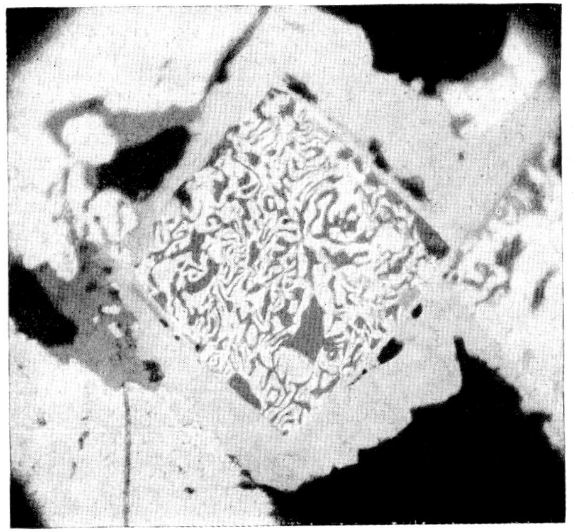

Fig. 249 500 ×, dry RAMDOHR
"Peru"

Myrmekite of native *silver* (white) and *fahlore* (light grey) in magnificent development as components of an Ag–Ni–Co-deposit. The myrmekite here is apparently a decomposition product of a complex minerai (cf. fig. 281)

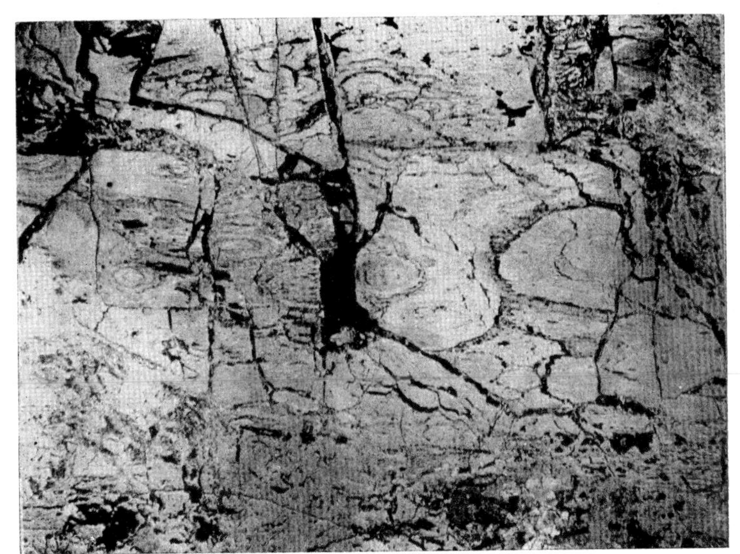

Fig. 250 8 × RAMDOHR
Matooster Mine near Rustenburg, Transvaal, S.Africa

Limonite which was formed via pyrite/marcasite from pyrrhotite still shows typically the "birds-eye" textures of the intermediate transformation of pyrrhotite into pyrite

IV. FURTHER POSSIBILITIES OF GENETIC INTERPRETATION OF TEXTURAL CHARACTERISTICS

In the preceding sections much has been said about the evaluation of the textures to the interpretation of deposits, e.g., as regards typomorphic textures, geologic thermometers, and relicts. In these cases it is often emphasized that the data are, and even have to be, quite defective. It is, however, desirable to present a number

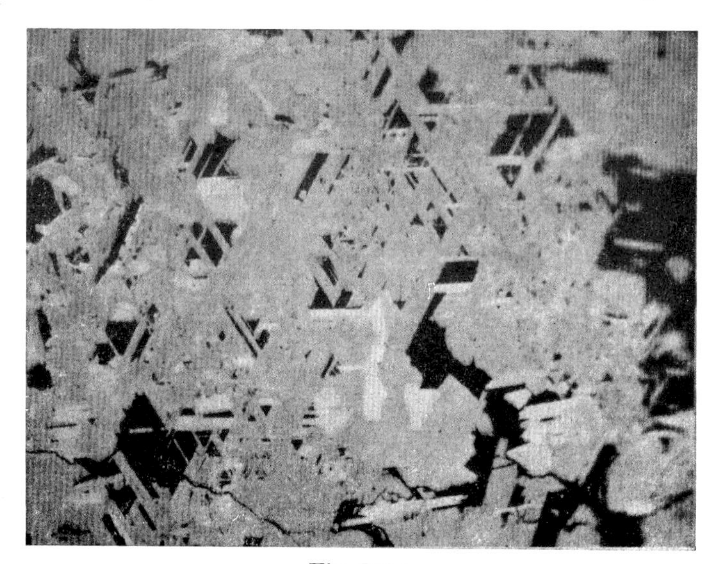

Fig. 250 a

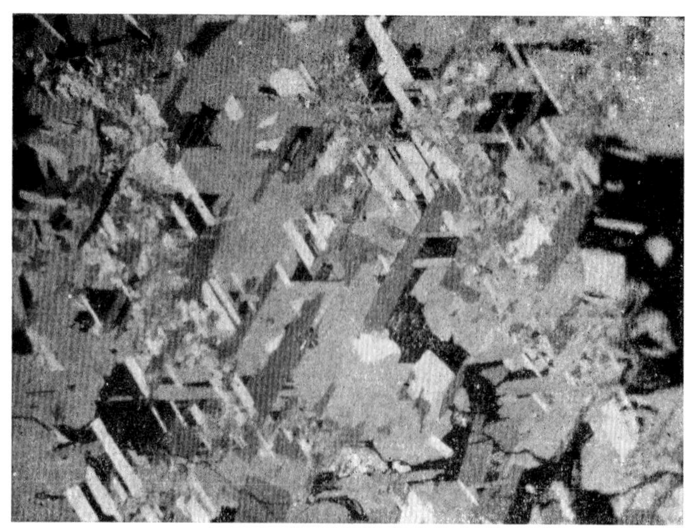

Fig. 250 a, b 250 ×, imm. Ramdohr
a) one nicol b) + nicol

Cut // (0001) of a former pyrrhotite, which has now changed to a triplet system of marcasite-lamellae, partly pyrite as a filling material

of possibilities of interpretations, which were not covered in the above sections, without special order and in random choice. Mentioned may be one case: The fact that the selvage of a vein sticks tightly to the country rock ("frozen to the walls") is in itself rather beside the point and not in connection with the gross genetic relations, but it is a very good indication of high temperature of formation. This is because at high temperatures no clay-like minerals are formed in the selvage. The margin of the

Fig. 251 325 ×, imm. RAMDOHR
Mantrain, Mine, Tarkwa, Ghana

A former hematite with excellent unmixing of *rutile* is entirely replaced by *chalcopyrite*. Rutile remains unaltered

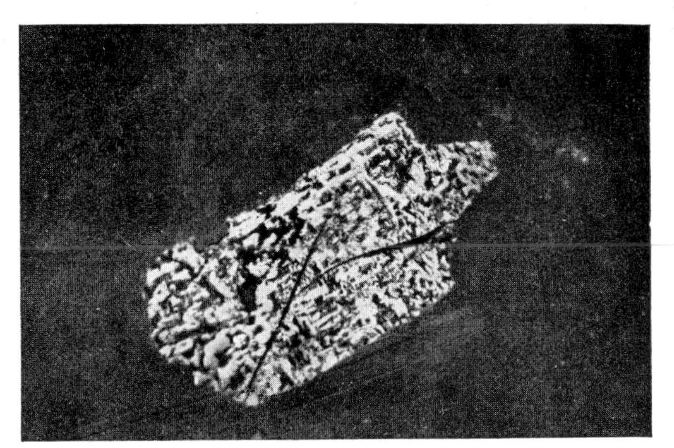

Fig. 252 225 × RAMDOHR
Rietfontein near Johannesburg, S.Africa

A relict of an "ilmenite-hematite" grain with exsolutions of ilmenite parallel to (0001) and new development of rutile normal to (0001). The hematite is now dissolved away and the ilmenite completely converted into *rutile* with very good preservation of the texture. Bright white is some *native gold*

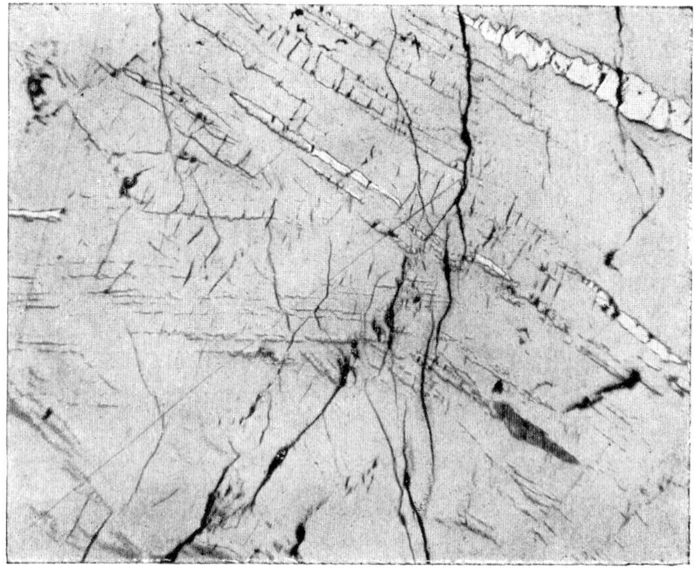

Fig. 253 170 ×, imm. Ramdohr
Jakobsbakken Mine, Sulitelma

Typical relict texture! Former lamellae of *cubanite* in *chalcopyrite* (white) are destroyed with the development of *pyrrhotite*, which, however, still allows the old form of the cubanite to be recognized

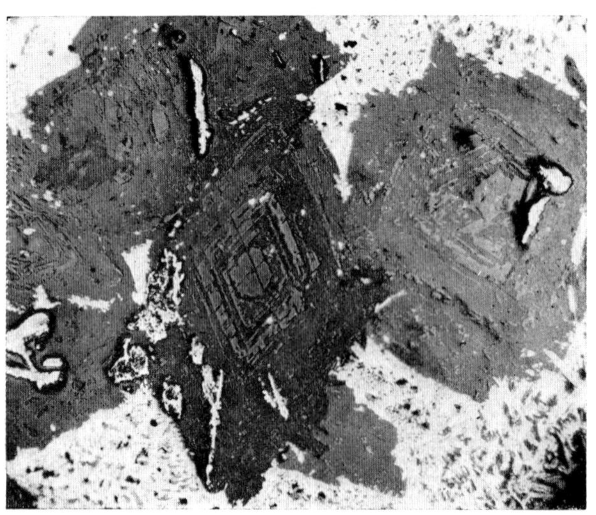

Fig. 254 25 × Ramdohr
Marienberg in Saxony, Germany

Ankerite (light grey, high relief) is replaced by — only partly in equal orientation — *calcite* (partly dark grey, partly light grey, softer). Beside it *jamesonite* (white) and secondary *marcasite* (white), with high relief From so-called "Weissgültigerz"

Fig. 255 170 × RAMDOHR

Baia de Aries, Siebenbürgen

Peculiar pseudomorphs after a thin tabular mineral in *quartz*. The former nature of the mineral is unknown, apparently nagyagite, even though the pseudomorphs now also consists of nagyagite. — The other minerals are alabandite and very little tetradymite

Fig. 256 170 × RAMDOHR

Garpenberg near Falun, Sweden

Minerals leached out by special weathering conditions. The former cement of roundish sphalerite grains composed of *galena* is pressed together to a sort of briquette through rock pressure

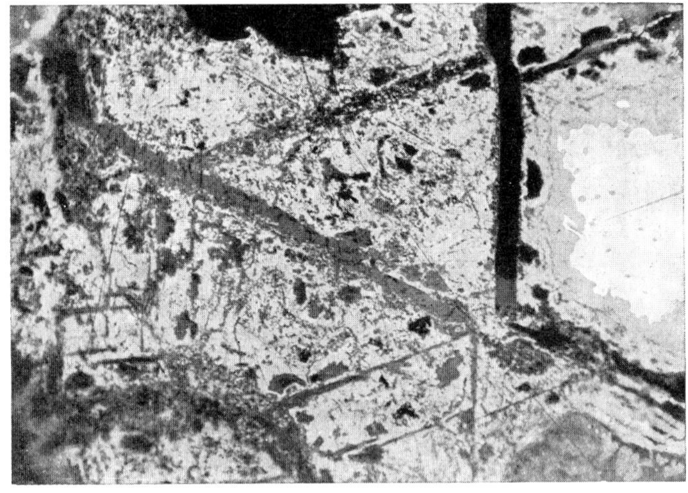

Fig. 257 25 ×, imm. RAMDOHR
Broken Hill, N.S.W.

Chalcocite (greyish white) and *cerussite* (dark grey) in place of coarse lath-shaped *galena* (pure white)

vein and the immediately adjacent rocks are strongly and mutually silicified. Thus there are inumerable individual observations in ore-microscopy, which, more or less "accidentally", can supply good genetic indications and which were not taken up in the preceding sections.

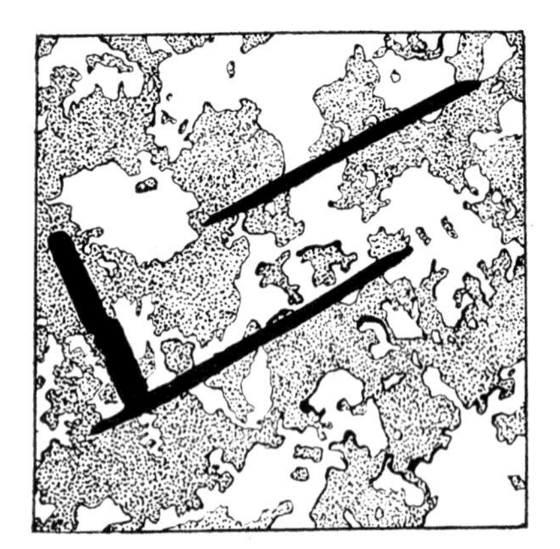

Fig. 258 A. SCHERBINA

Krennerite (pseudomorphous after *nagyagite*) in a mixture of previously homogeneous chalcocite and bornite. Chalcocite has even now the same orientation

Some cases should be cited: It can happen, that one can be in doubt about a metamorphosed sedimentary iron ore, whether it represents a former placer, or whether magnetite was formed from other ore minerals through metamorphism. Without any analysis an assiociation of exsolved titanomagnetites and common magnetite indicates, that the second interpretation is out of the question.

In many occurrences of fahlores, noble Ag-ores, etc. the hypogene-cementative interpretation is not easy. If we find here associated, e.g., galena in compact form, perhaps even intergrown with sphalerite containing chalcopyrite exsolutions, in one association then hypogene formation is probable, and galena in skeleton forms and other low temperature forms, cementation will be likely for the accompanying ore minerals. In a similar case: silver can be hypogene as well as cementative. The textures are little revealing, since it precipitates in both cases on less noble ore minerals. On the other hand it often has the tendency, even in definitely cementative precipitation not to coat or to enter the mineral causing precipitation.

If the deposit shows an auriferous silver (electrum), along with common silver, at least a large formation hiatus of both silver types exists, which means generally cementation.

Manganese ore minerals can give rise to extremely variable relations with regard to both texture and mineral content. This is in spite of the fact that ultimately most of them were of sedimentary origin. Iron can come out of complex Mn–Fe minerals, and conversely it can also enter into simple minerals. The study of certain associated minerals, dominantly from the group of the silicates, can supply certain or probable indications.

Certain magnetites, can regardless of origin, become very coarse grained. One will easily draw false conclusions about the complexity of the history and of the migration of material, if one sees such gigantic, homogeneous crystal aggregates in which the crystals are as much as 15 cm in diameter. Slight weathering or also etching shows the old, slightly wavy, strong parallel structure the individual layers of which are fractions of a millimeter thick. Very likely this is an original depositional texture. Thus only a minimal amount of material migration can have taken place during metamorphism.

The chosen examples are to be a stimulation which the experienced investigator of mineral deposits can supplement according to his inclination.

C. The relationship of ore textures to industrial minerals and beneficiation problems*

The microscopic study of ores and mineral products provides valuable assistance in mineral beneficiation, i.e. the processing required for the concentration of minerals form crude ores by:

the identification of minerals, i.e. the recognition of the compounds formed by the constituent elements which have been determined by partial or complete chemical analyses, and the understanding of the *pattern of the mineral intergrowths*, the *size of the intergrown particles* and the *nature of the boundaries between intergrown particles.*

The *mineralogical identification* is of importance because quite different mineral assoications can at times yield the same analytical results. This applies especially to associations of various "sulphosalts", the different oxides of Mn, and to members of the group: pyrrhotite — cubanite — chalcopyrite — bornite — chalcocite, in addition to the various oxide ores of iron and iron silicates. By contrast the influence of complex solid solutions is not nearly so troublesome as perhaps it is in the calculation of rock analyses. With a large number of mineral associations possible, flotation behavior, and in others, the magnetic response or gravity effects can be entirely different so that there must be a careful determination of *all* components. When rare elements are involved, especially the precious metals, the question always is, whether they can be enriched to a simple, clean high grade concentrate (for direct smelting or sale!), or whether their recovery is possible only as a "by-product" of the processing of other components. An example which has gained special significance is the recognition of renierite as a germanium-bearing mineral in many copper ores.

* See p. 281 for literature on this section.

Often a single, carefully studied polished section can establish whether an ore should be mechanically separated after coarse grinding, treated by fine grinding and floation amalgamated, leached, or smelted directly; whereas, without it, many useless, time consuming and expensive experiments are necessary. In respect to the precious metals a case in point is the process of gold recovery in which the chalcocite reacts with the chemicals of amalgamation or of cyanidation, thus rapidly making the process inefficient.

The pattern of the *mineral intergrowths* is important because it can give an immediate indication as to whether at some selected sizes the portions of the intergrown grains will be large or small. The size of the *intergrown particles* (liberation size), which naturally must be established statistically by numerous observations, directly gives the maximum particle size to which it is necessary to grind. The nature of the *boundaries between intergrown particles* discloses whether, as is desired but is seldom the case, the mineral grains will break at the boundaries or not. If such breakage does not occur, finer grinding will be required to obtain liberation. These three factors work closely together.

The problems mentioned above have become apparent following the advent of modern ore-microscopes and a large number have been studied. The results have not been widely published, however, since often they are only recorded as expert opinions or as plant memoranda and are in part held as confidential by the ore dressing companies. The way in which one or another happens to be published represents a fortuitous, but not always the best selection. The above remarks apply to Germany and indeed to Europe in general more than to America, where many good papers well provided with useful photographic illustrations are always available. In Australia an especially active and successful program has resulted in a series of publications of CSIRO.

A method which is very useful in average cases, was presented by GAUDIN (1957) and later by AMSTUTZ (1961). AMSTUTZ has also given a bibliography with over 200 references on the application of microscopy to ore dressing problems. In Germany KRAFT (1961) has published a good case study of concentration products from Freiberg. REHWALD (1954) gave an excellent general view of all possibilities in that materia.

Almost every ore dressing problem of this kind demands a special method of approach because universally applicable standard methods cannot be given — that which in one instance can be essentially absolute and conclusive can for another case represent completely useless work. The following statements, therefore, present only guiding rules which do not cover by far all possibilities. G. M. SCHWARTZ (p. 1938) presents a compilation which will be followed in its main traits in this tabulation:

Ore dressing microscopy can provide information on:

I. *Identification of minerals in the ore.*

a) Regarding ore minerals which gives information in:
  α Flotability
  β Solubility, especially in acids
  γ Reducibility
  δ Access to solution
  ε Tarnish, coating, etc.

ζ Magnetic properties
η Specific gravity
ϑ Complex minerals
ι Artificial components developed in sinters, slags, etc.

b) Valueless metallic minerals such as pyrite, pyrrhotite or oxides.
c) Gangue minerals.
d) Specially important objectionable minerals carrying sulphur or phosphorus in iron ores, or bismuth in lead ores.
e) Minerals with troublesome structures such as sericite, talc, bentonite.

II. *Fabric relations.*

a) Size and interrelations of various ore minerals.
b) Size and relation of valuable to valueless minerals.
c) Porosity, fractures, etc. allowing access of solutions, gases, etc.

III. *Quantitative data.*

a) Relative amounts of ore and gangue minerals.
b) Percentage of each ore mineral.
c) Percentage of metal values occurring in each ore mineral.
d) Approximate chemical composition calculated from mineral counts of unbroken or ground ore for comparison with average chemical analyses.
e) Percentage of the various size fractions, first in crude ores, and secondly in the mill products.
f) Locked and free minerals in each different product.

IV. *Special data.*

Recognition of origin of ore and changes to be expected with depth, with consequent changes in mill practice, etc.

The above outline, which I follow here, although I consider it a bit too detailed, will be discussed point by point.

First some remarks about the *polished sections*: On pp. 285-290 the principles described; special reference is made to the remarks of RAMDOHR and REHWALD in the LEITZ-Handbook (1954a). It is extremely important in the case of classifications (3a–3f) that the *number* of samples of *crude ore* be sufficiently large, and for relatively coarse material the size of the *sample* itself must be *ample*. It can happen that one can become misled because of the apparent monotony of mineral components and textures and one wants to save work and time, whereas the actual counting and measuring will often show that first impressions can be deceptive.* This applies particularly to crude ores with a relatively high proportion of gangue. The desire to see very many ore types — making it clear and necessary to establish all minerals of the mineral association [Paragenese] — can lead to an overaccentuation of the uncom-

* The writer remembers an ore in which only sphalerite, pyrrhotite, pyrite, galena and chalcopyrite, occasionally magnetite were known, where, however, an accurate study showed portions in which the following minerals were present in abundance: marcasite, fahlore, cubanite, mackinnwite, jamesonite, wittichenite, ?luzonite, argentite, silver, electrum, ilmenite!

mon* components. *Ore concentrates* are generally so fine that very few preparations of normal size are necessary to provide a proper statistical evaluation, provided that the polished sections are relief-free and otherwise flawless (figs. 259, 260). However, in this connection it is to be observed that in every type of preparation method, segregation of heavier or lighter minerals as easily occurs, so that the section should be cut perpendicular to the underside of the mount or this phenomenon must be taken into account in some other way. The writer has, e.g., observed amazing differences in the concentrates from shaking tables in the case of the Witwatersrand ores.

For crude ores especially when using mounts of sealing wax or of plastics for concentrates with a high proportion of harder ore minerals such as pyrite, (fig. 259) arsenopyrite, etc., the extremely fine intergrowths around and through the grains can only become recognizable with absolutely relief-free polishing. Also the broad black edges which occur around hard components give a false impression of their proportion; they are hard to count or to estimate and are, for the reason given, usually overestimated. The polishing defects in the hard ores should be entirely absent otherwise one may overlook actual porosity or fail to recognize small inclusions. The latter applies, e.g., to minute gold or argentite dust in many pyrites and arsenopyrites and the like (figs. 264a, 354).

For plastic mounts and also those of sealing wax or the much to be avoided dental cement, it may be necessary to elutriate the finest slimes especially kaolin, bentonite or sericite which can restrict uniform wetting and thorough mixing with the embedding material and cause uncontrollable brittleness of the mounts. The procedure is often unsatisfactory, as has been experienced in modern sedimentary petrography, because gangue minerals of the layer-lattice types, likewise those with strong cleavage tend to selectively settle out. This is also the case with ores which have very brittle and cleavable minerals, e.g., the silver carriers, fahlores and alaskaite which can be removed in large quantities by elutriation.

Flotation products commonly contain residual traces of flotation chemicals, especially oils as thin films which either impede the "bonding" of mineral grains with plastic or bakelite or prohibit it altogether. Thorough washing with a succession of benzene, acetone, and alcohol generally helps, but unfortunately not always.

It is self-evident that a comprehensive study is attained only by a comparison of the mill feed with all the products including the tailings.

In the examples mentioned in the discussion one notices that with this somewhat fine subdivision a particular item may often be placed in several positions. Attention should be paid to the captions of figures!

With reference to I:

α) Unexplainable losses or contamination can result where, associated with common minerals with known benefication characteristics, there are rare minerals which appear outwardly the same. SCHWARTZ (1938) reports an instance from his experience, where a black manganese ore always remained rich in SiO_2. It became obvious that the otherwise rare manganese silicate *bementite* was present. The writer can show, that in a copper concentrate in which a very obstinate zinc content persisted, the

* By way of comparison one thinks of the overestimation of foyaite in petrography!

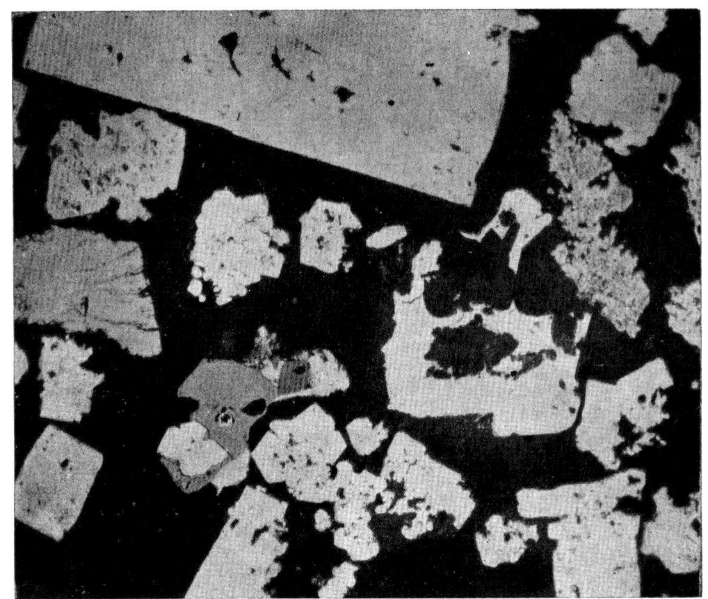

Fig. 259 a 170 × REHWALD

Gänsberg near Wiesloch, Baden, Germany

Concentrate (isolated from limestone with hydrochloric acid) predominant in *pyrite* (pure white), *galena* (grey white) and *fahlore* (grey). Note that the imbedding material fills all surface irregularities, facilitating the production of relief-free sections of extremely brittle grains with the machine

Fig. 259 b 600 ×, imm. REHWALD

Detail of (259 a)

Tennantite, grey, idiomorphic, with a somewhat darker border, *pyrite*, white, *sphalerite*, dark grey, and a trace of *galena* (light grey)

carrier of copper in addition to enargite and chalcocite was a zinc-rich tennantite, and thus, a lowering of the Zn-content was therefore not possible.

β) The flotation behavior of ores can be very poor or difficult to manage with ores containing a high content of troublesome oxidized minerals. The determination of the mineralogical nature of the latter often is sufficient to eliminate the difficulty. They can e.g., be soluble in acids and thus be removed, or they can be modified by admixtures and be made suitably flotable. Limonite and manganese oxide films on gold are an example of the first possibility (fig. 262a), and cerussite of the second. The presence of malachite in ores composed chiefly of sulphides results in high Cu-losses, unless this mineral is leached with acid and the resulting copper-rich

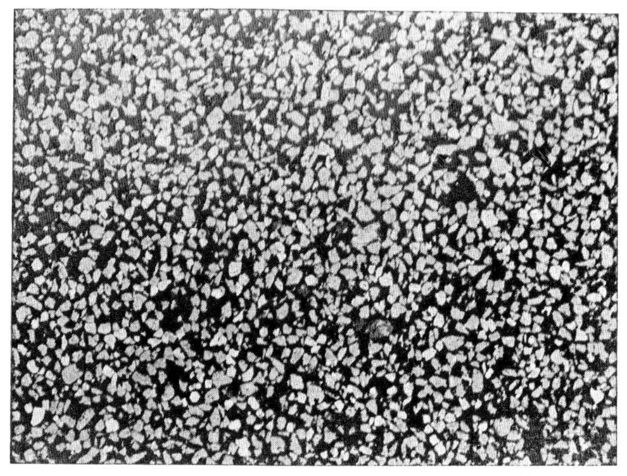

Fig. 260a

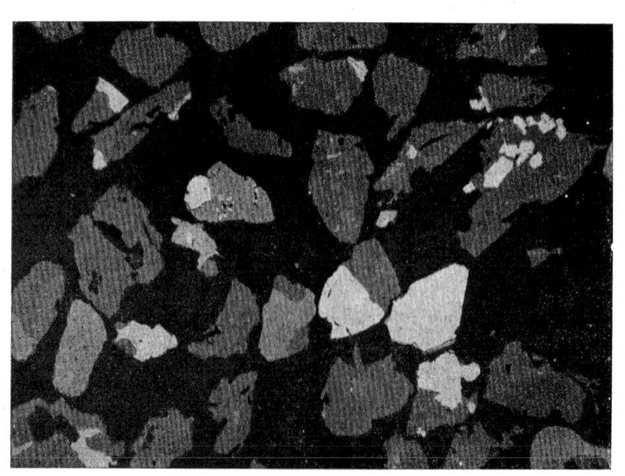

Figs. 260a and 260b 8 ×, 170 × resp. REHWALD
"Canada"

Machine-polished concentrate (plastic mounted). Predominant *sphalerite* (grey), *galena* (white) *tetrahedrite* (in b medium grey). Sporadic pyrite, some gangue and other things

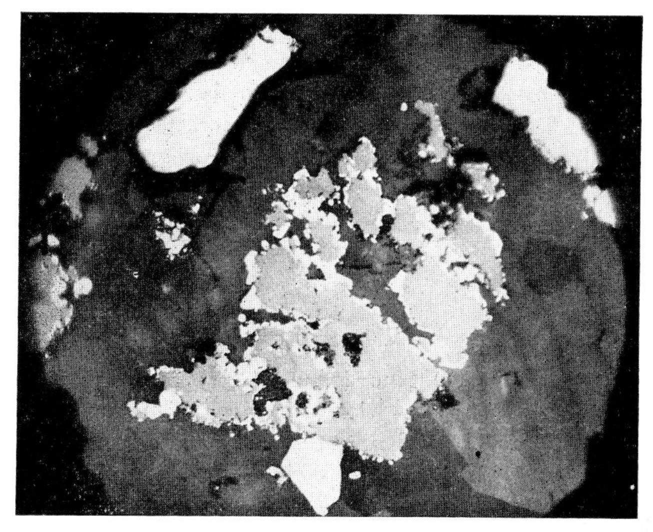

Fig. 261 400 × (dry) RAMDOHR

Keban Dere, Anatolia, Turkey

Sphalerite with small *chalcopyrite* granules surrounded with a cementative crust of *galena* (white) amidst *smithsonite* (dark grey). A little *chalcopyrite* (large white grain) and *pyrite* (strong relief)

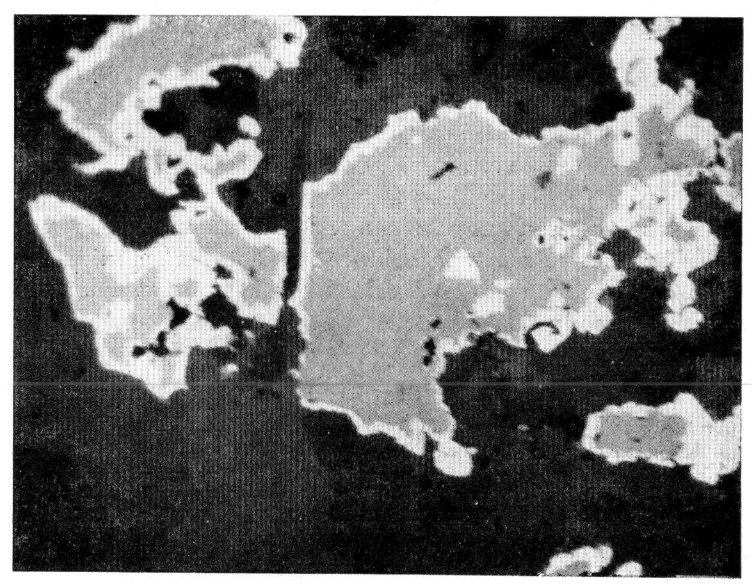

Fig. 261 E 450 ×, dry RAMDOHR

Mansfeld, formation in the so-called "Rücken"

Intergranular film of *chalcopyrite*, white, between *sphalerite*, medium grey, and *calcite*, nearly black

solution isolated and treated. HEAD (1928, 1932) has described such cases in several papers.

γ) Similar metal content, in fact perfectly corresponding overall chemical composition by no means implies similar metallurgical behavior. Hematite is more difficult to reduce than magnetite, and sphalerite with a high FeS content more difficult than varieties without this or with iron in the oxidized form etc. In chrome ores a Mg-content in silicate gangue must be evaluated in a different way from the same content within the mineral itself. Accurate knowledge of the mineral types, but also knowledge of the elements which make up each single one of them is therefore to be desired.

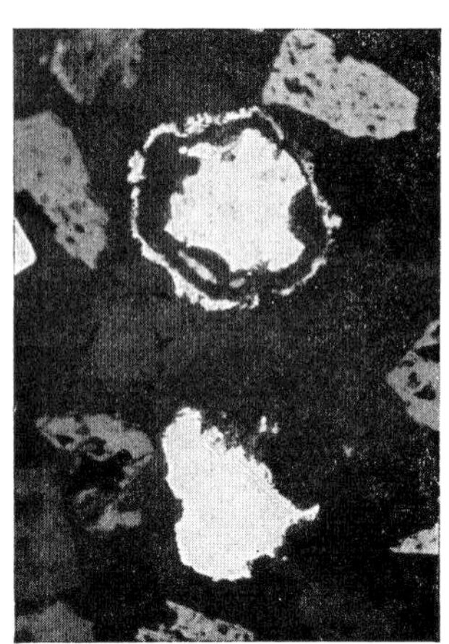

Fig. 262a RAMDOHR
170 ×
Las Medulas, Spain

Gold from a placer concentrate. A grain "armoured" with a complete limonite crust (and therefore under certain circumstances excluding amalgamation or cyanidation!), the other fully free. The remaining grains partly limonite, partly zircon, etc.

δ) Knowledge of cleavages and fissures is necessary for flotation, for leaching processes, and in part also for easier grinding. The microscope reveals these features and can, in their absence, occasionally show the way how they can be artificially induced by some appropriate process (e.g., heating and chilling, figs. 26, 29, 31).

ε) Unusual behavior in otherwise easily flotable ore or, by contrast in non-flotable gangue, is often easy to interpret microscopically. This can be caused by a "primary" encrustation with gangue, e.g., with sericite or kaolin minerals, or thin graphitic skins around quartz or felspar. This can also be caused by crusts of oxides not visible to the naked eye. Galena from a mine in the Ruhr Area responded to flotation in a variable and uncontrolled manner which appeared to be related to the preliminary treatment. This difficulty was shown to be caused by thin encrustations of cotunnite ($PbCl_2$) on fractures. Cotunnite is very differently watersoluble according to temperature so that on cold days it caused more difficulties than on warm days. HEAD (1937) was able to show that there can be considerable losses involved with gold ores in flotation, cyanidation or amalgamation due to encrusted minute gold particles which are not rendered sufficiently free. EDWARDS writes about a large number of similar cases of various ores (figs. 261, 262).

ζ) *Magnetic separation* can also be difficult. Progress can be made only by trial and error in those relatively rare cases where the same ore mineral is sometimes strongly and sometimes weakly magnetic as can be the case with ores containing platinum, franklinite, chromite or occasionally hematite. More important are mineral associations which involve easily overlooked magnetic ores or those occurring in places where non-magnetics are anticipated. SCHWARTZ (1938) describes an example where a high Cu-content was carried into the magnetic fraction and there became lost until it was found that the magnetic copper carrier *cubanite* was extensively intergrown with the non-magnetic chalcopyrite (fig. 437). — Strikingly divergent results can be found in the magnetic separations of magnetite ores in which is variably strong or full martitization. The paper of COOKE (1936, p. 82) affords excellent examples.

Ilmenite from a large ore deposit was sometimes strongly magnetic, sometimes not. The strongly magnetic ilmenite was found to contain many lamellae of magnetite (fig. 574) and in tectonically stressed portions, it carried twin crosses (fig. 576a) of newly formed magnetite. — Coarse grained "nickeliferous pyrrhotite" of the Sudbury type can be subdivided magnetically and the valueless pyrrhotite + magnetite can be separated from the valuable ores pentlandite and chalcopyrite. The ore-microscope shows that this separation can never be complete in a quantitative sense because of the occurrence of "pentlandite flames" in the pyrrhotite, even not considering the middling product, which is unavoidable, but here does not constitute great quan-

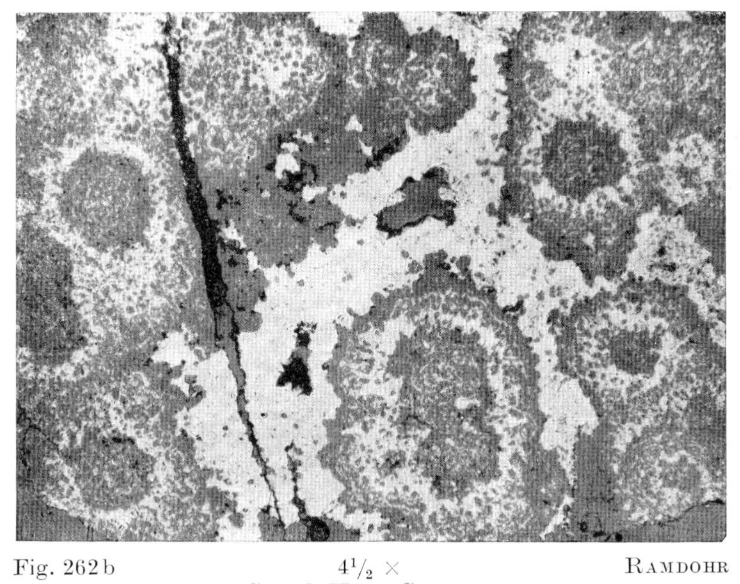

Fig. 262b $4^1/_2 \times$ RAMDOHR
Grund, Harz, Germany

Small rhythmic cockades of quartz-galena intergrowths in changing relative amounts Such ore prior to the introduction of flotation, was an insoluble beneficiation problem

tities (fig. 414). — In many instances complex concentrates of cassiterite and wolframite may be magnetically separated because cassiterite is non-magnetic whereas wolframite is magnetic. Where the separation fails, finely disseminated dustlike magnetite in the cassiterite can sometimes be detected (e.g., in some samples from Nigeria). Likewise cassiterite can be magnetic because of an exsolved content of columbite, while chemically similar compounds without exsolution show normal behavior.

η) The various carriers of one and the same metal can have very different densities, also regular fine intergrowth structures can cause intermediate specific gravities. COGHILL, HOWES and COOKE (1931) describe a case where from two manganese-bearing iron ores of similar appearance one yielded favorable results on the Wilfley Table and unfavorable results by flotation, whereas the other behaved in the opposite way. Microscopic identification of the mineral components showed that it could not be otherwise.

ϑ) Minerals of complex composition can produce special difficulties in industrial beneficiation. The silver content of galenas, e.g., is mostly fixed in mineralogically well defined "silver carriers", often in fahlore, but also in pyrargyrite, andorite, argentite, or polybasite; however, in other cases, it is dissolved in isomorphous compounds, e.g., existing as $AgBiS_2$. In the last case both it and the Bi-content remain microscopically nondetectable, however, it is perhaps regularly distributed in the galena, whereas otherwise — an old experience — the silver content in the various ranges of grain sizes of galenas can be quite variable. In the Upper Harz district, e.g., a relatively high Ag-content was found in the lead bearing portion of the coarse grained concentrate of the settling tanks, a low content in the finer grained portions, and the highest content in the slimes lost in earlier times in the settling ponds. The

Fig. 263 170 × Ramdohr
San José Mine near Oruro, Bolivia

Pyrite white, also *chalcopyrite*, light grey, *cassiterite*, dark grey and *quartz*, grey-black. — The section shows that with the polishing machine it is possible to produce relief-free sections of pyrite and cassiterite whereby all intergrowth structural characteristics are well exhibited. — Moreover it is clearly recognizable that idiomorphism commonly provides very little about the true paragenetic sequence

explanation is as follows: the small granules of tetrahedrite are situated at the edges, especially around the peripheries of chalcopyrite, but also of sphalerite, gangue minerals, etc. With moderate size reduction these grains remain in galena, whereas, with longer grinding they become progressively more and more removed (abraded) from the edges and find their way into the finest slimes. It is similar with the pyrargyrites in other ore deposits in which the Ag-bearing minerals occur especially in existing or recemented cleavage cracks and by progressive crushing they become enriched in the fines (figs. 31, 271). — With the development of flotation the losses in slimes have been reduced; nevertheless the peripheral positions of fahlore granules still cause these to report in the middlings (fig. 271). — "Sulphosalts" which are usually quantitatively insignificant, although carriers of Ag, Bi, As, Sb and also Au and Hg,

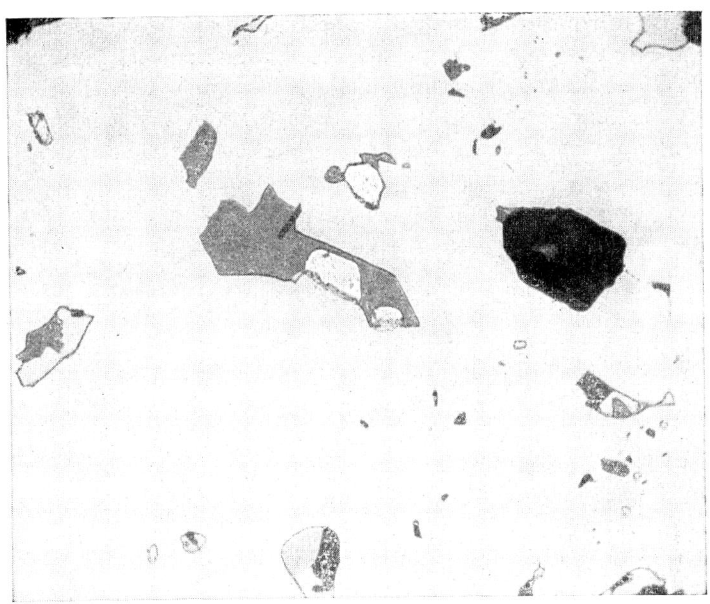

Fig. 264a 240 × RAMDOHR
Quemont, Ontario

Gold, two grains together with sphalerite, in addition *sphalerite* with chalcopyrite + pyrrhotite, gangue etc. — all in a large perfectly polished *pyrite* grain (entire field of view.) So-called "mineralized" gold is mostly only occluded in similar forms

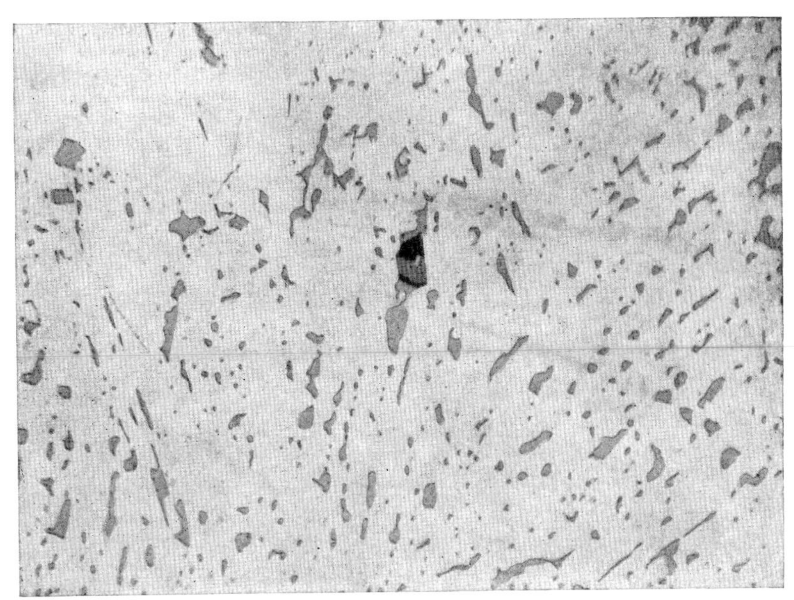

Fig. 264b 80 ×, imm. RAMDOHR
Pachuca, Mexico

Galena granules, doubtless in part oriented in a large *pyrite* crystal, perhaps an idioblast

are generally not taken into account in the course of ore dressing. In one lead mine, however, when ever Pb-recovery was satisfactory, a concentrate containing more than some 68–70% Pb could not be produced. The microscopic investigation showed that here a substantial part of the lead existed as jordanite. When the jordanite was collected during flotation the recovery of Pb was good; on the other hand, when only the galena was recovered, the concentrate was of outstanding grade, however, approximately 50% of the Pb was lost! The sulphosalts which in selective flotation are

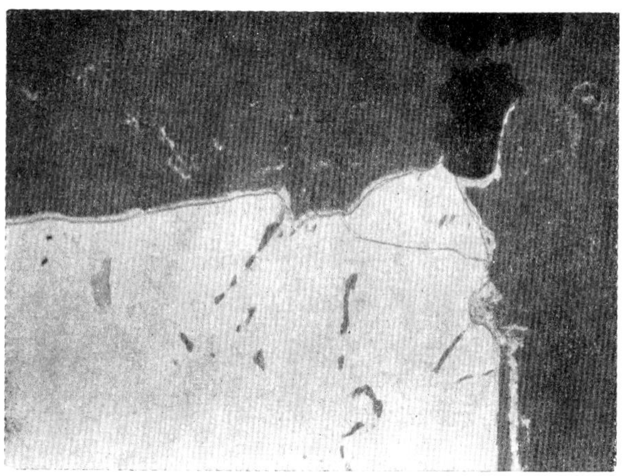

Fig. 265 170 × RAMDOHR
Pachuca, Mexico

Pyrite with a uniformly thin rim of galena against the *gangue*. This pale film would not be seen in poorly polished briquettes where it would be masked by the black relief-rim of the pyrite! — Under certain conditions of beneficiation the galena film would cause a large portion of the pyrite to enter into the galena concentrate

Fig. 266 500 ×, imm. RAMDOHR
Guelb Moghrein, Akjoujt, Mauritania

Electrum (pure white) as a thin, surely primary film between *chalcopyrite* (grey) and *cobaltite* (grey white). Also *arsenopyrite* (somewhat brighter)

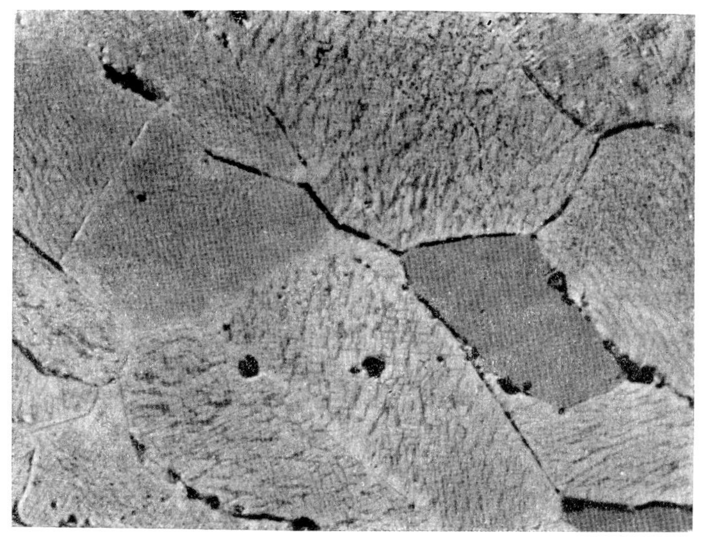

Fig. 267 600 ×, imm. RAMDOHR
Chihanji, Angola

Magnetite with fine (100) network of *ulvöspinel* (somewhat darker). Two grains of *ilmenite* are grey. The true value of the color difference is difficult to record photographically. — The high titanium content in the ulvöspinel cannot be separated from the magnetite; however, the slightly magnetic ilmenite may be easily isolated during beneficiation

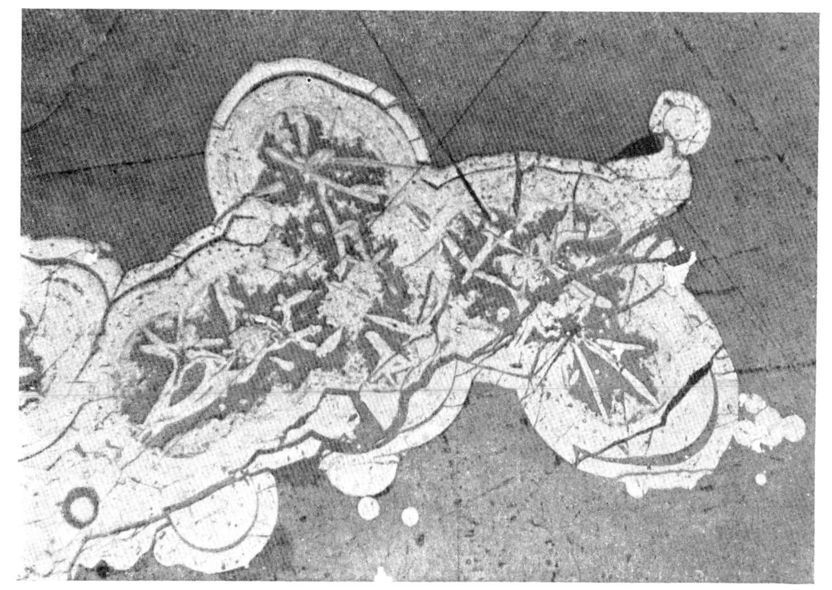

Fig. 268 80 × RAMDOHR
Kowary (Schmiedeberg), Silesia

Pitchblende of colloidal origin in *calcite* (dark grey). The radial fractures visible in fig. 129 now filled with younger pitchblende and with the interstitial parts later dissolved out are very characteristic. Very little *arsenides* (pure white)

recovered in part with galena, in part with chalcopyrite, and in a few instances, with sphalerite, can thus divert their valuable metal contents to various concentrates depending upon the mineralogical associations.

i) *Sinter* and *slags* have been repeatedly investigated with success by metallurgists (LUYCKEN, FABER, TROJER and AMSTUTZ, see AMSTUTZ, 1961) by use of ore-microscopy. Metallurgical slags were investigated, e.g., by EDWARDS. It is possible to establish to what extent the material is fused, which new components have formed and to

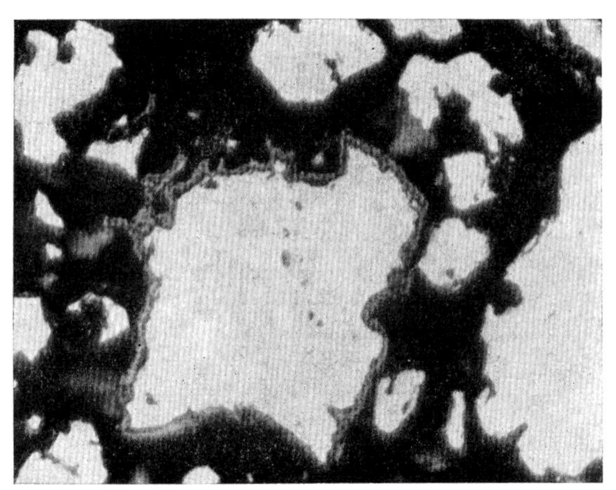

Fig. 269 250 ×, oil imm. RAMDOHR
Heath-Steele Mine, New Brunswick, Canada

Pyrite, with a thin cementation rim of *argentite*

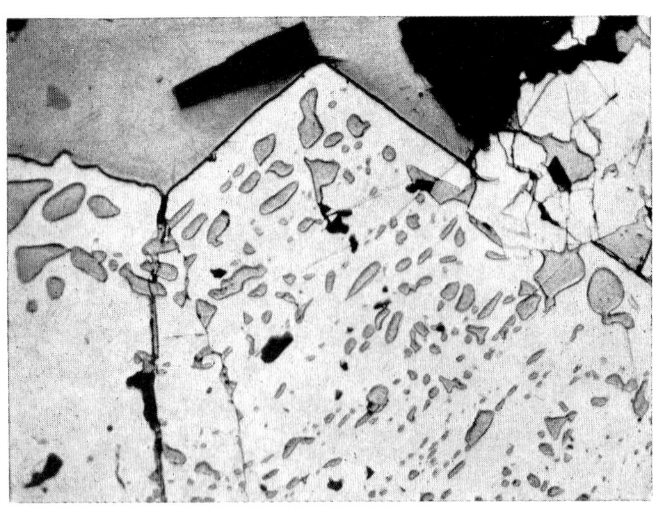

Fig. 270 250 ×, imm. RAMDOHR
Pachuca, Mexico

Porphyroblast of *pyrite* with numerous inclusions of *galena*. With beneficiation a substantial portion of the galena will necessarily go into the pyrite fraction and vice versa

what extent the intergrowth textures of these are favorable towards the reduction, whether the secondary ingredients (Mn!) go into silicates or oxides, etc.

I b) and c). The microscope, often also with transmitted light, can frequently provide information about "gangue minerals", "worthless" sulphides and oxidized ore. Thus it is frequently shown that the "worthless" tailings can also often be made useful, e.g., pyrite + pyrrhotite, or barite, fluorite or siderite, occasionally indeed, even quartz or calcite. In many gold-silver mines of the "subvolcanic type" rhodochrosite and rhodonite are avaible as abundant gangue. Both may often be recovered in adequate quantity and grade to guarantee their saleability and in certain circumstances to determine the economic success of the mining venture if they are favorably intergrown, which may only be established by microscopic investigation.

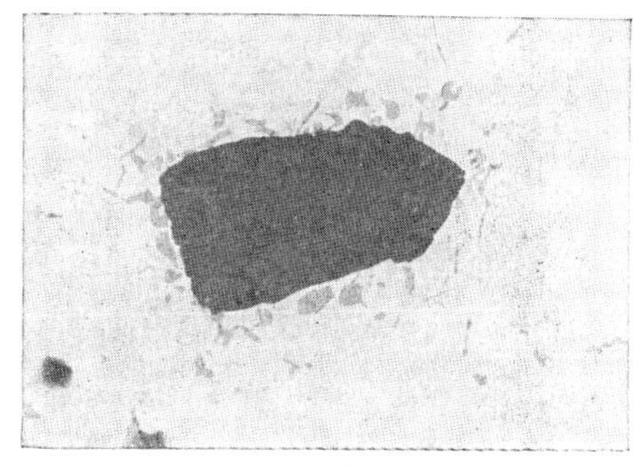

Fig. 271 80 ×, imm. RAMDOHR
Mine Pfannenberger Einigkeit, Siegen, Germany

Fahlore granules around a *sphalerite* grain in *galena*. This common form of occurrence of tetrahedrite signifies that the Ag-content which would be expected to occur in galena appears instead in the sphalerite

The microscopic investigation of intergrowths of the so-called "non metallic mine rals" determined the treatment of the apatite-nepheline rock from Kola, where in addition to apatite a high grade nepheline concentrate for glass manufactory is recovered. In the case of a nepheline syenite for ceramic purposes, as quoted by SCHWARTZ, aegirine and other Fe-bearing minerals can be entirely removed magnetically. Naturally in the case of such relatively cheap nonmetallics, the economic limit of the grinding costs is lower than in the cost of metallic minerals.

I d). The *removal of detrimental impurities*, e.g., the carriers of phosphorus and sulphur in iron ores, arsenic and tungsten in tin ores and bismuth in lead ores can sometimes be a problem, which is a task which lies beyond the actual scope of ore dressing. The often cited paper of COOKE indicates methods of removal of small phosphate contents from already very high grade Fe-ores in order to achieve Bessemer-grade ore. — SCHWARTZ quotes an example where the high S-content of completely weathered Fe-ores was not present as sulphides, but as the non-separable jarosite which is easily overlooked microscopically.

I e). Minerals, whose characteristics adversely affect certain beneficiation methods may be identified by a preliminary microscopic study, thereby the substantial costs of beneficiation testing are precluded; as examples may be mentioned kaolin, talc and sericite. can be very troublesome in flotation; chalcocite almost prohibits cyanide leaching, etc

II. *Textural and structural relationships*

a) and b). The first task, naturally, in each beneficiation program is to establish the grain size and intergrowth relationships (locking types) of the individual ore and gangue minerals. Each unnecessary grinding entails costs which could have been avoided, however, each inadequate grinding causes bad separation or excessive yields of middlings. Examples of both possibilities are common to those who have concerned themselves with these problems. Excessive amounts of middling products lead often to overgrinding. The optimum conditions capable of being attained with respect to economy and to the efficient and profitable extraction of values contained in the ore-body, demand extremely thorough microscopic supervision and a series of experiments often extending over months. They must include studies of crude ore, the con-

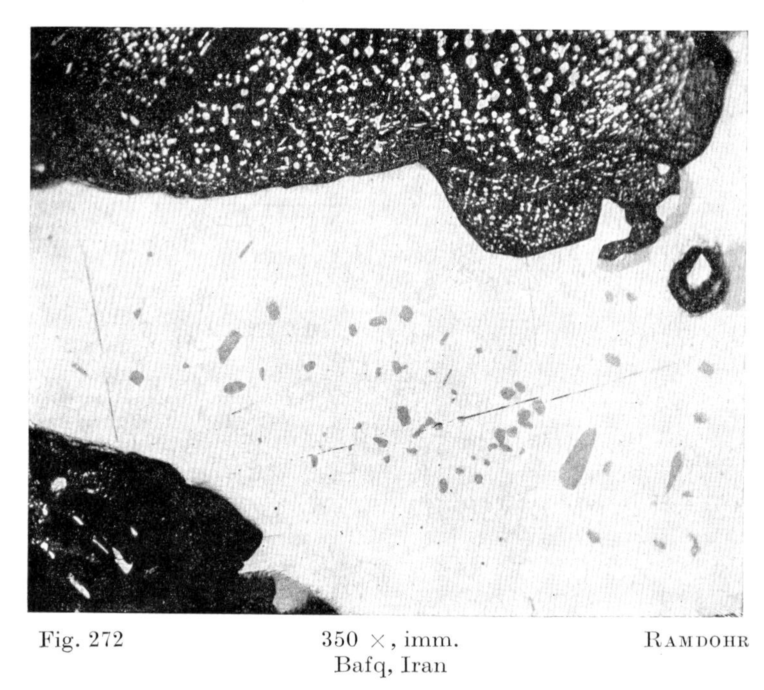

Fig. 272 350 ×, imm. Bafq, Iran RAMDOHR

Tetrahedrite and other "silver carriers" in *galena*. Next to it appears *sphalerite* full of *chalcopyrite* exsolution granules. Some *pyrite* is present

centrates and the tailings and must encompass statistical and intergrowth-textural data together with concurrent chemical assaying. A large report from HEAD, CRAWFORD et al. (1932) provides an example of this type of thorough control of the beneficiation at the Utah Copper Company. Recently KRAFT (1961) has published a good case study on Freiberg ores. Similar procedures have been accomplished with the Rammelsberg and Meggen ores, however, they have not been published in details. — As pointed out by SCHWARTZ, the possible processing of very low grade, but very thick and extensive ore deposits, as for example that of the "protores" of the Fe-ores of Minnesota, must be evaluated on the basis of very accurate microscopic interpretation of many hundreds of drill hole samples. The porphyry copper ores and the copper ores of the Northern Rhodesian types, etc., are naturally analogous examples.

IIb). A purely geometric classification of types of intergrowths has been established by AMSTUTZ (1960). It is reproduced in fig. 272a. Its advantage is that it allows to describe the types of intergrowth without any genetic connotations; this is the disadvantage of the terminologies of SCHWARTZ and others.

A geometrical clasification of basic intergrowth patterns of minerals

This classification is purely descriptive

1. for studies of rocks and minerals, deposits to present the real structures;
2. as an aid for ore-dressing microscopy and related intergrowths at all.

Naturally between most of these types gradational transition can be stated everywhere. When particle or grain-size data are given this figure can be a prerequisite of any accurate study of rocks, ore and mineral deposit.

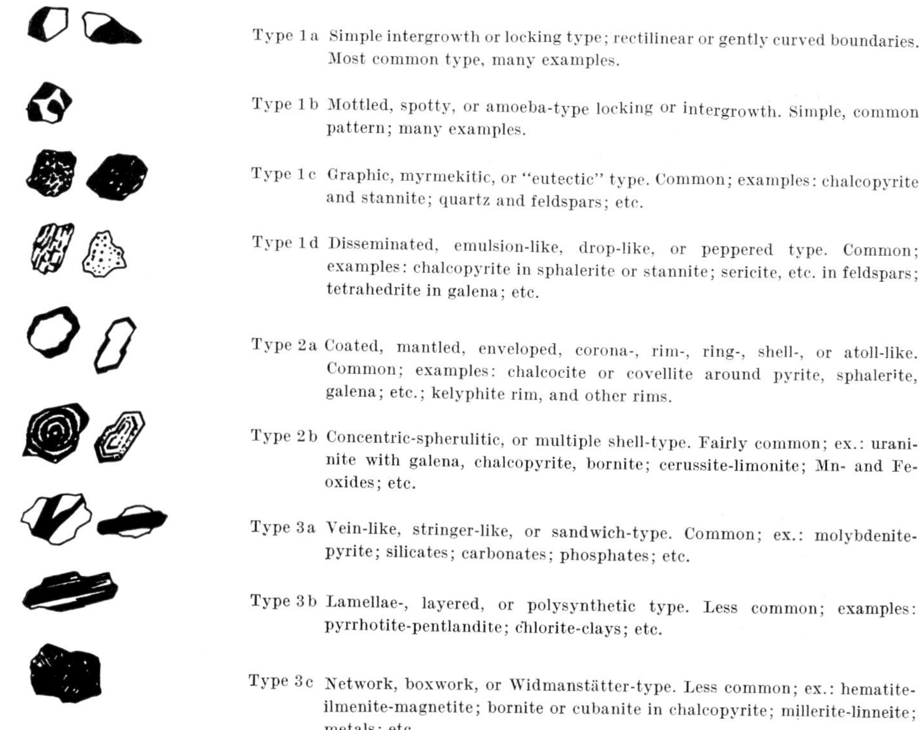

G. C. Amstutz — 1954, 1960

Fig. 272a

IIc). *Porosity* and *Fissures* are of great significance in regard to comminution procedures. The knowledge of local variations in the bonding relationships in the raw ore is closely related. The degree of fissuring and porosity is influenced by silicification, calcification, etc. in normally porous material and oppositely by solution of the bonding material in normally firmly cemented material. In many cases the milling practice recognizes these oddities, although the cause may not be clear. Here and in other less distinct cases the microscope can provide important clues (figs. 263, 268).

The importance of porosity and fractures in flotation and leaching has already been discussed.

III. *Quantitative data.* Quantitative estimations are to be made in all cases if possible, even though they are obtained only by accurate, tedious and time-consuming measurements. They are especially significant where the results apply to the correction of the conclusions drawn solely from chemical analyses. An example might be the case where the calculation of an analysis is inconclusive as to the form in which the valuable or worthless components exist in the raw material. Thus, iron losses in beneficiation might be attributed to the fact that not all Fe exists as magnetite or hematite but in addition as silicate compounds. With zinc ores it is especially important to determine whether the Fe-content is contained in the sphalerite itself, or exists as pyrrhotite, magnetite, etc.

IIIa). The relative amount of ore and gangue minerals plays a special role where the sulphide ore is not the specific subject of the recovery, but is only the carrier of the valuable substance, especially gold. Usually only the analysis of the precious metal is recorded. It is a matter of great importance, however, whether an ore with 20 g. Au/ton contains 1% or 5% or 20% pyrite, arsenopyrite, or similar ores, in which relative amounts of these minerals occur, and also whether the gold itself always occurs in the same form, e.g., as relatively coarse grained, unenclosed native gold.

IIIb). SCHWARTZ (1938) mentions that noble metals in consequence of their low quantity make difficulties, because it is improbable to get them in right distribution in a section. That is surely right as far as the grain size is large, but becomes increasingly less serious with smaller grain size: A gold ore with 10 g/t will show at a grain size of 1 mm in 200 sections (of 2 × 2 cm) one grain only, at 1/10 mm one grain in two, at 1/100 mm already 50 in one, at 1/1000 mm, the boundary of possible microscopic evaluation, 5000 in one! In spite of the objections to get the right distribution it is possible if enough material for study and a good statistic is available. This was demonstrated by the scientific staff of the Lake Shore Gold Mine (1936), even though about 80% of the gold existed as very fine flakes and the rest was in the form of calaverite, etc. Nevertheless, the path and the concentrating of the gold could be traced quantitatively in relation to its associate minerals from the mill feed to the concentrates.

IIIc). The complex ores, especially many copper ores with more abundant mineral associations must be traced quantitatively; since in these the components can behave quite variably in beneficiation and from the outset one must be clear as to what proportions copper goes into individual fractions. Sometimes it is possible to disregard the portion in a fraction; sometimes on the contrary further separation becomes necessary.

IIId). If it is possible during sieving to keep the grains so homogeneous that their counting is possible, then with a knowledge of the chemistry of all individual minerals the chemistry of the entire product can naturally be established. This presupposes a very critical consideration of all sources of error (cleavability, inclination to spindle or tabular fracture or also to concave fragments, etc.). Thus, in favorable cases this provides a really good method to follow the path of an element through processing steps including increasingly finer grinding. This will not be possible without shortcomings, since, in the same mine the following combinations may occur: old iron-rich and younger iron-poor sphalerite, chalcopyrite with mackinawite and cubanite inclusions, fahlores of various compositions, magnetite with zonally variable impurities — to mention but a few trivial examples. Unexpected deviation between micrometric analysis and the chemical analysis may commonly be traced to unexpected or overlooked minerals or to solid solutions.

Occasionally special difficulties occur. A concentrate showed a conspicuous discrepancy between chemical and micrometric analysis. It became apparent that approximately $^{1}/_{5}$ of the supposed sphalerite was in fact anatase!

IIIe). A quantitative counting of the particle sizes of each one of the crude ore components gives an indication of the most favorable grinding conditions (figs. 259a, b). It is commonly the case, indeed often even the rule, that carrier of certain metals (e.g., gold) are considerably finer grained (sometimes also coarser grained!) than the associated minerals. One must carefully consider which component one selects or which compromise one chooses. It is self-understood that a certain content of over- and undersize in the intermediate product must be taken into account, but one is often not clear about the consequences of particle size classification or about insufficient fine grinding or too much fine grinding. A generally valid principle cannot be given. A separation of the main components, e.g., galena, sphalerite, chalcopyrite can be sufficiently achieved by the selection of certain particle sizes, while the important accessory minerals, perhaps the silver carriers, still remain completely unenriched. The paper of COOKE (1936) shows problems of some iron ores. Very extensive size reduction can cause deleterious effects on the main components, not even counting the increased costs.

IIIf). Also after beneficiation, a separation at first seemingly very favorable is found to contain abundant intergrown particles in each product. Quantitative particle counts make possible the information as to whether further grinding (of the whole or a part), appears to be desirable. On the other hand particle counts and particle size recordings of the intergrowth textures can demonstrate for relatively low beneficiation yields whether with beneficiation procedures used so far, improvement on the whole is possible.

IV. *Special data.* Conclusions regarding the nature of the ore paragenesis and the amount of the ores with increasing depth, are commonly possible. The existence of rich masses of limonitic iron ore in contact ore deposits of magnetite, as pseudomorphs after pyrite, pyrrhotite, chalcopyrite makes it possible to say with surety that large sulphide masses exist at depth. Pseudomorphs after Fe-silicates indicate that there will be beneficiation losses in fresh ore. In copper ore-bodies similar predictions are possible and in the various ore-body types one is readily able to make correct and accurate recognition of "primary" depth distinctions. To explain the extrapolation of the paragenesis of the primary depths from that of the oxidized zone is not necessary because this is too trivial.

AMSTUTZ, G. C. (1960), *A geometric classification of basic intergrowth patterns of minerals.* Am. Geol. Institute Data Sheet **21**.

— — (1961), *Microscopy applied to mineral dressing.* Quart. Colorado Chool of Mines, vol. **56**, No. 3, 443–484. [37 figures, over 200 references.]

COGHILL, HOWES & COOKE (1931), *Mineragraphic aid in the concentration of manganiferous iron ores.* Eng. Min. J. **131**, 361–365.

COOKE, S. R. B. (1936), *Microscopic structure and concentrability of the important iron ores of the United States.* U.S. Bureau Mines, Bull. **391**.

DIETRICH, W. F., A. F. ENGEL & M. GUGGENHEIM (1937), *Ore dressing tests and their significance.* U.S. Bur. Mines Rept. Inv. **3328**.

FABER, W. (1954), *Mikroskopie der Metallhüttenschlacken.* LEITZ-Handbuch der Mikroskopie, 521–594.

HEAD, R. E. (1928), *Microscopic study of an ore as an aid in copper leaching.* Eng. Min. J. **126**, 13–15.

— — (1937), *Form and occurrence of gold in pyrite from a metallurgical standpoint.* U.S. Bur. Mines Rept. Inv. **3226**, 27–31.

HEAD, CRAWFORD et al. (1932), *Statistical microscopic examination of mill products of the Copper Queen concentrates of the Phelps Dodge Corporation, Bisbee, Arizona*. U.S. Bur. Mines, Tech. Pp. **533**.

KRAFT, M. (1961), *Zur erzmikroskopischen Untersuchung von Aufbereitungsprodukten am Beispiel von Blei–Zink-Erzen des Brander Reviers*. Zeitschr. f. angew. Geol., H. **10**, 517–523.

LUYCKEN, W. & L. GRAEBER (1931), *Untersuchungen über die Saugzugsinterung von Eisenerzen*. Mitt. Kais. Wilh. Inst. Eisenforschg. **13**, Abh. 192.

REHWALD, G. (1954), *Die mikroskopische Untersuchung von Erzaufbereitungsprodukten im Auflicht*. LEITZ-Handbuch der Mikroskopie.

— — (1965), *The application of ore microscopy in beneficiation of* ores of the precious *metals and of the nonferrous metals*. — In: Applied Ore Microscopy, ed. by H. FREUND 1966. The MacMillan Comp. New York.

SCHWARTZ, G. M. (1923), *Chalmersite at Fierro, New Mexico*. Econ. Geol. **18**, 270–277.

— — (1929), *Iron pre-sinter*. A.I.M.E. Tech. Publ. **227**.

— — (1938), *Review of the application of microscopic study to metallurgical problems*. Econ. Geol. vol. **33**, 440–453.

— —, *Minnesota Protore; only cited, apparently unpublished.*

"The staff" (1936), *Milling investigations into the ore as occurring at the Lake Shore Mine*. Can. Min. Met. Bull. **290**, 297–434.

ANNOTATION CONCERNING THE ARRANGEMENT OF MATERIAL

IN THE DESCRIPTIVE SECTION

Although the descriptive material is actually only a means to the ore-microscopist for the study of genetic relationships, geochemistry, classification of ore deposits, and applied problems, it has yet been the writers principal aim to work towards the thorough perfection of these means, and to this he dedicated many years of his life. Evaluating the large amount of material of observation collected over some 50 years, probably of unsurpassed perfection in the world, appeared to be a scientifical duty to him.

It is not easy to present the subject in a way being profitable to as many colleagues as possible. The compilation of a handbook which simply registers facts would be easier though it would be unlimited in scope. Critical study covering the literature as well as the writer's own notes on observation necessitates a strict *schematical subdivision of the material.* The subdivision made here is even stricter than the one in the "textbook" (Schneiderhön—Ramdohr 1931–33) but follows it very closely. The increased usefulness of the book, originally hoped for, seems to live up to its expectations. The desire that other writers might have used the same or at least a similar schema in analogous descriptions was fulfilled sometimes, though not too often. Let that be suggested again. The lack of a system in many papers makes a study next to impossible according to certain comparative points of view. A collection of material is often found with which one cannot build up at all, or only very uneconomically. The most serious consequence is that the lack of a system tempts to omissions. Thus, many papers miss stating the observed characteristics of some supposedly present mineral which perhaps is notorious for the difficulties in its determination. Thus, many data cannot be checked over which makes their value illusory. Of course, very often it is impossible to give all the data suggested here because many of them can only be obtained from more copious material.

In sections I–XII the schema deals with everything necessary for our purpose. Some recent methods of examination, still essentially in the development, have been omitted, such as the τ and ω values obtained by means of Berek's astigmatic tube analyzer or analogous equipment by Turner et al. (1945). Those data have been determined only for very few minerals and depend to such an extent on the individual case history (polishing etc.) that at the time being they are of no diagnostic value and will only be of such after the evaluation of the latest papers written by Berek. Also the print method ("méthode des empreintes"), very promising for certain purpose, is not yet developed. So are the ingenious experiments by Gaudin, dealing with the determination of elements in polished sections by producing induced radioactivity and photographic track recordings of it.

The method used as early as 1951 by CASTAING to apply electron rays for the use of fluorescence analysis (microsonde, microprobe) has become a most excellent aid for the ore-microscopy. It allows, best conditions supposed, quantitative determinations on objects of $2 \times 2\,\mu$. Very many problematics, inclusions recognized since 40 years as peculiar, but by no former method available for exact determination, could be solved. I mention the minerals: Roquesite, indite, mawsonite, castaingite, trüstedtite, wilkmanite, certainly many dozens of platinoid compounds, the discrimination of the very similar minerals mackinawite and valleriite, the recognition of new mixed-crystal-series and so on. — It is here not the place to describe the apparatus itself and the methods to work with.

Though strictly following the schema, several of the numbers will often be omitted or taken together, because nothing is yet known about it! This will frequently be the case in V (Physical chemistry), in VII (Special fabrics), and VIII (Diagnostic features).

Generally only safe data are stated. In chapter V and in the interpretation of the structures of course many things are hypothetical as long as further check is missing. Sometimes one must doubt the possibility to apply experimental data for natural occurrences.

Obviously the author made faulty statements in many places in spite of the fact that he had a material for comparison, collected during 55 years, surely not available at any other place. Very often one is constrained in the case of very rare material to rely on perhaps very old determinations of other people, and only step by step it will be possible to approve them by microprobe. Accordingly I ask in such cases not only for an indulgent criticism, but for informations concerning new data and if possible for reliable material for comparison.

I. General data. In this chapter the general physicographical data shall be given in a concise form, but up to date. In this compilation where no literary sources are quoted, it may be seen that many an important statement is very ancient and that a modern check-over should be desirable.

Mostly the international name is used; in some cases with an addition of miners' names, such as zincblende, peacock-ore, and hematite, which are actually more elucidating than sphalerite, bornite or specularite. Only seldom names of other languages are additionally given if they deviate strongly (e.g. mispickel–arsenopyrite), or some misleading names are mentioned. Here the most common name is heading and the less common names were added together with the German name, if they differed substantially.

Chemical composition. In accordance with the most recent research the most probable formula will be used for the more frequently occurring minerals.* Many an "isomorphic component" has been proven to contain a mechanically more or less accidental impurity, others to be products of exsolution from a high temperature phase. The

* The discussion of "sulphosalt compounds" will not be considered. To date it obviously cannot be distinguished clearly if, e.g., jordanite is $Pb_4As_2S_7$ or $Pb_{14}As_7S_{24}$ or perhaps $Pb_{14}As_4S_{23}$, whereby the "surplus of sulphur" of the first formula, which seems to correspond most accurately to the analysis, takes its place in some lattice positions of Pb or As of the second formula or vice versa, the second formula being perhaps better from a structural point of view.

formation of mixed crystals in sulphide systems seems to be much more common than thought before.

Crystallographic data, as far as the shape is concerned, are presented by a short description only (e.g., long needles, granular aggregates), whereas all the space groups, number of molecules per unit cell, and lattice constants are given completely with the exception of lattice points, coordinates, crystal-chemical data such as layer lattice, ionic bonding etc.* This is followed by *cleavage*, *density* = D, and *hardness* = H; then by optical characteristics provided that data other than the ore-microscopical ones are available. In particular also the somewhat vague but occasionally useful data important for the determination such as lustre, color, streak, discoloration by darkening with time, etc. Citations of literature are here mostly omitted.

Fig. 273 55 × Ramdohr-Ahlfeld
Vila Apacheta, Bolivia

Cassiterite (dark grey, scarred despite careful grinding) as a secondary product of the oxidation of *stannite* (greyish white). The white, hard grain in centre is *pyrite* which also has been altered

II. Polishing properties. Under this topic all the peoperties of each ore mineral will be given that can be observed during grinding and polishing. Since the introduction of excellent machines for this purpose, this chapter could be abbreviated very strongly, but these machines are expensive and surely not available in the field or camp. So a too strong reduction did not seem to be desirable. — Very long experience shows that the behaviour of ore minerals during polishing may vary very distinctly in different association, degree of oxidation, alteration by pressure and many other conditions. Therefore no uniform method in preparation of polished sections has developed. The best results can be obtained with the Graton-Vanderwilt apparatus (Graton 1937, Vanderwilt 1941 ?) and the much more improved construction of G. Rehwald (1960). The last one can supply much larger sections, of critical importance in the study

* The value given in Å (Ångström units) is to be understood in the modern sense, not as "kX" of the new definition. "Å" in the modern meaning is smaller than kX by 2‰.

of coarser intergrowth. The author could apply this method for about 6000 sections. Many times, but not with minerals of low or moderate hardness, easily polished otherwise, the method by TROJER (1952), polishing on a disk made out of a core of linden wood, gives excellent results. The classic polishing on a billiard table cloth results in a relief with components of different hardness which is unbearables for the researcher accustomed to the use of polished sections made by machine. In spite of that it may

Fig. 274 275 ×, imm. RAMDOHR-EHRENBERG
Mina Fabulosa near La Paz, Bolivia

Prolonged fine grinding and polishing on a pitch disk succeeded to some extent in showing *cassiterite* (dark grey) free of relief and holes against *stannite* (light grey) and replacement relicts of *chalcopyrite* (white)

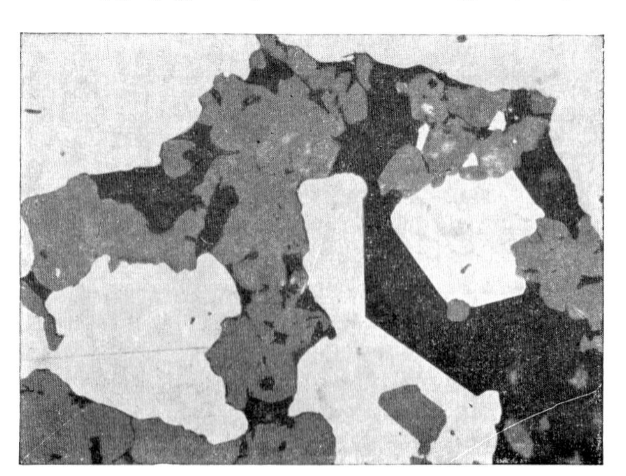

Fig. 275 350 ×, imm. as in Fig. 263 RAMDOHR
Polished with REHWALD's machine

Pyrite (white), *chalcopyrite* (light grey), *cassiterite* (very dark grey with internal reflexes), *quartz* (black in oil). Contrasts are even stronger than in fig. 263

be superior to the former in specimens with many cracks such as in chalcopyrite, sphalerite, fahlores in calcitic gangue, but without any other hard components. — The final grinding and polishing by hand on a pitch or rubber disk (RAMDOHR or VAN DER VEEN) is excellent but much too time consuming. — In any case an intelligent technician is of greatest significance.

The recently improved quality of polished sections is due to the replacement of older imbedding media by modern plastics. Adding a catalyser some harden at moderate*, some already at room temperature, without any sign of shrinkage. They have the advantage of being of very low viscosity at first, wetting ores and gangue uniformly and are after binding not attacked by oil used in the grinding process, neither by immersion oil nor by agents used in cleaning such as gasoline or benzene. In the preparation of polished sections of concentrates, using REHWALD's apparatus, those plastics yield results which were considered impossible at previous times (figs. 275, 260).

Especially "Araldite" ... manufactured by Ciba in Basel, Switzerland, and "Gießharz P 5" (Palatal) of BASF, the latter one because of its hardening at lower temperature, were useful. Difficulties still not solved give only ores containing free sulphur and some compounds giving off water at very low temperatures like sulfates and secondary uranium minerals.

In the old technique the preparation of polished sections larger than 2 × 2 cms was very time consuming if one did not want it to result in poor quality. The latest methods make it possible to prepare polished sections of 6 × 6 cm and larger. In Heidelberg we bed all polished sections in plastics using round bakelite containers. This saves the removing of edges and prevents the peripheral curvature at the rims of smaller specimens. Both procedures are a welcome improvement.

The necessity of *"individual" treatment of polished sections* is shown in each method by the time it takes for the rough grinding with the finest carborundum powder, the time for the polishing ,and the selection of the best material to polish with. Many cases of "subjective" attitude show up to a lesser extent if, as is always recommended, the specimen is cut by a diamond disk and if the grinding powder and polishing substances are carefully freed from oversized grains.

Grinding and polishing procedures are described by SCHNEIDERHÖHN (1952) and in the fundamental descriptions by REHWALD (1960).

Recently, diamond powder of different grain size is recommended. The author examined this method rather carefully and used partly prefabricated polishing laps or plates, partly he worked with nylon or silk clothes and different pastes. The success is at hard objects sometimes astonishing regarding quality and time. Against that the result at soft (not even "very soft") minerals and especially in intergrowths: "much hard — little soft" is often deplorable. — The method allows imbedding in plastics, but very carefully the heating during the polishing process, normally made with rather high pressure, must be avoided. Synthetic diamond composed of finest rounded octahedral crystals of exactly the same size is distinctly better than powdered natural diamond. Due to more recent methods and publications condemning a poor reproduction of polished sections the photographic pictures in publications have been improved. However, repeated attempts have been made to make use of poor preparates. A warn-

* Moderate temperature is essential because, e.g., stromeyerite will be altered at 93°, chalcocite at 103°, which results in very much disturbing changes in the structure.

ing has to be given that: a poor polished section may be worse than none at all, it causes self-deception!

In polishing two characteristic physical properties can be observed immediately: *hardness of polishing* and *cleavage of polishing.*

Hardness of polishing does not have to correspond with the hardness of scratching. The slightest differences in hardness of practically opaque minerals can be recognized by means of the light method used first by SCHNEIDERHÖHN as a diagnostic characteristic and explained by KALB (shifting of the light line into the softer mineral when the tubus is lifted). If one of the minerals should be transparent in the magnitude of microscopic thicknesses, then the observation of the light line is not always conclusive.

"Entirely without relief" a section cannot be made with any method. Such a state would not be desirable regarding the diagnostic value of the relief. The latter can be augmented, if necessary, by a very short polishing on a rotating cloth lap with magnesia.

The "polishing hardness" depends in some minerals rather strongly on the polishing liquid, i.e. water or oil, and may be relatively inversed at different minerals. Chalcopyrite e.g. is, using water, distinctly harder than galena, as we expect it from the V.H. (Vickers hardness) or scratching hardness. But using oil we observe just the contrary. The same happens occasionally for taenite: kamacite in meteorites. The beginner sometimes will be disturbed — a short repolish with MgO and water will reproduce immediately the properties on which the determination tabellae are based. — This difficulty was observed and mentioned in the German edition of 1960 by the author, alas, in a footnote only, STEINIKE (1965) observed it independently and described it more in detail.

Polishing cleavage is an important characteristic, but, if absent, it should not be considered as a criterion. Cleavage recognized megascopically cannot always be detected in a polished section, the cleavage observed in a coarse aggregate may be missing in a fine grained aggregate. Striking differences between megascopical and microscopical cleavage partly may be due to smearing of the surface.

The smearing of the surface by a "polish film", "BEILBY layer", or something similar, is often disturbing with certain minerals, mostly with such as have a low hardness and strong ability of translation. To some extent it may always be present and is a prerequisite of a good polish. Strong polish films cause a decrease of the effects of anisotropy, show many scratch marks visible with crossed nicols only, and may induce quick tarnishing. The smearing of the surface, as is shown by A. F. TURNER et al. (1945), may change the optical characteristics completely.

The mounting of polished sections in general is not dealt with. For permanent use, the writer mounts his polished sections with plasticine on a labelled and numbered glass slide. Others often prefer to preserve them upside down in small boxes, on a bedding of velvet or cotton, well protected from the influence of dust. Either method has its advantages and disadvantages. Certain ore minerals are sensitive to mounting on common plasticine (chalcocite, native copper, stromeyerite, etc.); they have to be mounted on bees wax. Many species of ore minerals or gangues that were precipitated in the form of colloids, as well as pyrolusite, extract oil from the plasticine, thus turning it brittle and loosening the mounting. Mounting on wax is required here too. — The best way is to reduce polished sections to a few standard sizes by bedding them in plastics. We use imbedding in Araldite or "Gießharz P 5" and ring- or square-sized forms of bakelite. That allows a standardized storing and mounting and especially numbering. Immediate imbedding in bakelite is impossible because it requires high pressures and temperatures 150°, both destroying many structures etc. Sometimes we use "Gießharz". In the first case the casings may be numbered so that no mistake arises and in the second case even pencilled labels may be put on the reverse side of the specimen and cast in the mould. For permanent use, standard forms of bakelite in which the specimen is bedded are good looking but the bedding in bakelite requires considerable temperature (> 150°C) and high pressure which ruins many minerals or aggregates entirely. It should be restricted to concentrates or avoided at all.

Alterations of the surface of finished polished sections are to be observed in many cases. To avoid gross delusions their forms and causes of formation have to be known. There are many transitions from a slight change in color in pyrite and chalcopyrite, over a simple tan in bornite, to complete and partly deep reaching surface decomposition such as in native arsenic. These alterations may take place immediately after the polishing as is the case with lead, or after hours, days, or months. Slight alterations are common in nearly all minerals, thus making it useless to study a specimen of an apparently perfect polish after one year, without giving it a short repolishing. This also counts for minerals in collections which have been mentioned to be stable (galena, magnetite). In these the physical-chemical condition of the polished surface layer obviously causes alterations. — In many cases the moisture of the air is responsible for alterations. It may cause complete decomposition in some ores in a relatively short time (argentopyrite, bravoite, weathered pyrrhotite, loose marcasite). For such specimens careful drying after preparation and preservation in a desiccator over P_2O_5 is recommended. To cover the polished surface "Zapone lacquer" may preclude the tarnishing sometimes. The varnish of course has to be removed at each new examination.

Talmage hardness. St. B. Talmage (1925) gives a scale of seven letters for the scratch hardness which he determines by means of a sklerometer; "+" or "−" signifies a little above or below the average value. These data generally parallel the hardness differences of scratch marks and, with rare exception, are excellent. The scale should following the right statement of Young & Millman be enlarged for about 5 steps.

The method using measurement of the size of an impression ("indentation"), made by a standardized pyramid of diamond at distinct pressure, are very similar (Vickers, Bernhardt-Hanemann, Grodzinski, Knop et al.). They were used by many scientists but alas just in the case of many minerals deviations of the results are so high that in the result the method remains valuable only when connected with other tests. Especially the very careful examinations of E. Pärnamaa (1963) and Young & Millman (1964) made that evident. Pärnamaa especially explains the deviations connected with discussions of the data of other workers and so on.

The main influence regarding the deviations have small, sometimes very small differences in compositions, conditions during formation (e.g. structural defects) and differences in orientation. Further on, polishing techniques, duration of indentations, weight, sometimes temperature and more or less care in the methods may be essential. So, it is better to give the range of deviations than middlings which can suggest higher accuracy. In very careful work differences of 100% (in relation to the lower value) can occur, those of 30–40% are common. Value spreads for only 10% or so are mostly to be explained by too few measurements. Comparing values of different authors the spread may be still higher (e.g. 63 and 184 resp.), without a distinct trend, in other words, where one author would have consequently the higher or lower values. But notwithstanding, the results go roughly parallel to the MOHS and TALMAGE values — which is not so astonishing regarding the fact that the steps in the Mohs Scale differ by about $2^1/_2$ times. Consequently the diagnostic value should not be overestimated; it is not much higher than the Talmage-data. Thus I can omit here the indentation values. But I refer for them to the papers of PÄRNAMAA and YOUNG & MILLMAN.

That I myself did not say too much in the judgement of hardness measurements was approved by a paper of JAIN et al. (1972) who refer to 119 iron meteorites, shocked more or less, which hardness values between 100 till 350. The Ni-content has here very small influence (the highest and lowest value referring to Om-iron with about 7,5% Ni.

The data are often given thoughtlessly. Minerals, well known as complicated mixed crystals, and actually prototypes of those are examined in complicated and timevasting measurements. One example: a monoclinic mineral such as brannerite ($\sim UTi_2O_4$) is treated as isotropic, besides that, the formula allows high replacements for U by Ca, RE, Th, Fe^{II}, for Ti by Fe^{III}. Fresh brannerite obviously will vary in unforeseeable amounts. And further: with geological age brannerite besomes isotropized and that isotropization can mean a decrease of reflectivity of at least 30% and that variable in different parts of the same crystal and variable with temperature of the rock and depending perhaps on reheatings. To give here numerical data is actually nonsense!

III. Reflection behavior. "Impression of color and reflectivity" as seen for the specific mineral beside its common associates was the first qualitative property we mentioned in our "textbook" (SCHNEIDERHÖHN & RAMDOHR 1931). "We always stated this property though it may be quite misleading very often." The first impression of the properties causes prejudging, which is hard to dispel. An ore mineral of medium light color such as galena may be of a weak whitish grey beside silver, gold, or bismuth, whereas its appearance may catch the eye as a pure white color when associated with sphalerite, pitchblende, or some gangue. A moderately colored mineral such as pentlandite or even chalcopyrite may show up as nearly white adjacent to minerals of strong colors; adjacent to white minerals, however, it may well be of decidedly colored appearance. As the mineral of the lightest color and to some extent the most colored mineral are responsible for the general color impression one gets in a polished section, there is a tendency of comparing other minerals with it. Thus chalcopyrite 38% reflectivity) beside gold (90%) looks approximately like sphalerite (16%) adjacent to chalcopyrite! Even an eye with unusual sensitivity for distinguishing colors and the greatest experience does not protect from errors. In spite of all this, the present chapter was not given as much weight as it had in previous editions. The difficulties are so well known and commonplace that it is enough to mention them here in general, which should do for most cases. Observations of this kind are no longer clearly separated from those dealing with the general effect of reflectivity.

A *table* is given, showing the true reflectivity and the color. Both these characteristics are given for observation in air and observation in an immersion oil (always Cedarwood oil of $\sim$ n 1·51) because of their specific differences. Above all, reflection and color are compared not only with those of common standard minerals which show very little variation such as galena, sphalerite or chalcopyrite, but also with special associates in the mineral assemblage. If necessary a decisive comparison of color and reflection may be obtained by using a comparison microscope. The use of such an instrument is highly recommended for the study of strange mineral associations. The beginner as well as those with an insufficient ability to distinguish colors (there are many stages up to total color blindness) will frequently make use of it because otherwise very slight variations in the color of ore minerals cannot be distinguished. The attempt was made to give as clear a description as possible, — perfection being a goal beyond limits.

The *quantitative data* of optical characteristics offer great difficulties that lie in the theory and partly in the nature of the technique of measuring and polishing. The reflectivity of isotropic, strongly absorbent minerals depends on "n" and "$\varkappa$" which are for each wave length of different and independently varying magnitude. The measurements can be made independent from the knowledge of "n" and "$\varkappa$" which in fact are known only for very few opaque ore minerals. The best method is measuring the percentage ratio of the reflected to the incident light. The variation of the single wave length gives a curve of dispersion for the reflection intensity, causing the subjective impression of color. A moderate increase from red to blue commonly results in the color impression from white to bluish white, maximum value in red with decrease towards blue being reddish, maximum in yellow being yellow or yellowish white etc. The same color impression may result from a complicated dispersion curve. It has to be considered that the subjectively brightest color impression e.g. of yellow, may not be immediately evident when measured. In anisotropic minerals "n" and "$\varkappa$" are different for different directions. Random sections furnish mean values which are hard to appraise. Naturally, all optical orientations have to be considered and use has to be made of the polarizer. For the optical principles refer to Vol. I, 1. of the Lehrbuch der Erzmikroskopie, SCHNEIDERHÖHN & RAMDOHR 1934, and various publications by BEREK (1931, 1937), GALOPIN & HENRY (1972). It has to be mentioned that even very objectively written publications contain some elementary errors. To measure the reflectivity quantitatively, three principal methods have been suggested: 1. Photometric comparison against a standard mineral under the comparison microscope, using a "*grey wedge*". 2. Measuring with the slot photometer ocular by BEREK. 3. Application of photoelectric cells of various construction. For polished ore sections the first method has not yet been worked out systematically. At the suggestion of SCHNEIDERHÖHN, the slot photometer with several slightly altered models was used by H. FRICK (1930) and by A. CISSARZ (1932) at the Institute of H. SCHNEIDERHÖHN. The values obtained with red, orange, and green filters which essentially correspond with the spectral lines *C*, *D*, *E*, are quoted for air and for oil and if necessary, for different optical directions. It has to be worn in mind that some minerals of considerable pleochroism in the reflection (see below) have not yet been systematically evaluated.

Re. 1.: Polishing defects always weaken the reflectivity. In using a slot photometer the error can be reduced subjectively, which cannot be done in the objective measurements with the photoelectric cell. Re. 2.: Polishing films at times reduce the reflectivity considerably whereby both isotropization and partial oxidation, the development of the latter partly being favorably accounted for by the former, result in a reducing effect, which in turn seems to be balanced by the smearing. Re. 3.: An ideal reference base should be the white noble metals as silver, platinum, rhodium. They cannot be polished without showing considerable defects. Such mirrors, produced by other methods always have lower values than calculated theoretically. Quartz and diamond which yield an ideal polish and allow mathematical control have much too low values to be useful for comparison because every error becomes multiplied. The formerly frequently used minerals galena and pyrite have not proved to be stable. Now standardized mirrors of metal carbides are used. — Other sources of error are partly surprising: Different polish-liquid can be up to 2% absolute raise or lower the values. — The method of measuring itself could be reproduced up to an accuracy of 1/10% and farther, but the object not! The ranges of variation of 4% absolute, i.e. e.g. 8% relative, are quite common with the same method (cf. loose Tables) and still higher ones with different methods. The values of the Sovjetic colleagues are almost always higher (see below).

In spite of all scepticism (see p. XVII of the preface) against the trustworthiness of the quantitative results of the reflection-values in literature, a scepticism, which should also be more familiar to the young microscopists as it is recently used, I gave here many data from the literature. But I emphasize: The single measurement e.g. of nm 589 does not mean anything — 2 or 3 wave lengths not much more! More useful are data across the whole visible spectrum (i.e. until the mostly neglected deep red) since they reflect, in spite of the rather strong deviations at least the tendency in the color-variation and possibly even the fine tints of the differences in air and in oil. The importance of that is perhaps recognized for the first time by several Russian colleagues, e.g. VJALSOV (1973) and a team under the guidance of M. S. BESMERTNAJA (1973) and shortly later by Western scientists. Data of both are given many times. Other ones are taken from literature directly, partly from text-books, such as the International Tables of the Commission of Ore Microscopy of the IMA or from UYTENBOGAARDT & BURKE. For the latter ones no other literature is mentioned. It is regrettable that here and elsewhere the values in oil are mostly missing. Of course, obviously grave errors are omitted (there are some with more than 100%, even 200%).

The small tablelike summary by MCLEOD & CHAMBERLAIN of the Geological Survey of Canada (Paper 68-64, 1969) gives some good statements but contains also much ballast by not controllable (e.g. isotropized minerals) or obviously wrong values.

The *comparison* of the *reflectivity in air* with that in *oil* proved to be of extraordinary diagnostic importance. Brightness and color changes are characteristic both in their qualitative subjective impression and the quantitative measurement.

I refer to the preface where was, briefly and therefore a bit drastically expressed that many data of reflectance were given carelessly and inconsiderately published. Though excellent methods were used, object which are from various points of view unfitting were chosen and therefore valueless and sometimes coarsely misleading. Only a few chemically well defined and well polishable material with using very many measure-points across the whole spectrum are really useful and fitting for reproduction. There are naturally minerals which follow more or less these requirements. But astonishingly they show that even the best authors come to very differing values.* In this case the imperfection of the structure, especially abundance of faults, seems to be the reason. It often shows that the reflectance curves in their entire lengths are extremely similar but are sometimes 2 to 5, even 10% and more higher or lower, whereby

* Please compare the values of VJALSOV or BESMERTNAJA for native gold, or native Antimony of BESMERTNAJA and LEONARD.

the nuances of the color impression can then be very constant, constant at least as far as the human eye can see with the same illumination.

All tabellae which give data on complicated mixed-crystals are valueless if they don't give in the same paper exact analyses of every grain. Absolute nonsense are everywhere data on isotropised or partly unisotropised minerals.

Inspite of this at least a part of the data, especially of BESMERTNAJA, the COM-Committee, VJALSOV and the French School should be given. The former reflectivity values, back to the old measurements of CISSARZ, EHRENBERG, FABER, ORCEL and FOLINSBEE are by no means obsolete. Often they lay in the middle of most recent data. Of course these values are inadequate because they give only one wave-length and only for green-orange-red. But more could not be expected with the insufficient apparatus of 45 years ago.

Alas, the values of VJALSOV, BESMERTNAJA, as well as the COM-group (HENRY and FONT-ALTABA) are incomplete: Some trivial minerals are completely missing, against that they contain data on extremest rarities, which 1) are strongly to be doubted, and 2) are not controllable. Often I didn't give them.

In the sequence of the book I bring now a part of the values by VJALSOV or BESMERTNAJA of different single-numbers of about 40 or 50 nm. It is tecnically not possible to give complete dispersion curves — favorable as it would be! Which author I prefered as more reliable is the result of my personal experiences — which do by no means exclude gross mistakes. Two authors I have mentioned where evident differences exist. Correction, e.g. in the form of (?) or a short remark I have made only in a few cases.

The suffices PIOR resp. BLESK in the tabellae refer to the used apparatus. Missing of these data means that the authors used other apparatus or that the data are taken from another reference.

It has to be emphasized over and over again that whoever shuns the use of oil immersion misses an important diagnostic tool and will never see hundreds of details described in this book.

Reflection pleochroism, or as priority demands, "*bireflection*" can be measured quantitatively very easily by the methods just mentioned. Even without being measured, it is one of the most characteristic properties which in many cases makes possible the recognition of all textural characteristics. In the recognition of delicate differences in brightness and color any excessively strong source of light is of disadvantage. Important also are a uniform illumination of the field and the use of the same light source. The reason for using the term pleochroism of reflection in this book (as nearly everywhere in literature dealing with ore-microscopy) lies in the obvious similarity to the pleochroism of non-opaque minerals.

The diagnostic importance of the bireflection ranges within wide limits in its distribution and variations, that is, for a trained and color sensitive eye. The writer principally observes with the polarizer nicol. The contrasts (particularly if "$\varkappa$" is small) often become considerably intensified by using an oil immersion. The color phenomena described under "reflection pleochroism" have been stated as correctly as possible, the description of very minute color changes being very difficult.

Considerable theoretical and practical difficulties become apparent in the behavior of anisotropic ore specimens under crossed nicols, the effects of "*anisotropy under +*

+ *Nic.*". Nevertheless, to the experienced observer this procedure offers important diagnostic data, above all the criterion of isotropic or anisotropic. It is the best method known to make visible grain boundaries, twin lamellae, zonar structure and other structural features, without destroying the polished surface by etching. For the observation a very strong source of light is necessary, a tungsten projection lamp, a uniformly operating electric arc, a Xenon high pressure lamp or something similar. To recognize weak effects of anisotropy, e.g., that of anomally ansiotropic magnetite is only possible by those means. For stronger effects, the contrast in intense and in moderate light is as different "as black and white". A small arc light would be ideal without doubt. Unfortunately such a lamp burns too irregularly in spite of a clock work or electromagnetic control. The well suited but very sensitive Nernst lamps are no longer manufactured and the only means available is the projection bulb*. Caution has to be taken to avoid getting in the eye the intense light of this lamp undamped by a glass filter and without crossed nicols. Not to speak of the damage to the eye, the glare is so strong that it takes hours to recover from it. The glare of a well sun-lit room is already considerably strong.

Principal effects of anisotropy have been observed in almost all non-cubic ore minerals and further anisomalous anisotropy in cubic minerals. Very often the effect is very weak so that it becomes visible only by applying special tactics.** Theoretically most reliable is the "conoscopic" observation, that is the observation of the exit pupil of the ocular at the proper focal distance or by means of a lens and a card board tube. Then, of course, no more details can be distinguished but all the brightness is concentrated at one tiny spot. A different procedure consists in moving the analyzer or polarizer by 1–2–3 degrees from the crossed position. The effect becomes much stronger and more apparent by sudden color changes when rotating the stage. Errors result when comparing different minerals, further when the diagnosis does not depend only on the effect of isotropy–anisotropy, or the observation of grain boundaries. The method given by GALOPIN (1936, 1947), i.e. the measurements being made by deviating a defined amount, e.g., 4 degrees from the 90 degree standard position, should be feasible and acceptable in practice because of its higher light intensity though the mathematical interpretation becomes very difficult.

* The writer temporarily used with success a Xenon high pressure lamp. The apparatus is so expensive that the saving of time scarcely pays off.

** In a position of precisely crossed nicols the light intensity amounts to a few per mils in comparison to the common observation and only in very few cases 1–2%.

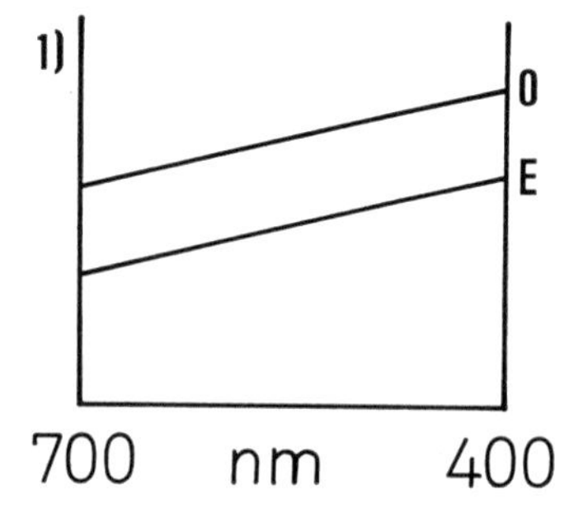

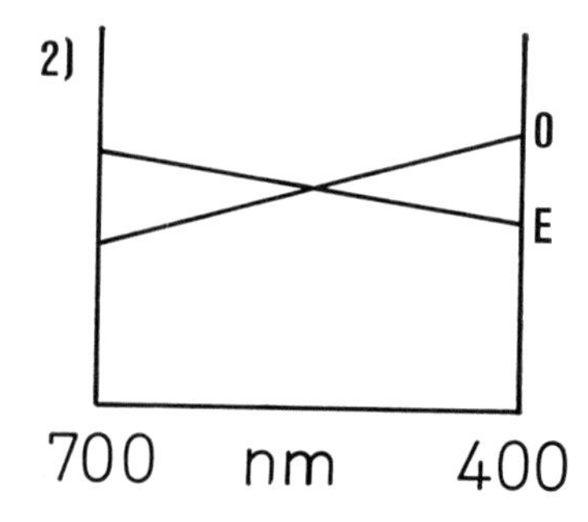

Due to its complex nature the quantitative measurement of the effect of anisotropy seemed to be a hopeless affair until 40 years ago. Investigations by BEREK (1931, 1937) however, showed the way. Statements in publications up to the present which deal with "only one position of extintion under crossed nicols", "interference colors", and effects of particular colors are either grossly erroneous or purely descriptive and haphazard. BEREK showed that a variation of 1/10 of a degree (!) in crossed nicols, which is a common error that can be eliminated only difficultly with the present instruments, gives in certain minerals fast changing effects which are hard to control. In monochromatic light, under precisely crossed nicols and known position of the bisectrix, extinction results at the 90° positions and a maximum of brightness at the 45° positions. In white light the color effects are characteristic and can be reproduced.* For common practical application not much can be gained from this fact.

Strong color effects, though, may indicate strong dispersion, and under approximately crossed nicol position certain colors may be predominant and characteristic (e.g., bright leather brown to red brown in covellite). The term "approximately crossed nicols" is flexible and is more limited for one particular mineral than it is for another. Therefore all statements on color effects have to be made under greatest reservation and very many are omitted at all (cf. GREEN 1952). Many sources in the literature on this subject are misleading!

The remark "with + N at no position absolutely dark", should always be enlarged by "using white light", when it concerns minerals of low symmetry. Using monochromatic light the extinction can be perfect.

CAMERON (1961), gives for many minerals as a diagnostic help the "rotation properties", characteristic angles, which can be measured with a rotationable analysator and a likewise rotationable mica plate rather quickly for different wavelengths (comp. his tables 272). Unfortunately these values differ rather often from the theoretical ones so strongly, that probably the method contains some unknown sources of error, and so at the moment the method must be handled with much caution.** According to the footnote I omit these values determined by CAMERON and his pupils with much care and time.

As is the case with the reflection pleochroism, theoretically the effects of anisotropy in oil should be considerably more striking than in air. This can be seen clearly when observing a specimen first under a dry medium magnification objective and then by using the same magnification for observation with oil immersion. On the contrary, when using the dry objective No. 3, then oil immersion objective 1/7 a, strong depolarization is caused by the strongly convergent light so that the effect of anisotropy in oil appears to be weaker than in air. Reduction of the opening of the aperture iris, which cuts off the oblique rays of light yields the characteristic effect, though of course very weak.

---

* The most recent microscopes (about since 1958) have a device that by means of degree circle and vernier both nicols can be moved and reset.

** The report of the I.M.A.-Commission of ore-microscopy says: "The discussion of the "rotation properties" (Berek, Cameron), which can be regarded as an indirect method for the measuring of the bireflection, did not bring a result. The indirect experimental values do not agree mostly with the direct ones. The reasons of that are still unknown ... "Rotation data at present cannot be regarded as reliable".

IV. Etching. A. *Diagnostic etching.* The chemical characteristic of etching by certain reagents was once considered important and still is today. With a specially developed small pipette a drop of the reagent is put on the freshly polished surface of the section and the reaction of the mineral, mostly within a fixed time, is observed, such as the effervescence or the less vehement development of gases and vapors. The tarnish of the minerals is observed, and whether it is a cohesive coating or may be rubbed off, or whether the mineral is dissolved without showing any tarnish. Finally the form and kind of influence on the particular surface is registered. J. MURDOCH (1916) developed as early as 1916 a book on ore-microscopical determination based on the attitude against 8 reagents. In 1920 DAVY & FARNHAM extended and improved the data by MURDOCH (1920). The booklet was issued in its second edition in 1932 and instead of five standard etching reagents, seven were used. In 1931 M. N. SHORT in exceedingly hard work restudied the etching effect of all accessible ore minerals and supplemented and extended his data in his second edition in 1940. This book has become a well known standard publication on ore-microscopy. His data are given everywhere in my book and only in relatively few cases a critical comparison has been made with those by MURDOCH, DAVY & FARNHAM, and others. Certain reasonable abbreviations were made; common data, such as the stronger reaction of some mineral faces, or such as say very little and give way to errors, were omitted. The reaction with aqua regia has been presented only where in striking contrast with that of HNO_3.

It has been found, however, that a schematic application of diagnostic etching yields gross errors in the determination (cases are known where out of 12 minerals of a mineral association only four were correct!). Such insufficiencies of course discredit the method. The writer alone and together with SCHNEIDERHÖHN has repeatedly warned against the application of diagnostic etching as the sole method and he uses it very rarely. Nevertheless, it would be wrong to reject it as useless. It can supply very useful clues, but has to be substantiated by optical methods, microchemical tests, and above all by X-ray studies or microprobe. Last but not least its principles and sources of error have to be known.

A prerequisite of etching is, of course, that the mineral itself is soluble under the conditions given, such as kind and concentration of the etching reagent, temperature, mineralogical association and intergrowth. In a fine powder the reaction may be instantaneous, even where on a polished section scarcely any etching or none at all may become evident.

In detail we shall consider as sources of error in diagnostic etching:

1. *Orientation of the crystals.* In crystals, the rate of going into solution depends on the crystallographic direction; one lattice plane may be dissolved hundreds of times more rapidly by an etching reagent than another lattice plane. In turn, the latter may be attacked faster than the former by a different reagent. Thus in diagnostic etching it is possible to determine the same mineral differently, depending on the orientation. The basic data contained in tables of determination may be insufficient, particularly if the mineral is relatively scarce and preferably occurs as coarse grained aggregates.

2. *Crystal chemistry.* The chemical solubility is a property depending on the perfection of the lattice, i.e. slight irregularities in the lattice may have a strong effect. Whoever had in a collection the "opportunity" to observe water damage in such a "pure" mineral as pyrite, is not surprised at all. Pyrites of the same appearance, the

same coarse grain, and without any intergrowth may become completely destroyed in one case and remain completely intact in an other. A tiny imperfection such as the missing of a cation or anion may furnish the point for an attack, and once the reaction has started, it may proceed at the speed of a chain reaction. It is the same if there are lattice imperfections caused by components non related to the lattice, in part in the group of anomalous solid solutions; by exsolution bodies, perhaps those that are not yet visible microscopically; by mechanical strain in the lattice, or strain already set free by cataclasis, translation, or gliding. Also here the differences between an ideal and a real crystal may be enormous.

3. *Solid solutions.* A great number of ore minerals form solid solutions, and certainly more often than has been assumed up to now also in the group of sulphides and suphosalts, to various degrees depending on the conditions of their formation. The resulting compounds may show different reaction to chemical attack, the mixed compounds presenting all transitional stages in a way which cannot be controlled. Take for instance solid solutions of silver and gold and their attitude against nitric acid or the "tarnish" of solid solutions of $(Fe\text{-}Ni)S_2$. Part of the astonishingly variable etching properties of magnetite are due to solid solutions which is also the case with chloanthite, fahlore, and sphalerite.

Even though already mediocre preparats and photographs in some rare cases can afford an insight into ore structure problems, unfortunately the equipment of many ore dressing papers with pictures is deplorable till now in most countries. Thus, I must confine myself for the most part to literature (American, Australian, German) available by fortunate circumstances. But mainly the examples mentioned here are taken from problems which I have been able to examine myself.

4. *Influence of the grain boundaries.* As can be seen very often, the grain boundaries are the starting points of the etching reaction. A single grain may be completely indifferent against the reagent whereas grain aggregates are attacked. All the more, reaction may take place if the neighbouring grain is of different composition. The causes vary. Small pores allow the reagent to penetrate and they confine it, which is purely a mechanical process. The grain boundaries may have random orientation, whereby the systematic saturation of the lattice planes is disturbed. Finally a formation of galvanic chains may take place, the electrical potential accounting for chemical reaction where the etching reagent alone would be of no effect.

It is practically quite impossible to survey all those causes of disturbance (1–4) in each mineral grain. To some extent, some of them can be taken care of (e.g., avoiding all intergrown grains), others are beyond control.

5. *Influence of the etching reagent.* Further causes of disturbing effects are due to the etching reagent itself. Traces of impurities which find their way into the reagent while it is being applied may work in an unexpected way like a catalyzer. Slight changes in the temperature may at a strokelike increase the solubility many times. Effects as are known of iron against concentrated HNO_3 (complete inertness up to 75°, then sudden reaction) may show up in ore minerals in a modified form. Even here an exact control will be difficult because the above effects (1–4) have to be considered.

On the whole, diagnostic etching is not to be recommended to the "occasional" ore-microscopist, that is, a scientist who in course of investigations of ore deposits or petrographic work has to determine some less frequently occurring ore minerals. The routine specialist will make use of it, but under constant control by X-ray methods.

It is the specialist whom one will consult if there is no other way left. It is less embarrassing if rare sulphosalts or rare tellurides are not recognized and is less of a burden to the literature rather than wild determinations which afterwards may be proven to be erroneous.

In the application of diagnostic etching the various writers do not use the same technique. The temperature at which the etching is applied and which is of great influence, is rarely stated. That is why all published data have to be viewed with reservation. *In general*, etching is applied for 60 seconds and the reaction observed under the microscope. This procedure is not without any objections. A description of the result of the etching very often cannot be given in a few words. It was not possible to give in this book every detail of it, as mentioned above.

B. *Structure etching*. The aim of structure etching is different. It has been taken over from metallography where the similarity of bodies is closer than in ore-microscopy. Further there the help given by the use of polarized light has not yet become too familiar. The structure-etching was introduced into ore-microscopy by SCHNEIDERHÖHN in 1921; of course, there were earlier occasional applications and reports on excellent etching patterns.

SCHNEIDERHÖHN attempts to give, where possible, for each ore mineral some etching reagent which develops the following details:

1. Internal texture of individual grains (twin lamellae, zoning, deformation). This is called *internal etching of grains* [*Korninnenätzung*].

2. The vectorial reaction in different crystallographic directions, in order to recognize the orientation of the grains in an aggregate and the shape of the grains. It cannot be clearly separated from 1. It is called

Surface etching of grains [*Kornflächenätzung*]. 3. The grain boundaries of the single grains of a homogeneous crystal aggregate. This is called

Grain boundary etching [*Korngrenzenätzung*]. Internal etching of grains and surface etching of grains are the result of the vectorially different reaction of crystallized bodies. It is more difficult to understand the grain boundary etching of grains. The assumption shortly sketched above, where non saturated points in the crystal lattice at the grain boundaries give way to easier reaction may serve as an explanation. Metallographers developed here in part superfluous hypotheses about an intercalated plane or something similar.

A further application of structure etching is to be found if different ore minerals that show little contrast are needed for subjective observation or for photographic purpose. *Contrast Etching* may be of importance principally to the less color sensitive eye for the technical purpose of counting or measuring the components of a complex aggregate.

The writer admits that he has not taken pains in further systematical extending the contents of the "textbook" (SCHNEIDERHÖHN-RAMDOHR, 1931, 1934). In some of it he succeeded by chance. If homogeneous aggregates of isotropic minerals are not concerned, he prefers to work without the knowledge of certain structural details rather than exposing to deep destruction a polished section which took hard work to prepare. It would be desirable to have suitable structure etching reagents for each ore mineral. Also in anisotropic minerals zonal structure can often be made visible only by etching.

Other means of making structures recognizable have been applied by the writer only on some occasions. Here, in his statements he depends largely on published sources. Some of the described methods are simple, nearly trivial, others are complicated or methodically very elegant. To these belong annealing by heating (it has to be warned against because of possible transformation!); deep etching for the production of etch patterns with oriented reflections; the print method as applied by GAUDIN (1930), GALOPIN (1936), and HILLER. Furthermore recent procedures where one element in the mineral is transformed into a radioactive isotope by use of the cyclotrone after which the element in question can be detected photographically because of its radiation (GAUDIN).

V. Physico chemistry. The physico chemistry of relatively few ore minerals is known as thoroughly as that of almost all rock forming minerals. The difficulties presented in experiments by the vapor pressure of S and O, have not yet been always overcome. Nevertheless, a microscopic investigation very often gives an idea of the stability of a mineral or a mineral assemblage under increased (or decreased) temperature or pressure conditions. If for a specific mineral more is known than on an average, then a special chapter might be useful.

It contains published data of the *temperature of transformation, temperature of decomposition* at given pressures, *melting points or ranges,* solid solutions and exsolutions, as well as all conclusions possible from ore-microscopic examination. Recently the number of references has increased enormously. It is impossible to give only a small part of all data. Pyrrhotites may give an example!

In many cases it proves appropriate, to give data, e.g., on exsolution in section VI together with considerations on its consequences, and to make only a short reference in section V. For very many minerals data are still missing. It should be pointed out explicitly that in no way any effort was made to exhaust the subject, and all data of physico-chemical character that are unimportant for natural ore occurrences, have been omitted.

VI. Fabric. The chapters "*fabric*" and "special *fabrics*" give information on all relationships of the single mineral grain, of the aggregates, and of intergrowth patterns with associated minerals. It is clear that much more can be said on frequently occurring ore minerals and those that have a wide range of formation, rather than on rarely occurring ores or those that are limited to few and similar occurrences. According to the amount of material available for observation a more or less extensive subdivision will take place. For rare minerals the scope of this book imposes limits. Section VI, "*fabric*", deals with the "*internal properties of grains*", *textures,* i.e. the properties of the single grain, and fabric, the orientation of the individual grains in aggregates and in intergrowth. As the general section of this book deals especially with textures, only an outline will be given here.

Section VII "*special fabrics*" emphasizes a few very important and characteristic fabric patterns which could be discussed under "fabric", but deserve special treatment. Among others there are replacements, myrmekitic texture, intergranular films, recrystallization, and occasionally some common important textural characteristics. Rare minerals are not specially dealt with in section VII.

Internal properties of grains. The minerals of a homogeneous or heterogeneous aggregate may be well formed crystals which are intergrown or grown into open spaces, or crystal grains of random boundaries, which are called "*crystallites*" by metallographers. The first case is less frequent than the second. The individual crystals may be internally homogeneous or they may have inhomogeneities of diverse kinds.

Twinning boundaries are the most frequent. The formation of polysynthetic twin lamellae is very common in ore minerals, much more than simple twinning. This characteristic may be a wide variation: the same twinning law may be after several planes of equal value, while several different twinning principles or simultaneous translation may change the picture. — As to the origin the following distinction has to be made: *growth twinning,* meaning a formation of twins during the growth of the crystal, which is often the sign of hidden pseudosymmetry of the lattice. *Pressure twinning,* which is often caused by external pressure exerted on the already full grown crystal or more commonly on the aggregate. *Inversion twinning* which is also to be found in already fully developed crystals, but was caused by an allotropic reversible or monotropic transformation. Transitions are possible in the two latter groups.

Zoning is also very frequent. It is due to isomorphic layering in a solid solution series as it is to be found very frequently in ore minerals formed at relatively low temperature, or to interrupted growth. In etch tests old surfaces can be etched with particular ease. It is also due to a zoned arrangement of inclusions or other minerals. Judging the change of the habitus as it is often visible in zoned crystals has to be done with caution because the elementary geometrical considerations require the knowledge of the crystallographic plane of the section and its distance from the centre of the crystal. Many data found in publications are erroneous.

Deformations are very common. They often give evidence in the same mineral in many different ways: length of the period of stress (gradual or instantaneous), temperature, purity of the object, and a wrong treatment of the specimen in the preparation etc. may cause plastic deformation in one case or cataclasis of very different kinds in another case. Undulatory extinction, bending, translation, formation of "gliding twins", and cataclasis show the lattice to be under strain or that the "free surface" has become very large. Therewith the prerequisites for *recrystallization* are given which in the different minerals starts more or less well and is influenced by the case history and temperature. The signs of deformation may already have been wiped out completely in one mineral while in another associated mineral they still may be seen clearly. This is dealt with, just as the fabric orientation, in section VII.

Intense deformation of an ore deposit, together with evidence of recrystallization, often causes in ore minerals what is known in metallurgy as rolled structures, and is called "*fabric orientation*" [Gefügeregelung] in crystalline schists.

Evidence of *inversions* is found in the above mentioned inversion twinning. In other cases transformation may cause intricate structures which are hard to explain.

Exsolutions are also considered under "internal properties of grains". Since SCHNEIDERHÖHN mentioned in 1921 for the first time their wide distribution in ore minerals, more and more of it has been found (cf. pp. 165—194). It has to be mentioned that many cases of unmixing may become wiped out by "collection crystallization" without leaving any traces. On the other hand exsolution textures may be simulated by diffusion or partial replacement, primary oriented intergrowths, and crystallized gel textures.

Oriented intergrowths have a very close crystal-chemical relationship to exsolutions, showing the same mutual lattice relations. They may be linked by transitions. The "sphalerite stars" described for the first time by the writer may be exsolutions as well as primary (rarely) oriented intergrowths.

The decomposition of solid solutions or compounds is often observed. Some are clearly interpreted, such as allargentum-silver, "maldonite", now gold + bismuth, schapbachite-galena; others still are of doubtful nature and certainly have a more complicated mechanism of decomposition, such as fahlore into arsenopyrite, gudmundite, chalcopyrite and sphalerite, etc.

Patterns as observed for the first time in minerals of the ore from Broken Hill by STILLWELL in 1926, *autometamorphic* or *synanthetic* (SEDERHOLM) intergrowths may be classified among this group, but rather should be discussed under *replacements* (see below).

Structures and Textrures. The properties mentioned above, (unmixing, decomposition) already lead to what is called "*fabric*"; that is the entity of the geometric development of the components of an aggregate and the spatial orientation and distribution of the components therein.

In Europe, different from America, it is customary to understand by "Struktur" the properties based on the shape development, size and mutual boundaries of individual grains, whereas "Textur" encompasses characteristics of spatial orientation, distribution, and packing. The latter expression comes close to (but not quite) the term of "fabric" in American publications.

In order to classify a system the following points are dealt with:

Grain shape, grain size, grain bonding within an aggregate. The data are, at first, purely observational, i.e. descriptive; but some conclusions become immediately apparent.

The *grain shape* in a monomineralic aggregate is hard to determine without etching. Some experience, however, will help to do without etching in most of the cases, even with isotropic minerals (relief, cracks at the grain boundaries, etc.). The variety is so wide that in a description the petrographic nomenclature is not sufficient. Statements of the *grain sizes* are needed for technical reasons. They may be very important, since the tendency of development either in coarse or fine grained crystalline patterns seems to be a characteristic of the mineral species, or in some other cases, a characteristic of the kind of origin.* Grain bonding relations (intergrowth or locking types) which are of great technical importance are also of genetic and crystallographic interest because compounds with related crystal lattices tend to intergrow more easily than others.

A particular part is played by *gel structures* which in no way originated at such low temperatures as is suggested by some limonite occurrences. In many cases the impurity and viscosity are decidedly of importance; characteristical *rhythmic precipitation* therefore is here common.

* Caution is recommended. Loellingite which commonly is fine-grained is known to occur in Kaatiala/Finland in grains that weigh many kilograms.

VII. Special fabrics. True but still questionable *patterns of replacement* are shown here; the question has to be considered, whether the replacement is of ascendent nature in the normal sequence of the formation of ore minerals, or whether it took place independently later on, or whether it is of descendent nature in the zone of oxidation and cementation.

In many cases there will be an extensive discussion as to whether structures are due to replacement or a different process. Many a true replacement can only be recognized by relicts of characteristic associates which may be very scarce; other apparent "replacements" are most likely primary intergrowths. The publications give the impression that some writers are too subjective in their personal views whereas others simply do not say enough.

Intergrowth with other minerals that cannot be explained as replacements are discussed next. The widely distributed types of "myrmekitic", "graphic", "subgraphic", and "eutectic" intergrowths belong to this. How complicated they may be genetically is to be seen in the general section of this book. *Intergranular films* which are genetically very complex; intergrowth of "*mutual boundaries*" which mostly is taken as a "proof" of simultaneous origin, but which is not necessarily the case; *relict textures*; *palimpsest textures* and many others are discussed.

Also included are the very frequent and important features of "*rolling out*" with *fabric orientation* [Gefügeregelung] and recrystallization in all stages though part of it has already been discussed in the chapter "grain shape", "grain size", or "deformation". In spite of the enormous abundance of occurrences (there is scarcely a lead deposit where we do not find "galena tails" („Bleischweif"), i.e. rolled out and recrystallized galena) little work has been done on "Bleischweif" though it has always been properly interpreted by mining men. Metallurgy also found microscopic methods of studying the degree of rolling and fabric orientation in opaque and isotropic minerals. They are very tedious and today they are everywhere replaced by X-ray methods. In ore minerals the conditions for microscopic determination are to some extent easier (cf. textbook of SCHNEIDERHÖHN-RAMDOHR, 1934, Vol. I. pp. 265ff.). Publications by SCHACHNER-KORN, HUTTENLOCHER, and the writer (1945b) show the first attempts of systematic research.

"*Recrystallization*" very often wipes out the external form of the grain in an oriented fabric. The orientation, though, as seen crystallographically, may become exceedingly strong. Many ore minerals recrystallize so easily that it is not to be recommended to boil mechanically strained specimens in sealing wax, or simply embedding them in it, because this alone may obscure some fabric features by recrystallization (in lead e.g. as metallographers know, this process takes place when the specimen is exposed to room temperature). Many new observations can be made here.

Among *special fabric types* very often *weathering textures* of ore minerals are counted, which to some extent have been dealt with in the discussion on replacements. Many textural features, such as cleavage and zoning become better developed by weathering than by any etching. The formation of secondary sulphides, that is by transformation or new formation, (marcasite, bravoite, millerite, covellite) or oxides can be recognized excellently though very often analytically no change has yet taken place. Many strange phenomena shown by ores in ore dressing can be thus explained and predicted. The many characteristics of descending cementation in rich copper and silver ores that were studied microscopically are discussed here as well as their relationship to the latest hydrothermal phases. Proper evaluation can be of great economic importance. It has to be emphasized that correctly many ore parageneses that had formerly been interpreted as [supergene] cementation, recently have been recognized as ascendent [hypogene].

In many weathering textures, even after complete destruction of the primary ore, the original nature of the latter can be recognized more or less easily, but definitely. This is of importance in the knowledge of "gossans" ["Eiserner Hut"].

VIII. Diagnostic features. In many excellent books on determinative mineralogy are contained all the data which make a definite determination possible. The application of those tables in practice, if everything is taken into account, is very time consuming. This is felt by everybody and is the reason for working things out "roughly" that is omitting some of the testing methods. This is how misinterpretations occur. Why? Because, among the given data are not mentioned the particularly dangerous traps that lead to wrong determinations. If those traps are known, then even with rapid work, the probability of errors becomes less. The chapter "diagnostic features" or "possibilities of misinterpretation, and determination" will therefore be of particular importance.

Points of view of the most heterogeneous sort may facilitate a diagnosis and make a positive determination possible, even in cases of extreme similarity according to the tables. Among such points are those of general geology, physical and optical characteristics, etch reactions, microchemical tests, internal structure, and the mineral association in particular.

Though the pattern of presenting the subject becomes thus a little disorganized, I hope, that it still remains clear enough.

Concerning the less frequently occurring minerals the data show many gaps. Efforts were made in vain to supply plenty of data of comparison for rare minerals. In the statements of the locations of occurrence of less common minerals all the investigated occurrences are quoted. This should be evaluated in the sense that it may improve the diagnostic possibilities.

It happens at times that in more common minerals some rare possibility of erroneous determination was not considered, though it might be a specifically dangerous one. It will always be like that in mineralogy; a lucky change makes possible for the careful observer the observation of a rare and formerly bypassed mineral.

IX. Paragenetic position. The large amount of the material for research and its variety that served as a base for the textbook to SCHNEIDERHÖHN and the writer, as well as the considerably more extensive material the writer was able to work over during the last 40 years in cooperation with his students, makes possible to report with a certain statistical reliability on the occurrence, quantity, and distribution of ore minerals in the single groups and subgroups of classified ore deposits.

For those data fairly clear, but not too strongly subdivided classification becomes necessary. This is discussed in this book in connection with other things.

Part of the data are commonplace, known for centuries, and were the reason for the classification of the "vein deposits." It has been discovered, however, that certain rare minerals, very often such minerals as contain rare elements, seem to be leading and to be characteristic in many classes of ore deposits. Thus, ore-microscopy may offer many valuable clues to pure and applied geochemistry. On the other hand, many minerals important as name giving play a lesser part in ore deposits.

In many cases particular emphasis is put on whether a minerals is "*ubiquitous*" [*Durchläufer*] in several or in many ore deposits or if it is "*typomorphic*", that means, limited to few or only one type of ore formation. Those data are independent of the amount of the mineral concerned, though it is obvious that a common mineral tends to be a more frequent companion of different parageneses than a rare one. In this connection, very complicated geochemical and crystal-chemical relations play an important role. Conditions of temperature and pressure as well as the sensitivity to alterations are phase limits for the formation and stability of a mineral.

For minerals which are commonly considered as type minerals in specific ore deposits, but occasionally may occur in different association, such cases are referred to. Not knowing them may result in an erroneous classification and appraisal of an ore deposit. "Lievrite" is an important characteristic mineral in contact metasomatic formations in limestone with an introduction of iron. It may occur, however, as a late magmatic product of reaction between a calcium rich plagioclase and a Fe-rich olivine and apparently also magnetite. It is easily missed in the latter case. — Covellite generally is a leading mineral in the upper parts of the cementation zone, but may also be found in S-rich hydrothermal copper veins and in volcanic exhalations. Very rarely it also occurs in parts of contact metasomatic or contact metamorphic formations which are very low in iron. In the first case one has to beware of a geochemical conclusion and in the second the lack of iron is very striking.

Of course all information is given for cases where certain minerals "avoid" each other or are incompatible. Some of those rules of exclusion are commonplace, others are a surprise, e.g., the marked decrease of the formation of martite in magnetites which are connected with more than a trace of sulphides etc.

Naturally all statements in this chapter leave gaps and some points are more extensively dealt with than others. On the whole, the chapter may present an important collection of material and much of which is new.

X. Investigated occurrences. Supplementary to IX, and also to show to the reader the reliability of the statements, the *occurrences studied*, mostly mining towns, have been given. If a rare mineral is known from only one mine and has possibly been investigated in only one polished section, then even the optical data may be incomplete. It may be that under circumstances a non-typical specimen was selected for an example, that the data on textures, intergrowth etc. are lacking perfection in spite of great care taken. If perhaps 20 occurrences have been considered then the statistical average is sufficient to recognize specifically some differing specimen and to give more definite statements as to textures, intergrowths and mineral associations.

If there is only one occurrence known but with many specimens available, then occasionally the number of the specimens is quoted. That means that the data in themselves may perhaps be incomplete, though exhausting the particular occurrence.

For the common minerals the occurrences have been mentioned with limitations. If possible those have been preferred that formed the base for more extensive research without any regard to economic importance. That is why, e.g., St. Christoph near Breitenbrunn (Germany) has been mentioned in the occurrence of sphalerite because there is a particularly dark species of iron rich sphalerite with exsolutions of pyrrhotite and chalcopyrite, the nature of which has also been established geologically. The occurrence is of no economic importance today.

The locations of occurrences, where many are listed, are quoted according to their geological position and further according to the geographical distribution, particularly for the rarer minerals.

It might have been of advantage to use here again the same spacing in the printing as used in the textbook of SCHNEIDERHÖHN-RAMDOHR (1931, 1933). This would have resulted in larger volume or in a shortening of the contents.

XI. Literature. On such a comprehensive topic it is difficult to give a proper selection of literature which satisfies both writer and reader. It is impossible to comply with all the demands of the specialist. In the textbook efforts were made to get some perfection but in the present book it was not attempted in order not to lose the general line. The incompleteness of many publications issued during the war and on the other hand the important and excellent Soviet literature, as can be seen from single publications and occasional quotations, made the decision easier. Many points presented in older editions were not repeated here. Those editions and the textbook (SCHNEIDERHÖHN-RAMDOHR, 1931, 1934) may be consulted. This will hardly be necessary, however, for, all the publication which in turn have a literature index, are given whenever possible.

In this English edition, for the literature quoted and listed in the bibliography, the name of the writer(s) is always given with the year of publication of the specific paper. The difficulty arising in finding a specific quotation in some publication has to be coped with.

In quoting some frequently repeated known reference or textbooks such as those of MURDOCH, DAVY & FARNHAM, SHORT, VAN DER VEEN, SCHNEIDERHÖHN, UYTENBOGAARDT, sometimes only the writer's name was used. With some rare exceptions no reference has been made to the textbook of SCHNEIDERHÖHN-RAMDOHR (1931, 1934) and to former editions of the present book.

It is clear that for frequent and much described minerals less emphasis was put on perfection than on less frequent ones. It was mostly omitted to mention and to correct the frequent erroneous data in the literature, necessary as that might be.

Very many data of literature especially of recent time consider only the last ten years (or less!) and bring often references, showing that the author has only copied — and that often wrongly. I try to go back to the sources. "The olds" observed sometimes more thoroughly!

XII. Powder diagram. The DEBYE-SCHERRER method became increasingly important in the determination of minerals and other substances, and forms today the most important criterion of any ore-microscopic determination. X-ray data therefore should not be missing in a book dealing with ore-microscopical diagnosis.

The writer has been using the method for many different purposes during the last forty years and most of the time he got satisfactory results. Most of his collection which amounts to several hundred comparison diagrams unfortunately was lost after the war and so he is compelled to quote foreign publications. Fortunately collections of data are available, in the first place by HARCOURT, MICHEEV, HANAWALT, and NOVITZKY. The many gaps and errors existing in those lists were corrected by further data from publications and the writer's own determinations. Especially useful were for that purpose the data of ASTM, BERRY & THOMPSON, and CHUCHROV. In some cases the data for simple lattice configurations can be calculated easily from the lattice constants.

Methods: A very small quantity of substance of 1/1000 mg and less may be sufficient for a reliable diagnosis. With a hand or machine operated drill the bit of which is a steel pin, a small amount of fine powder is recovered under the microscope. If done slowly and smoothly the operation yields grains scarcely larger than those from which LAUE spots result. The powder is picked up by using a pig's bristle to which some glue has been applied and the sample is brought into the axis of the camera. If possible, it should be rotated during exposure. The radius of the camera should not be less than 2·86 cms because of the accuracy desired. A larger radius requires too long an exposure and is only necessary in precision measurements and in the search for reflections close to the primary beam. — The recovery of fine samples by drilling is a matter of clean work, a steady hand and a little bit of consideration. It has to be borne in mind that the eye sees in two dimensions while the drill recovers material from the third. In the area to be drilled, that is a rough rule, a distance of more than the width of the drill bit has to be left between the drill bit and the grain boundaries. Then at most 1/9 (11%) of contamination is to be expected, which is about the tolerable limit.

There is no general method as to the kind of radiation, voltage, filters, etc. Many sulphide ores, particularly those that contain iron and many layered lattice patterns give poor diagrams with general darkening. The cause may be the process of grinding in many ore minerals that show good translational properties. It is less disturbing if a drill is used. In some cases the cause is not known. Most researchers seem to work with iron radiation partly because of little blackening effect and partly because of the relatively great wavelength which does not give too many lines. Everywhere, the "d" values* and the relative intensities for 3 or 4 and in many cases more of the strongest lines of the diagrams are quoted. With rare exceptions this is sufficient for the positive determination of a mineral.

A difficulty arises because in publications some writers denominate the strongest line by 10 while others use some kind of an empirical personal scale wherein the strongest line is marked as I = 5 or even less. At first sight the former method seems to be simpler and more reasonable (the writer mostly used it) while the latter indicates right away the quality of the diagrams. The "quality" may depend not only on the mineral but also on the experimental conditions such as grinding and wavelength. Further it may depend on conditions of a very subtile kind (e.g., sulphides ground in solid carbon dioxide-dry ice-yield better diagrams) so that at times it is hard to make a choice.

The sequence of the degree of intensity in *each* diagram gives no difficulties. If some writers give the sequence of the degree of intensity differently for the same mineral, then there is some reason for it. A large camera radius also supplies lines which in spite of strong intensity may get completely lost in a small radius setup in the black zone near the primary spot. Very thin samples show the large d-values stronger than the small ones; thicker samples also show the latter very clearly. After some ex-

---

* The d-values are calculated from $d = \frac{\lambda}{2 \sin \vartheta}$.

perience conclusions can be drawn by comparison as to the technique of the writer in question (e.g., of the different tables for selenium in NOVITZKY, 1957).

It often is of advantage to give additional lines or to draw the attention to some peculiarities of some diagram. The scope of this book does not allow to give tables of line values.

It has to be mentioned that in checking the HARCOURT data some errors have been found. It is to be expected that some more will be found in cases that have not yet been checked, which suggests to be cautious.

Considering the weaknesses of the HARCOURT values the above mentioned latest collections of data have been used to supply additional information and mostly more than three principal values were quoted. The most frequent references are:

HANAWALT: HANAWALT, J. D., RINN, H. W., FREVEL, L. K. (1938), *Chemical analysis by X-ray diffraction.* Ind. Eng. Chem. **10**, 457–517.
MICHEEV: MICHEEV, W. I. (1957), *Determination of Minerals by X-ray.* Moscow (in Russian).
ASTM: *X-ray diffraction patterns.* Amer. Soc. Test Mat., Philadelphia 1943, supplements 1944, 1950 etc.
B. + TH. = BERRY, L. G. & THOMPSON, R. M., X-ray powder data for ore minerals: The Peacock Atlas. Geol. Soc. Amer. **Mem. 85**, 1962.

For quite a number of cases new determinations were made at my institute (e.g., for nearly all Mn-minerals).

In rare cases it may happen that positively different minerals yield nearly the same diagrams. That means similarity of the lattice dimensions and analogous distribution of atoms of the same relative size. It often indicates the possible formation of isomorphic crystals where it certainly was not expected. HARCOURT gives some examples. To the writer of this book the similarity geocronite–jordanite appears to be most striking.

ELEMENTS AND INTERMETALLIC COMPOUNDS

COPPER

I. General data. *Chem.* Cu; with very little Ag in solid solution. Admixtures of As constitute transitions to "whitneyite". Solid solutions of Cu_2O in Cu which are possible at high temperatures within certain limits, do not occur in nature.

Cryst. Isometric; O_h^5, $a_0 = 3{\cdot}61$ Å. The crystals are often very strongly distorted and it is difficult to interpret them. — ⋕ lacking. H = $2^1/_2$—3; can be hammered and pressed. D = 8·9. Opaque. $n_{Na} = 0{\cdot}64$! (DRUDE), $\varkappa_{Na} = 4{\cdot}09$. Very high metallic lustre in freshly polished or cut pieces; if still fresh, delicately rose colored, but darkens soon to "copper red", later dull and dark brown.

II. Polishing properties. Polishing is relatively easy, but scratches due to strange inclusions are difficult to avoid. For the technique of polishing, cf. the section on silver. Loose, porous masses must be well impregnated with artificial resin, otherwise they may easily shatter or rupture the polishing cloths or scratch the lead lap. No cleavage

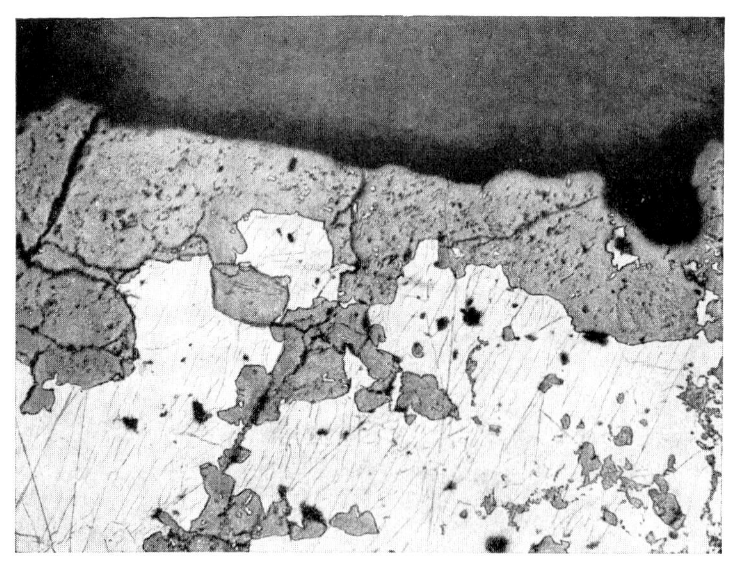

Fig. 276 40 × RAMDOHR
Otjosongati, South-West-Africa

Native copper (white) is being replaced by *cuprite* (grey, in part indicating idioblastic forms). The replacement starts at a portion of very weakly reflecting Cu-oxide minerals (almost black) and follows partly the cleavages

Fig. 277 E 180 × RAMDOHR
Mina Copreasa, Mexico

Native copper surrounded by *cuprite*. The platy crystals of about the same color are *delafossite*. The dark grey minerals are *quartz* and *malachite*

pattern can be observed. The polishing hardness is noticeably higher than that of chalcocite and chalcopyrite, but lower than that of cuprite and most of the limonites.

Talmage hardness: B.

III. Reflection behavior. *Color and reflectivity.* The color of the very freshly polished surface is a brilliant deep rose, which darkens to redbrown if the polished surface is not immediately protected with a coating of immersion oil. The following data thus refer to freshly polished surfaces.

	in air	in oil
general	extremely high, rose-white; if isolated "copper -tint" is not conspicuous	not noticeably changed
intergrown with		relationships similar; differences still more marked
silver	noticeably darker and reddish	
chalcocite	much brighter, chalcocite is dull bluish grey	
cuprite	much brighter; cuprite is dull bluish green to grey in comparison	

	air				oil			
R. nm	470	546	589	650	470	546	589	650
	38·4	42·5	68·2	78·4	32·7	39·9	65·6	73·2

in oil: not essentially different.

Other data deviate in part strongly. Here for nm 650 seem to be low. Values of VJALSOV are, beginning in yellow till 20% higher – Calculated from data of DRUDE, the reflectivity for Na results in 73·2%

It is isotropic, however, it will not turn completely dark under + nic. Old, nearly vanished scratches can easily give false effects.

The reflectivity is distinctly reduced by As in solid solution; at the same time, the material will quickly become more brittle (cf. algodonite–whitneyite).

IV. Etching (accord. to DAVY & FARNHAM, MURDOCH and VAN DER VEEN). *Positive*: HNO_3; quick solubility and roughening; concentrated HNO_3 dissolves instantly, tarnish slightly colored. Even vapors produce coating. KCN; slightly brown coating. Concentrated HCl; slight coating. $FeCl_3$; immediate blackening with etch cavities. KOH; weak coating. $HgCl_2$; quick blackening, iridescent coating. *Negative*: HCl.

Structure etching: 1 part 50% CrO_3 and 1 part concentrated HCl, mixed immediately before use, 10–20 seconds, then wash off with one drop of concentrated HCl to remove the coating, gives excellent etching. The "Heyn's Reagent" (solution of copper-ammonium chloride in ammonia water) as used in metallography, is not quite as satisfactory; $H_2O_2 + (NH_4)OH$, also reveals the structure.

Etching by air starts within a few hours and is quite disturbing; it rarely reveals the texture and it varies with the occurrences.

V. Physico chemistry. The small atomic radius implies that under natural conditions mix crystals of copper with Au and Ag can form to a very limited extent only.

Very rare high temperature mix crystals of Au–Cu (Mooihoek-Hortonolith-Pipe), break down to Au and $AuCu_3$. Mix crystals, however, are formed with up to 12% As by weight (whitneyite) which differ in their properties from those exhibited by the pure Cu of other deposits.

VI. Fabric. *Internal properties of grains.* Similarly to gold and silver, copper also exhibits, after etching, almost always lamellar twin-formation in coarser or finer lamellae after (111). One part is formed through growth, the bulk through pressure often with simultaneous translation (bending of the lamellae).

Zoning does not appear to be too unusual in upgrown crystals. *Exsolutions* are very rare, which is not surprising considering the generally very low temperatures of formation. Only in one instance (cf. under V. above) exsolution of a mix-crystal was noted. A similar occurrence is described by SCHOLTZ (1936).

Structures and texture. Copper is mostly granular, often strongly interfingered. The grain size varies greatly; usually it is rather small, though generally uniform crystals can be observed in supergene porous masses. Many occurrences precipitated as finely dispersed material, similar to botryoidal goethite, become coarsely crystalline, regardless of this concretionary texture; similar to many instances of native arsenic. Good pictures showing the texture of this are published by VAN DER VEEN. Recrystallization phenomena are well known in metallography and occur also in nature. In the masses formed by descending solutions skeleton-like forms are common; often they become finer, moss or sponge like. Copper occurring in the Lake Superior area and in similar deposits, shows no such development. Furthermore, dustlike impregnations are known to occur; e.g., in tuffs of the Coro-Coro region, where they may have originated from ascending solutions, or in granite at Reichenbach in Odenwald where they formed from descending solutions (figs. 276, 277).

VII. Special fabrics. Through lack of cleavage, characteristic replacement forms are rare; though there is no doubt that copper often replaces chalcocite or cuprite, and is in turn replaced by these and other copper minerals from the oxidized zone, particularly by carbonates.

The manifold pseudomorphs known from mineral collections are of minor importance for ore-microscopy.

Through the decomposition of the very rare unstoichiometric relatives of chalcopyrite, e.g., in hortonolithdunite, copper often forms in finest sheets.

Rhythmic textures with limonite are excellently represented, e.g., in the occurrence of the Wingertshardt Mine near Siegen. In other low temperature occurrences, these structures become identifiable only after etching (fig. 278).

VIII. Diagnostic features. Due to its enormous reflectivity and its characteristic color this mineral is rarely misidentified. Minute particles occasionally show similarity to tarnished silver or bismuth, but even here the paragenesis protects mostly from errors.

IX. Paragenetic position. Extensive occurrences of copper are rare; there are few economically significant deposits. According to origin several fundamentally different groups can be distinguished.

1. In the lowest parts of the oxidized zone of chalcocite deposits where already formed secondary enriched copper ores (chalcocite, bornite) become subject to oxidation. In such cases plentiful cuprite also develops beside native copper. Cuprite is often apparently found alone; though fine flakes of native copper always are present. In the process of precipitation, colloid-chemical reactions often play an important part. Deposits of this nature have temporarily produced large amounts of native copper, particularly in some of the arid regions.

2. The deposits of "arid basins" often contain native copper, which may have been formed from chalcocite, as described under part. 1, but in other instances copper has probably directly precipitated. Sometimes relatively pure copper is found, in other instances it occurs as (As-containing) copper, described under algodonite–whitneyite. The Coro-Coro occurrences are often classified as "arid basin" type of deposits by some, (Entwistle & Gouin 1953) and as hydrothermal-hypogene deposits by others (Geier, Singewald, Ahlfeld).

3. Mention must be made here of the deposits precipitated by carbonaceous substances in the sandstones of the Nonesuch Formation. These also have been interpreted in at least two different ways.

4. The most noteworthy occurrence, the native copper deposits of the Keewanaw Peninsula of Lake Superior, is as yet genetically not clearly understood. The latest interpretation considers these deposits as of hydrothermal origin and the precipitation as a result of oxidizing conditions. Perhaps this can be supported by the occurrence of very numerous, though generally unimportant occurrences in basalts or melaphyres. Another explanation considers previous stratabound origin, mobilized and redeposited in various, in part impregnative or veinlike form.

5. Unusual and as yet not of clear origin is the occurrence of copper in marble close to manganese ores at Jakobsberg and Långban in Sweden and a similar occurrence at Franklin in New Jersey, which was described by Ries & Bowen (1922).

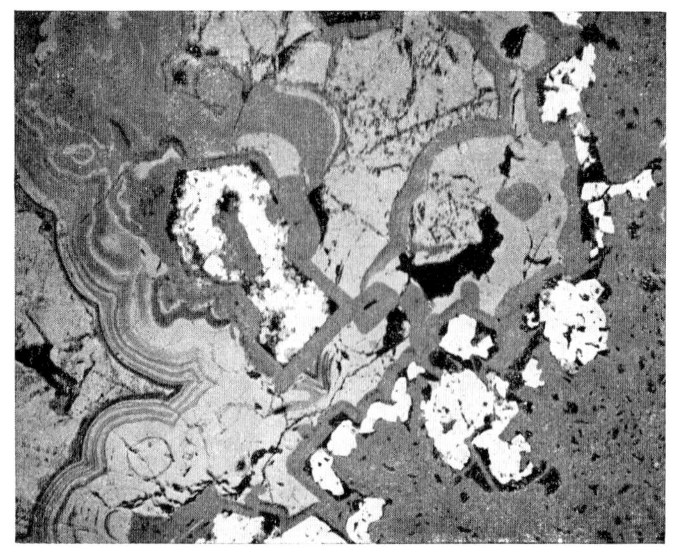

Fig. 278 10 × Ramdohr

Käusersteimel Mine, Siegerland

Cuprite (light grey) partly intergrown with very nicely, rhythmically botryoidal *goethite* (dark grey). Pure white, in part corroded, is *copper* in more or less well developed crystals

Fig. 278a 150 × Ramdohr

Calumet, Michigan

Native *copper*, (pure white) with an equally thick rim of cuprite (dark grey) and gangues

6. Without doubt of "primary magmatic" origin is the copper in the pyroxenes of the diabases at Doros, South-West-Africa, where it fills cavities, in some platinum occurrences (especially the "Pipes") in the Bushveld, and in many parts of the Skaergård intrusive in East Greenland (VINCENT & PHILLIPS, 1954). The paucity of sulphides in all these cases points to the extreme deficiency of sulphur within the host rocks.

7. An example of metamorphic deposits is the occurrence near awaruite and heazelwoodite in epimetamorphic ophiolitic serpentines.

8. Small quantities of native copper are present in very many stony meteorities, some iron meteorites and in the Moon-rocks.

X. Investigated occurrences. Some of the almost monotonously similar type 1 deposits are Mechernich, Eifel; Rheinbreitbach; Bogoslowsk, Ural; Bisbee in Arizona; Otjosongati, South-West-Africa; Tetiuhe, East Siberia; Burra-Burra and Mt. Isa, Australia. No. 2 type deposits: "Gouvern. Perm"; Coro-Coro, Bolivia. Type 4 deposits: Lake Superior, U.S.A.; Faröer; Annisead Valley, Nelson, New Zealand. No. 5 type deposits: Jakobsberg and Långban, Sweden; Franklin, N.J.; No. 6 type deposits: Doros, South-West-Africa; Skaergård, East Greenland. No. 7 type deposits: Josephine Co., Oregon (cf. with awaruite page 396 and artificially produced types).

A content of native copper in kimberlites may be caused by the breakdown of chalcopyrite into copper and pyrite at very high pressure. As curiosity it may be mentioned that in concentrates of ore dressings products of most different origin often tiny flakes of native copper can be observed. It derives from the detonator capsules. They show sometimes excellent shatter cones.

XI. Literature. The ore-microscopic determination was originally established by MURDOCH, DAVY & FARNHAM, in more detail by SCHNEIDERHÖHN and in still more detail by VAN DER VEEN.

XII. Powder diagram (B. & TH.). (10) 2·078, (6) 1·807, (4) 1·278, (5) 1·090, (5) 0·830, (5) 0·809 Å.

SILVER

I. General data. *Chem.* Gold and silver display unlimited solid solution, yet alloys with 0–1%, then again with about 20%, and finally with more than 60% gold are particularly common. Some occurrences may contain up to 12% Hg; for hither contents of Hg see there.

Copper up to 1% and bismuth up to 8% can form solid solutions with silver. Concerning mix crystals with silver see gold. Concerning mix crystals with Sb, see under Dyscrasite.

Cryst. Cubic, O_h^5, $a_0 = 4{\cdot}078$ Å. Cleavage not observed. H = 2–3, D = 10·5, when pure. *Opt.* Opaque, $n_{Na} = 0{\cdot}181$, accord. to DRUDE, 0·27 (white) accord. to KUNDT. Absorption index = 20·3 for Na, 19·5 for red light accord. to DRUDE. Very strong silvery lustre, white, very soon reddish through tarnishing.

II. Polishing properties. Uncontaminated silver polishes easily with high brilliancy, retains, however, extremely stubborn scratches and porelike cavities. Fine grinding and polishing is best achieved by using a slowly rotating, slightly moist disk. No cleavage in polished sections. *Polishing hardness* is higher than that of galena and of all "noble" Ag ore minerals, similar to chalcopyrite, lower than that of fahlores and much less than that of sphalerite. Talmage hardness: B.

III. Reflection behavior. *Color and reflectivity.* Color of polished surface is shiny, glowing white when fresh. Exposed to air, it soon gets darker; various colors appear such as: clear yellowish, also red, blue and other colors of the thin lamellae formed through surface films. On such older polished surfaces the color tint may be similar to that of tarnished copper. The formation of surface films varies in speed in individual occurrences, even where no chemical differences are present.

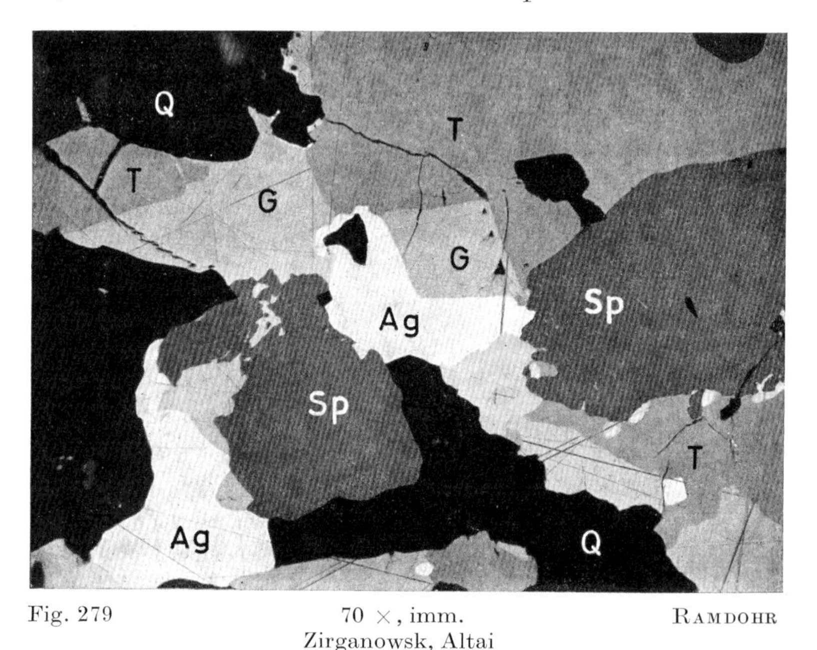

Fig. 279 70 ×, imm. RAMDOHR
Zirganowsk, Altai

Native silver, probably rich in gold, white (Ag) enclosed in a matrix of *tetrahedrite* (grey (*T*)) *sphalerite* (dark grey (Sp)) *galena* (greyish white (G)) and *quartz* (black (Q)). A distinct sequence cannot be observed

Isotropic. Scratches, though nearly eliminated by polishing can, nevertheless, cause false effects through depolarization.

IV. Etching (accord. to MURDOCH, DAVY & FARNHAM, VAN DER VEEN). *Positive*: HNO_3, light short effervescence, roughened surface; fumes cause coating. $FeCl_3$, iridescent coating, remains after rubbing. $HgCl_2$, grey coating, after rubbing brownish grey. *Negative*: KOH. The data vary for KCN and KCl but in any event the effect is very minor. *Texture etching*: Accord. to VAN DER VEEN, the best is CrO_3 with HNO_3 in 1 to 5 seconds, also HI. The solution of H_2O_2 and NH_4OH used in metallography is not quite as effective.

V. Physico chemistry. Concerning miscibility with Au, Cu, Bi, Sb, see above. Silver precipitates easily from solutions with less precious metals and their compounds.

	in air				in oil	
general	white with fair yellow tinge. Reflectivity the highest known				not noticeably different	
compared to copper compared to platinum compared to bismuth compared to iron	white, somewhat brighter creamish tinge, brighter similar in color, distinctly brighter distinctly yellow				— analogous	
BESMERTNAYA nm	450 83·0	500 87·2	550 88·5	600 90·0	650 91.0	700 91·5
Photocell ORCEL 460 mμ MOSES/FOLINSBEE	 86·5% 86·0–97·0/93·8%					
	Accord. to the specifications of DRUDE for n and $\varkappa$ the reflectivity in Na is 95·3%; thus the above reflectivity measurements fully correspond with these figures					

This process is even now very often attributed exclusively to descending solutions, an assumption which is not based on any known facts.

VI. Fabric. *Internal properties of grains.* Many crystals particularly those in coarser grained occurrences, show twin lamellae as a result of growth or compressive stresses. The lamellae are frequently bent as a result of simultaneous translation. *Zoning* in dendritic silver is occasionally well defined (good illustration by VAN DER VEEN, fig. 100). Around the margins some occurrences show finely porous portions which are difficult to polish and which probably originated from original encrustations of argentite.

Exsolutions are generally absent. Those of "dyscrasite" spindles see there. *Textures and structures*: The shape of the grains is greatly variable. Idiomorphic development is rather rare and occurs only occasionally in crystals grown into openings and in dendritic and skeletal forms (fig. 280). In many "knitted" growths, the oriented network does not originate from silver, but from other replaced minerals. In general the aggregates are fine-grained and allotriomorphic with the individual grain intimately interlaced. Regarding the peculiar textures of grains formed from argentite as "silver teeth" and "curls" see VAN DER VEEN (1925). The grain size within sheets, bands and wedge shaped fillings which silver very often forms, also varies greatly.

VII. Special fabrics. *Replacements.* Considering the paragenesis of silver, replacements are often to be expected. Even from ascending solutions silver is formed generally rather late and can therefore replace fahlores, chalcopyrite, galena, "noble" silver ores, but particularly cobalt and nickel ore minerals, such as niccolite and maucherite, as well as zones within smaltite and bismuth.

Extensive literature concerning these occurrences deals in particular with the typical occurrences of Cobalt, Ontario (BASTIN, 1950; CAMPBELL & KNIGHT, 1906; WHITEHEAD, 1920), Silver Islet (CHADBOURN), Jáchimov (Joachimsthal) (ZÜCKERT, 1925), Nieder-Ramstadt (RAMDOHR, 1923). The replacements generally occur in rather coarse

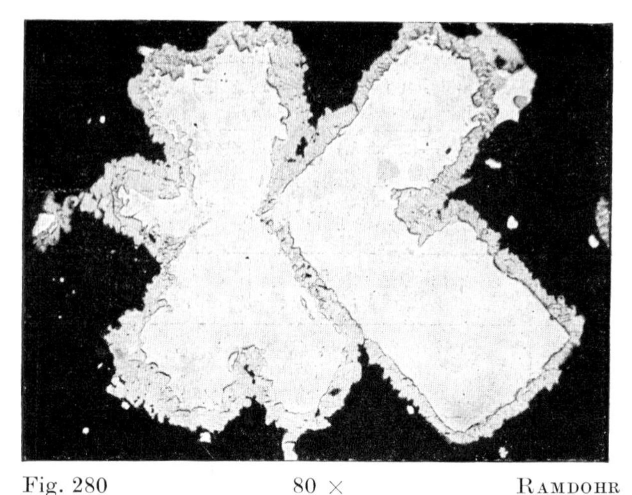

Fig. 280 80 × Ramdohr
Cobalt, Ontario
Silver with a thin rim of *arsenides*

The skeletal forms, which are remarkably often five cornered, permit occasionally the recognition of crystal faces (cube!). The exsolution of "Oleander-leaf" flakes of dyscrasite is readily recognizable

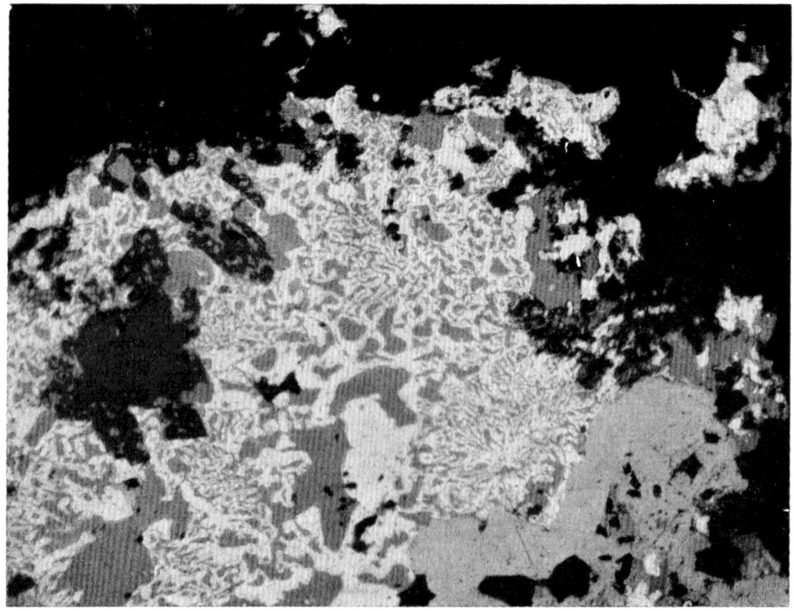

Fig. 281 500 ×, dry Ramdohr
Colquijirca, Peru

Myrmekite of *silver* (clear white) and *fahlore* (grey) in an envelope of *chloanthite* (white, a shade darker than silver). The illustration shows the remnant of a skeleton-like mineral, enveloped by chloanthite and subsequently subjected to pseudomorphosis

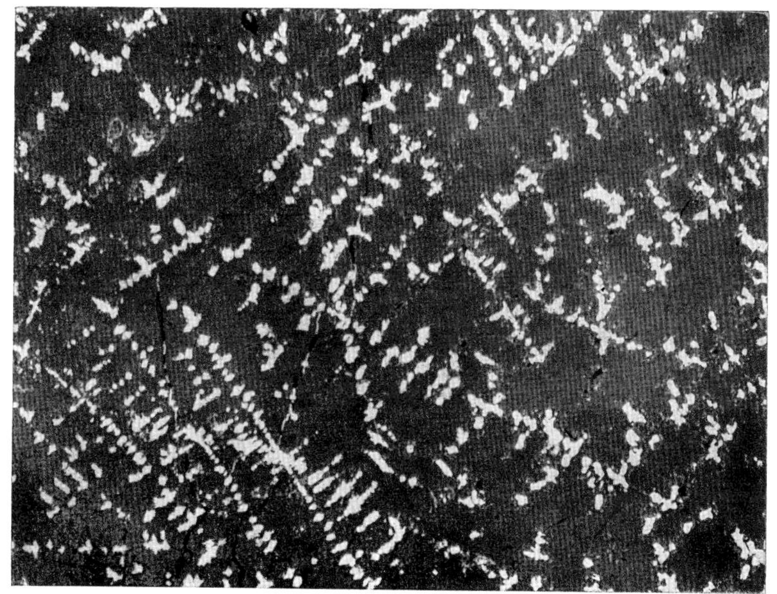

Fig. 282 about 10 × RAMDOHR
Freiberg, Saxony

Native silver, "knitted" in quartz. Magnificent crystal skeletons parallel to the cube edges

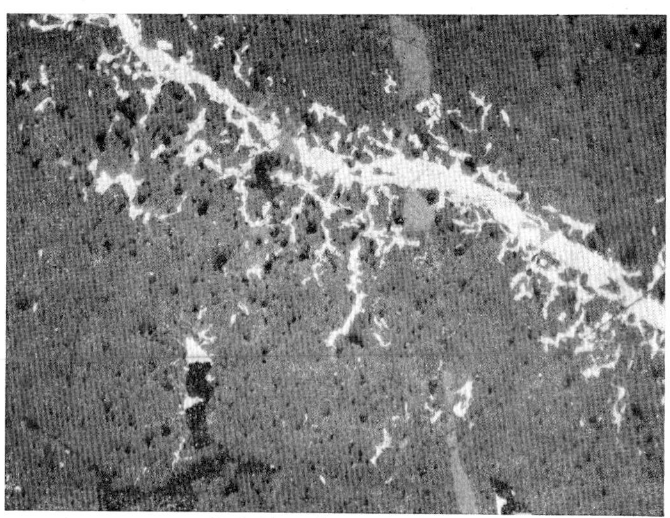

Fig. 282a 150 × RAMDOHR
Umberumberka, N.S.W., Australia

A fissure filled with finely branched *native silver* in *siderite*, besides that a bit *sphalerite*

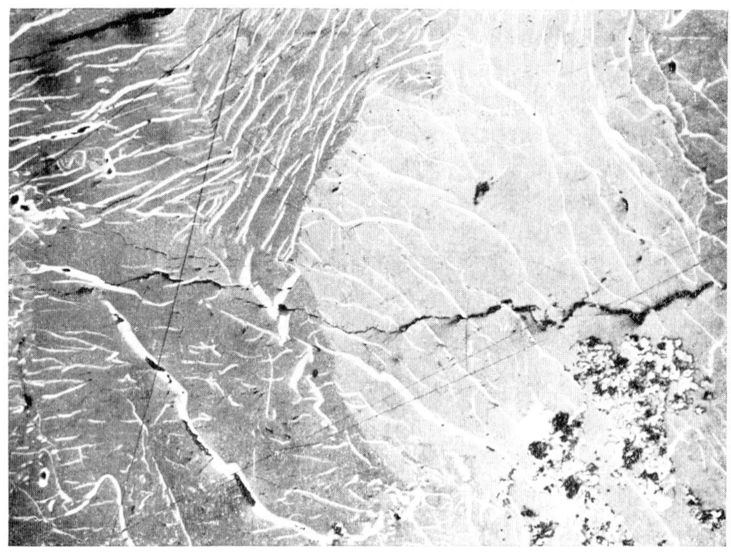

Fig. 283 110 × RAMDOHR
Colquechaca, Bolivia

Silver (white) as cementing mineral in fractures of *wolfsbergite*

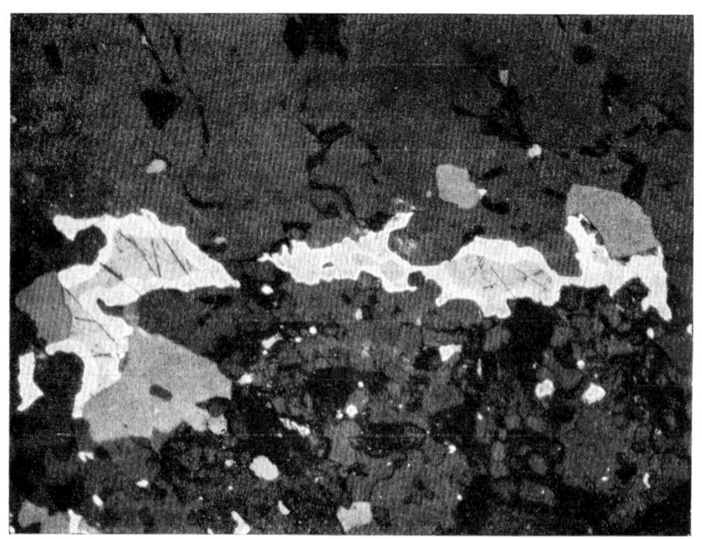

Fig. 283a 150 × RAMDOHR
Mina la Prieta, Mexico

Silver (bright white) surrounds and replaces *argentite* (greyish white and badly scratched). The gangue is mainly *quartz* (dark grey)

form. Ascensing decay of complex sulphosalts, with or without introduction of material can develop good myrmekites (fig. 281).

Supergene silver replacements in the zone of cementation generally form thin surface films and fissure fillings. These are particularly common as cementations of galena, sphalerite, fahlore, bornite and chalcocite. They favor cleavage cracks, fissures and grain interstices and consequently also cataclastic zones within these minerals (fig. 283). Doubtful cases: In many instances the question whether ascending or descending solutions are responsible for the depostion will be impossible to answer on the basis of the fabric.

Replacements, due to descending solutions, which can partially be observed (Aspen) at great depths, are so extensively described that they cannot be dealt with here in detail; see, e.g., in FINLAYSON, DOLMAGE, GUILD (1928), BASTIN, and many others.

Peculiar relationship exists between silver and argentite; in cases of ascending as well as of descending origin, transformations of silver to argentite and vice versa, can be observed. "Silver-teeth" derived originally from argentite are, e.g., in turn again transformed into argentite, after the formation of this new argentite at temperatures about 175°. Stromeyerite can as well break down simultaneously precipitating native silver, perhaps mostly through descending solutions, and form attractive "replacement" textures. The exsolution products resulting from the breakdown of Ag–Sb crystals are a special case (cf. with dyscrasite).

Originally eutectic intergrowths with copper are reported to be scarce (cf. VAN DER VEEN, fig. 97). Similar material was not available to the writer.

Rhythmic precipitations in typical form have not been observed to date but can well be expected to occur occasionally.

VIII. Diagnostic features. On freshly polished specimens, silver may be confused with relatively few ore minerals, i.e., mostly with native metals with high reflectivity; in tarnished, older polished specimens, caution is highly recommended. Bismuth is considerably softer, darker and also clearly anisotropic; antimony does not tarnish, but neither does some of the native silver; it is also weakly but noticeably anisotropic and occasionally also shows cleavage; dyscrasite is hard to differentiate from cubic and hexagonal Ag–Sb solid solutions. It is, however, more anisotropic, and in comparison to native silver distintly duller; platinum can be very similar, it is somewhat less yellowish, does not tarnish and is not attacked by almost all reagents; pyrite is much harder, confusions are possible only in smallest grains.

The differentiation between a few very rare silver minerals is without microprobe not always possible though in general they are distinctly anisotropic.

The differentiation of pure native silver from white silver-gold alloys (electrum) is generally easy, due to a lack of tarnish, the more stable chemical composition of the latter, and often also by the mineral association.

IX. Paragenetic position. Extensive literature exists on occurrences of native silver; because native silver was known for many thousand years. However, many details are often contradictory and apparently at least in part incorrect and result from preconceived opinions.

There is no doubt that silver has been precipitated from ascending solutions, i.e. as a product of different hydrothermal solutions, as well as from descending solutions as a product of transportation in surface waters. The resulting textures and intergrowth patterns can in either instances be megascopically as well as microscopically so similar that in determinations the greatest care is advisable and it is suggested that the mineral association should always be taken into consideration.

The silver of *ascending origin* occurs particularly in:

1. Type Schneeberg—Joachimsthal—Cobalt, widespread throughout the world, and responsible for a substantial portion of the silver production from this mineral (Cobalt City, Great Bear Lake, Schneeberg in the past, etc.). Accompanying minerals are the Co, Ni, Bi minerals and in part pitchblende, as well as many "noble" Ag-minerals; gangue minerals are: calcite, heavy spar, quartz. In the closely related *Kongsberg* type, silver strongly predominates over subordinate argentite in carbonate gangue, although accord. to LIETZ (1939), the mineral association is not quite as monotonous as originally assumed.

2. In the Freiberg-type of deposits, to which some of the richest silver occurrences in Mexico closely resemble, native silver is generally very rare (rich, e.g., at Parral, Mexico), the main minerals are tetrahedrite, in part freibergite, sphalerite, galena. pyrite and many "noble" silver minerals, partly in a gangue predominantly of calcite ("noble calcite formation") and partly in quartz. The occurrences prove to be of epithermal origin through their manifold pseudomorphs, incrustations and open cavities.

3. Within the typical subvolcanic "*young gold silver veins*", most of the silver occurs in sulphides, but sometimes important amounts occur in native form or alloyed with gold as electrum of ascending origin.

4. Also in some other vein types silver may occur occasionally in subordinate amounts. Sometimes it is clearly of ascending origin (Boliden); usually, however, the differentiation whether it is of ascending or of descending origin is difficult. It could be a lateral-secretionary formation out of black shale (in the ,,Kupferschiefer").

Sedimentary formations contain native silver in various groups.

5. Silver is most widely distributed in the zones of oxidation and of secondary enrichment associated with many sulphide deposits, often in large amounts, even in cases where the primary ore is relatively poor in silver. In the zone of oxidation it is associated with silver halides and limonite, with the zone of enrichment particularly with argentite, with corroded, almost powdery appearing galena, with "precious silver minerals" (of ascending or descending origin) or with fahlore. It has been previously noted that the differentiation between silver, originating from ascending or descending solutions is not always easy. Even the "silver content" of the cerussites often appears to be present as finely disseminated native silver.

6. In some deposits of the sulphur cycle, especially in the Kupferschiefer of Mansfeld, silver precipitated most likely through reduction by organic substances. The native silver which is found next to copper in the carbon-rich sandstones of the Nonesuch-formation of Michigan (cf. K. NISHIO) and in the detrital beds of arid basins such as the "Red beds" has been formed by similar reduction processes.

Many occurrences of native copper contain silver in the form of incrustations or on-grown crystals which were formed as electrolytic precipitates.

Very strange is the formation of silver on fahlore which is found in a crushed state in the midst of "galena tails" (steely galena) (Müsen near Siegen). These forms cannot be attributed to enrichments caused by ascending or descending solutions. In these examples any signs indicating cementation processes are completely missing.

X. Investigated occurrences (small selection)

1. Nieder-Ramstadt near Darmstadt; Wittichen, Schwarzwald; Schneeberg, Saxony; Jáchimov (Joachimsthal); Great Bear Lake, Gowganda and Cobalt; Kongsberg, Norway.
2. Andreasberg, Harz; Freiberg, Saxony; Parral and Zacatecas, Mexico.
3. Pinos, Mexico; Chanca near Coquimbo, Chile; Salida and Mangani, Sumatra.
4. Boliden, Sweden.
5. Silverton, Colorado; Zacualpan, Cuale, Bramador a.o., Mexico; Smeinogorsk and Semenowskoy, Altai.
6. Mansfeld; Göllheim, Palatinate; Frankenberg, Siegerland.

XI. Literature. Naturally, a considerable amount of literature pertaining to the metallurgy of the silver exists. Also, much has been published on studies under the ore-microscope. A longer, excellent compilation was published by VAN DER VEEN (1925). Individual occurrences are dealt with by CAMPBELL & KNIGHT (1906), GUILD (1917, 1928), BASTIN (1939), WHITEHEAD (1919, 1920), DOLMAGE, HOFFMANN, RAMDOHR (1923), ZÜCKERT (1925), KIDD & HAYCOCK (1935), KRIEGER (1935); their publications contain important genetic and ore-microscopic data.

XII. Powder diagram (B. & TH.). (10) 2·359, (4) 2·042, (6) 1·444, (8) 1·232, (7) 0·937 Å.

The hexagonal silver "allargentum" (Ag, Sb) is discussed under dyscrasite.

GOLD

I. General data. *Chem.* Gold, generally with some silver, which even in natural occurrences always forms unlimited solid solutions with gold. Mix crystals containing 30–45% Ag are called electrum; they are almost pure white. Those with about 80% Ag are designated with the rarely encountered name "küstelite". Gold often contains small amounts of Cu (see there), also Bi, Pt, and Hg. Some occurrences are mix crystals with Pd (porpezite). *Cryst.* Cubic; O_h^5, $a_0 = 4{\cdot}070$ Å. The relatively rare crystals (111, 100, 110) are often distorted. No cleavage. — $H = 2^1/_2$–3, $D = 19{\cdot}3$ when pure; generally, however, lower. — *Opt.* opaque, $n_{Na} = 0{\cdot}366$, $\varkappa_{Na} = 7{\cdot}71$ (DRUDE), $RV_{Na} = 85{\cdot}1\%$. Golden yellow with variable tints, highest metallic lustre, finest powder brown.

II. Polishing properties. Pure gold is easy to polish; scratches, however, are difficult to avoid unless slow moving polishing disks and very weak pressure are used. Great pressure when prepolishing by hand as well as when polishing by machine must be carefully avoided, because hard polishing material embeds itself into the soft metal, which then stubbornly resists any polish (fig. 294). This is particularly true for polished sections of "nuggets". No cleavage; low hardness. The polishing hardness is similar to that of chalcopyrite, lower than that of fahlore and sphalerite, higher than that of galena. Talmage hardness: B.

III. Reflection behavior. *Color and reflectivity characteristics.* The color of the polished surface is a luminous golden yellow, much variable with the silver content, the reflectivity is enormously high. Compared to gold even the brightest sulphide ores appear very dull; chalcopyrite in comparison appears dirty yellowish-green, within larger chalcopyrite masses the color of small gold particles can only be described as "white", due to their high reflectivity.

Isotropic, it never becomes completely dark with crossed nicols, but rather peculiarly green. Even on smooth surface under + nic. all scratches become visible under the polishing film.

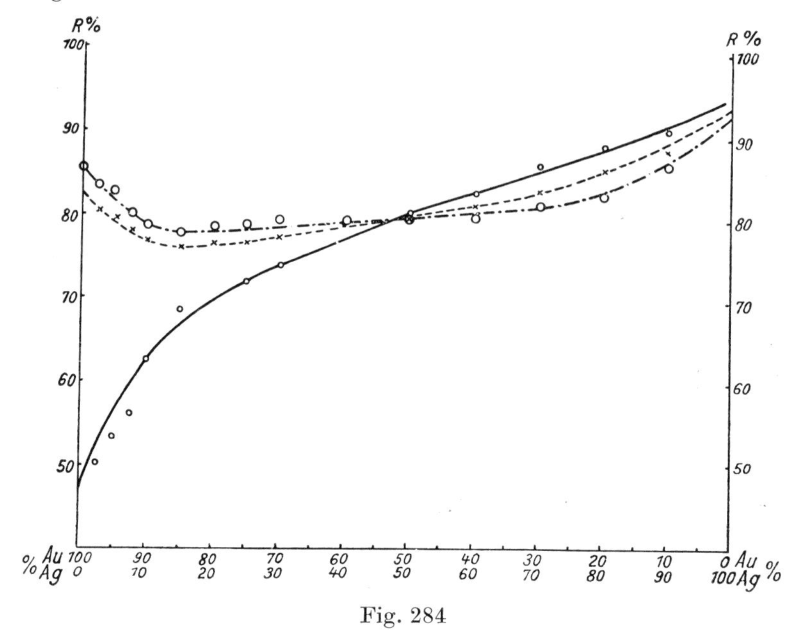

Fig. 284

Reflectivity curves of gold–silver alloys. Green: ○ — Orange: × – – – Red: ₒ.

The *reflectivity* of the Au–Ag alloys has been studied many times. The gradual changes show particularly well in a series of measurements by R. FRICK undertaken at the suggestion of H. SCHNEIDERHÖHN. The corresponding graph is reproduced in fig. 284. Foremost characteristic is the fast increase in the value of green and therewith the approximation to white as the total color impression.

IV. Etching (accord. to MURDOCH). *Positive*: KCN, instantaneous blackening with formation of rough surfaces; generally poor, only seldom good development of textures. Hg gives white amalgam. HI etches occasionally. *Negative*: HNO_3, HCl, KOH, NaOH, etc., also aqua regia in the usually applied etching times.

Texture etching: Accord. to VAN DER VEEN (1925) with aqua regia good structure development, also with CrO_3 in HCl and CrO_3 in aqua regia. Time of etching varies with Ag content. EDWARDS obtained good results with diluted aqua regia, BETECHTIN in case of copper-rich gold with aqua regia vapors and in general with KCN, if fluorspar is added.

V. Physico-chemistry. At high temeprature, e.g. crystallising from melt, Au and Cu form mixed crystals, which during slow cooling break into the intermetallic com-

	in air	in oil
general	very high, color from deep yellow to whitish-yellow	likewise
compared to all		
sulphide ores	much lighter	likewise
bismuth	yellower and brighter	
platinum	very much yellower	
silver	more yellow, tarnished silver in one stage can closely resemble gold	likewise
Photometer ocular pure, artificial gold (see further below)		
green	47·0%	42·5%
orange	82·5%	80·0%
red	86·0%	83·5%

The values of BESMERTNAYA & VJALSOV differ in blue and green very strongly (10%) but not very much of those of FRICK.

Photocell		
MOSES/FOLINSBEE	64·8/73·4%	
BOWIE (orange)	80·6%	
Reflectivity	Directly determined 85·1% from n_{Na} and $\varkappa_{Na}$ (DRUDE), which to some degree corresponds with the above determinations	

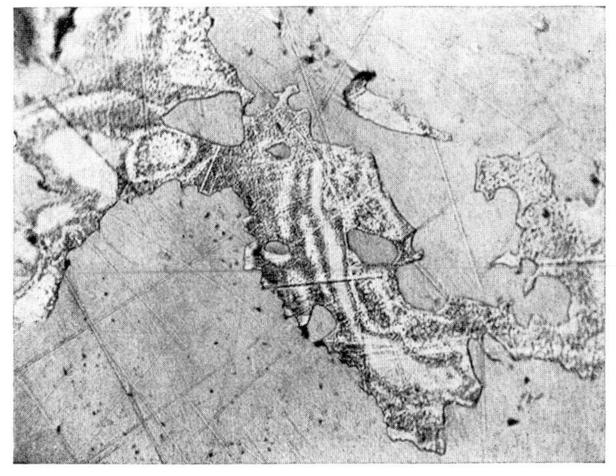

Fig. 285 30 ×, etched with AqR RAMDOHR
Smeinogorsk, Altai Mts.

Gold, rich in silver is strongly zoned. It forms a younger vein filling in tetrahedrite (unetched)

pounds $AuCu_3$ and AuCu. Under very restricted conditions — Cu normally is too chalcophile — they can form even hydrothermally. An occurrence of Mooihoek (Transvaal, from a hortonolite dunite) shows gold and $AuCu_3$ in an obvious unmixing structure (slender skeletons of Cu-rich lamellae // (100)). A similar observation has been described by SCHOLTZ (1936) from Insizwa. It is surprising, that the natural Au–Cu-compound from hydrothermal gold–quartz deposits mostly seem to show Au:Cu = 1:4·5 (= 40% Au), whilst stoichiometric $AuCu_3$ (= 49·3% Cu) not yet has been analysed (see auricupride, pg. 344). Gold forms, with Bi, the compound of Au_2Bi; as a natural mineral this is known as maldonite (see page 336); with Ag and Pt gold forms unlimited mix crystals, which, however, only in part are known as natural minerals.

VI. Fabric. The *internal properties of grains* have already become fairly accurately known through older studies such as those of LIVERSIDGE, ZEMCZUZNY and VAN DER VEEN. They have, however, not fully assessed the multitude of forms. Twinning in the form of growth lamellae // (111), which are often bent through stress is very common. *Zoning* is widespread in freely grown crystals, particularly in those of epithermal origin, as well as intergrowths of gold of different colors and therefore of different age are known (RAMDOHR, 1932; TOLMAN & AMBROSE, 1934). *Deformations* are frequent; they can be explained by the great toughness and the translation characteristics.

Texture and structure. The grain shape is mostly equant and more or less serrated. In some colloidally deposited pieces botryoidal reniform structures are well developed, occasionally with rhythmic banding (fig. 286). Within these bands the individual grains are very small or sub-microscopic. Coarse grains are, as VAN DER VEEN showed, by no means an indication of hypogene gold, as re-deposited gold can become quite coarse grained (1 cm!). It must, however, be considered that fine grained, concentric shell-like textures can be obliterated by "collection crystallization".

The individual grain within other minerals is generally completely xenomorphic. Only within galena and antimony and, of course, also in vugs and open pore spaces, idiomorphic crystals are found. Quite peculiar is the very frequent case, that within individual grains of galena, sphalerite or chalcopyrite, gold occurs in the form of individual, quite irregularly formed granules. All other metal grains in the environs are completely devoid of gold (see fig. 290). Therefore an evaluation of mineral deposit suspected to contain gold must always be based on the examination of a large number of polished sections. Quite frequent are thin sheets, which occur as binding material in cataclastic pyrite and arsenopyrite and more rarely in tetrahedrite and sphalerite. In similar manner gold also occurs in the interstices or cracks of quartz grains. This gold has the same age as tetrahedrite, chalcopyrite, galena and others and in association with them intrudes the older ore minerals. Part of the gold contained in the gold quartz veins is older and is found in the form of granules and drops, very often microscopically difficult to recognize. This gold is undoubtedly contemporaneous within quartz or pyrite and arsenopyrite. The question as to where to seek the gold in some ore minerals is difficult to answer; some ore deposits — and indeed often without regard to their genesis — contain all their gold in the form of microscopically identifiable "free" grains, others, sometimes very rich deposits, show hardly a trace of free gold, even in cases when no gold tellurides are present. SCHNEIDERHÖHN describes, e.g., pyrite-rich portions from vein 25 of the Bredisor area near Brad, where in spite of the most careful examination no free gold could be detected in pyrite and quartz.

Fig. 286 a 170 × RAMDOHR

Cooktown, Queensland, Australia

Gold, bright white, next to *limonite* (grey) and gangue, mostly *quartz* (very dark grey). Certainly cementative enrichment

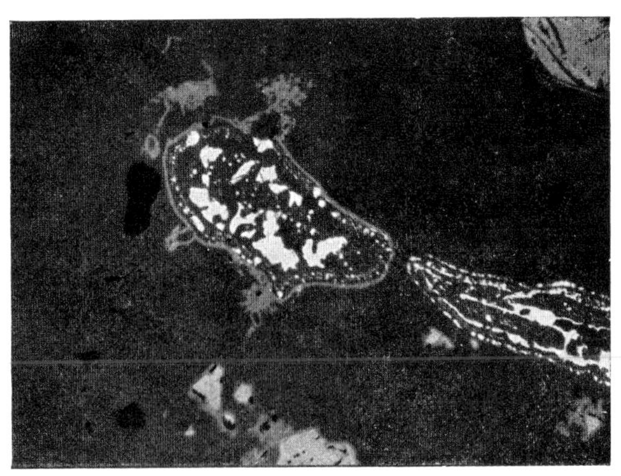

Fig. 286 b 30 × RAMDOHR

Denuncio Mine, Minas, Uruguay

Gold (white) as cementation product replacing an entirely decomposed chalcopyrite crystal. The fine gold particles are completely encased in a thin *limonite* crust, which, under certain circumstances, hinders the recovery of the gold. Small aggregates of γ-FeOOH (in pseudomorphs after pyrite, left top and left center). The main mass is quartz

HÜTTENHAIN (1932) and BÜRG made similar observations. The latter did succeed in such instances to liberate the gold from the pyrite through careful heating with the blowpipe. Later STILLWELL & EDWARDS (1946) treated this problem very thoroughly. They found in the ores of Fiji up to 1000 g/t without visible free gold. Soviet scientists (IWANOFF, 1951), however, ascertained with electronic optical methods that here free gold is present as well, but with grain sizes of 10^{-6} cm (0·01 μ).

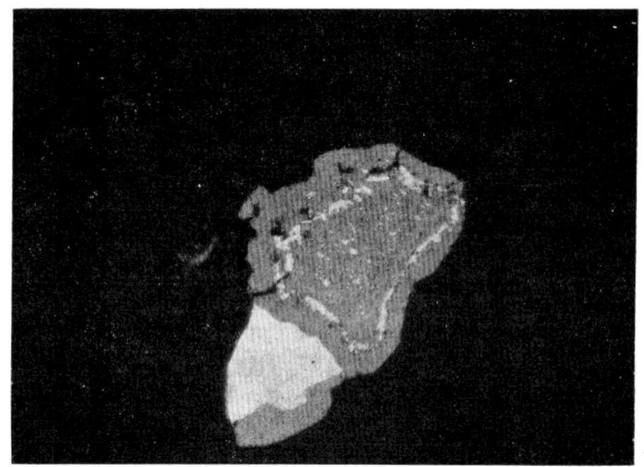

Fig. 287 400 × — dry RAMDOHR
Mt. Morgan, Queensland

Gold rhythmic in *limonite* (grey). Nearby coarser gold grains. The matrix (almost black) is *quartz*

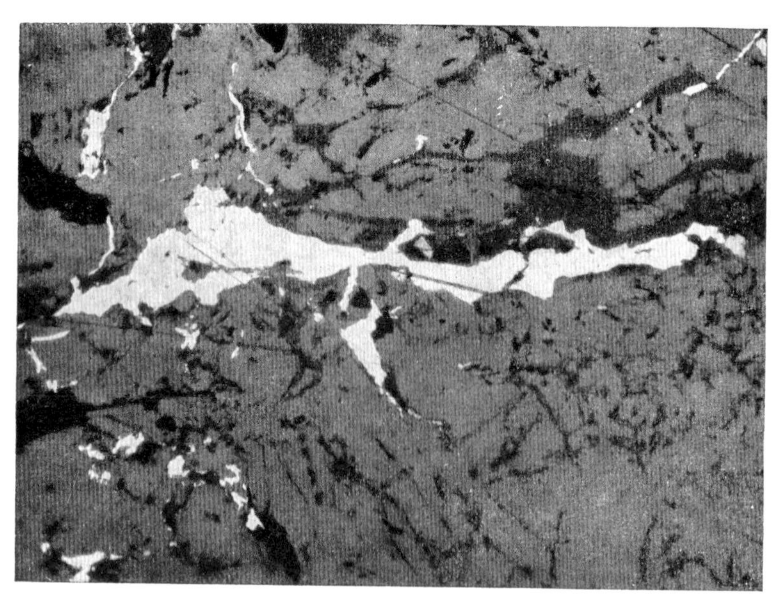

Fig. 288 170 × RAMDOHR
Bou Azzer, Southern Morocco

Cataclastic *brannerite* (grey, in fissures and interstices somewhat discolored and slightly darker) is cemented by much *gold* (clear white) with a little *galena* (grey-white) and quartz (dark grey)

VII. Special fabrics. Gold replaces and cements other minerals in many different patterns (figs. 288–294). Whether these replacements are of ascending or descending origin is very often difficult to decide from one specimen. Since gold precipitates easily from solutions, transportation in descending solution over longer distances is rather unlikely. Therefore, gold found far below the ground water table, certainly originates from ascending solutions. Very extensive literature exists dealing with transport possibilities and precipitation; experimental work in this regard was undertaken by PALMER & BASTIN (1913) and GROUT (1913). Further data concerning redeposition will be given below.

VIII. Diagnostic features. The high metallic reflectivity, together with the very low hardness and the bright color exclude most mistakes. Very small granules of pyrite or chalcopyrite in gangue can, particularly without immersion, appear similar and can give occasion for mistakes. In these instances treatment with an $AgNO_3$ solution (VAN DER VEEN, 1925) is recommended which attacks other metals but not gold. Or, one may use, as SCHNEIDERHÖHN recommends, one drop of mercury on the polished surface which amalgamates gold and which dissolves it under prolonged reaction. Native silver can in the early process of tarnishing appear similar to gold but is etched by an HNO_3 solution.

IX. Paragenetic position (with association and *typomorphic characteristics*). Gold is found very widespread, however, in small quantities. In economically unimportant quantities it is present in various rock types in finely disseminated form, namely in volcanic rocks, in sediments and in crystalline schists. Perhaps some of the placer gold originates from such deposits.

1. The important accumulations occur mostly in veins, and indeed in a wide range of conditions of formation. Within the intrusive-magmatic sequence it extends from the pegmatitic-pheumatolitic formations to the last phase of the hydrothermal stibnite veins (figs. 288–294). The mineral association is therefore very variable, even often within the same ore deposit. In many types of transitional ore deposits it is possible that, e.g., the oldest gold has the same age as wolframite, tourmaline, rutile, and the youngest gold the same as lead–antimony–sulphosalts, or even as zeolites. Several generations of gold are often visible, even when most of the gold was introduced with chalcopyrite, fahlore, galena or bournonite only. The sometimes expressed opinion, that the late gold originated from a special, more or less independent mineralizing process is not tenable and at best only true once in a while locally. Gold of metasomatic origin, particularly in the pegmatitic-pneumatolytic stage of limestones and silicates, is not rare; it occurs here together with chalcopyrite, sphalerite, bismuth and bismuth minerals, often also with bismuth tellurides and skarn minerals. A compilation of all ore deposits and of the literature pertaining to this group is not possible here. Mention must be made, however, of the comprehensive publications of F. BUSCHENDORF (1931), UGLOW & JOHNSTON (1923), LINCOLN (1911). Uncommon and certainly an extremely low temperature formation is the gold occurrence of Corbach described by the writer, where gold is found in a knitted pattern with clausthalite and in other strange patterns.

2. In the ore formations of the subvolcanic group, gold is often mineralized as a telluride, extremely rarely a selenide, but more often occurs here also in its native state. The usually much higher silver content in these ore deposits causes an elevated

Fig. 289 80 ×, imm. RAMDOHR
Chiliani, Bolivia

Gold with *antimonite* and *quartz*. Antimonite shows variable brightness due to the varying orientation of the sections

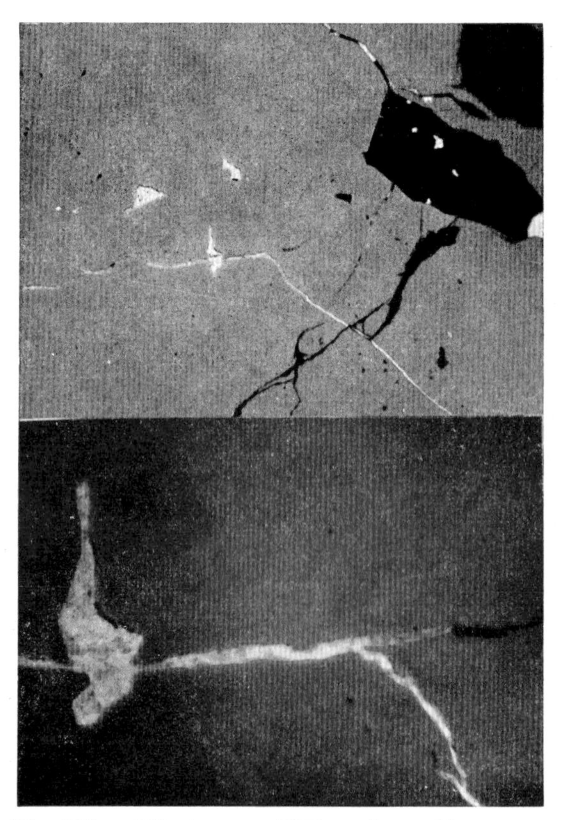

Fig. 289a 250 +, resp. 1200 ×, imm. RAMDOHR
Homestake Mine, Lead

Gold, filling stringers and tiny holes in *arsenopyrite*

silver content and a brighter color of the native metal. Nevertheless the gold is yet in most cases present as "free gold" and not as a "gilded" silver or as "electrum". The major portion of the silver is contained here in sulphides or in sulphosalts. The association is very variable and rich in varieties with pyrite, chalcopyrite, bornite, silver-rich fahlores, sphalerite, stephanite, pearceite, ruby silver, etc. Within the gold tellurides free gold is often of primary origin. In the altered occurrences, particularly in those due to weathering, the tellurides very often produce secondary gold, frequently in fine powder or moss-like form ("mustard gold").

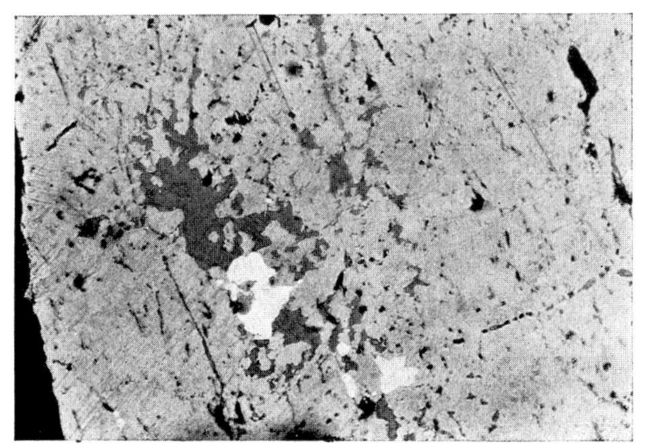

Fig. 290 225 ×, imm. RAMDOHR
Nassfeld near Böckstein, Austria

Gold (clear white) in *chalcopyrite* (light grey) next to *galena* (here dark grey, due to appropriate filter)

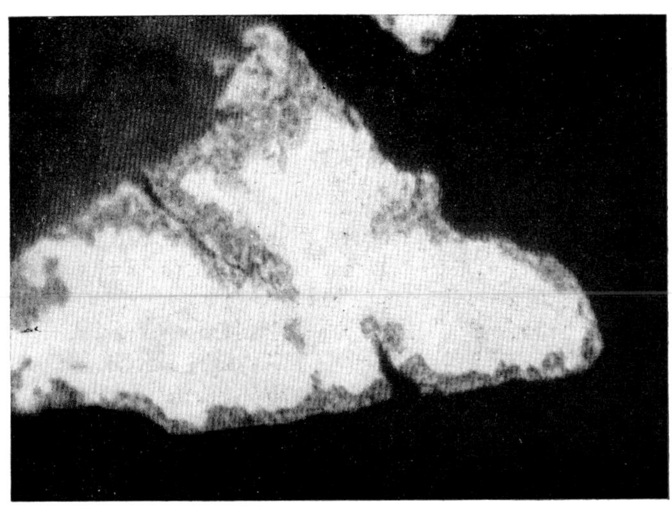

Fig. 290E 500 ×, imm. RAMDOHR
Rheingold

Silver-rich gold, of which in the outer part the silver content is leached out by the transporting water

3. Within the sedimentary sequence gold occurs predominantly in placers. The gold originates from destroyed deposits of another kind. Whether or not these rearrangements are due strictly to mechanics or whether at that time occurred as a result of dissolution, is in individual cases often not clear. That gold can be dissolved through rather heavy concentrations of ferrisulphate, and also as a chloride in the presence of MnO_2 and NaCl, through sulphurous acid, is a proven fact and certainly occurs quite frequently. The extent and range of these changes are generally known with very

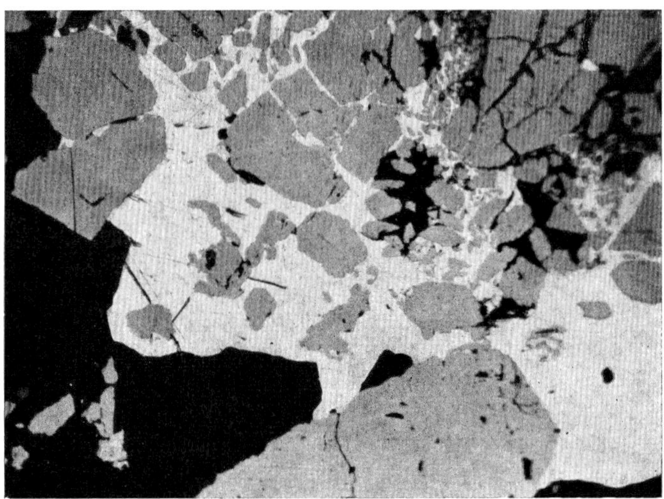

Fig. 291 250 ×, imm. RAMDOHR
Stawell Mine, Victoria, Australia

Similar to fig. 292, *gold*, however, more plentiful

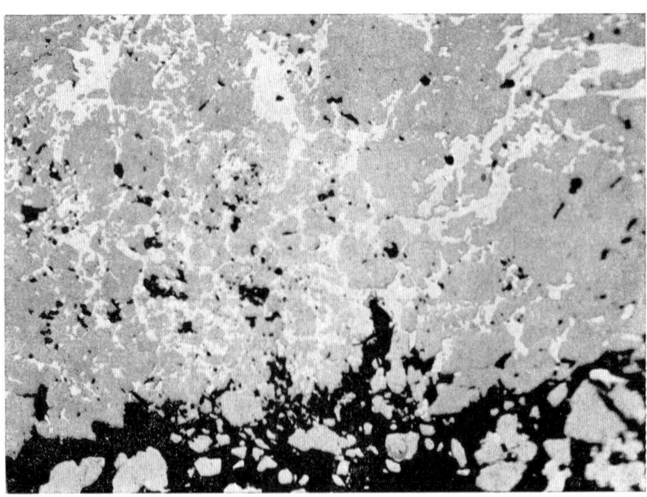

Fig. 292 250 ×, imm. RAMDOHR
Stawell Mine, Victoria, Australia

Gold (white) cementing and replacing badly broken *arsenopyrite* (light grey). Black is quartzy gangue material

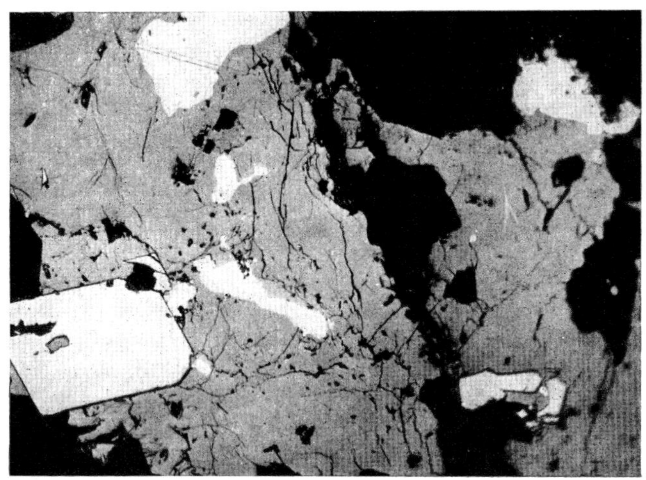

Fig. 292a 250 ×, imm. RAMDOHR
Fürstenzeche near Brandholz, Bavaria

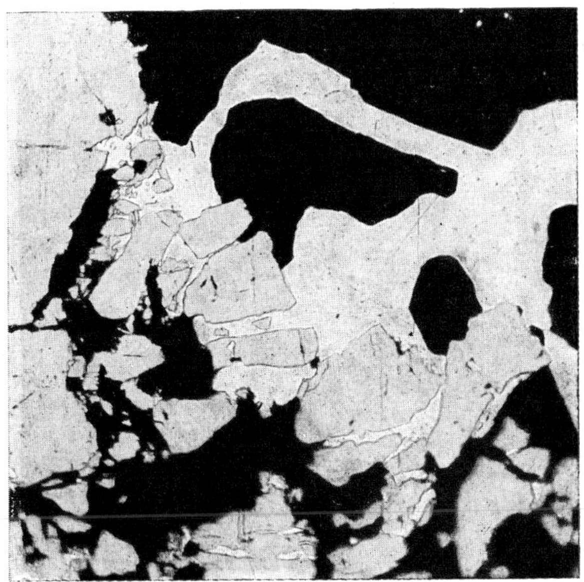

Fig. 293 170 × RAMDOHR
Stawell Mine, Victoria, Australia

The gold content here is particularly high.
(The excellent polish of the gold, which is normally very difficult, is achieved by means a of very short, dry, secondary polishing by hand with dry MgO on a folded handkerchief)

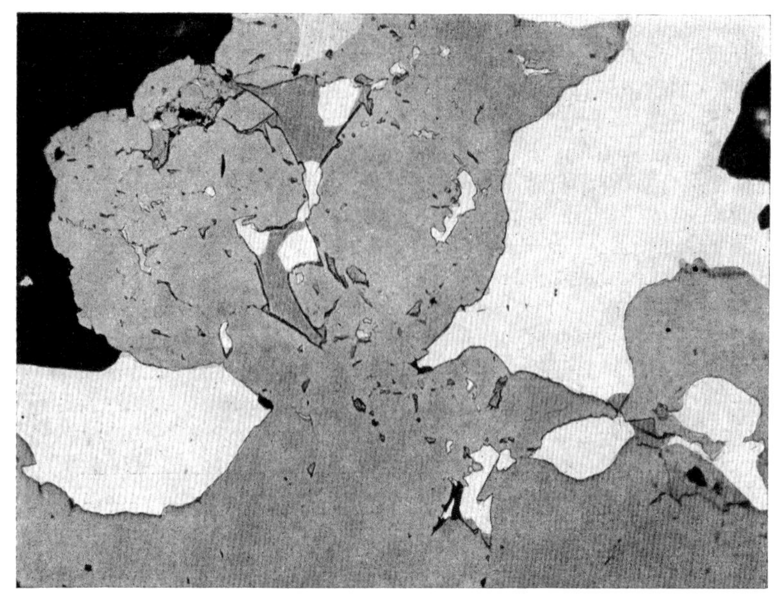

Fig. 294 250 ×, imm. RAMDOHR
Crown Deep Mine, Witwatersrand

Gold (clear white) next to older and partly corroded *pyrite* (greyish white) and *chalcopyrite* (grey); black is gangue. Outstanding machine polish!

Fig. 295 250 ×, imm. RAMDOHR
Astano near Lugano, Switzerland

Native gold (clear white) and *galena* (dark grey) infiltrate without essential replacement the *arsenopyrite* (to achieve good contrasts it is necessary to employ very contrast-rich developing!)

little certainty. Out of such solutions gold can be precipitated by means of reducing substances (organic remnants, sulphides, ferrosulphate) or even by "autoreduction". The transport can continue when the precipitation resulted through $FeSO_4$, because $Fe(OH)_3$ acts as a protective colloid. This last occurrence is surely of great significance in the oxidized and cemented zones of gold ore deposits (fig. 286). Many botryoidal specimens and some quite clearly developed crystals without any pressure strains have certainly continued to grow in the placers.

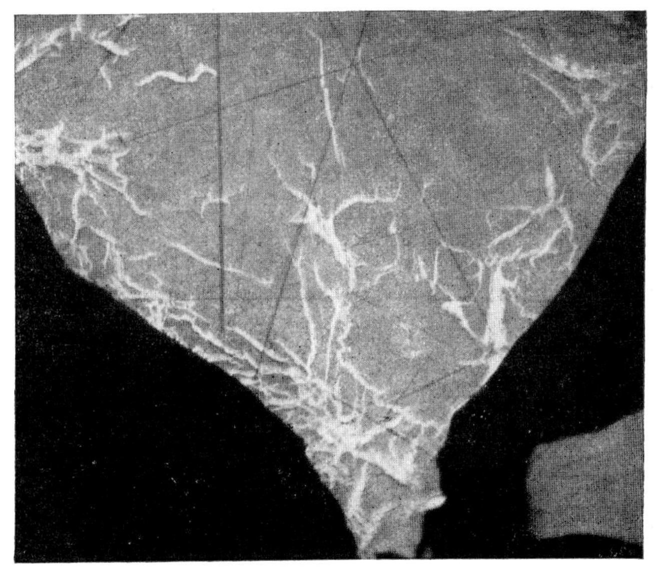

Fig. 295a 250 ×, interference contrast MEDENBACH
A partly idiomorphic photograph crystal of pyrite, "Witwatersrand"

This special technique shows — in unetched section — excellently minute differences in hardness. Neither chemical differences nor porosity can be observed at all

Placer gold is almost always of deeper color than "mountain gold", this is due principally to the superficial leaching of the silver, which is soluble as Ag_2SO_4 or Ag_2CO_3, but can locally also result from the aforementioned continuation of growth (fig. 290E). Films of limonite, silver chloride and others are disturbing features in the milling process; they are dealt with by STILLWELL & EDWARDS (1941) and by HEAD (p. 282) (fig. 262). Part of the "rusty" gold is also coated with a superficial film of oxidized antimony minerals.

4. In the outcrops of primary gold deposits (see above) the gold generally does not migrate far. Nevertheless in some cases enrichment zones of great importance originated through this process. Much that was said before under the heading of placers also applies here (figs. 286a, b).

5. The Witwatersrand assumes a very special position; its economic importance is enormous, at present some 65% of all gold production originates here. Originally a very large placer deposit was located here, whose gold content, however, was redeposited

in a manner as yet not known. The ore deposit has gone through high temperatures ($> 250°$), since chalcopyrite, with enclosed cubanite, was newly formed. The gold content is mostly present in the form of fine granules and films on the interstices of the cementing materials of the large quartz pebbles, only a subordinate portion is associated with the plentifully present pyrite. Even in the very rich portions the gold can easily be missed in microscopic observations using second class sections.

X. Investigated occurrences (small selection).

1. Brandholz, Fichtelgebirge; Schallgaden, Salzburg; Böckstein, Hohe Tauern, Austria; Boliden and Falun, Sweden; Bor, Yugoslavia; Beresovsk and others, Ural; Trail Creek, B.C., many mines in Ontario, e.g., Quemont, Caramora, etc.; Grass Valley, California; Neuras near Rehoboth; Pilgrimsrest and Barberton, Transvaal; Que-Que and Cam and Motor, Rhodesia; Kalgoorlie and Ballarat, Australia. Associated with uranium ore from Shinkolobwe, El Sharana, and Hüttenberg.
2. Schemnitz, Carpathians; Fata Baie and Brad, Siebenbürgen; Guanacevi, Mexico; Cripple Creek, Colorado; Comstock Lode, Nevada; Simau Sumatra; Edie Creek, Papua.
3. Many gold sands and concentrates from such (fig. 14).
4. Minas, Uruguay; Mt. Morgan and Cooktown, Queensland.
5. Witwatersrand (abundant material!)

XI. Literature. Due to its economic importance, a great amount of publications exists dealing with the microscopic characteristics of gold, most of which is cited in the text. Mention should be made particularly of the ore-microscopic summaries by SCHNEIDERHÖHN (1922), VAN DER VEEN (1925) and ZEMCZUZNY (1926), and the paragenetic-microscopic studies, of BUSCHENDORF (1931) and ÖDMAN (1939, 1941). The accompanying minerals are comprehensively dealt with by G. M. SCHWARTZ. Much work has been done by the writer on the Witwatersrand ores (RAMDOHR 1954 and1958).

XII. Powder diagram (B. & TH.). (10) 2·354, (7) 2·039, (6) 1·442, (8) 1·229 Å.

AURICUPRIDE and AuCu

I. General. *Chem.* Au and Cu form the intermetallic compounds $AuCu_3$, and AuCu, perhaps some other ones. In nature $AuCu_3$ ("auricupride") has been observed several times, AuCu rarely and not fully approved. It is noteworthy that the natural auricupride carries 40% Au instead of the theoretical 50·8%. — *Cryst.* Cubic face centered, where Cu forms the centers, Au the corners. — No cleavage, H $\sim 3^1/_2$, distinctly harder than pure Au, D 11·5. Lustre highly metallic, yellow with a distinct tint to red. — *Polish* much better than gold; scratches in the latter often do not enter goldcupride.

III. Reflection properties. Typical auricupride is distinctly darker than gold, and has in oil a clear lilac tint. Since gold and auricupride often occur together, the difference can be rather pronounced — without an immediate comparison some care is necessary.

For Rozhkovite (acc. RAZIN)			
R. nm	460	580	1110(!)
	47·5	63·0	83·6

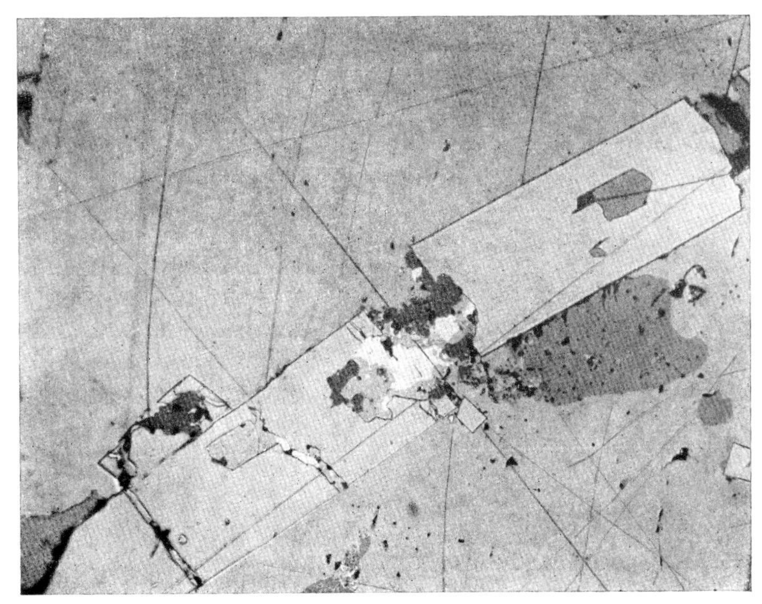

Fig. 296 150 × RAMDOHR
Bayerland Mine near Waldsassen, Bavaria

Large crystal of *arsenopyrite* embedded in *chalcopyrite* and partly replaced by it. At the fracture point *native gold* (white) and some *sphalerite* (dark grey). Upgrown on the arsenopyrite is a large portion of *tetrahedrite* (medium grey)

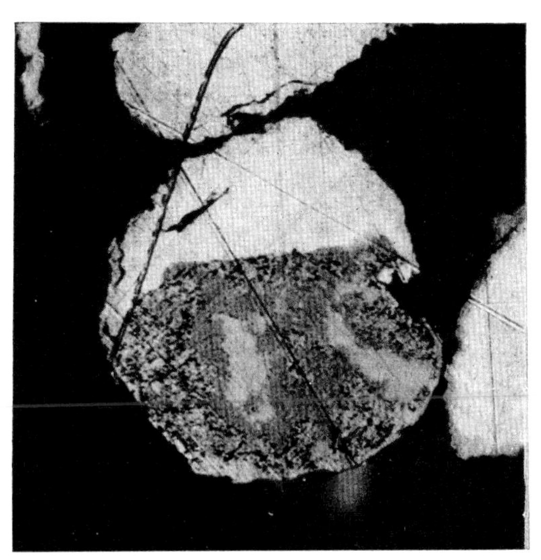

Fig. 297 500 ×, imm. RAMDOHR
"Rheingold", Steinmauern, Baden

Auricupride, small remnants only, is, starting from the surface, altered in a "browny" *gold* full of holes. The crystal is primarily upgrown, perhaps oriented, and surrounded by Ag-rich gold which, in its turn, shows a small leached rim

Besides auricupride a compound, again with a reddish tint but distinctly brighter and more similar to gold, can rarely be observed (e.g. in concentrates of "Rheingold"). It behaves in the weathering process very similarly to auricupride. For the time being I take it to be AuCu.

IV. Etching. No data! In any case rather resistant against weathering and reagents.

V. Phys.-chem. Cf. gold (pg. 324).

VI. and VII. Structural properties. Normally the occurrences in gold quartz veins show in their ramified forms much similarity with those of normal gold (Victoria, Austr.; Kerr Addison Mine, Ontario; Tankavaara, Finland). Concentrates of "Rheingold" contain rounded monocrystalline grains, sometimes idiomorphic against normal gold (Fig. 296a).

During the river transport Cu leaches out, analogous to the silver content of Ag-rich gold. Corresponding with the very high Cu-content at first a dark brown, loose spongy mustard gold is formed, mostly changing soon to "normal" gold by the cold working of the river pebbles. All intermediate transitions are known, remnants are mostly preserved. — The probable AuCu has similar but thinner crusts of pure gold. — Because of its hardness auricupride becomes during the river transport less platy than normal gold.

IX. Paragenetic position. The known occurrences — many more may have been overlooked — belong to hydrothermal veins and are probably formed in the stability field of $AuCu_3$, i.e. below 390°. An exception can be Mooihoek (see p. 324). Regarding the highly chalcophile properties of copper the deposits must have been very low in sulphur. Otherwise the knowledge of that mineral is very restricted.

X. — Localities are mentioned in VI and VII.

XII. Powder diagram. — No data!

Remark: Regarding the intermetallic low-temperature alloys of gold and copper, resp. gold, silver, palladium and copper during the last years a rather long list of data, esp. by Russian authors, has appeared, which are sometimes contradictory. As already told, $AuCu_3$ = Auricupride seems to be the most common one, AuCu, recently called *Rozhkovite,* o'rhombic, distinctly anisotropic doesn't seem to be rare. Besides that Au_2Cu_3, perhaps stabilized by Pd, occurs. For the last one (from Talnakh), again o'rhombic, RAZIN gives R-values and P.D. too.

MALDONITE

I. General data. *Chem.* Au_2Bi. *Cryst.* Cubic; C_{15}-Type, a_0 = 7·97 Å; in natural occurrences predominantly decomposed as Au + Bi.

II. Polishing properties. It polishes excellently, much better than its associates in nature, gold and bismuth. Hardness somewhat greater than that of gold. Brittle.

III. Reflection behavior. High reflectivity, however, distinctly less than that of gold or bismuth. Rough estimate: 50–60%. The color has a light bluish, after longer exposure more greenish-grey color tint, particularly in oil.

IV. Etching not investigated.

V. Physico-chemistry. Through fusion one can readily produce Au_2Bi. In nature the conditions of formation are obviously rarely achieved. Gold and bismuth occur relatively frequent in close proximity, without any indication of forming any compounds; not even compounds which are again decomposed.

VI. Fabric. The texture of the unaltered mineral of Maldon can hardly be reconstructed. In the present state it is found decomposed into a predominantly coarse myrmekite of Au and Bi with ~ 1:1 surface proportion. Smaller and larger intact remnants occur only in wedges; they are recognizable by their color and sometimes by their cleavage. The occurrence of Gilgit forms tetrahedrite-like grains in the placer. They show all crusts of secondary gold.

IX. Paragenetic position. The formation of maldonite is apparently limited to a narrow margin within the high temperature range. The classical occurrence (Maldon Mine, Victoria) is a particularly high temperature gold-quartz vein. Likewise of high temperature origin is the occurrence in the garnetiferous skarn of Baita Bihorului. With reservations only should be mentioned the high temperature occurrence in a gold rich löllingite at Ingram in Ontario. Pictures of gold-bismuth myrmekites are also shown in a Japanese publication (ISHIBASHI). In the meantime, it has been found unreplaced as an inclusion in the arsenopyrite of Akjoujt and also in a form with incipient weathering in the sands of the Indus.

XI. The **Literature** is scanty. The writer has described the occurrence in which the mineral was originally discovered (1952) (by utilizing personal information from STILLWELL). S. KOCH shows pictures of typical decomposed myrmekite from Rezbanya (1950).

XII. Powder diagram (B. & TH). (4) 2·818, (10) 2·404, (5) 2·296, (6) 1·534, (5) 1·409 Å.

LEAD

I. General data. *Chem.* Pb, allegedly with traces of Ag and Sb. *Cryst.* Cubic, O_h^5, a_0 — 4·94 Å. # missing. H = 1–1½, exceedingly malleable. D = 11·4. Freshly cut, it is metallic white, but coats immediately grey, later earthy white coating.

II. Polishing properties. Due to its low hardness and malleability it cannot be polished without scratches; in addition, exposed to air it immediately tarnishes making observation almost impossible.

The polishing hardness is very small, much lower than that of galena. Talmage hardness: A.

III. Color and reflection behavior. The reflectivity in freshly polished surfaces is quite high, but decreases immediately. The color is grey-white. Color and reflectivity do not change noticeable in oil.

Quantitative measurements are very difficult. LAWRENCE gives for lead from Red Cap, Chillagoe, Queensland: green 82%, orange 74%, red 74%. The values seem very high compared with the calculated 62·1%. — Isotropic.

IV. Etching (accord. to VAN DER VEEN). *Positive*: Concentrated HNO_3 gives texture etching. HI gives yellow coating. *Negative*: HCl, H_2SO_4.

VI. Fabric. *Texture and structure.* Aggregates of rounded grains, which occasionally show twin-lamellae; occur as individual crystals. No other textural characteristics were observed.

VIII. Diagnostic features. The very low hardness and the immediate tarnishing after polishing preclude misidentifications.

IX. Paragenetic position. Intensively as well as extensively very rare; approved occurrences are very few, all of which originated under simultaneous scarcity of sulphur and oxygen. The richest occurrence was Långban (Wermland), where it is accompanied mainly by manganese oxides in a granular limestone. Similar conditions prevailed at Franklin where lead occurs in rhodonite in very finest lamellar inclusions. The conditions required to form lead are also found, as an exception, within oxidation zones (Hailey in the Wood River District, Blaine, Idaho). — In Cheleken, USSR, native lead in rather high quantities is deposited recently from hot brines.

X. Investigated occurrences. Pajsberg, Långban, both in Sweden. Artificial products.

XII. Powder diagram (B. & T.) (10) 2·856, (7) 2·476, (6) 1·750, (9) 1·492 ... (6) 0·783 Å.

MERCURY

I. General data. At ordinary temperatures native mercury is a liquid and therefore *not suitable* for examination under the ore-microscope. At −39° it crystallizes into rhombohedric crystals. Associated are frequently the two "amalgams" *moschellandsbergite*, Ag_5Hg_8 with about 70% Hg, and *kongsbergite* a mix-crystal with more than 40% Ag with silver structure. Both are cubic, the first one forms good crystals, the latter almost always is anhedral. Both are brittle, particularly the moschellandsbergite; no #; H ~ 3, D = 15 to 12. Metallic lustre, silver white, kongsbergite tarnishes much less than native silver.

Polish of moschellandsbergite good, better than that of precious metals; that of kongsbergite is the same as that of silver (perhaps better!).

The amalgams are *microscopically* not yet sufficiently well known, to give definite data.

Remark: Recently a third member of the Ag-amalgam-family was described:

Schachnerite. I. General data. Chem. β-phase in the system Ag–Hg with 40% Ag. Cryst. hexagonal, D^4_{6h}, a_0 c_0 = 2·97, 4·84 Å, Z = 2 ($Ag_{1.1}Hg_{0.9}$). D = 13·52. Seldom big crystals.

III. Reflectivity. White, metallic. R_{589} = 72%. Anisotropism very weak, in oil also.

VIII. Diagnostic feature: Schachnerite has been observed only in the vein "Vertrauen auf Gott", Moschellandsberg; *Paraschachnerite*, o'rhombic, similar properties.

XII. Powder diagram (SEELIGER and MÜCKE, from whom are also the other data) (5) 2·420, (10) 2·273, (5) 1·268, (5) 0·954, (6) 0·8595 Å.

PALLADIUM

I. General data. *Chem.* Pd, according to scarce information concerning its natural occurrences; it occurs either pure or intergrown with Pt. *Cryst.* O_h, a_0 = 3·89 Å. — H = $4^1/_2$–5, D = 11·9 — opaque, white, strong metallic lustre.

II. Polishes easily, notably softer than platinum.

III. Reflection behavior. Reflectivity is very high, very similar to freshly polished silver (but darker), compared to platinum, somewhat yellower. It does not darken like silver or bismuth.

Isotropic.

	in air	in oil
Photometer ocular		
green	69 %	66%
orange	70 %	66%
red	71·5%	67%

IV. Etching — not well known, similar to that of platinum; etched, however, by aqua regia and particularly fast by chromic-hydrochloric acid.

VI. Fabric. Structural characteristics are hardly known. Small octahedral individual crystals occur in Brazil. Xenomorphic and often with platinum concentrically overgrown palladium is found in Transvaal (SCHWARTZ & PARK, 1932). In Choco, Columbia, it envelopes partly the Pt-sulphides embedded in platinum. Small portions are precipitated through weathering from stibiopalladinite.

VIII. Diagnostic features. Regarding the diagnostic features cf. the table under stibiopalladinite.

IX. Paragenetic position. Paragenetically almost nothing is known concerning the formerly described occurrences in Brazil, St. Domingo, and the Ural. It occurs in greater quantity in the outcrops (within the weathering zones) of the platinum deposits of the Transvaal, where it originates from oxidation of the Pd-bearing sulphides and of stibiopalladinite, as described by SCHNEIDERHÖHN (1929). In Columbia it is contemporaneous with platinum and shows no indication of secondary formation. The great similarity, both, macroscopically as well as microscopically, to platinum makes an identification next to

Fig. 298E 325 ×, imm. RAMDOHR
Sudbury, Canada

Rounded idiomorphic crystal of *platinum* (pure white) included in a mixture of *pentlandite* (greyish white), *pyrrhotite* (light gery), and *magnetite* (dark grey). A distinct age relation cannot be established

this but without direct comparison, very difficult. Meanwhile a lot of Pd-compounds has been described.

XII. Powder diagram (HANAWALT). (10) 2·23, (5) 1·92, (3) 1·37, (3) 1·17 Å. (B. & T.) (10) 2·36, (7) 1·364, (5) 1·273, (6) 1·233 Å.

Remark!

Allopalladium Allopalladium, for more than 100 years regarded as a "good" Mineral has been recognized recently by GENKIN & TISCHENDORF (1977) to be really Stibiopalladinite. It must therefore disappear at this place and the text for stibiopalladinite was a bit enlarged.

The two figures — formerly 297 and 298 will now appear as fig. 332a and fig. 332b. Actually the old name should have priority, but regarding the wrong definition, the name stibiopalladinite is preferable.

IRIDIUM

I. General data. Ir, cubic O_h^5, a_0 = 3·83 Å. Small, loose crystals, very rare. H = 7(!) D = 22·6. Very high metallic lustre, silver white.

II. Polishing is relatively easy when iridium is found as inclusion in native platinum, polishing hardness is considerably higher than that of Pt.

III. Reflection behavior and color not different from that of platinum.

IV. Etching. The unusual chemical stability makes etching with known reagents impossible.

V. Physico chemistry: Ir and Pt should form unlimited mix-crystals, according to synthetic experiments; in nature, however, distinct exsolution particles of Ir in Pt may be observed; cf. figs. 300, 303 to 305. Extremely slow cooling periods are probably necessary for such exsolution processes.

VI. Fabric see platinum.

VII. Special fabrics. Occurrence, hardness, and resistance to chemical attack make mistakes almost impossible.

IX. Paragenetic position. Observed microscopically only as exsolution particles in platinum, and here associated with occurrences derived from chromitites.

XII. Powder diagram (WYKOFF). (10) 2·21, (6) 1·91, (6) 1·36, (8) 1·15 Å.

PLATINUM AND FERRO PLATINUM

I. General data. *Chem.* Pt, always with Fe (up to 20%), minor Ir, Cu, Pd, Rh, Ru. Most of the analyses are based on very impure material which has not been investigated microscopically, with the result that it is not possible to determine which of the accessory elements are present in solid solution, and which as inclusions or gross impurities. Accord. to VYSOTZKI (1925) the Ural platinum can best be classified as "ferroplatinum" (dark grey to almost black) with 71–78% Pt and 16–21% Fe, or "polyxene" (light steel grey to silvery-white) with 80–90% Pt and 6–11% Fe. Further subdivision, based on iridium content, does not appear to be justified according to microscopic study

Cryst. Hexoctahedral; in placers mostly rounded xenomorphic masses; in some South African occurrences, good crystals. *Lattice*: facecentred cubic: $a_0 = 3{\cdot}92$ Å.

Cleavage not observed.

$H = 4–4^1/_2$, frequently apparently greater; hackly fracture. D of natural material = = 14–19; that of pure synthetic crystals, 21·5. *Opt.* opaque, $\varkappa_{Na} = 20{\cdot}6$, $n_{Na} = 4{\cdot}28$; reflectivity for Na-light is $\sim 70{\cdot}1$%.

Highly metallic lustre, white to steel-grey; the rounded grains are tin-white to quite dark grey.

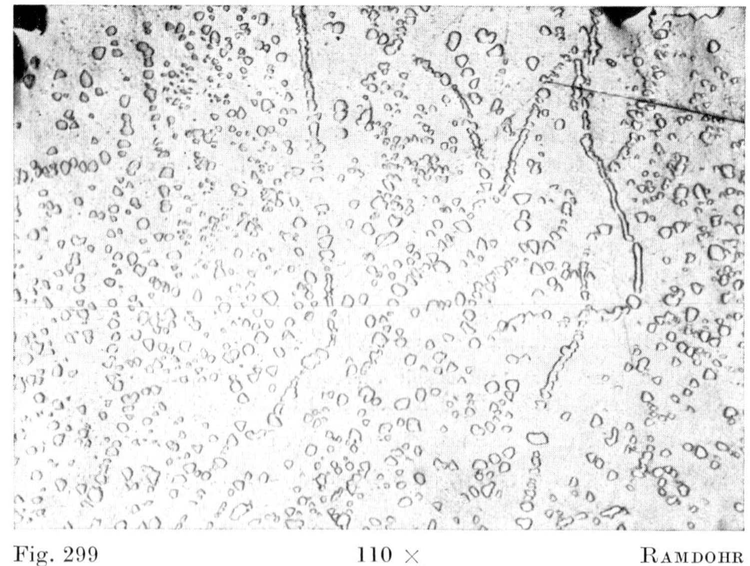

Fig. 299 110 × RAMDOHR

Nishne Tagilsk, Ural

Similar to fig. 300. *Iridium* forms more individual grains

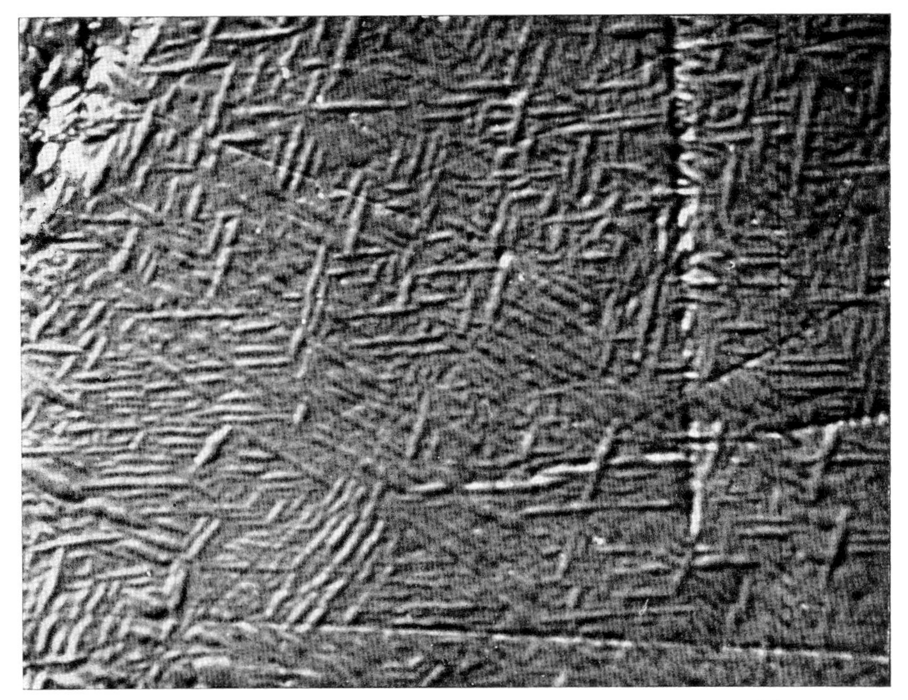

Fig. 299 a 300 ×, Interference contrast MEDENBACH
Nishne Tagilsk, Ural

"Platiniridium", an exsolved mix-crystal // (100) fine lamellae, spindle-like, are iridium, in contrary to the equally thick lamellae or iridosmium (Fig. 307), which are enclosed // (111)

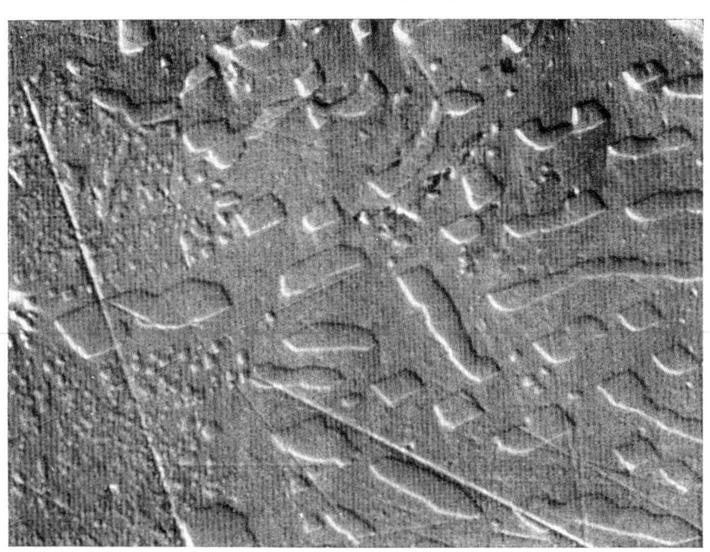

Fig. 299 b 250 ×, Interference contrast MEDENBACH-RAMDOHR
Nishne Tagilsk, Ural

Large exsolution bodies of iridium from platinum. There are signs that this bodies originated from the fine bodies (right) of the lamellation of Fig. 299 a

II. Polishing properties. After good pre-grinding, an outstanding polish is easily obtained, but in some cases (perhaps especially with Fe-poor samples?), very per-

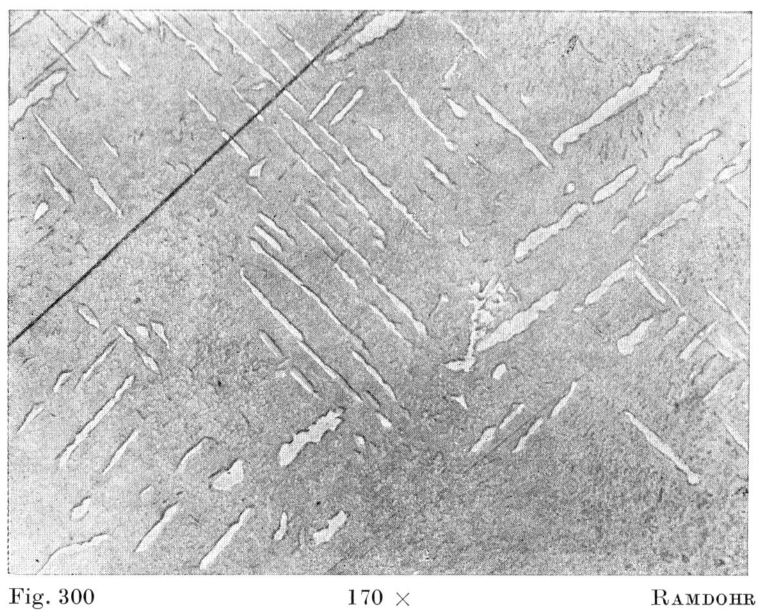

Fig. 300 170 × RAMDOHR
Nishne Tagilsk, Ural

Coarse unmixing of hard *iridium* // (100) in *platinum* (slightly darker). Also fine exsolved bodies with rounded forms

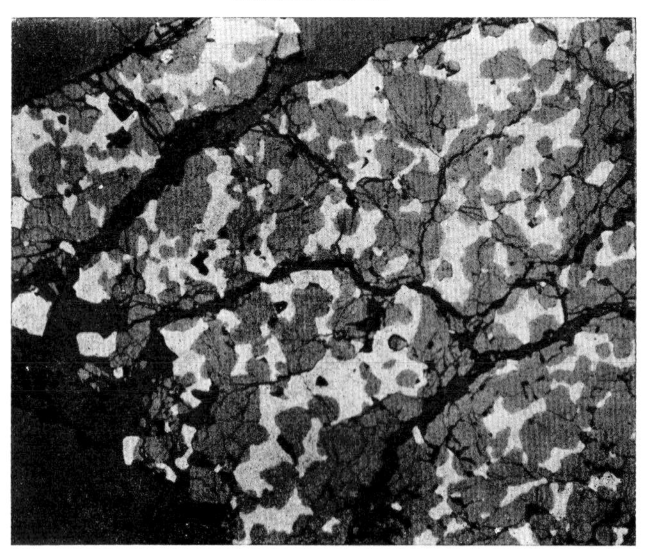

Fig. 301 a 8 × RAMDOHR
Nishne Tagilsk, Ural

Native platinum cementing *chromite* grains. The chromite portion has been severely fractured during fluvial transport, but the nugget has been held together by the tenacious platinum

sistent scratches remain. No cleavage discernible. The hardness varies appreciably with the chemical composition and geological "pretreatment". Naturally occurring placer platinum is appreciably harder than commercial platinum with the same composition; the difference is due to "cold working" during fluvial transport, as pointed out by ZEMCZUZNY (1920). The hardness is usually between that of sphalerite and pyrrhotite. Talmage hardness: not given.

III. Reflection behavior. Color and reflectivity impressions: pure white, with a tint varying from bluish to yellowish, depending on the accompanying minerals. This tint depends largely on contrast impressions, but also on varying chemistry. Reflectivity is very high.

	in air	in oil
general	very high; exceeded only by few other metals. White to bluish, somewhat variable	change is unusually slight so that the contrast to accompanying minerals is increased
against silver	without creamy yellow tint	silver is decidedly creamy
against palladium	similar, sometimes a little more bluish	trace more bluish
against iridosmium	appreciably yellower somewhat brighter	appreciably yellower; distinctly brighter
against sperrylite	appreciably brighter; similar color tone	much brighter; similar color tone
against iridium	trace more yellowish	distinctly yellower

VJALSOV (PIOR) R nm	460	500	540	580	620	660	700
	76·5	77·3	78·3	79·7	80·1	80·3	81·0

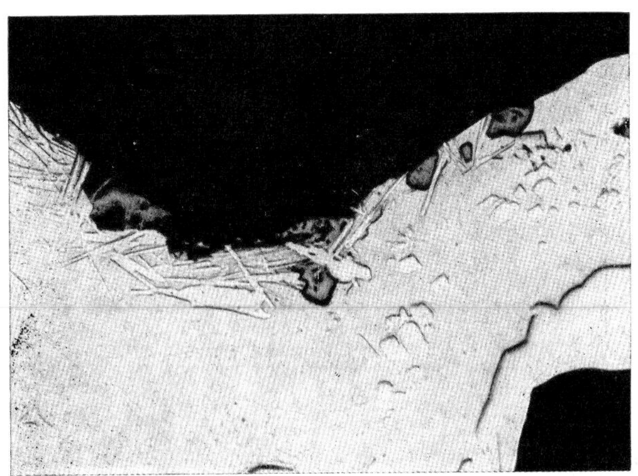

Fig. 301 b 700 ×, imm. RAMDOHR
Nishne Tagilsk, Ural

Native *platinum* (principal component) with thin lamellae of iridosmium, one large and many small grains of *iridium*, together with a number of grains of an extremely hard (H ≫ iridosmium) mineral, probably an opaque oxide. The almost black mineral is *chromite*

Anisotropic effects have not been observed; however, complete extinction is not attained with + nic. at any position.

Differences in color and brightness of chemically different samples are not sufficiently high to make them easily distinguishable.

Minute differences in iron contents of platinum (4.7% and 10.7% resp.) can lower the reflectivity for the whole spectrum for ~ 4% (TARKIAN & STUMPFL 1975).

IV. Etching (accord. to VAN DER VEEN). Very resistant to etch reagents; is attacked only by aqua regia and chromic acid with hydrochloric acid.

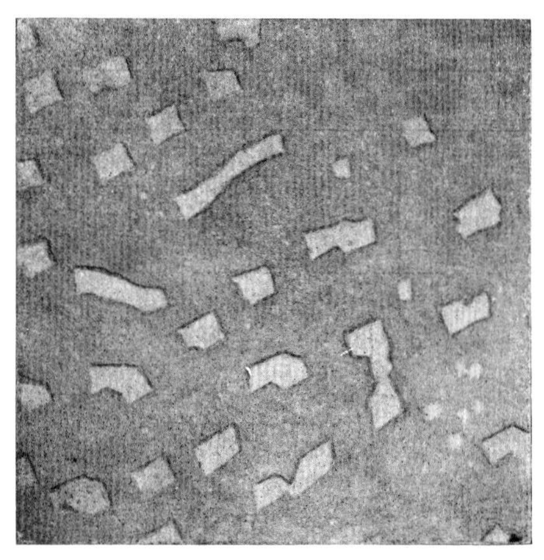

Fig. 302 350 +, imm. RAMDOHR
Nishne Tagilsk, Ural

Iridium in perhaps octahedral exsolution particles in platinum. Also, later fine-grained exsolutions

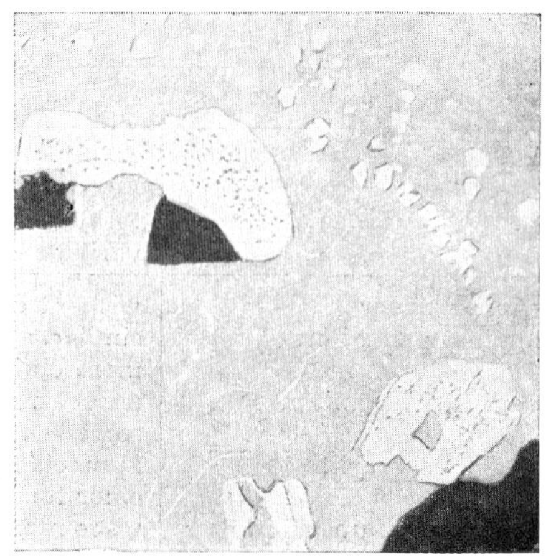

Fig. 303 600 ×, imm. RAMDOHR
Nishne Tagilsk, Ural

"Sub-exsolution" of *platinum* in an *iridium* exsolution grain in platinum

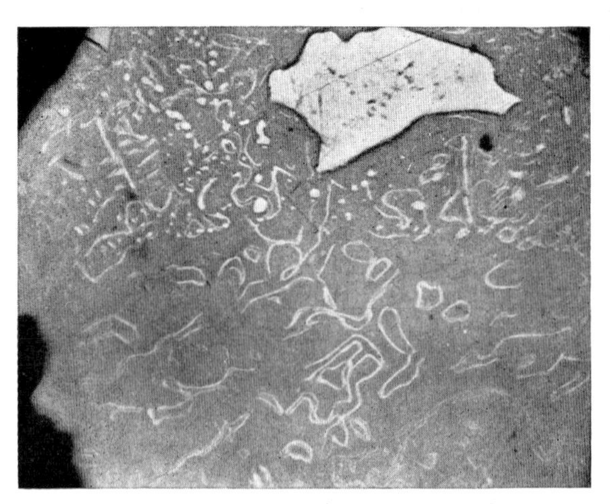

Fig. 304 350 ×, imm. RAMDOHR
Nishne Tagilsk, Ural

"Snake-like" exsolution of perhaps an intermediate Pt–Fe compound in *platinum* beside one large and many small rounded bodies of iridium itself

Texture etching. Aqua regia or chromic acid with hydrochloric acid. Time required for etching is variable; Pt–Ir alloys are much more resistant than those of Pt–Fe; Pt–Pd and Pt–Cu alloys are most easily etched. — Iridosmium is not attacked by aqua regia. — BETECHTIN describes a (111) etch cleavage.

V. Physico-chemistry. The natural platinum forms complicated mixed crystals and other intermetallic alloys with Fe, Ir, Rh, Ru, etc. not all fully understood. Some of them seem to remain in complete solid solution, Ir unmixes as metallic Ir and probably an intermetallic compound (figs. 302, 303, 304) — both in spite of examinations in laboratory. Os + Ir unmix in form of tabular iridosmine. Iron may unmix in the form of the low temperature intermetallic compound FePt with the tetragonal superstructure (a_0 c_0 = 3·84, 3·78 Å), as described by GENKIN & VASOVA (1965b). The d-values (5) 1·888, (7) 1·142 of it deviate rather strongly those of platinum.

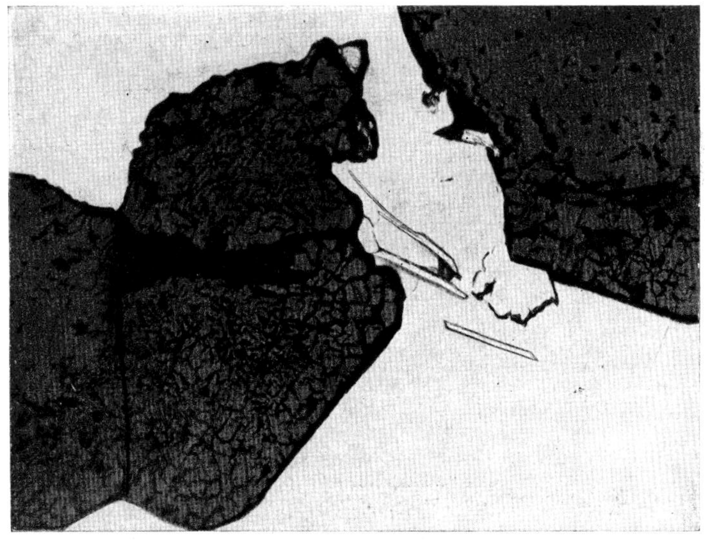

Fig. 305 170 × RAMDOHR
Nishne Tagilsk, Ural

Platinum with *chromite* and *iridosmium*. The highly idiomorphic chromite here exhibits a hitherto unnoticed fracturing texture

VI. Fabric. *Internal properties of grains.* Twinning along (111) is noticeable after etching, in the form of numerous lamellae, undoubtedly formed by mechanical deformation (see fig. 133 in VAN DER VEEN, who, however, interprets the lamellae as cleavage). Zonal growth, accord. to R. BECK (1909), is caused in granular aggregates by purer platinum forming the inner, and more Fe-rich, lower melting alloys, the outer portions. ZEMCZUZNY (1920) gives good illustrations of this. Accord. to ISAAC & TAMMANN, a complete solid solutions exists in the Pt–Fe system, with a melting point minimum at 1529°. *Cuproplatinum,* which frequently forms seams between actual platinum and chlorite, results from a high temperature Pt–Cu solid solution which, although it exsolves at ~ 700°, often persisted (BETECHTIN, 1935).

Exsolution of an unknown substance is probably represented by systems of lines or dashes // (111) which may be visible as cavities, or appear by slight etching. These phenomena, accord. to ZEMCZUZNY, who first described them, are due to recrystallization, which, however, is quite unlikely judging from his accompanying photograph.

A number of samples (nuggets) from Nishne Tagilsk exhibited peculiar exsolution of the appreciably harder iridosmium in the form of fine tablets, grains, and also very delicate skeletons (figs. 300, 304, 305). Iridosmium also occurs as an exsolution in extremely small tablets // (111) (fig. 307).

The fine garland like white lines of fig. 304 and poorly visible in 303 may be the exsolved FePt of GENKIN or another low temperature intermetallic compound.

Zoning is widespread, and is occasionally recognizable by a difference in hardness between the core and the softer shell.

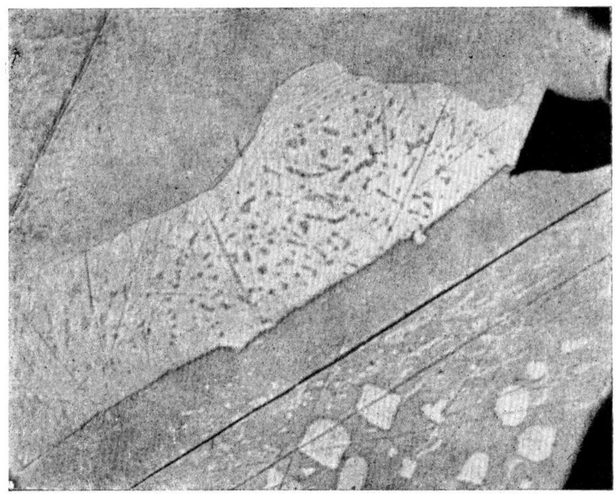

Fig. 306 350 ×, imm. RAMDOHR
Nishne Tagilsk, Ural

Coarse exsolution body of *iridium* (white, hard) in *platinum* exhibits fine "sub-exsolution" of platinum

Structure and texture. Grain form: Where platinum occurs in coarser aggregates with chromite, as is frequently the case in the Ural, it forms xenomorphic polygonal grains. In the hortonolite dunites of S. Africa well-developed cubic crystals with edges up to 4 mm are found beside quite irregular larger grains with tattered-like outlines, or together with finegrained scales. The latter form is here, as also elsewhere, probably the most abundant one. In most cases, certainly in the hortonolite dunites, there are two generations, a question which could be investigated more thoroughly if abundant material were available. In the hortonolite dunites the two generations of platinum appear to correspond to two generations of chromite.

The platinum originating from the decomposition of platinum arsenides or sulphides or from the meagre platinum content of other sulphides (Ni-pyrite, pentlandite, etc.) has a completely different form (see SCHNEIDERHÖHN, 1929). It occurs as reniform or wart-like concretionary masses which exhibit a concentric shell structure after etching, and sometimes even without etching. They appear to have been recognized first by HUSSAK, but are probably not rare. The Pd content is unusually high, corresponding to the abundant presence of this element in sulphidic platinum ores; for this reason they can be etched with aqua regia with particular ease; accord. to SCHNEIDERHÖHN, they even have occasionally a core of pure palladium. The platinum in the quartz veins in the Waterberg district also consists of Pd-rich and Pd-poor shells.

VII. Special fabrics. *Inclusions.* In spite of the work of ZEMCZUZNY, VAN DER VEEN, and BETECHTIN, not nearly enough systematic work has been done on platinum. Practically every nugget of different origin exhibits new, unexplained phenomena.

Very little is known about temperature of deposition; the melting temperatures of the pure material, and that of Pt–Fe mix-crystals are out of the question. It can be

Fig. 307 500 ×, Interference contrast MEDENBACH
Nishne Tagilsk, Ural

Platinum with exsolved iridosmium; tablets with (0001) // (111) of platinum

established that platinum in chromitites and dunites is relatively younger than the chromite and olivine, and in the hortonolite dunites the larger grains are older than hortonolite but younger than the chromite. Accordingly, chromite crystals are quite frequently found in platinum, and drop-like silicate inclusions have been found someimes. Very complex ,surely originally molten, droplets of sulphide-magnetite are rarely present (fig. 307 E).

The hexagonal mineral iridosmium is frequently embedded as simple tablets or bundles sometimes // (111), as determined by VAN DER VEEN (figs. 307, 305). P. A. WAGNER has also described iridosmium inclusions in the platinum from the Onverwacht pipe, which the writer can confirm from observation on abundant material. In addition there are widespread lamellae of another, appreciably softer platinoid embedded // (111). Doubtlessly, other hexagonal platinoid minerals besides stibiopalladinite do exist; thus a nugget from Kuschwa exhibits an inclusion of a more bluish anisotropic metallic ore // (111). — Platinum from the Onverwacht pipe also contains large, partly corroded, inclusions of sperrylite (fig. 308), cooperite, braggite laurite, and crusts of stibiopalladinite.

Natural platinum of unknown origin contained three unidentifiable highly reflecting minerals as inclusions; material from Choco, Columbia contains five, of which palladium, stibiopalladinite, braggite, and sperrylite could be determined with good probability. This Choco platinum, according to the type of intergrowth with palladium, cannot be interpreted as a magmatic, or even high temperature formation. The platinum of the platinum quartz veins of Rietfontein contains specular hematite inclusions, frequently in rhythmic alternation, as well as a pseudo-tetragonal, very light brown platinum mineral (in addition to overgrowths of stibiopalladinite).

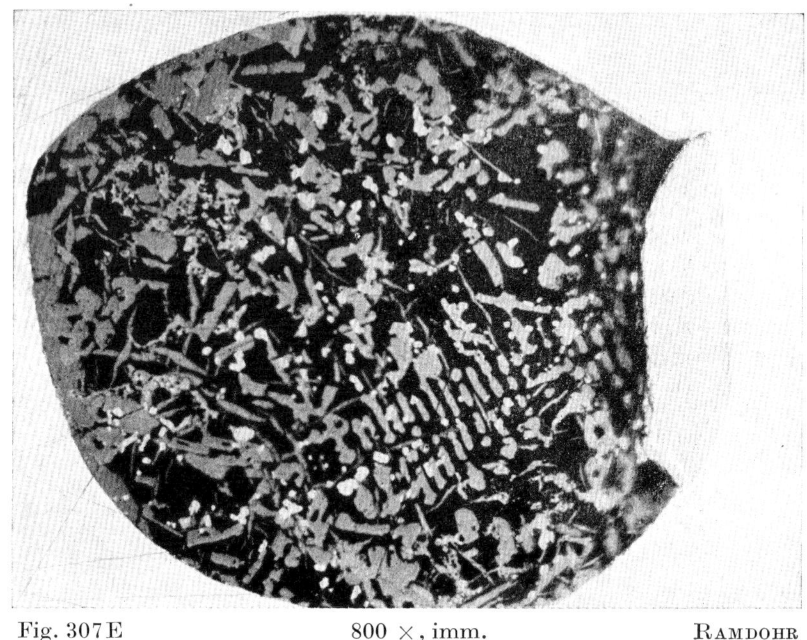

Fig. 307E 800 ×, imm. RAMDOHR
Nishne Tagilsk, Ural

A globule of *bornite* nearly black etc. included in *platinum* (the white surrounding) and *iridosmium* (tabular, a trace greyish, on the right). The bornite contains skeleton formed *magnetite* (medium grey) and some *chalcopyrite* (white)

The aforementioned nuggets from Nishne Tagilsk exhibit a very *peculiar mineral inclusion*. It was so hard that it stood out in strong relief and was very poorly polsihed in a section that was ground with great care on a pitch lap and which, e.g., produced almost ideal relief-free chromite (fig. 301b). Its reflectivity was similar to that of hematite; it is isotropic and opaque, or nearly opaque. Apparently it is a natural carbide or oxide; it does not seem to be identical with eskolaite, as has been suggested.

Replacements are known only to a limited extent. Platinum replaces sperrylite and the rarer Pt minerals both in hypogene and supergene processes. In iron-rich samples an impression of "replacement" may be given by the solution and oxidation of the iron to form limonite.

Gel textures with an extraordinary variety of forms are exhibited by the platinum of platinum-quartz veins. There are, e.g., 1. long threads, somewhat branching, frequently with a hollow interior; 2. tiny cockades surrounding grains and circular thin-walled cells; 3. finely dispersed to reticulated, frequently concentric, intergrowths with hematite; 4. tufts having cauliflower or chrysanthemum-like forms; 5. forms similar to the "mineralized bacteria".

VIII. Diagnostic features. Ferrite, cohenite, awaruite, native palladium, iridium, and osmiridium are similar. Iron, cohenite, and awaruite are much more strongly attacked by acids. Palladium is slightly more yellow, and is more easily attacked by chromic acid – hydrochloric acid. Iridium reacts more slowly with cold aqua regia, while osmiridium remains practically unattacked. Both are appreciably harder. Iridosmium is tabular and anisotropic. Against many of the recently discovered platinoid minerals being mostly much rarer, the diagnostic features are not yet examined. Very probably wheathering products of them will deliver platinum of different textures.

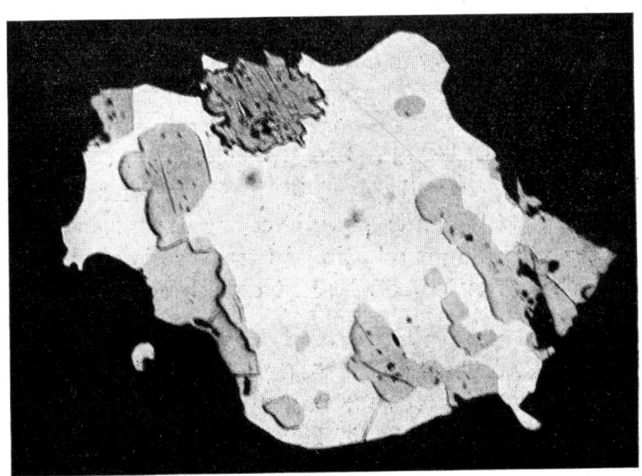

Fig. 308 130 ×, imm. RAMDOHR
Onverwacht Mine, Bushveld, Transvaal

Platinum with replacement remnants of *sperrylite* (light greyish-white), *laurite* (grey, extremely hard and poorly polished), and in addition, *stibiopalladinite* (color similar to sperrylite, but much softer). Grain from a very rich, fresh hortonolite dunite (contrasts greatly enhanced photographically!)

IX. Paragenetic position. Geochemically, platinum belongs chiefly to the siderophile and, to a lesser degree, to the chalcophile elements. Recently it has been shown that platinum can be transported under pneumatolytic–pegmatitic and even under hydrothermal conditions. Especially the latter one was recently emphasized by GENKIN (1965), STUMPFE (1962), BETECHTIN (1961). – The Pt content of meteorites is camouflaged in the iron. – In the crustal rocks, native platinum is found in the chromite schlieren of dunites, more rarely, in the dunites themselves, and in placers derived from them (figs. 301 a, 305). The latter, till about 1930, were the only deposits of commercial importance. The liquid magmatic sulphide segregations contain platinum in the native state, dissolved in certain Fe and Ni sulphides, and in sperrylite, braggite, cooperite and many others. In some cases, the native platinum only occurs as the weathering product of sulphides. The platinum of the hortonolite dunites in the Transvaal, which was probably precipitated at a relatively low temperature ($\sim 800°$), is unusual. The platinum from the now commercially unimportant platinum–quartz veins in the Waterberg district of the Transvaal, where aggregates of manifold forms (see above) are contained in a quartz vein of fairly low temperature of formation together with fine-scaly hematite, has surely been transported hydrothermally. – At the Boss Mine in the Yellow Pine district, Nevada, from which I unfortunately have not had a sample, the Pt (as well as Au and Pd) has been concentrated by descending solutions, together with oxidation minerals of bournonite.

X. Investigated occurrences. Placer Platinum from chromitites from Nishne Tagilsk, Kuschwa, Solowiew, and other localities in the Ural and Altai; Choco, Columbia. The Mooihoek, Onverwacht and Willemskopje (Driekop) pipes in the Lydenburg district, Tweefontein in the Potgietersrust district, and Waterval in the Rustenburg district, Transvaal. In unweathered nickeliferous pyrrhotites from Fäoy, western Norway, and the Mouat Mine, Montana; Sudbury, Canada, and others.

XI. Literature. Much work has been done with regard to the occurrence of platinum, and some evidence was established by early microscopic studies, e.g., by R. BECK (1901). Several works of L. DUPARC summarize the Ural platinum deposits, while the South African ones have been treated exhaustively by P. A. WAGNER (1929). The ore-microscopy of platinum is discussed especially by S. F. ZEMCZUZNY (1920), VAN DER VEEN (1925), BETECHTIN (1935), and a chapter of the WAGNER volume by H. SCHNEIDERHÖHN (1929). Very voluminous additional literature is given by VAN DER VEEN, ZEMCZUZNY, and N. VYSOTZKY (1925). Recent data brought BETECHTIN (1961) and GENKIN.

XII. Powder diagram (B. & TH.). (8) 2·235, (3) 1·936, (4) 1·369, (10) 1·167, (8) 0·889, (8) 0·866, (7) 0·791 Å. GENKIN gives for "tetragonal platinum" PtFe a_0 c_0 = 3·84, 3·78 Å: (5) 2·41, (10) 2·182, (5) 1·919, (5) 1·888, (5) 1·343, (6) 1·154, (7) 1·142 Å.

OSMIRIDIUM AND IRIDOSMIUM

Newjanskite and *Sysserskite**

I. General data. *Chem.* Iridium and osmium in isomorphic mixture, sometimes with considerable Ru; very Ir-rich members are lacking. *Cryst.* Iridosmium with prevalent Os dihexagonal bipyramidal. D_{6h}^4, hexagonal closest packing; $a_0 = 2{\cdot}72$, $c_0 = 4{\cdot}32$ Å, c/a = 1·568, Z = 2. # (0001) nearly perfect. H = 6–7, D = 19–21. Opaque. Osmiridium is cubic and similar in its properties to iridium.

Tin-white to steel-grey, highly metallic lustre. The lighter osmiridium was also known as "newjanskite", iridosmium is a trace more blue, and is essentially the socalled "sysserskite". The following data concern mostly iridosmium.

II. Polishing properties. The small crystals, which are recovered mostly as tinsel-like particles with the crude platinum, have to be polished as preparats of concentrates. Because of their excellent results the cleavage // (0001) of iridosmium is often visible. If, which will be a rare case, iridosmium is present in a polished-section of a compact ore, the polish remains bad; it has always a high relief. It is distinctly harder than pyrite.

III. Reflection behavior. The observations are a bit contradictory perhaps caused by a content of Ru, My preparats (I had many hundreds of grains), have an extreme high reflectivity (not essentially altered in oil) with a faint yellowish tint. In placer material sometimes at the boundary that tint is altered to bluish and the reflectivity lowered. Similar observations are made by KOEN (1966), but not agreeing fully with mine. VJALSOV's values

* These mineral names are given as *nevyanskite* and *sysertskite* in HEY (1955), chemical index of minerals.

are surely for ~ 8–10% too low. CISSARZ' values were green 67·5, orange 66, red 67% in air, 64, 64, 58 in oil.

IV. Etching. Both minerals are extremely indifferent against etching and not affected by cold Ag.R. A small attack is sometimes visible in form of a fine dentation. Some etching in polish-sections can be done with blowpipe by evaporation of some perosmic acid.

V. Physico-chemistry. For experimental work the extremely slow adjustment of equilibria in the solid state results in very high differences of the data of different workers and shows that a comparison of the — again differently interpreted — natural occurrences (RAUL & BUSS (1940) and of the author in this paper) brings many difficulties.

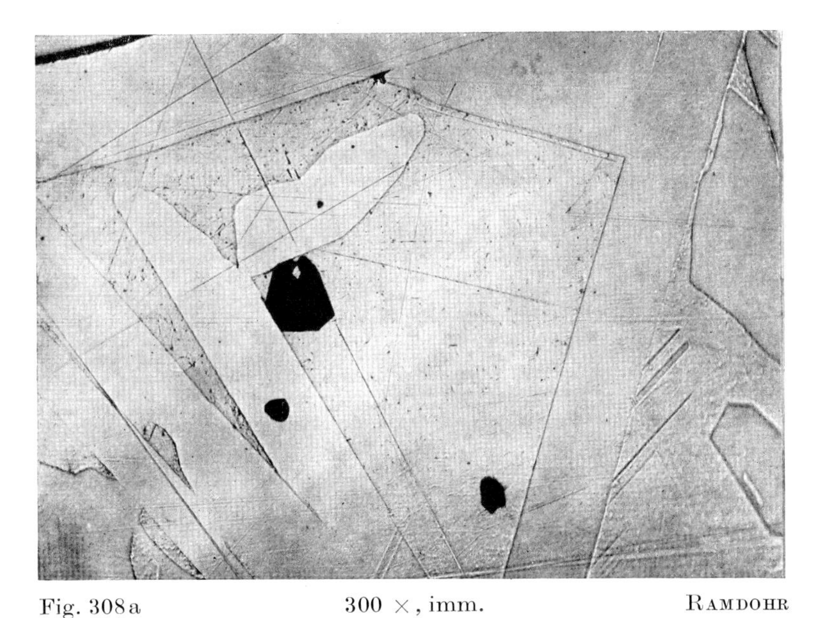

Fig. 308a 300 ×, imm. RAMDOHR
Choco, Rep. Columbia

Platinum, white, a bit rough, with a large quantity of iridosmium, coarse crystals, harder, but better polish

Perhaps some discrepances can be understood if we assume that the composition of the solution and high pressures affect the form of the diagrams. At pressures of only a few bars exists a diagram with two mixed crystals, Ir–Os and Os–Ir resp. and perhaps an eutecticum where the mixed-crystal regions on both sides first are rather broad and at room-temperature on the Irside goes still till about 39% Os and the Os-side till 21% Ir. The intermediate field should contain two phases. With this assumption many structures esp. on the Irside can be well explained: Ir shows often unmixing of thin lamellae of iridosmium // (111).

VI. Fabric. Basal tablets, rarely over 1 mm in width and of variable thickness which, accord. to ZEMCZUZNY, are frequently twinned (according to the illustrations they are similar to tridymite twins). Because of their hardness, they are idiomorphic even in placers. — During movement in placers the grains are frequently "cold worked", and exhibit beautiful fine-grained peripheral recrystallization. Quite common in the form of inclusions in platinum, either irregular or oriented // (111). Some samples are quite commonly encrusted by a very hard, somewhat lower reflecting, bluish white metal, which probably represents a leaching product. Sometimes it is undoubtedly older than platinum. For inclusions in platinum, see there.

VIII. Diagnostic features. Easily distinguished from native platinum by its anisotropism, appreciably greater hardness, and its resistance to aqua regia. Yellowish white and anisotropic in contrast to iridium. It is not likely to be mistaken for any other minerals because of its chemical inertness and its hardness.

The identification against some of the (much rarer) recently described platinoid minerals does not seem possible without microprobe.

IX. Paragenetic position. As a magmatic segregation together with platinum in the chromite schlieren of peridotites, at many places in the world. The amount is almost always exceptionally small, so that it is recovered only from placers, and there mostly as a by-product of platinum production. It also occurs in the Witwatersrand ores.

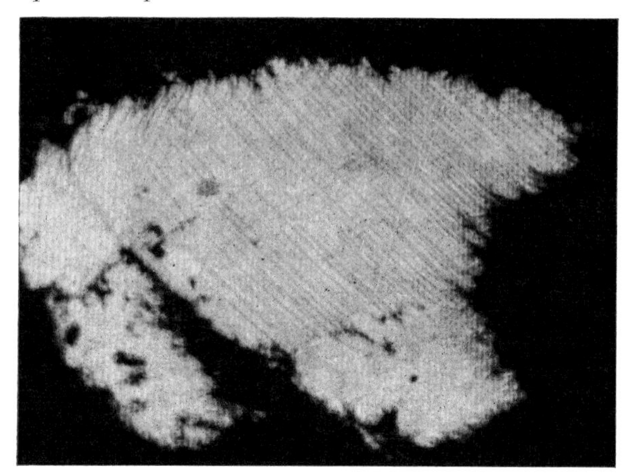

Fig. 308 b 250 ×, imm. RAMDOHR
"Mint" of Philadelphia,
from alluvial gold of California

Unmixing of abundant lamellae of *iridosmium*, a bit harder and darker, in three directions in a matrix of osmiridium (about the same quantity) a trace brighter

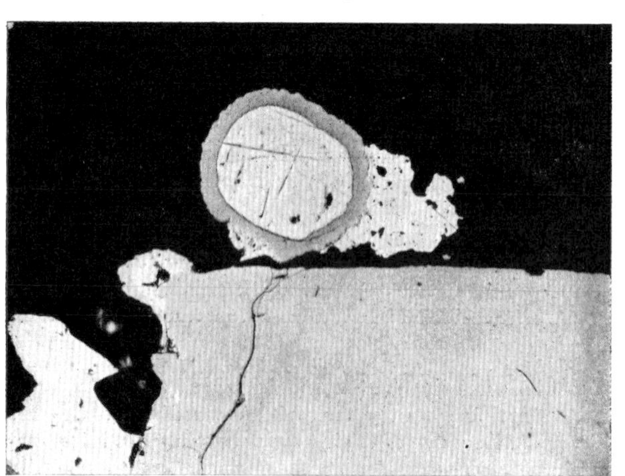

Fig. 308 c 250 ×, imm. RAMDOHR
Gvt. Areas Mine, Transvaal

A rounded grain of Osmiridium (white), scratched as a rolled pebble is overgrown in the deposit by pyrite. The large light grain (below) is also pyrite. Pure white with some scratches is gold, black gangue

X. Investigated occurrences. Toimanoff Mine at Kyschtym and another occurrence in the Ural; gold placers in the Sacramento Valley, California. Rustenburg and Witwatersrand, Transvaal.

XII. Powder diagram (B. & Th.). (10) 2·162, (4) 2·068, (5) 1·229, (7) 1·081, (5) 0·912, (6) 0·872, (8) 0·812 Å.

IRON

(α-Iron) + Taenite

(only terrestrial occurrences)

I. General data. *Chem.* Fe; sometimes with appreciable amounts of cohenite (= natural cementite) Fe_3C. In meteorites, which will not be discussed further, with a content of 6–15% Ni.

Cryst. Hexoctahedral; natural crystals are as good as unknown. *Lattice*: the iron stable at ordinary temperatures (customarily called α-Fe) has a body-centred cubic lattice with a = 2·87 Å; the modification stable above 906° (γ-Fe) has a face-centred cubic lattice with a = 3·63 Å. A content of about 12% Ni stabilizes the γ-Fe (taenite). Cleavage // (100) is occasionally visible. *Phys.* H = 4–5, depending on treatment history; greater hardness values are due to a Fe_3C content. D = 7·3–7·6. Magnetic. The refractive index n_{Na}, accord. to Drude, is 2·36. Kundt's values of 1·81 for red; 1·73, white; 1·52, blue are appreciably different.

Metallic lustre, white to grey, rapidly tarnishing to grey.

II. Polishing properties. Easily takes a good polish after appropriate pre-polishing treatment. Pre-grinding on dry emery laps, as practiced in metallography, achieves the desired purpose more quickly than the wet method used for ores, but is not satisfactory, in inves-

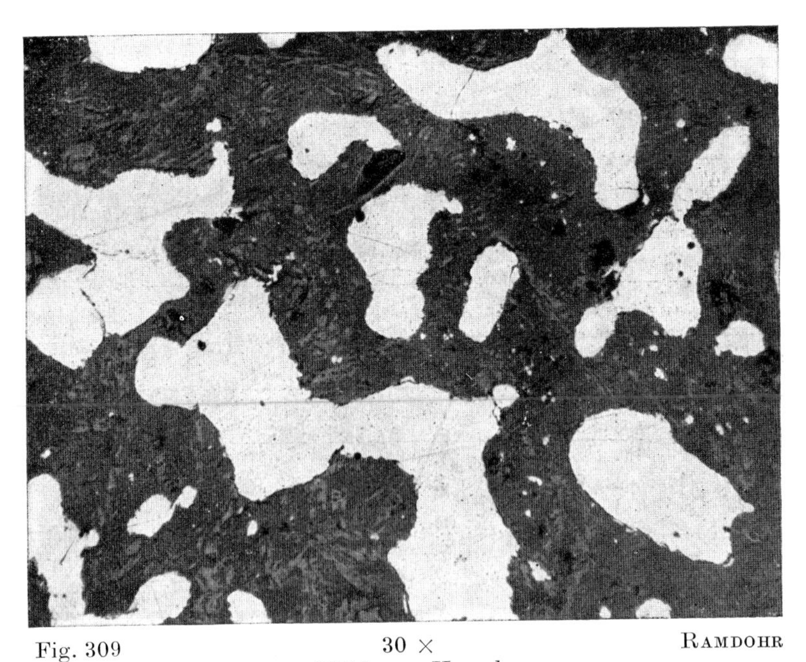

Fig. 309 30 × Ramdohr
Bühl near Kassel

Coarse spongy *iron* in glassy basalt. The iron was melted, or at least strongly sintered

tigations using polarized light. Cleavage not observed. The polishing hardness is lower than that of magnetite, and distinctly less than that of cohenite. This fact permits the recognition of cohenite (= cementite) in iron (= ferrite) without etching.

Talmage hardness: not given.

III. Reflection behavior. Color impression: Without a comparison object, pure white, with a high metallic reflectivity. Tarnishing of the section after polishing can be prevented for years by lacquering.

Isotropic, remains completely dark in all positions between + nic. In contrast, cohenite (cementite) is distinctly anisotropic and is noticeably pleochroic in oil.

	in air	in oil
general against silver against cohenite against platinum	very high, white light grey light bluish very similar; darker at close comparison	slightly changed analogous
Photometer ocular (for meteoritic iron from Atacama) green orange red	 64% 59% 58%	 50 % 51 % 47·5%
calculated reflectivity	Accord. to DRUDE's values for $\varkappa = 1{\cdot}36$ and $n = 2{\cdot}36$, $RV_{Na} = 56{\cdot}1\%$	

R. acc. BESMERTNAYA

nm	450	500	550	600	650
	60.5	62	59.5	58	60.2

	air				oil			
RV. nm	470	546	589	650	470	546	589	650
R_p	46.0	44·6	43·9	43·5	31·1	30·7	30·9	28·6
R_g	46·7	44·8	44·2	44·8	31·5	30·9	30·2	28·8
accord. CERVELLE & CAYE								

The values of BESMERTNAYA for R_g are ∅ 15% higher and ΔR is for a multiple (till 80 times) higher. I myself think that for fresh and well polished sections BESMERTNAYA is better!

MOSES gave R_g Na 52·8 a more reasonable value than C. & C.

IV. Etching (accord. to VAN DER VEEN). *Positive*: HCl, dilute H_2SO_4, dilute HNO_3, $CuSO_4$. *Negative*: concentrated HNO_3.

Cohenite is not attacket by dilute HCl, but by sodium picrate.

Textural etching: Picric acid in alcohol (1:100) etches iron (ferrite), but not cementite. Iodine in alcohol has similar effect. Awaruite remains unattacked in both instances. A kind of air etching, produced by the moisture in the air, attacks basaltic iron particularly rapidly.

V. Physico chemistry. The physical chemical relationships of iron and its alloys are very complicated, but have been studied thorougly. The iron-carbon and the iron-nickel diagrams are important for the natural occurrences. Carbon governs the appearance of the

Fe_3C — cohenite, whose presence depresses the melting point of the iron at its eutectic to 1140° with $5^1/_2\%$ C, so that melting can still occur in ordinary basalts; sometimes a natural sponge iron may have been formed without melting. The appreciable solubility of Fe_3C in iron decreases rapidly below 900°, which results in the segregation of "needle-shaped" (lamellar!) cohenite (Disko) (fig. 19). The maximum cementite content — which varies greatly even in samples from the same occurrence — attains 25%, and even appreciably more (Ovifak, Greenland); in general, however, natural iron is markedly lower in carbon.

Terrestrial native iron does not exhibit any of the complicated break-down phenomena observed in meteoritic iron. These will not be discussed here.

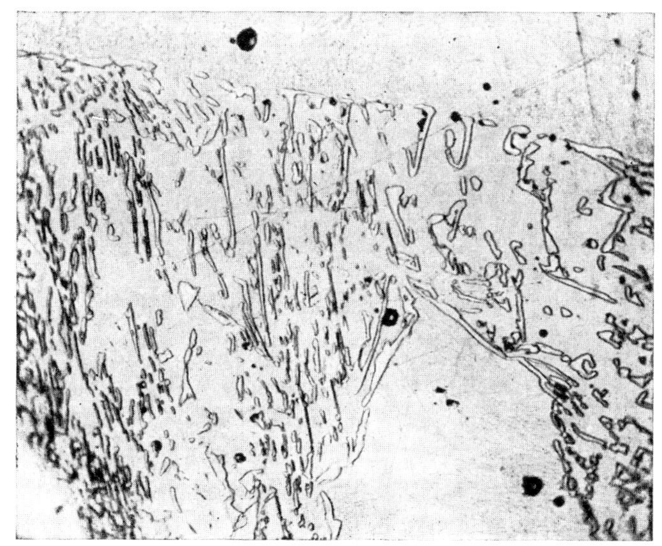

Fig. 310 400 × imm., Ramdohr-Ehrenberg
etched with iodine in alcohol
Bühl near Kassel

Iron is principal component, etched. Enclosed in it is unreacted *cohenite* (= cementite) in vermiform aggregates

VI. Fabric. *Internal properties of grains.* Twin lamellae are well known in commercial and meteoritic iron; they are produced by pressure // (211). Neither twinning, nor zoning, has yet been observed in terrestrial iron. The allotropic transformation γ-Fe → α-Fe is not recognizable under the ore-microscope.

Structure and texture. Droplets or spongy aggregates in basalt, which are usually very uniformly oriented. In the rare case of granular development, the grains are polygonal and quite intimately interlocked.

VII. Special fabrics. Most of the natural occurrences are of low-carbon iron, in which cementite (cohenite) is not prominent. A clear development of eutectic perlite textures is not common even in the terrestrial irons higher in carbon. The cementite is distributed as rather irregular festoon-like or speck-and drop-like particles (fig. 310). A high cementite content produces a quite different picture (figs. 19, 311). In weathered samples of Greenland iron, the tabular cohenite is easily discernible due to selective weathering; the same may occur in the Bühl iron. Frequently thin shells of ilmenite crystals have been deposited in a seam-like fashion around the iron droplets in Ovifak and Bühl. Mellemfjord (Disko) material exhibits beautiful textures of oriented intergrowths of iron and magnetite.

VIII. Diagnostic features. The metallic reflectivity, color, and isotropy restrict the possibility of error, particularly when the occurrence is taken into consideration. Platinum may be quite similar, and some samples may even be magnetic. The inertness of platinum to almost all reagents is diagnostic, whereas iron is easily attacked, and awaruite more slowly. "Ferrite" and "cementite" are readily distinguishable by etching with picric acid in alcohol; with a little practice cementite can be easily distinguished by its light yellow color, somewhat greater hardness, and its anisotropy.

IX. Paragenetic position. Apart from its occurrence in the inaccessible core of the earth, which is to some degree analogous to meteorites, and apart from a few doubtful or unusual reports, masses of terrestrial iron are known in three modes of occurrence:

1. In highly bituminous sediments and coals, e.g. at Chotzen, Č.S.S.R., very pure native iron has been found repeatedly, but its genetic interpretation is still open to question.
2. As a "natural smelter product" in places where eruptive rocks have broken through reducing sediments, particularly coal seams, thus giving rise to the reduction of iron ores in the basalt, or of inclusions. Of the very many occurrences of that type, those of Bühl near Kassel and of Ovifak in West Greenland have become famous. There it is present partly as natural "wrought iron" and partly as natural "steel". The Ni content is either low or absent. EITEL (1920) assumes from the Bühl iron that it originated from pyrite, through pyrrhotite and magnetite. Like VAN DER VEEN, the writer cannot agree with EITEL. — A variety of 2. is represented by the iron from Leigh Creek (BAKER), which originated naturally from a burning coal seam.
3. Some redox conditions during the serpentinisation of olivine may yield iron besides awaruite, wairauite, heazlewoodite, and magnetite at low temperatures (cf. p. 80).

X. Investigated occurrences. Bühl near Kassel; Ovifak, Greenland — abundant material from both occurrences; Muskox, Canada, and similar occurrences.

XI. Literature. There is a vast amount of metallographic literature, and much work has also been done on meteorites. Microscopic data on terrestrial iron are quite scarce. VAN DER VEEN (1925) has complied and enlarged them. LÖFQUIST & BENEDICKS (1941) have conducted detailed investigations into the Ovifak iron, while the Bühl iron has been studied by IRMER and EITEL (1920). Furthermore the Bühl iron was investigated by the writer (1952). The meteoritic irons are covered by an exhaustive, superbly illustrated monograph by PERRY (1944).

XII. Powder diagram (HULL). (10) 2·05, (5) 1·43, (6) 1·16 Å.

Remark: ε-Iron with hexagonal closest packing is a highest pressure modification of Fe ($>$ 170 kb). It seems to be occasionally stabilized in the condensation globules of the impact cloud in the surrounding of the Canyon Diablo crater and in many iron meteorites.

AWARUITE (JOSEPHINITE, SOUESITE, BOBROVSKITE)

I. General data. *Chem.* Ni_3Fe, at low temperature, as in natural occurrences, a wel defined intermetallic compound. *Cryst.* Cubic and, contrarily to low-iron, face centred $a_0 = 3{\cdot}60$ Å. No distinct crystals, tiny splinters common in many serpentines, larger grains sometimes as a placer component. — H = $5^1/_2$–6, D = 8 (actually, caused by silicate inclusions, essentially less, down to $5^1/_2$). — White metallic, not staining, ductile, magnetic.

II. Polishes exceptionally well without effort. Sections do not tarnish; no cleavage.

III. Reflection behavior. High, metallic reflectivity (estimated 75–80%), yet lower than that of the precious metals, and that of the red of copper, but brighter than iron. — Pure white, isotropic.

Very faint color differences; youngest external portions are a trace more bluish.

IV. Etching. No satisfying data!

V. Physicochemistry. The formation of Ni_3Fe at low temperature is not fully understood. Surely the metal content derives from material mobilized during serpentinisation. According BETECHTIN (1961) and others we mostly assume, that the reduction happened by hydrogen, freed during serpentinization from the fayalite content and water: e.g. 6 $(Mg_{1.5}Fe_{0.5})SiO_4 + 7\ H_2O = (3\ Mg_3Si_2O_5(OH)_4)) + Fe_3O_4 + H_2$.

VI. Fabric. Extraordinarily variable!

1. Granular masses of grains which form a loose tissue intergrown with abundant silicates. Evidently a granular aggregate of another mineral has been pseudomorphosed.
2. Narrow veins in magnetite and silicates.
3. Compact masses with inclusions of serpentine and chlorite, locally intergrown with native copper and with up to eight sulphide minerals.
4. Rhythmic botryoidal masses, sometimes alternating with native copper.
5. Pseudomorphs after magnetite and perhaps other minerals in various forms (see fig. 247).
6. Finely disseminated dust in silicates — and still others!

VII. Special fabric. Here too, a multitude of features could be mentioned.

Replacements. Awaruite, in various forms, replaces a variety of minerals, silicates, sulphides, but apparently especially magnetite, possibly trevorite. The latter is perhaps actually the point where the awaruite formation starts, and probably the illustrated pseudomorphism can be attributed to it. However, sulphide minerals high in nickel (heazlewoodite, niccolite) should not be overlooked as a source.

Recrystallization has been observed exceptionally well in a few instances in the outer portions which have been strongly affected by the fluvial transport.

VIII. Diagnostic features. The paragenesis is so unique, and the metallic properties are so typical, that apart from tiny grains of magnetic platinum, wairauite, and pure iron, an error in diagnosis is quite unlikely.

But for distinction of native iron in the same occurrence, the microsonde should always be used.

IX. Paragenetic position. Natural nickel-iron with its abundant local names is always found in regionally or dynamically metamorphosed serpentine masses, and is late there, having been formed during or after metamorphism, probably at about $< 200°$. That the temperature was quite low is shown by the lack of mix-crystal formation with copper, and even by the occasional rhythmic botryoidal intergrowth with the latter.

X. Investigated occurrences. Very abundant material from Josephine County in Oregon, typical material from Awarua in New Zealand and from British Columbia, and also some from a number of Alpine occurrences, particularly from Poschiavo.

XI. Literature. The awaruite from Poschiavo, which is associated with heazlewoodite, has been investigated by DE QUERVAIN (1945); the writer has thoroughly investigated the properties and genesis, particularly of samples from Oregon (RAMDOHR, 1950b). Soviet occurrences have been treated by BETECHTIN et al. (1958). More recently CHAMBERLAIN et al. (1965), NICKEL (1959), and KANEHIRA et al. (1964) gave new occurrences.

XII. Powder diagram. (10) 2·044, (6) 1·775, (4) 1·255, (5) 1·070 Å.

Remark

Relict structures after ε-iron are present in many ironmeteorites.

WAIRAUITE

(Data mostly acc. to CHALLIS & LONG (1964) and CHAMBERLAIN et al. (1965).

I. *Chem.* CoFe with traces of Ni. *Cryst.* cubic, $a_0 = 2{\cdot}856$ Å (α-Fe resp. CsCl-structure). Strong tendency to form idiomorphic crystals (100) and (111). H $\sim 4^1/_2$, VH ~ 250, i.e. $>$ Fe, $<$ awaruite. D = 8·2, strongly magnetic. — Good polish.

III. Reflectivity. Wairauite has been discovered accidentally with the microprobe during an examination of awaruite. Once recognized it shows against the latter a bit lower reflectivity (green 54% in air). Isotropic.

V. Physical chemistry. Wairauite corresponds with the artificial intermetallic compound CoFe in the system Co–Fe.

VI. Structures. Very tiny grains (1–10 μ), mostly overlooked, in the occurrences of awaruite (see pg. 356). The grains are mostly idiomorphic, octahedrons or cubes. Sometimes it is, as in the original observation, included in awaruite.

VIII. Paragenetic relations. Wairauite belongs to the association of serpentinization (see pg. 79), which contains e.g. awaruite, iron, heazlewoodite etc., mostly connected with plentiful secondary magnetite The formation probably is caused by hydrogen in state nascendi with the traces of Co in the former olivine.

For wairauite and the normal iron, not rare in this association, should be examined (microprobe) but awaruite is much more common than both.

X. Localities. The author saw a wairauite in ores from Muskox, Northern Canada; Selva near Poschiavo, Switzerland; Ôko near Koti, Japan. Elsewhere it may be overlooked.

XII. Powder diagram. No data!

COHENITE* – CEMENTITE OF METALLURGY

I. General data. *Chem.* Fe_3C with some % Ni and Co instead of Fe. *Cryst.* o'rhombic, D_{2h}^{16}, a_0 b_0 c_0 = 4·53, 5·08, 6·75 Å, Z = 4 in the high form. All cohenites seem to be optically twinned in very fine lamellae actually not allowed in the o'rhombic system (s. below). Imbedded in natural iron, cohenite is partly coarsely crystalline with tendency to be euhedral, partly myrmekitic, partly lamellar (wrongly "needle like"). Cleavage // (100), (010), (001), perhaps a prism too. — H = $6^1/_2$, that is much harder than Fe, D = 7·68. — Tin white, a tint yellowish against Fe.

II. Polish, more slowly than Fe, but easier to get without scratches. The higher hardness than Fe is always visible. The distinction of the different cleavages in polished sections is impossible, but (100) is the most distinct one.

III. Reflection behavior. First impression is highly metallic white. Only in comparison with α-Fe a faint yellowish tint becomes visible, much less than schreibersite, but higher than taenite. Quantitative data are not available (estimated 58% for orange).

Bireflection is visible in oil and at boundaries only; the anisotropism is weak, visible without difficulty in oil and at grain boundaries.

(acc. Cervelle etal.)

Cohenite nm	400	440	480	520	560	600	640	680	700
R_p	49·8	51·2	52·3	53·3	54·5	55·7	56·8	57·9	58·4
R_g	47·7	51·2	53·6	55·3	56·7	57·9	59·1	60·5	61·1

IV. Etching, with cuprous ammonium chloride, which does not attack iron, or vice versa, with iodine in alcohol. Careful observation makes etching superfluous.

V. Physico-chemical. compare iron (p. 353). The author (1964) observed a peculiar twinning which he logically must ascribe to a lower symmetry modification. The twinning can be observed in all cohenites natural and artificial ones. F. Heide (1966) could give an astonishing explanation: this is not a crystallographical twinning but

* Cohenite (Weinschenk 1889) has priority over cementite (H. M. Howe 1890).

the boundary between magnetic domains. It can be made excellently visible by floating of finest powder of magnetite in immersion oil. The magnetite migrates at once to the "twin" boundaries. As far as I know, here for the first time magnetic properties could be observed by optical anisotropism. I admit that already PEPPERHOFF (1963) could show the same effect with the help of reinforcing foils. — It remains difficult to explain that sometimes recrystallisation in naturally shocked material starts distinctly at those boundaries.

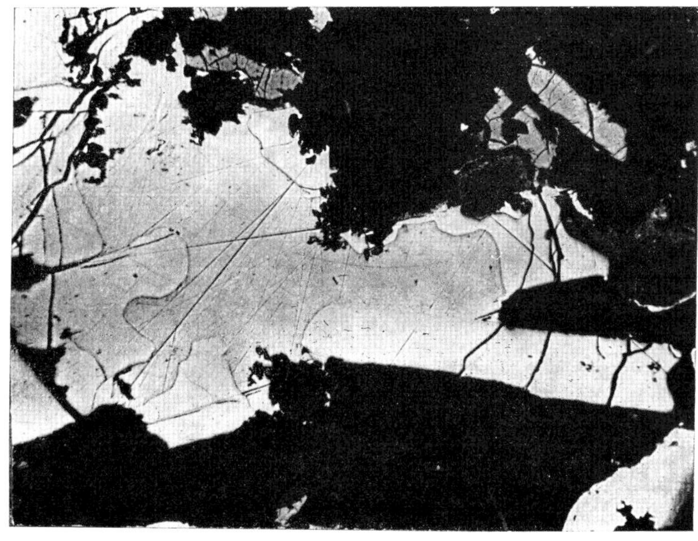

Fig. 311 170 × RAMDOHR
Ovifak, Greenland

Cohenite (cementite) tabular, idiomorphic crystals beside *ferrite* (centre, softer, somewhat more scratched). Silicate minerals are dark grey

VI. Fabric. The distribution of cohenite in iron (terrestrial, cosmic, technical) can be observed by using hardness and anisotropism, for routine work by contrast etching.

The most common form, the myrmekitic "perlite", can be seen fig. 310. Often it becomes easily visible by quicker weathering of iron (e.g. Bühl and Ovifak). A second form, the so-called, "needle cementite" is characteristic by its fine distribution of thin platelets (not "needles"!) of cohenite in iron. The formation can be explained by unmixing. Terrestrial occurrence is noted from Ovifak only.

A third, rare type can be observed where cohenite is abundant. Rounded grains of iron are enclosed in a thick crust of cohenite which, in its turn, can form well developed crystals up to 2 mm at the boundary to glass (fig. 19, 311). — Cohenite is extremely rare in stone meteorites, in irons it is often abundant in the surrounding of graphite and troilite nodules. Here it forms partly euhedral grains in the kamacite.

The magnetic areas described below can look exactly like twinning lamellae but they fork into a very characteristic form: a sequence is produced thin a, thick b, thin a, thin b, thick a, thin b, thin a, thick b . . .

VII. Special fabrics. The breakdown to form iron and graphite, occurring not too rarely in artificial cementite, is typically observed once in cohenite from Ovifak. Weathering of cohenite in iron meteorites may have a similar effect.

XI. Literature. Cohenite — cementite and its relations are described in much detail by metallurgists and "meteoritologists" hundreds of times.

XII. Powder diagram. (HANAWALT) (8) 2·38, (10) 2·10, (9) 2·02, (9) 1·97 Å.

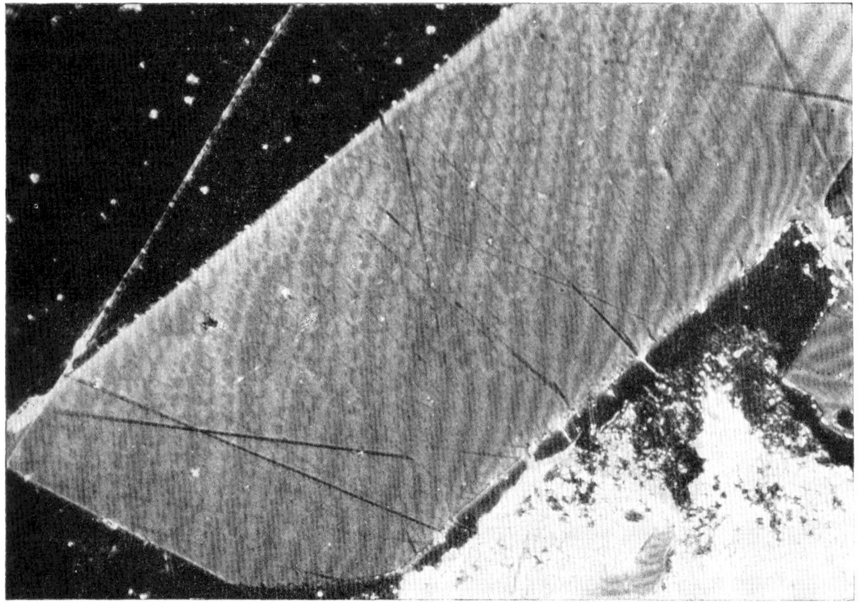

Fig. 311a 250 ×, imm. Nic. + MEDENBACH
Ovifak, W. Greenland

A distinct crystal of *cohenite* shows at nearly crossed polarizers without any further treatment, the so-called *Bitter*-lamellae, here characteristically each time a broader stripe rimmed by two fine stripes who in themselves look granulated. The stripes are equally thick

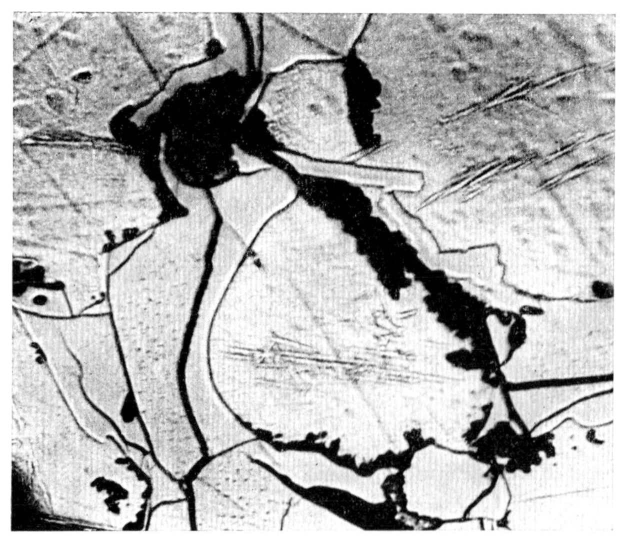

Fig. 311b 300 ×, imm. MEDENBACH
Disko, Greenland

An "iron" very rich in carbon, which consists of nearly equal parts of "*ferrite*" and "*cohenite*" (cementite). *Cohenite* encloses rounded grains of ferrite, which contains fine lamellae of cohenite ("Needle"-cementite). In the middle (above) an idiomorphic crystal of cohenite

SCHREIBERSITE AND RHABDITE

I. General data. *Chem.* (Fe, Ni, Co)$_3$P, where acc. to MERRILL and GALOPIN iron may go till Fe_6P. — *Cryst.* Tetragonal, perhaps S_4^2, $a_0 = 9{\cdot}03$, $c_0 = 4{\cdot}43$ Å, Z = 4. Schreibersite — much more common — forms xenomorphic, partly tabular masses and grains, rhabdite idiomorphic prisms. EL GORESY discovered in meteoritic schreibersite a spindle-formed twin lamellation, perhaps analogous to the cohenite "twinning". Schreibersite and rhabdite are very common in iron-meteorites, schreibersite widespread but rare in stone meteorites. The terrestrial iron of Ovifak contains a little of the latter. H $\sim$ 6 (lower than cohenite), D = 7·1. Color compared with iron is distinctly brownish, reflectivity lower. The mentioned "golden yellowish tarnish" is not observed by the author. Very brittle. — *Polish* good, H > Fe, < Fe_3C. Very often conchoidal cracks.

III. Reflection behavior. Very high reflectivity, few % lower than Fe; in oil, has a strong brownish-pink tint. Schreibersite, in direct comparison with iron, is very little pinkish and (in oil) a trace darker. Reflection pleochroism is sometimes noticeable, anisotropic effects then distinct. Schreibersites of many meteorites are actually isotropic.

Schreibersite (acc. CERVELLE et al.)

R_O nm	45·0	46·8	48·3	49·8	51·2	52·6	54·05	55·4	56·1
R_E	47·0	48·4	49·8	51·4	52·8	54·3	55.8	57·2	57·9

IV. Etching. Etching must serve primarily to facilitate the distinction of rhabdite and schreibersite from nickel-iron (actually for well trained eyes no problem!) and has therefore been investigated from a totally different viewpoint than in the case of other ore minerals. 50% HCl is negative (positive for Fe), acetic acid (Fe^+); copper ammonium chloride (Fe^+) and acid copper sulphate quickly deposits copper on schreibersite and rhabdite, much more rapidly than on Fe. Dilute HNO_3 etches immediately (more rapidly than on Fe); a brief etch develops the texture well.

VI. Fabric. "Rhabdite" forms sharply idiomorphic short prisms to long needles. The forms observed in section vary accordingly. In "schreibersite" the forms are more irregular, frequently tablets // (001), sometimes strongly corroded by iron, sometimes much coarser. Especially massive schreibersite is enormously cataclastic forming "in situ breccias".

Both rhabdite and schreibersite are occasionally oriented in Fe, but the law does not yet appear to be known.

Rhabdite commonly forms clumps with an approximately uniform distribution in iron, while schreibersite varies widely.

Among themselves, rhabdite needles are arranged in swarms which are roughly perpendicular to each other.

VIII. Diagnostic features. The anisotropism, hardness $\gtrsim$ Fe, < Fe_3C, and the light brown color tone permit rhabdite and schreibersite to be easily recognized in Fe, and to be distinguished from cohenite and troilite.

IX. Paragenetic position. Schreibersite and rhabdite appear to be restricted predominantly to meteorites, particularly holosiderites. — The above description, taken principally from R. GALOPIN, is based on the holosiderite from Union, Chile. — Since LÖFQUIST & BENEDICKS (1941) have demonstrated the occurrence of schreibersite in native iron from Ovifak, it must be recorded here.

XI. Literature. The abundant references in the metallurgical and meteoritic literature are not listed here, and attention is drawn to only two works, those of MERRILL and GALOPIN. Recently EL GORESY and the author did additional work.

XII. Powder diagram (CHUCHROV). (10) 2·18, (8) 2·10, (8) 2·01, (10) 1·959, (9) 1·755, (9) 1·268, (9) 1·194 Å.

NATIVE NICKEL

Native metallic nickel is formed in nature in minute quantities only but in many different occasions. Heazlewoodite from New Caledonie contains it in very pure form, probably formed by selective oxidation, in minute crystals (100) and (111) and spiderlike on grain boundaries. In the Henbury Impact-glass droplets of "iron" have been described, actually containing more than 92% Ni. Cosmic globules of magnetites from deep-sea-dredgings and in manganese nodules there contain often "iron" cores, actually enriched in Ni to 80%. They represent burning rests of iron meteorites. Finally moon rocks contain "iron" from 0%–95% Ni.

Ore-microscopically it is not much known. Nickel is a trace more yellowish-wite than iron. The reflectivity may be a bit higher than that of iron (for nm 548 52.8 against 50% in iron — both values for pure metal). Nickel will be discovered only accidentally or in systematic examination with the microprobe.

It is found as an alteration product of awaruite from Jerry River, New Zealand.

POTARITE

I. General data. *Chem.* PdHg. Inclusions partly account for deviations to $\sim Pd_3Hg_2$. *Cryst.* Cubic; an accessory component which is sometime present (perhaps Pd_2Hg) is anisotropic. $a_0 = 5{\cdot}21$ Å, $Z = 4$. T^4 (FeSi structure). BERRY & THOMPSON give D_{4h}, $a_0 = 3{\cdot}02$, $c_0 = 3{\cdot}71$ Å, $Z = 1$. — In any case — my specimen was isotropic! $D_{X\text{-ray}} = 14{\cdot}33$, D_{exp} variable — 14·8.

II. Polishing properties. Because of its low degree of hardness, it does not polish very well, at least in granular preparates. The polishing hardness is low. The accessory component s distinctly softer. Cleavage is not discernable. Talmage hardness: C +.

III. Reflection behavior. *Color and anisotropic effects.* The reflectivity is very high and is similar to native palladium. The included component has an appreciably lower reflectivity. The color impression is pure white; the inclusions are light grey in contrast.

Reflection pleochroism cannot be discerned; with + nic., the principal component appears to be isotropic, while the inclusions are distinctly anisotropic without pronounced color effects.

IV. Etching. (THOMPSON in PEACOCK). *Positive*: HNO_3 tarnishes slowly; $FeCl_3$, etches rapidly. *Negative*: HCl, KCN, KOH, $HgCl_2$.

VI. Fabric. The principal component is made up of small, loosely packed grains whose interstices are filled with the included component. The included component is frequently dissolved peripherally, and adjacent to these pores, the principal component is also corroded.

IX.–XI. Paragenetic position and literature. The few grains, which have been found up to now, form small nuggets in the Potaro River in Guyana, from where a small sample was available which was investigated by CISSARZ (1930). PEACOCK was able to supplement the data appreciably, and to determine the structure.

XII. Powder diagram (ASTM). (10) 2·31, (7) 2·11, (6) 1·50, (7) 1·391, (10) 1·265 Å.

TIN

I. General data. Native tin is reported to have been observed occasionally as a placer mineral; this is doubtful, and in view of a possible confusion with tin–palladium and platinum compounds, it cannot be considered authenticated. Only very recently it appears to have been substantiated by an observation by J. F. B. SILMAN (1954), who found it with pitchblende, hematite, calcite, and many other minerals in the "Nesbitt La-Bine Uranium Mines", Beaverlodge, Saskatchewan. All the data are taken from his publication.

II. Polishes scratch-free with great difficulty, and is about as soft as argentite.

III. Reflection behavior. Few data. White, highly metallic. Exhibits moderate anisotropic effects between + nic.

IV. Etching. *Positive*: HNO_3 effervesces, etches, and forms a brownish black tarnish. Vapors form a coating. HCl slowly causes a dark tarnish. $FeCl_3$, rapidly light brown; $HgCl_2$, rapidly dark brown. *Negative*: KCN, KOH.

VI. Fabric. Grains up to $1^1/_2$ mm in size extending in tongue-like fashion into fractures in the accompanying minerals; completely xenomorphic.

IX. Paragenetic position. In the mine noted above, the tin was found in three veins in the following paragenesis and age relationships: Early calcite and quartz, hematite, pitchblende and pyrite; later calcite and hematite; chalcopyrite, bornite, chalcocite, sphalerite, galena; native tin; finally calcite.

Very recently native tin in larger quantities was described from an obviously high-temperature pegmatitic vein from Oued Berkou in Algier besides columbite, xenotim and triphylite.

XII. Powder diagram (HANAWALT). (10) 2·91, (9) 2·71, (10) 2·01 Å.

Recently a large number of compounds of the type platinoid (± Cu + Ni)–tin has been described, or in other words stannides. The similar lead-rich compounds come essentially from the Siberian deposits Norilsk and Talnakh, the tin-rich from the Bushveld and Insizwa, but there is no doubt that Pb and Sn can vary in all relations. All these minerals are rarities. CABRI (1972) mentions about 15 of them, which are not named till now, of them some may not be properly independent, because obviously Sn and Pb can partly be replaced by Sb, Bi, and perhaps Te. Since the platinoids can be furthermore replaced by iron, nickel and Co the complication becomes still higher. Only a few members will be treated here. The reflectivity values in such mixed-crystals series are of little importance.

NIGGLIITE

I. *Chem.* PtTe, perhaps a mixed crystal with isostructural PtSn. *Cryst.* Hexagonal, a_0 c_0 = 4·11, 5·45 Å, Z = 2. – Bright white, variegated shiny. – *Polish* good; soft but brittle.

III. Reflectivity. Very high; without nic. "white", actually by superposition of O and E. *Bireflection* very high, shiny creamish white (O), blue white and a bit darker (E). Anisotropism extreme.

VI. Textures. Originally known in tiny grains from the Insizwa deposit (and perhaps Rustenburg), the original specimen* from that locality did show the mineral in the original association. It shows polygonal grains, partly twinned and partly with pressure effects.

IX. Association, Niggliite obviously is connected with the youngest parts of the deposit Waterfall Gorge, Insizwa, S.Africa, which contain a very complicated association (~ 20 components).

XII. Powder diagram (SCHOLTZ). (5) 3·56, (5) 2·98, (10) 2·16, (5) 1·780, (10) 1·490, (10) 1·206 Å, which fits for PtTe and PtSn likewise.

SVIAGINTSEVITE

I. *Chem.* $(Pd, Pt)_3(Pb, Sn)$ with approx. 55 Pd, 7.5 Pt, 25 Pb, 12 Sn, but obviously varying considerably, Cryst. cubic, a_0 = 4·02 Å. Grains partly idiomorphic. – Polish difficult to get scratch free.

* Kindly dedicated to the author by Prof. SCHOLTZ.

III. Reflection behavior. Pure white, against electrum grey to brownish. — isotropic.

IV. Etching (C. a. T.), positive: HNO_3, dull brown, HCl and KCN very weakly brownish, $FeCl_3$ quick brown-black. Negative: $HgCl_2$, KOH.

VI. Fabric. Rather large areas with partly idiomorphic single-grains enclosed in pentandite. Often thin crusts of ferroplatinum.

R. (accord. GENKIN)

nm	470	520	550	580	600	650	700
	63·7	65·2	66·8	67·6	68·6	69·4	67·6

i.e. distinctly maximum in orange red, effecting a pink tint alike stibiopalladinite.
Accord. CABRI & TRAIL this tint depends strongly from small differences in chemism.

IX. Paragenetic position. Besides many other platinoid-minerals in the relatively Pb-rich nickel–copper-paragenesis of Norilsk, Siberia.

XII. Powder Diagram (GENKIN). (10) 2·315, (8) 2·011, (7) 1·418, (9) 1·20 Å.

Remark. GENKIN (1968) describes briefly $(Pd, Au)_3Pb$, cubic, with a rather similar powder diagram like Sviagintsevite. Similarly mentioned RAZIN & BIKOV: $(Pt, Pd)_3Sn_2$ with powder diagram (10) 2·29, (8) 1·984, (7) 1·404, (6) 1·196 Å. A further mineral with the same P.D. seems to be $(Pd, Pt)_7(Sn, Pb)_2$.

Paolovite. *Chem.* Pd_2Sn.
Reflectivity. Lilac-rose, bireflection distinct.
The optical character changes in visible light. Talnakh, Siberia

PLUMBOPALLADINITE

I. *Chem.* Pd_3Pb_2; *cryst.* hexagonal, NiAs-Type, $a_0\,c_0$ = 4·470, 5·719 Å.

III. Reflection behavior. White with pink tint.

Photocell (in air)	nm	440	480	580	660	740
	R_g	50·3	52·6	58·6	62·6	68·7
	R_p	46·2	49·5	57·0	61·7	65·4

Bireflection in oil is distinct.

IX. The original deposit Talnakh near Norilsk.

XI. All data accord. GENKIN et al. 1970.

POLARITE

I. *Chem.* Pd(Pb, Bi), similar synthetic. PdBi. *Cryst.* O'rhombic, $a_0b_0c_0$ = 7·19, 8·69, 10·68 Å. H ~ chalcopyrite.

III. In the section white to yellowish, weak anisotropism.

Photocell (in air)	nm 460	540	580	660
R	56·8	59·2	59·6	61·2

IX. From the Cu–Ni deposit Talnakh besides $CuFeS_2$, Cubanite, Talnakhite etc. especially in lumps rich in platinoid-minerals, such as Pd_3Pb, stannopalladinite, native Pt.

XII. (accord. GENKIN). (10) 2·65, (5) 2·25, (9) 2·16, (5) 1·638 Å.

STANNOPALLADINITE

I. *Chem.* Pd_3Sn_2, but by no means stoichiometric and with a considerable amount of Pt and Cu. *Cryst.* hexagonal, a_0 c_0 = 4·40, 5·66 Å. – Polish good, H ~ ferroplatinum.

III. Reflection behavior. Pink-white metallic. Accord. GENKIN (1968):

nm	460	540	580	660	
R_g	48·5	54·0	55·5	60·0	in oil no data
R_p	46·2	53·0	54·0	57·0	

VI. Fabric. Single crystals, almost myrmekitic droplets in ferroplatinum, both in chalcopyrite of the Norilsk-deposit; ferroplatinum is here lighter, harder, and whiter. Besides that in single grains in alluvial Ural-platinum. A component of the Bushveld concentrates (Driekop, Rustenburg) is at least very similar.

X. All data accor. GENKIN (1968).

XII. Powder diagram (GENKIN). (10) 2·30, (10) 2·16, (3) 2·01, (2) 1·440, (2) 1·222 Å.

ZINC

According to the writer's opinion, zinc was not known as a mineral, and its occurrence as such is improbable. Even the very positive data of BEDER and OLSACHER appear to apply to an artificial product. – Only recently it is mentioned by BOYLE as formed from exceptionally cold brines.

VI. Greyish metallic platelets on an in partly oxidized siderite and powdery limonite with Mn-oxides in an ice lense. Partly in finely clustered form. Often connected with native silver, oxidation products like anglesite and cerussite, and remnants of sphalerite.

IX. To form native zinc in nature, conditions must be very peculiar. Very probably in Keno was a coworking of reduction by organic compounds, weakly alkaline solutions and very low temperature.

R. W. BOYLE, Native Zinc at Keno Hill (Yukon). Canad. Mineral. *6*. 692–694, 1961.

NATIVE ARSENIC

I. General data. *Chem.* As; contains quite commonly Sb, which was originally contained in isomorphous admixture or in a mixed gel, or formed a now-decomposed compound with it. *Cryst.* Ditrigonal–scalenohedral. c/a = 1·401*; thick, botryoidal crusts, or rarely agglomerations of crystals. *Lattice*: D_{3d}; a type structure, similar to a deformed halite lattice. a_{rh} = 4·15 Å, α = 54° 16′. Perfect (0001) cleavage, $(01\bar{1}2)$ indistinct. H = $3^1/_2$, D = 5·7. In fresh fracture it is metallic tin-white, quickly darkening to dark lead-grey even black.

* The axial ratio and indexing of As, Sb, Bi are based on the pseudo-cubic cell!

II. Polishing properties. Polishes very well, and is fairly easily made scratch-free, much more readily than the true metals. — After a few days the sections tarnish to a dull brownish grey. The tarnish cannot be rubbed off, but can quickly be removed by a wet polish.

The cleavage // (0001) is also frequently visible in section. Slightly harder than dyscrasite, distinctly harder than antimony, much harder than bismuth.

Talmage hardness: C.

Fig. 312 300 ×, imm., + nic. Schneeberg, Saxony RAMDOHR

Native *arsenic* in divergent sub-parallel development with numerous twin lamellae

III. Reflection behavior. *Color and reflectivity impression.* White; a high, but not distinctly metallic reflectivity.

Anisotropic effects under + nic. Very distinct; strong in immersion, but without striking colors; stronger than those of antimony. In diagonal position (in oil), yellowish-brown; light grey to yellowish grey. Between + nic., the scratches below the smeared polished surface are easily visible.

IV. Etching (accord. to MURDOCH, DAVY & FARNHAM, VAN DER VEEN). *Positive*: HNO_3 colors immediately (brown to black), effervesces slightly. $FeCl_3$ colors brown immediately. *Negative*: KCN, HCl, KOH.

Structure etching. Air etching for only 2 to 3 days usually develops the structure well. ZÜCKERT recommends a short etch using $KMnO_4 + H_2SO_4$; ORCEL, a 2-minute immersion in 3% H_2O_2 solution; VAN DER VEEN, $K_3Fe(CN)_6$ for 10–20′. — All yield very good results.

VI. Fabric. *Internal properties of grains.* Twinning on $(01\bar{1}2)$ is very abundant in exceptionally delicate lamellae. Doubtless many cases represent glide twinning, since lamellae sometimes form from preparing the sample with the hammer. The twinning

Fig. 313 225 ×, imm. RAMDOHR-EHRENBERG
Gabe Gottes Mine near Ste Marie aux Mines, Alsace

Native *arsenic*, granular texture with pressure-twin lamellation; bottom left, fracture pits. Otherwise invisible scratches become visible in a number of places between + nic.

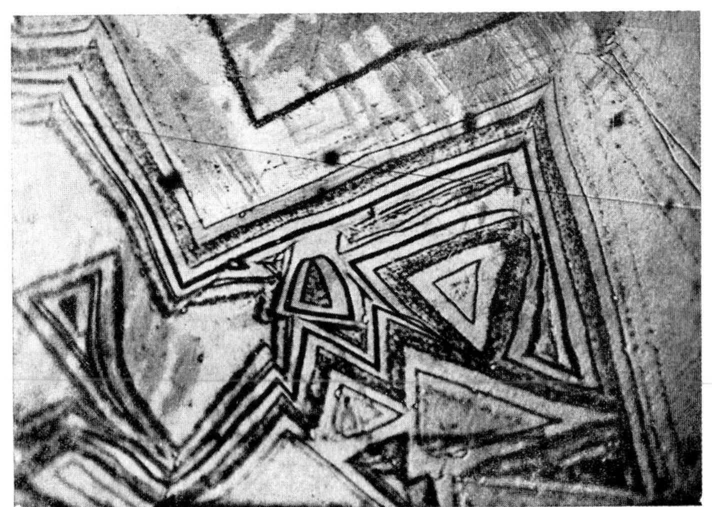

Fig. 314 500 ×, imm. RAMDOHR
Kowary (Schmiedeberg), Silesia

Native arsenic with exceptional zoning which has been developed by air etching within two days, and twin lamellae which are recognizable by their bireflection may be quite different in adjoining grains (fig. 313). Tridymite-like twins occur in arsenic which has been formed from the decomposition of geokronite.

	in air	in oil
general	high, white	greatly reduced
against galena	trace lighter, more creamy	almost exactly = galena
against smaltite	darker, slightly grey	distinctly darker
against antimony	darker, toward grey	distinctly darker
against bismuth	appreciably darker, toward grey	much darker
against silver	much darker, less "yellow-white"	

	air				oil			
RV. nm	470	546	589	650	470	546	589	650
R_p	46·0	44·6	43·9	43·5	31·1	30·7	30·9	28·6
R_g	46·7	44·8	44·2	43·8	31·5	30·9	30·2	28·8

accord. CERVELLE & CAYE — surely too low.

The values of BESMERTNAYA for R_g are ∅ 15% higher and ΔR is for a multiple (till 80 times) higher. I myself think that for fresh and well polished sections BESMERTNAYA is better!

MOSES gave R_gNa 52·8 a more reasonable value than C. & C.

Bireflection at grain boundaries distinct

(in air)	(in oil)
O lighter, white	O grey white to yellow
E darker, grey white	E white grey to bluish

Zonal structure is rare, but occasionally striking (fig. 314); on the other hand, rhythmic structures are common (see below), sometimes in layers alternating with arsenides (fig. 103).

Deformation evidently produces a (0001) translation; crumpling and flexure are found in (0001) and in the twin lamellae.

Strukture and texture. Because of its late formation at low temperature, arsenic very often shows a striking concentric shell structure ["scherbenkobalt"], and only very rarely distinct crystals. In the shell structure, which may be pronouncedly rhythmic owing to the variable content of Sb and others, and which becomes very easily visible by etching and by its divisibility, grains can extend across many shells (fig. 312). Individuals are very often arranged ⊥ to the surface, in the form of sub-parallel sheaves or ice ferns. Everything points to a crystallized gel. Compact, not-botryoidal masses form irregular patterns of fine polygonal grains (maximum grain size 1 mm; fig. 313). In the botryoidal varieties the outermost shells are frequently scarred, suggesting crystal development. In some localities (Miedzianka [Kupferberg] in Silesia), arsenic forms fine-grained massive aggregates, with interlocking, irregular, frequently sieve-like, intergrowths and with very pronounced crumpling and deformation of the individual grains (mechanically moved).

VII. Special fabrics. Intergrown with stibarsen, oriented in various ways. Drop-like inclusions, fine vermiform veining (or rather, "like the borings of bark beetles"), and striking intersertal intergrowths. All these are called "allemontite", although (see under antimony), this name should be reserved for the latter. The others have developed from unmixed gels. Replacements are relatively rare. R. ZÜCKERT (1925) describes arsenic as a replacer of bismuth. The parageneses of Andreasberg are exceptionally variable; here arsenic is finely intergrown with very many other ores, without the age relations being easily recognizable. Inclusions of dyscrasite with enveloping concentric layers of "scherbenkobalt" are common; these intergrowths have frequently given rise to the formation of younger minerals (e.g., of ruby silver, sternbergite, and many others) which tend to retain the original textures. Idiomorphic grains of arsenic, which tarnish exceptionally quick in polished section, are found as the decomposition product of geokronite in a number of deposits.

VIII. Diagnostic features. Rapid tarnishing in air within a few days is quite characteristic, as are the usually concentric structures. Antimony is brighter, less anisotropic, and does not exhibit air etching. Bismuth is much brighter and softer. Also, arsenic is easily etched by H_2O_2.

IX. Paragenetic position. Native arsenic is found as a very late mineral in veins of the "noble calcite formation" and in the Co–Ni–Ag–U veins, both of which belong to the intrusive magmatic sequence, as well as in similar formations originating from the sub-volcanic sequence, and then usually at shallow depth. It is also mentioned in a variety of other deposits containing abundant arsenic minerals. Its occurrence is often interpreted as the result of gossan formation and cementation processes, which is true in some cases. However, if VAN DER VEEN (1925) and DAVY (1920) consider this interpretation as the only possible one, then they are surely wrong. Although arsenic is frequently late, it is certainly "ascendent" as, e.g., in Andreasberg where, at the greatest depths, ruby silver crystals were found frequently in arsenic which had been leached. Also Kowary (Schmiedeberg), e.g., has primary arsenic of the same age as safflorite. Associeted minerals include quite frequently the noble silver ores (ruby silver), dyscrasite, sternbergites, arsenopyrite, löllingite, lautite (Markirch), niccolite, etc. The arsenic liberated from the decomposition of geokronite is of particular nature.

X. Investigated occurrences. Andreasberg and Wolfsberg, Harz; Marienberg, Saxony; Kowary (Schmiedeberg); Markirch, Voges; Bayerland Mine at Waldsassen; Jáchimov (Joachimsthal), Č.S.S.R.; Sulitelma, Norway; Sala, Sweden; Les Chalances, Dauphiné; Caracoles and Copiapo, Chile; Akadanimura, Japan.

XI. Literature. Arsenic has been studied quite early under the ore-microcope. The work of VAN DER VEEN is particularly important. Further data are given by RAMDOHR (1924), MURDOCH (1916), ZÜCKERT (1925). For allemonite, see there.

XII. Powder diagram (HARCOURT). (8) 2·74, (5) 2·04, (6) 1·867 Å, a diagram with very many lines; other strong lines are: (4) 3·14, (5) 1·53, (3) 1·283, (4) 1·195 Å.

ARSENOLAMPRITE

I. *Chem.* As, with a reported content of several percent Bi. *Cryst.* o'rhombic; foliated, "hexagonal" or hypo-hexagonal masses, frequently somewhat radiating-rosette-like. — H ~ native arsenic, but appreciably more brittle. The existence of these minerals was confirmed in 1960 by JOHÁN: $a_0 = 3{\cdot}63$, $b_0 = 4{\cdot}45$, $c_0 = 10{\cdot}96$ Å, $Z = 8$, $D_{x\text{-ray}} = 5{\cdot}577$.

II. Polishes very well, but exfoliates somewhat // the cleavage, and then breaks out. Polishing hardness is quite similar to native arsenic, slightly higher.

III. Reflection behavior. Color impression and reflectivity in a freshly polished section are quite similar to arsenic. Reflectivity is high; generally appreciably lower in oil.

Reflection pleochroism is barely perceptible in air, and is considerably increased in oil. // (0001) the brightness remains very similar to that of arsenic, while ⊥ (0001) it is sharply reduced. In any case, the pleochroism is quite weak even then.

It is a peculiar observation, that arsenolamprite, without any recognizable air etching, turns dark under an oil immersion coating one day after polishing; in oil, O becomes light greyish-bluish-white, while E darkens to a duller, darker, non-bluish tint, somewhat toward brown.

Anisotropic effects between + nic. are detectable but weak, and are increased appreciably in oil. If the nicols are moved very slightly from the crossed position, the extinction direction quickly becomes ill defined, and the brightest settings at the 45° position are no longer equally bright. This is particularly noticeable in oil.

IV. Etching. *Positive*: HNO_3; *negative*: $FeCl_3$ (as contrasted to the common arsenic!) HCl, $HgCl_2$, KOH.

Air etching in section is very noticeable even after one day; close observation indicates that ordinary arsenic, which penetrates the foliated arsenolamprite as lamellae, is the carrier of this rapid etching. Shortly after this, the arsenolamprite, which has only darkened, is also attacked whereby As_2O_3 octahedra are formed as in the case of arsenic.

V. Physico chemistry. Compared to the rhombohedral arsenic, arsenolamprite is certainly metastable, but yet durable enough to be slowly replaced like a completely foreign mineral.

The reasons for the development of a different structure are not known; one can scarcely assume that the small amounts of Bi, which are not visible microscopically, and are therefore probably in isomorphous solution, would be responsible for this, since the Bi and As lattices are so similar.

VI. Fabric. The single sample available from the original locality shows a foliated, sometimes rather radial aggregate with perfect cleavage. The tablets, which themselves are thick, are abundantly exfoliated and somewhat bent, and are replaced to varying degrees by ordinary arsenic. The bending, the tabular development, and the cleavage are all earmarks of a typical layer lattice, presumably a hexagonal one.

Against the younger arsenic which surrounds and also replaces it, arsenolamprite is idiomorphic, but very poorly developed.

VIII. Diagnostic features. Arsenolamprite is extremely rare, but should be recognizable with relative ease because it is accompanied by the rapidly tarnishing arsenic.

IX. Paragenetic position. Arsenolamprite has heretofore been described only as a rarity from a few deposits. It now appears that its independent existence has been erroneously discredited for a long time. In the examined sample it occurred in arsenic ("scherbenkobalt") having the usual properties (Palmbaum Mine near Marienberg in Saxony) and constituted about 10% of the entire sample. Accompanying minerals were native silver and a cockade-like aggregate of sternbergite with some safflorite and traces of secondary pyrite.

XII. Powder diagram. Before JOHÁN's work (1960), a pure diagram could not be obtained because of its intimate intergrowth with native arsenic (JUNG has therefore erroneously discredited arsenolamprite!). The known diagram of arsenic has to be subtracted in every case. Naturally, coinciding lines were easily mistakenly eliminated in this way, yet relative intensities provided positive support. In 1960, Johán gave the following values: (10) 5·44, (7) 2·740, (10) 2·72, (7) 1·731, (10) 1·115 Å.

"ALLEMONTITE" – STIBARSEN

I. General data. The "mineral" allemontite lies between arsenic and antimony and was formerly considered to be a mix-crystal of these elements in varying proportions. The first ore-microscopic investigation showed the presence of a beautiful decomposition structure consisting of two components which were thought to be arsenic and antimony. Further investigation, particularly by Quensel and Wretblad, showed that the intergrowth, depending on the variable chemistry, was composed of arsenic and stibarsen ("Allemontite III"), or stibarsen and antimony ("Allemontite I"), and that finally, the compound stibarsen (AsSb) can also rarely occur alone ("Allemontite II"). All the properties of stibarsen are between those of arsenic and antimony, but it is a welldefined compound. This is noteworthy, since the laboratory melting curve does not show any intermediate compound, and there is no unmixing.

Apart from these "allemontites" with decomposition structures, samples of native arsenic with the typical "scherbenkobalt" form have been found which contain fine vermiform particles of stibarsen. This probably represents the unmixing of a mixed gel. Further, native antimony is also found with rounded drops of arsenic (or stibarsen ?), which can be concentrated in a cloudy fashion. — The latter forms have been treated at As or Sb, while "allemontite" has been used for the three varieties mentioned above.

II. Polishing properties. Exceptional polish. The slight differences in hardness tend to make the intergrowth forms easily recognizable.

III. Reflection behavior. The reflectivity of freshly polished stibarsen is between that of Sb or As. However, since stibarsen intergrown with antimony darkens rapidly, as does arsenic intergrown with stibarsen, stibarsen may be confused with arsenic or antimony depending on the case. The darkening is probably the beginning of air etching.

IV. Etching. The different forms of intergrowth influence the etch behavior in exactly the same way as air etching: stibarsen beside antimony behaves quite differently than stibarsen beside arsenic. — The results given by Wretblad (1941) are fairly reliable:

$FeCl_3$	Sb very weakly	AsSb strongly	As not attacked
HCl with H_2O	Sb brownish-grey	AsSb adjacent to Sb, weakly; adjacent to As, strongly	As not attacked
HI	Sb etched strongly and deeply	AsSb tarnished adjacent to Sb; adjacent to As, dark attack	As not attacked

For structure etching, air etching of several days duration suffices.

VI. Fabric. The decomposition texture is easily recognized in figs 315 and 174 (p. 178). The type of intergrowth shows that originally quite large mix-crystals of uniform composition must have existed.

IX. Paragenetic position. "Allemontite" has so far been known: 1. in hydrothermal veins, of the Co–Ni–Ag–As–Bi type; 2. in some pegmatites in coarsely foliated masses. "Allemontite III" is the most widespread, while the Sb-richer "Allemontite I" and "Allemontite II" have so far been restricted to Allemont and Varuträsk.

X. Investigated occurrences. Allemont, Dauphiné; Varuträsk, North Sweden; Přibram Č.S.S.R.; St. Catarina (so-called silaonite).

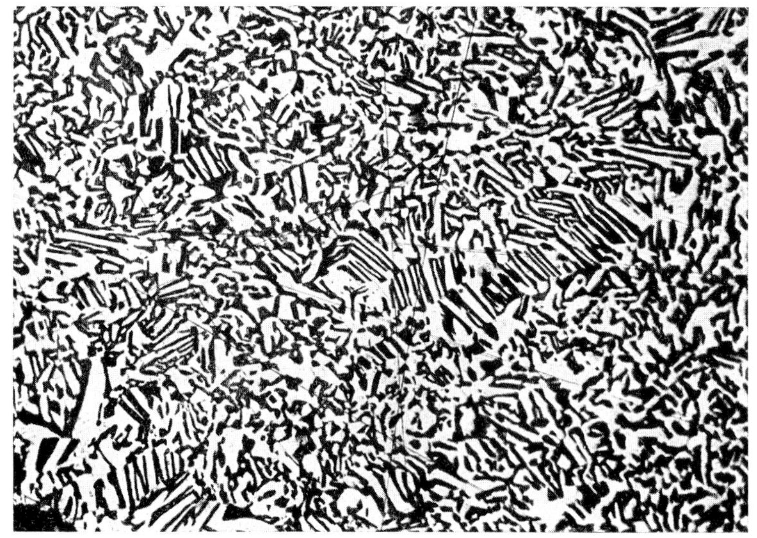

Fig. 315 225 ×, imm. RAMDOHR—EHRENBERG

Allemont, Dauphiné

Allemontite, air etched; black = arsenic, white = stibarsen. Decomposition of a former mix-crystal

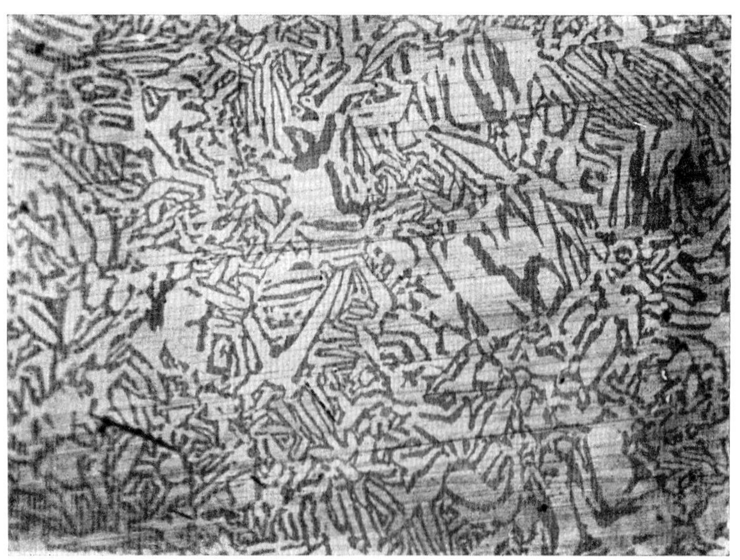

Fig. 315 a 250 × RAMDOHR

Allemont, Dauphiné

A cut very near to // (001) but otherwise very similar to 315

XI. Literature. The first data on the allemontite intergrowths were reported almost simultaneously by VAN DER VEEN (1925) and KALB (1926). ORCEL (1928) and R. J. HOLMES have dealt with them later. QUENSEL and his student WRETBLAD (1941) then gave the present, probably unequivocal explanation (1941).

XII. Powder diagram (B. & TH. for stibarsen). (10) 2·92, (6) 2·13, (7) 2·01, (4) 1·656, (4) 1·278 Å.

ANTIMONY

I. General data. *Chem.* Sb; the As content is almost completely exsolved at ordinary temperatures. — For allemontite, see p. 371. *Cryst.* Rhombohedral crystals are rare, generally as massive, spathic masses. c/a = 1·324. *Lattice* like that of arsenic. a_{rh} = 4·50 Å, α = = 57° 2′. Z = 2. # (0001) perfect, (02$\bar{2}$1) distinct. H = 3–3$^1/_2$, D = 6·7. Opaque. Tin white, metallic lustre with a very light yellow tint.

II. Polishing properties. Polishes exceptionally well and scratch-free. Cleavages are recognizable near fractures and on the edge of the section; the hardness is appreciably greater than that of bismuth, and slightly less than that of arsenic, stibarsen, and dyscrasite.

III. Reflection behavior. Brillant white, very high reflectivity.

Reflection pleochroism very slight; in oil, slightly more distinct, and recognized positively only at grain boundaries. Lower than that of bismuth, and especially that of arsenic. Similar to dyscrasite.

Anisotropic effects under + nic. In air, distinct but quite weak; in oil, lively, but without striking color effects at diagonal position. The anisotropy permits grain structures and twin lamellae to be recognized adequately.

	in air	in oil
general	very high; white with a light yellow tint	lowered but not as striking as that of arsenic
against silver	distinctly lower	appreciably lower
against dyscrasite	somewhat brighter	very similar
against arsenic	brighter	now appreciably brighter
against bismuth	distinctly lower, less creamy	lower
Photometer ocular		
green	67·5% } probably too low	60%
orange	58 % } probably too low	57%
red	55 % } probably too low	55%

Photocell, for air, freshly polished acc. LEONARD et al. (1971)

nm	470	546	489	
R_g	65·2	65·2	64·5	(BESMERTNAYA's values are much higher)
R_p	58·8	61·6	69·3	

MOSES, FOLINSBEE, BOWIE — for orange 62·6, 74·6, 72·3 resp.

IV. Etching (accord. to VAN DER VEEN and DAVY & FARNHAM — abbreviated). *Positive*: HNO_3, brown or iridescent coating; $FeCl_3$, slight brown tarnish; $HgCl_2$, light brown. Concentrated HNO_3, black; after the addition of a drop of water, a heavy white deposit, $KClO_3$, H_2SO_4, $FeCl_3$ in alcohol, etch with blackening effect; HI. Dilute and concentrated HCl almost negative, as also KOH.

Negative. KCN, $FeCl_3$ in H_2O (+ accord. to DAVY & FARNHAM).

Structure etching. Concentrated K_2S yields exceptional results.

In allemontites the Sb portion behaves somewhat differently than pure Sb; $FeCl_3$ in water etches more distinctly.

VI. Fabric. Mostly xenomorphic aggregates, rarely idiomorphic cube-like crystals.

Internal properties of grains $(01\bar{1}2)$ twinning is common, and is frequently polysynthetic lamellar, but is entirely lacking in some occurrences. In some cases the twinning is due to pressure (glide twinning). Cataclasis does occur, but rarely. Zonal growth has not been observed.

Structures and textures. In natural occurrences, mostly very fine-grained serrated aggregates; overgrown masses also very coarse-grained. Also botryoidal concentric but, as such, mostly enclosed in native arsenic.

Replacements of stibnite have been observed to a limited extent; QUENSEL & ÖDMAN have described native antimony as a decomposition product of a microlite mineral. Dyscrasite crystals from Andreasberg are often encrusted by a skin of antimony.

VIII. Diagnostic features. Against arsenic, the reflectivity of antimony is higher; and against bismuth, lower. Antimony exhibits weaker anisotropic effects than either, and is etched by K_2S. Dyscrasite is very similar, although if tarnishes somewhat to a brown color, is slightly harder, and does not have as distinct a cleavage. Antimony is also "whiter", while dyscrasite has a light brown tint. The isotropic AgSb mix-crystals related to dyscrasite are also very similar, except for their isotropism. Silver, itself, — freshly-polished — has a distinctly higher reflectivity. Tellurium is quite similar, but is much rarer and appreciably bluish.

IX. Paragenetic position. Native antimony is quite rare, though its occurrence is more extensive than assumed heretofore. It is almost never of economic importance.

It is restricted to hydrothermal formations, especially veins. It is common in pure stibnite veins as well as in stibnite-rich gold veins, and more rarely in Ag–Co–Ni veins and in those of the sub-volcanic silver group.

Associated minerals are extremely variable; arsenic, stibnite, smaltite, ruby silver, galena, arsenopyrite.

As to its origin, antimony seems to have formed often from stibnite by descending solutions. This is shown in many cases by its association with kermsite. Other occurrences appear to be of an ascending nature, however, and even of high temperature.

X. Investigated occurrences. Andreasberg, Harz (many mines). Brandholz, Fichtelgebirge; Jáchimov (Joachimsthal), Č.S.S.R.; Les Chalances, France; Allemont, Dauphiné. Seinajoki, Finland; Caes de Sobreira, Mizarella Mine near Coimbra, Portugal. Buena Vista, Bolivia; New Brunswick, Canada; Borneo; Broken Hill, N.S.W. — Artificial.

Many stibnites of widely differing origins contain antimony in very tiny specks.

XI. Literature. Except for short occasional reports, only a few writers have worked on antimony. Of importance is the relevant paragraph in VAN DER VEEN (1925), then the publications by G. KALB (1926) and J. ORCEL (1928) (both especially on allemontite), P. RAMDOHR and R. ZÜCKERT.

XII. Powder diagram (B. & TH.). (10) 3·10, (8) 2·25, (8) 2·15, (6) 1·774, (5) 1·370 Å.

BISMUTH

I. General data. *Chem.* Quite pure Bi; perhaps with very small amounts of As and Te. *Cryst.*: Ditrigonal-scalenohedral. c/a = 1·304, usually spathic compact masses. *Lattice* like that of arsenic, a_{rh} = 4·74 Å, $\alpha = 57° 26'$. # (0001) perfect, $(02\bar{2}1)$ moderate. H = 2–2$^1/_2$, somewhat ductile. D = 9·8. Opaque; n (white), accord. to KUNDT = = 2·26. Very white metallic lustre on fresh surface, and yellowish on an older surface.

II. Polishing properties. Bismuth alone polishes moderately well in spite of its low hardness, but it is often severely scratched when accompanied by hard ores, and in most cases it is severely polished out. Strong relief can be avoided if a polishing machine is used. Heating should be avoided regarding a (possible) transformation at $\sim 75°$.

Cleavage // (0001) is occasionally visible, especially in samples in the initial stages of weathering (to bismuth ochre).

Polishing hardness is lower than that of all associated minerals, including bismuthinite, although its hardness is given as only 2. It is also less than all minerals which could be mistaken for bismuth. Talmage hardness: A^{+}.

III. Reflection behavior. *Color and reflectivity impression.* Shining white, with a creamy tint. After polishing, the material soon tarnishes, at first without an appreciable decrease in reflectivity, and becomes more distinctly yellow with a reddish tint. This property, which is strongly emphasized by VAN DER VEEN, is not as striking as one would assume from his description.

Anisotropic effects under + nic. are distinct, more striking in oil. In imperfect sections, frequently severely masked by scratches; in small grains which are highly concave due to polishing, frequently not visible because of depolarization.

	in air	in oil
general	creamy white; later white, toward pinkish brown; very bright	somewhat reduced, in contrast to silver
against silver	somewhat less bright and a trace more yellowish	silver is distinctly brighter
against antimony	appreciably brighter	
against niccolite	always much brighter	
Photometer ocular		
green	67·5%	52·5%
orange	62 %	51 %
red	65 %	48(?)%
	Acc. to Cissarz	
Photocell	EHRENBERG / DESSAU	
red	55·5 ± 2·0% / 58–61%	42·2 ± 2·0%
yellow	65·5 ± 2·0% / 62–69%	46·3 ± 2·0%
green	59·5 ± 2·0% / 61–65%	47·3 ± 2·0%
MOSES/FOLINSBEE	65·2/71·3%	

BESMERTNAYA	nm	450	500	550	600	660	700
	R_g	60·5	63	65·5	69	72	72·5
	R_p	35·2	56	58	61·5	65	65

	in air	in oil
Reflection pleochroism	rather slight, recognizable only under favorable circumstances	slight, but noticeable O = brighter, creamy white E = darker, creamy white with a light grey cast

Fig. 316 250 ×, imm., nic. nearly + RAMDOHR
Lancelot Tin-Mine, Herberton, Qld.

Native bismuth, large grain, showing excellent twin lamellae following three directions (one poorly developed!). It is extremely difficult to get bismuth with such a good polish

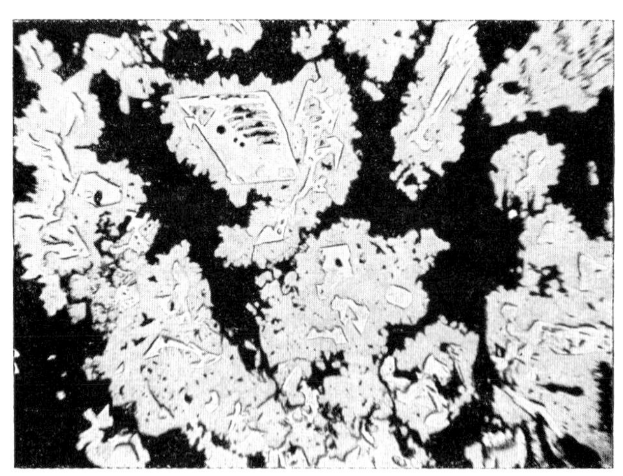

Fig. 317 55 ×, + nic. RAMDOHR—EHRENBERG
Mte. Romero, Sevilla, Spain

Bismuth, irregular grain boundaries. The individual grains exhibit the typical (? — transformation twinning). The individual bordering grains with high relief are *safflorite*

IV. Etching. *Positive*: HNO_3, immediate blackening, with vigorous evolution of gas; when rubbed off, a light grey tarnish; HCl, slow dark discoloration, solution; $FeCl_3$, immediate etching, which varies greatly with direction. *Negative*: KCN, KOH.

Texture etching. Accord. to VAN DER VEEN, HI works exceptionally well in 1″. Accord. to SCHNEIDERHÖHN, concentrated HNO_3 for 1–2″ is preferable especially since smaltite, safflorite, and niccolite, which frequently occur with it in Co–Ni ores, develop a good texture etching at the same time.

Etch cleavage // (0001) is well developed by etching with HNO_3.

V. Physico chemistry. Bismuth, like ice, belongs to the few substances that contract during melting. Its solidification point (~ 269°) is therefore depressed by increased pressure. Since, at this low temperature, bismuth is frequently precipitated in the form of drops, an effect resembling, "frost blastings" may develop occasionally in the accompanying minerals enveloping the drops, as, e.g., in a number of scheelites, and rarely in arsenopyrite (important geological thermometer).

VI. Fabric. *Internal properties of grains*. Twinning on $(01\bar{1}2)$, mostly in the form of lance-like laths in lamellar, frequently parquet-like arrangement, is almost always present. The assumption that it can be attributed to transformation at 75° is very uncertain. A portion can usually be designated as the "main crystal" (fig. 316). Only a few occurrences do not exhibit twinning; the lamellar structure in these cases has probably been partly obscured by recrystallization.

Growth twins, e.g., very fine penetration twins, do occur, but are generally rare.

Zoning has not been observed. *Deformations* apparently are frequently the cause of twinning ($K_1 = (011)$, $K_2 = (100)$ – rhombohedral symbols!). At the same time translation can occur parallel to the base. – Unmixing has not been observed. In individual cases the grains precipitated from the melt exhibit a sector-like structure.

Structures and textures. Grain form rounded, usually very serrated around the edges. Grain size is variable, often unexpectedly coarse. Skeletal masses in dendritic safflorite (the dendritic "smaltite" in most collections is almost always safflorite) (fig. 317), sphalerite, etc. are commonly homogeneous, frequently to the extent of several centimetres. Such skeletons have been found particularly in the Jáchimov (Joachimsthal) occurrence described by R. ZÜCKERT (1925), but they occur also in similar deposits. They are sometimes quite difficult to interpret genetically.

Where bismuth is present only in minor amounts, it forms small rounded grains in midst of the other ores or, more rarely, as interstitial fillings, which permit it to be immediately recognized as the youngest mineral in the paragenetic sequence. In many occurrences the individual bismuth grains are separated by very thin crusts of safflorite, löllingite, and smaltite.

VII. Special fabrics. The relationship of bismuth to the other ores is complicated and is still partly unexplained, in spite of abundant reference material. Some peculiar "replacements" can probably be explained by the bismuth being originally present in the form of drops, and later crystallizing with dilation. However, true replacements have been observed not infrequently; they occur particularly well in some zoned smaltites. A special position is occupied by the relatively abundant cases where Bi has been formed by the decomposition of bismuth compounds, especially lead–bismuth sulphosalts sometimes under decreasing temperature, sometimes with the beginning of

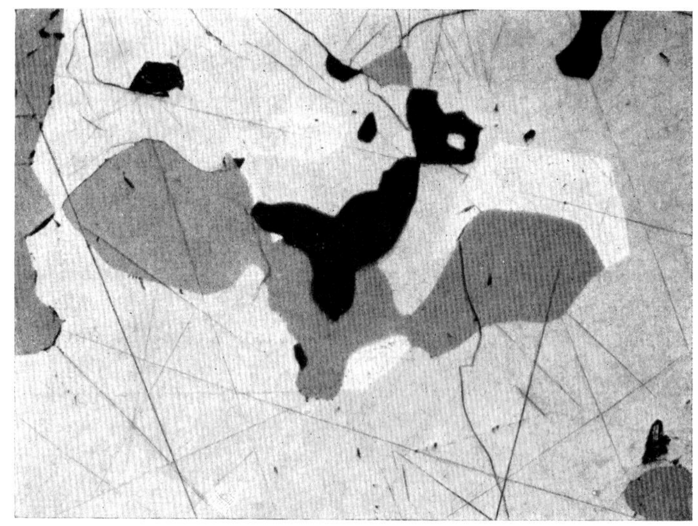

Fig. 317a 325 ×, imm. RAMDOHR
Yxjöberg, Dalarne, Sweden

Native bismuth, pure white, beside *pyrrhotite,* dark-grey, and *gangue,* black, in *chalcopyrite,* medium grey

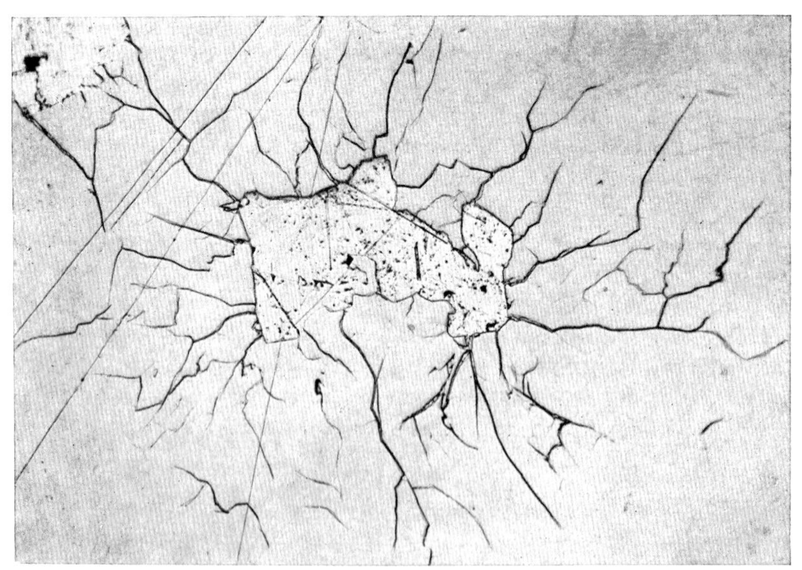

Fig. 318 250 ×, imm. RAMDOHR
Bieber, Hesse, Germany

Skutterudite (smaltite), with inclusion of an aggregate of *native bismuth,* is fractured by its crystallization with the formation of radial cracks. Grain boundaries in the bismuth are barely recognizable by reflection pleochroism

weathering. In this way, cosalite, galenobismutite, klaprothite, and wittichenite often have rims which contain abundant fine drops of bismuth.

The fracturing of chloanthite shown in fig. 318 can be attributed to liquid bismuth deposited in negative forms and expanding during crystallization, or to crystallized bismuth contracted less compared with chloanthite during cooling. Similar illustrations are quite abundant!

Some galenas contain bismuth in negative cube forms. This bismuth, like tetradymite and bismuthinite in other, more common cases, is responsible for the notable (111) cleavage of these galenas. In one case, idiomorphic cube-like crystals (well oriented) were enclosed by chalcopyrite (Kompaneno).

Myrmekites occur with a number of ores (see p. 113). Weathering results in a large number of transparent oxidation ores ("bismuth ochre"), especially a number of oxides with n $\sim$ 2. Oxidation generally proceeds along an irregular front; occasionally follows the (0001) cleavage.

VIII. Diagnostic features. Its high reflectivity prevents confusion with most minerals. Only silver, antimony, dyscrasite and a few very rare ores are similar. For its recognition, the important features are its low hardness, its anisotropy, and the rather common lamellar structure; against silver, also its noticeable darkening, but the absence of actual tarnish. Silver, in fine grains, can be very similar.

IX. Paragenetic position. Native bismuth is the most widespread (not most abundant) naturally occurring metal. In its occurrence it is a typical "throughrunner" (ubiquitous mineral) in the broad range of deposits belonging to the magmatic succession.

1. Bismuth is extensively deposited already under pegmatitic-pneumatolytic conditions of formation. Therefore it is usually found associated with scarce sulphide masses, but also without them in pegmatites of various kinds, e.g., the cassiterite and wolframite-scheelite pegmatites. It is very widespread, though hardly ever in commercial amounts, in contact metamorphic formations, where it is actually never lacking. Here it generally occurs in chalcopyrite aggregates, more rarely in arsenopyrite. According to the entire paragenesis, it must have been deposited in the liquid state. The possibility of its precipitation in the form of drops has been doubted from many sides, but it must surely have occurred like this, as proven by the occasionally observed "frost blasting" due to dilation upon solidification (e.g., in the scheelite from the Brusius fields at Omaruru).

1–2. Bismuth is almost universally present in sulphide deposits of the deeper basement complexes, of the Scandinavian type (Falun, Skutterud, Ljusnarsberg, Yxsjö, Kaveltorp, Tunaberg, Håkansboda, and many others), which formed under conditions intermediate between pneumatolytic and hydrothermal processes. The amounts are usually small, but they are very widespread.

2. The chief habitat of native bismuth is in vein deposits of the intrusive hydrothermal type, which frequently grade without a sharp division into those listed last under 1. Some of the deposits have characteristics of the "extrusive" succession (Bolivia). Principal bismuth carriers, which were, or are still bismuth producers, include the cobalt–nickel–silver veins (Schneeberg and Annaberg in Saxony, Jáchimov (Joachimsthal) (ZÜCKERT, 1925), Wittichen, Mte. Narba, Sardinia, Temiskaming

(Campbell & Knight, 1906), Kerr Lake, and many others; the occurrences at Turtmanntal and Eifischtal, which have been altered by mountain building, can also be classified here), the "Mansfelder Rücken" occurrences which are closely related to these formations, and finally the tin-bismuth–silver veins of the Bolivian cordillera (Uncia, Chorolque, Tasna), which have been supplying the world markets for a long time.

3. Bismuth is generally unknown in the sedimentary cycle; it occurs in traces at the Rammelsberg. River placers in Bolivia (and elsewhere!) contain large nuggets which have escaped weathering.

In group 1, the accompanying minerals are: chalcopyrite, pyrrhotite, arsenopyrite, pyrite, bismuthinite, molybdenite, sphalerite, galena (rather subordinate), wolframite, cassiterite, as well as cubanite and cobaltite. In group 2: smaltite, safflorite, löllingite, bismuthinite, and complex Bi sulphosalts of Cu and Pb, silver, niccolite, and breithauptite, and also a variety of rich silver ores, pitchblende, and others. Here the principal gangue minerals are calcite and barite. In the Bolivian occurrences, which extend from pegmatites to extrusive-hydrothermal impregnations, the paragenesis, except for the ever-present bismuthinite, is extremely variable.

X. Investigated occurrences. Because of its wide distribution, it is not even remotely possible to list all the investigated localities (in order, according to groups).

1a. *Pegmatites*: Hundholmen at Tysfjord and Iveland, Norway; Muro alto, Serra Cabreira, Portugal; Goodhouse, Lesser Namaqualand; Mina Fabulosa, Bolivia.

1b. *Tin ore veins*: Geyer and Altenberg in Saxony; many mines in Cornwall. Schlaggenwald and Zinnwald; Lancelot Mine, Queensland, Australia.

1c. *Contact metasomatic*: St. Christoph near Breitenbrunn; Blekka Mine near Silfjord, Norway; Fontana Raminosa, Sardinia. Stiepelmann Mine at Arandis, Brusius-claims and Feld Simson near Omaruru, South-West-Africa; Tetiuhe, Eastern Siberia.

1–2. *Sulphide occurrences* in the basement complex: Querbach; Silberberg near Bodenmais; Röros, Norway; Ljusnarsberg, Yxsjö, Los, Kaveltorp, Tunaberg, Långban, Falun, Håkansboda all in central Sweden; Boliden, northern Sweden.

1–2. *High temperature gold veins*. Cobar, N.S.W.

2. *Veins of intermediate temperature*. Eisleben; Bieber in Hesse; Wingertshag near Wissen; Wenzel Mine near Wolfach and the Daniel Mine near Wittichen, Baden; Markirch in Alsace; Johanngeorgenstadt, Schneeberg, Freiberg in Saxony; Jáchimov (Joachimsthal); Hüttenberg, Carinthia; Mte. Narba, Sardinia; Villa Nueva de Cordoba, Mte. Romero at Huelva, Spain; Wershne Sejmchanskoje; Gowganda and Cobalt, Ontario; Mina Salvadora at Uncia, Animas at Atocha, Chorolque, and many others in Bolivia.

3. *Sedimentary-metapmorphic*. Rammelsberg, Harz.

XI. Literature. van der Veen and R. Zückert have conducted a systematic study, of bismuth; short contributions have been made by P. Ramdohr, Davy & Farnham and Murdoch. — The microscopic aspects of bismuth have been treated briefly in many places from a general viewpoint or from a genetic angle, the first time by Campbell & Knight (1906).

XII. Powder diagram (B. & Th.). (10) 3·30, (7) 2·37, (8) 2·28, (5) 1·873, (6) 1·447 Å.

SULPHUR

I. General data. *Chem.* S, quite pure. *Cryst.* o'rhombic; many crystals apparently sphenoidal. a:b:c = 0·813:1:1·904. *Lattice* a = 10·61, b = 12·87, c = 24·56 Å; 128 atoms per cell. Indistinct (111) and (110) cleavages. H = $1^1/_2$–$2^1/_2$, D = 2·07. *Opt.* Refractive indices for nm 589: α = 1·950, β = 2·038, γ = 2·241. a = a, b = b, c = c. *Opt.* +. Light yellow, highly transparent, high greasy lustre.

II. Polishing properties. It is very difficult to polish sulphur scratch-free with any method and, because of its low hardness, polishes very concavely when accompanied by other minerals (except gypsum). Samples with abundant sulphur are not bound by most artificial resins, or only with great difficulty. A directional variation of hardness, which would be expected from the reported "H = $1^1/_2$–$2^1/_2$", cannot be observed. The polishing hardness is very low, and is less than 2, and distinctly below gypsum. — Grinding cleavage has not been observed.
Talmage hardness: B.

III. Reflection behavior. *Color and reflectivity impression.* A dull greyish white, but still distinctly brighter than the brighter gangue minerals. No noticeable colored tint.

	in air			in oil	
general	low, greyish white			very greatly lowered, quite dull grey	
against sphalerite	appreciably darker; not too conspicuous for direction // c in good sections			much darker in all positions	
against cerussite	generally somewhat brighter, but similar for certain directions			quite similar; appreciably brighter only // c	
Photocell	$\frac{a + b + c}{3}$ = 12·1	red 10·6 yellow 11·6 green 12·5		red 1·99 yellow 1·97 green 1·75	
Reflectivity, calculated for Na light	// a 10·5	// b 11·7	// c 14·7	// a 1·6	// c 3·7

Reflection pleochroism. Distinct in air, very striking at grain boundaries, but frequently also recognizable in the individual grains. Brightest // c; darkest // a. Sometimes somewhat masked by a poor polish. The relative difference between a and c is increased considerably by oil; the strong internal reflections and the really very low reflectivity frequently make the effects less striking than would be expected, especially in very coarse-grained and clear material.

Anisotropic effects under + nic. cannot be observed at all because of the omnipresent internal reflections. They are only clearly visible when the nicols are turned appreciably (5–10°) from the crossed position.

Internal reflections: Great numbers, white to light yellow.

IV. Etching. Has not yet been investigated systematically. Its easy solubility in carbon disulphide and the high inflammability of samples with only a small amount of S, are characteristic. In sections kept in oil for a relatively long period of time and under intense illumination, a slight air etching (or rather, "vapour pressure etching") takes place.

VI. Fabric. The technically important sulphur occurrences are usually quite coarse-grained if the deposit does not contain many clay impurities. The more important occurrences, from the standpoint of ore-microscopy, where sulphur was formed by the weathering of sulphides, are fine-grained, the grains are rounded, not serrated, polygonal; sometimes tiny, very highly faceted rounded crystals are formed. Also as interstitial fillings and rhythmic crusts.

VIII. Diagnostic features. Sulphur is easily recognized by its very poor polish, low hardness, striking internal reflections, etc. In its reflectivity it resembles some highly reflecting oxidation ores, e.g., cerussite, but the latter generally has a much better polish.

IX. Paragenetic position. The most common and for the industry most important type of occurrence which is associated with gypsum, calcite, aragonite and frequently salts, as well as hydrothermal deposits in areas of waning volcanism, are of no importance in ore-microscopy, since there sulphur is almost always easily recognizable. However, sulphur is not uncommonly found in the cementation zone of sulphide deposits, as a newly formed mineral on galena, sphalerite, pyrite, and others, where polysulphide-containing solutions probably encountered highly acid solutions. In such cases sulphur may form excellent crystals in cavities, or compact, inconspicuous masses and interstitial fillings.

X. Investigated occurrences. Mooseck near Golling, Salzburg; Rio Tinto, Spain; Kiuchta, East Daghestan; Els, Moravia.

XI. Literature. Except for a compilation by the writer, which has been reproduced almost word-for-word here, there are no reports in the literature, apart from occasional references.

XII. Powder diagram. (B. & TH.). (4) 7·76, (5) 5·75, (10) 3·90, (4) 3·48 etc. Å.

SELENIUM

I. General data. *Chem.* Se, very pure. Isomorphous miscibility with Te is possible, but has not yet been observed in natural occurrences. *Cryst.* Trigonal, D_3^4, $a_0 = 4{\cdot}34$, $c_0 = 4{\cdot}95$ Å, Z = 3; until now, only thin columnar, rather skeletal crystals* with $(10\bar{1}0)$ and $(10\bar{1}1)$, and fine-grained aggregates. c/a = 1·134, H ~ 2. D = 4·81; crystals somewhat pliable. *Opt.* +, $n_O = 3{\cdot}0$, $n_E = 4{\cdot}04$. In the very thinnest splinters it is transparent red. Red streak.

II. Polishing properties. Polishes easily to a lustrous surface, yet in single crystals scratches are unavoidable: better in aggregates from Pacajake.

III. Reflection behavior. The reflectivity in air is fairly high, yet appreciably lower than that of galena; the color, on first impression, is white. In oil, appreciably darker, white to brownish-grey.

Reflectivity calculated from n, without considering the not unimportant $\varkappa$, especially for E.

VJALSOV — air	nm	460	500	540	580	620	660	700
	R_E	38·8	36·8	36·7	36·7	36·3	35·5	34·6
	R_O	25·5	24·2	24·5	25·6	27·1	27·3	26·3

The *reflection pleochroism* is distinct even in a single crystal in air. In aggregates it is, of course, much more noticeable, and is appreciably increased in oil, since R_0 is decreased a relatively greater amount. In air, E is bright, creamy white, and O is darker, toward brown; in oil, E is greyish-white to blue, and O is dull brown. The color impression in general is not dissimilar to that of tenorite.

* I am grateful to my late colleague C. PALACHE for having sent me several crystals from this very rare occurrence.

The anisotropic effects between + nicols are very strong, both in air and oil. Dark red internal reflections are definitely recognizable, but with difficulty.

IV. Etching has not been investigated.

VI. Fabric. In material resulting from a mine fire in the United Verde Mine, selenium forms thin single crystals, sometimes with selenium glass. As a weathering product of selenium minerals it forms divergent crystal aggregates, columnar single crystals, and typical gel forms, sometimes with rhythmic deposition. It is very often accompanied by limonite, and embedded in it. The primary ore is almost always clausthalite.

Selenium in Colorado Plateau – type ores forms single crystals often with strongly "nibbled" prisms or aggregates as a matrix of sandstone.

VIII. Diagnostic features. Difficult, if no unweathered selenium minerals have been recognized. The total impression is very much that of tenorite. If, however, primary selenium minerals have been determined, then identification is easy.

IX. Paragenetic position is indicated by its semi-artificial development in ores affected by mine fires, where PALACHE first recognized selenium positively; then in tiny crystals sitting on sedimentary (red bed type) U, V, and Cu ores (e.g., of Thompsons in Utah), where it was recognized by HILLEBRAND, MERWIN and WRIGHT, and where it was meanwhile found in dozens of other deposits. Here, on the Colorado plateau, the monoclinic modification was also found as a mineral. Finally, is it probably most widespread as the weathering product of natural selenides, especially clausthalite, where it is associated with cerussite and others (RAMDOHR, 1937). In one instance it was found to be enclosed by unweathered pitchblende (Pinky Fault, Canada). Some more occurrences are given by SINDEEVA (1964).

XI. Literature. Of the studies mentioned, only that of the writer contains ore-microscopic data, which are essentially given here.

XII. Powder diagram (quite different from various writers!): (10) 4·29, (10) 3·79, (8) 3·36, (10) 3·01, (8) 2·20, (8) 2·07 Å. B & TH.. do not have, e.g., 4·29, and 3·36 Å.

TELLURIUM

I. General data. *Chem.* Te, some analyses with isomorphous Se. *Cryst.* trigonal, with *lattice* D_3^4, $a_0 = 4{\cdot}44$, $c_0 = 5{\cdot}91$ Å, $c/a = 1{\cdot}33$, # $(10\bar{1}0)$, (0001) less perfect. $H = 2\text{–}2^1/_2$, $D = 6{\cdot}2$. Opaque; white, with high metallic lustre.

II. Polishing properties. By itself, it polishes well, but with adjacent harder minerals, it is difficult to get it scratch-free. Cleavage can usually not be seen in section, but sometimes it is easily visible. Softer than chalcopyrite. Talmage hardness: B ±.

III. Reflection behavior. *Color and anisotropic effects.* White, very high reflectivity.

IV. Etching (accord. to DAVY & FARNHAM and VAN DER VEEN). *Positive*: HNO_3, blackening; $FeCl_3$ iridescent; $HgCl_2$, some spots light brown, others, negative. *Negative*: HCl, vapors weakly positive, KCN, KOH.

Texture etching: HCl + CrO_3, $KMnO_4$ + HNO_3 (1′).

VI. Fabric.' (Structure and texture.) Twin lamellae or zoning have not been observed. Xenomorphic granular aggregate, whose texture can be seen satisfactorily between + nic. Further observations have not been made on the sparse amount of material available.

	in air				in oil		
general	pure white, with a light creamy tint				distinctly lowered against native metals, tint becomes bluish		
against platinum, silver, gold	light grey; slightly bluish in contrast				appreciably darker		
against galena against calaverite against hessite	brighter and less yellow } much brighter } brighter, somewhat creamy				distinctly brighter		
VJALSOV — air nm	460	500	540	580	620	660	700
R_E	68·1	68	67·5	57·1	56·7	65	63·7
R_O	59·4	59·7	59	59·3	58·3	57·6	56
Reflection pleochroism	medium, but appreciable at grain boundaries O = white E = very light brown				more distinct than in air brown tint of E disappears		
Anisotropic effects between + nic.	quite strong; no pronounced color effects				likewise		

VIII. Diagnostic features. Can be confused with the other anisotropic metalloids. It is not twinned, and is brighter than arsenic. Much more rare.

IX. Paragenetic position. Tellurium is mostly restricted to the near-surface portions of the subvolcanic hydrothermal gold-silver veins, where it is found in association with quartz, chalcedony, barite, and rhodochrosite, with tellurides of Au, Ag and Pb, and also with pyrite and galena. Here it is of primary ascending origin, but of quite late formation. In Cripple Creek and in Porcupine it has been described as an intermediate oxidation mineral of tellurides. — The Frood Mine in Sudbury contains small amounts of Te in partially pegmatitic portions.

X. Investigated occurrences. Samples from Faţa Băii; Boulder Co., Colorado, and four localities in Japan; Frood Mine in Sudbury; Vatukoula, Fiji.

XI. Literature. Only occasional references by DAVY & FARNHAM (1920) and V. LEHNER; in more detail by VAN DER VEEN.

XII. Powder diagram (HARCOURT). (10) 3·22, (8) 2·33, (7) 2·22, (6) 1·820, (6) 1·610 Å.

GRAPHITE

I. General data. *Chem.* C; analytical deviations are due to mechanical impurities. *Cryst.* Hexagonal, D_{6h}^4, sometimes D_{3d}, $a_0 = 2{\cdot}46$, $c_0 = 6{\cdot}79$ Å, Z = 4, a pronounced layer lattice. Tabular-foliated crystals. Since many natural and synthetic "graphites" are more loosely packed because of an increase of the c dimension, there are small differences in all properties (semi-graphites or pregraphites). — Very perfect cleavage // (0001), H = 1*, D = 2·2. Opaque; dark blue in the very thinnest sheets. Leadgrey, characteristic dull metallic lustre; distinctly duller in rarely observed cross-fractures (visible in coke, e.g.). The refractive index of ~ 2, given by SENFTLEBEN & BENEDICT, is, accord to RAMDOHR (1928c), 1·5 for n_E, and for n_0 probably $\lesssim 2$ (with appreciable $\varkappa$). — Streak lead grey.

II. Polishing properties. It is very difficult to polish because of its low hardness and its well-known tendency to smear in coarsely foliated aggregates. When fine-grained, however, where the (0001) translation which causes the smearing cannot develop, the polish is quite good if the associated minerals are not too hard. Sections parallel to (0001) nearly always flake out. After very careful pre-grinding on a glass-or pitch plate, it should be polished briefly on a slowly-moving lap, if a polishing machine is not used.

Cleavage in polished section. In good sections, the (0001) cleavage is usually easily recognized. During polishing, it is seen that the polishing hardness is much higher than has been assumed from the scratch hardness. It is slightly greater than that of chalcopyrite. Talmage hardness: not given.

III. Reflection behavior. *Color impression*: In polished section, light brownish-grey, decidedly "ore-like".

According to literature reports, the reflectivity in air, calculated from $\varkappa$ and n, is quite close to 15%.

	in air	in oil
general	whitish-grey, with an orange tint on O; E, bluish-grey with much lower reflectivity	the reflectivity is not strikingly lowered for O, and the color is more brownish. On the other hand, E is extraordinarily much darker, so that the color appears to be satiny black, almost like a hole
against chalcopyrite and pyrrhotite	slightly darker, with a light brownish-grey tint	decidedly darker and more brown

Photometer ocular	in air O	in air E	in oil O	in oil E
green	22·5%	5 %	n.r.	n.r.
orange	23·5%	5 %	16%	±2%
red	23 %	5·5%	n.r.	n.r.

BESMERTNAJA — air nm	450	500	550	600	650	700
R_O	19	20·8	21·5	22·5	24	24·5
R_E	7	7	7	7	7	7

The reddish tint in R_O becomes evident!

* This very low hardness is only apparent, due to the very easy translation. It is actually much higher, and for pressure $\perp$ (0001), is even similar to diamond. But // (0001) it may also be > 4, if translation can be prevented.

The *reflection pleochroism* is enormously high, and is attained by only a very few minerals, It is the most striking characteristic of graphite (fig. 319), especially in oil, which practically eliminates errors in identification (see valleriite, mackinawite, and chalcophanite)!

Anisotropic effects under + nic. are very high in air and oil, and they also belong to the greatest known. The color effects, however, are moderate; in diagonal setting in air, very light brownish-yellow; in oil, quite similar (fig. 323).

Internal reflections have not been observed; graphite soon becomes transparent in infra-red.

Fig. 319E 250 ×, imm. RAMDOHR
Råna, Ofoten-Fjord, Norway

Graphite, crystals, differently bent, in different brightness accord. to orientations up-grown on *pyrrhotite* which contains smaller flakes of graphite and some *chalcopyrite* (white). — Dark is gangue

IV. Etching. Negative against all etch reagents!

V. Physico chemistry. At geological temperatures and pressures graphite is almost always the more stable modification compared with diamond and chaoite.

VI. Fabric. *Internal properties of grains. Zoning and twinning* have so far not been observed with certainty, but strikingly regular creasing near coarse scratches produced during polishing, may be attributable to twinning. Molybdenite exhibits quite analogous behavior. *Deformation* by (0001) translation, and flexure in all directions in this plane, are observed frequently. Bending and compression almost always result in exfoliation. All deformations are easily recognized by the reflection pleochroism (fig. 320a, 320b).

Structures and textures. The individual grains are tabular, and vary greatly in size, depending on their origin. In the homogeneous aggregates the grains are rarely intimately intergrown and hypidiomorphic. Graphite flakes tend to idiomorphic development

Fig. 320a 70 ×, imm. RAMDOHR

Namib Mine near Swakopmund

Graphite, thick tablets in marble, exhibits strong reflection pleochroism and crumpling as a result of bending

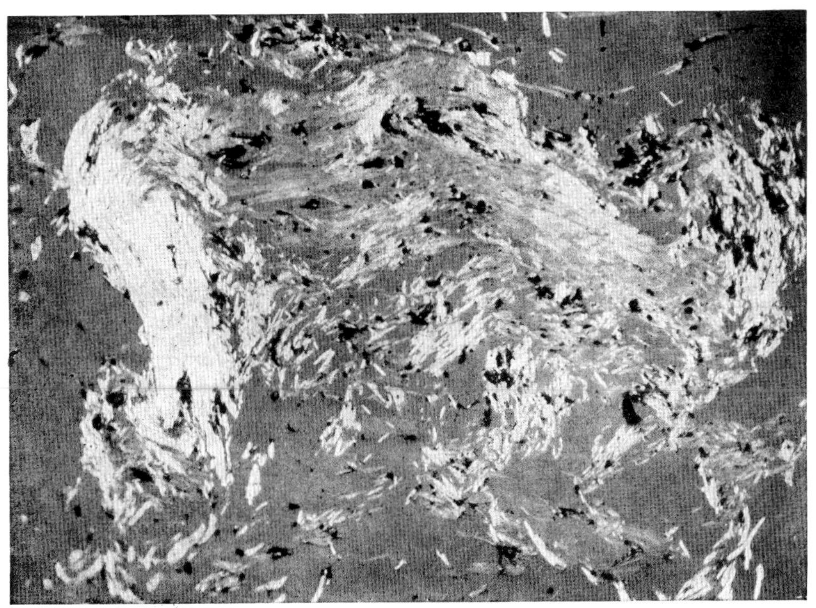

Fig. 320b 70 × RAMDOHR

Buch near Lindenfels, Odenwald

"Curl" of *graphite* in a metamorphic quartzite. With different orientations, using one nicol the reflection pleochroism from white to dark grey is very distinct when adjacent to other minerals, and are commonly arranged parallel to the rock texture, but sometimes they are very irregularly distributed and even bent in quartz and calcite (fig. 320). The texture of the aggregates is easily seen with one Nicol because of the strong reflection pleochroism. Although this is very rare, it sometimes exhibits, in spite of the high temperature of formation, concentric botryoidal forms which are quite similar to retort graphites (see below). The meteorite from Smithville and contact metamorphic material formed from hydrocarbons, at Doberlug, exhibit divergent, palmleaf-like forms with E // to the fibre length.

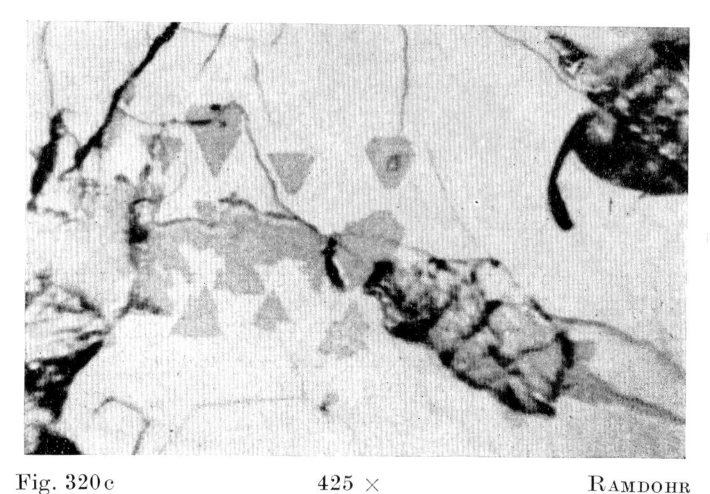

Fig. 320c 425 × RAMDOHR
Meteorite Mundrabilla

A bundle of //c elongated crystals of graphite is cut exactly //(0001) in trigonal sections. It is one of the trigonal polytypes of graphite in this meteorite. A very small deviation of the (0001) section acts

Graphites in a number of "titano-magnetite differentiates" exhibit spherical aggregates up to about 0·3 mm in size, in which, in contrast to the retort graphite, (0001) is arranged radially and sheaf-like , not tangentially. Similar structures are exhibited by graphite from a quartz diorite from Tottijärvi, Finland, and a hortonolite dunite from Mooihoek. On the other hand, some granites of sedimentary origin contain small graphite plates. Radial spheres adjacent to concentric ones occur in Akjoujt. There they line joints in dolomite and form films around magnetite.

Graphite in meteoritic irons is often needle-formed // c and shows several polytypes (RAMDOHR 1972) fig. 320.

VII. Special fabrics. Replacements (active or passive) are apparently very rare in natural occurences. An occurrence at Otjimbojo exhibits graphite as an oriented replacement of a mica mineral!

Some regional metamorphic graphites, in spite of extreme deformation, still exhibit local organic cell structures. — Very different and striking structures are shown by some graphite nodules from basalts, e.g., Storö, Umanak, Greenland.

VIII. Diagnostic features. Polishing behavior, anisotropy, and reflection pleochroism scarcely permit it to be confused with other minerals. Mackinawite is distinctly more metallic, and chalcophanite has and entirely different paragenesis and does not exhibit any cleavage. Both, as well as molybdenite and valleriite are much brighter than graphite, especially in oil.

IX. Paragenetic position. Graphite is generally a fairly rare mineral in ore deposits. It is widespread in contact and regionally metamorphosed sediments, therefore also in ores of all metamorphic grades. It occurs "orthomagmatically" in hortonolite dunites and titanomagnetite differentiates in gabbros — and therefore undoubtedly is of high temperature origin. It can also be present in large amounts in the primary formations of pegmatitic–pneumatolytic parageneses. It is also widespread in nickeliferous

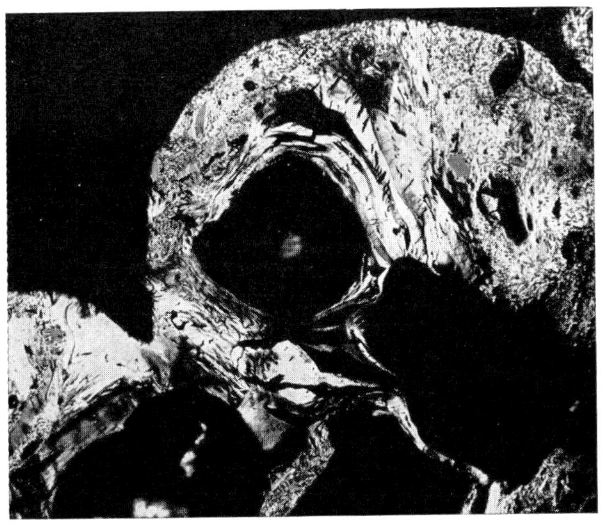

Fig. 321 125 ×, imm., one nic. RAMDOHR

Graphite in colliery coke, Eschweiler Mining Company, near Aachen. Graphite skins are moulded around coke pores. Upper right, small remnants of fusite. The reflection pleochroism easily permits textures to be recognized

pyrrhotites. In some of these magmatic parageneses it provides evidence of magmatic resorption of bituminous rock, but certainly not always. On the other hand, it is almost entirely lacking in hydrothermal veins, with very few exceptions such as Silver Islet; this, of course, is apart from slickensides derived from material in the adjoining rock. — It is typical that coarsely foliated graphites tend to occur in high temperature pegmatitic–pneumatolytic, contact-metasomatic deposits, while fine scaly or "earthy" types commonly occur in metamorphic deposits.

IXa. Artificial graphite. Graphite is the principal component of industrial cokes, in which it is present as a fine-grained ill ordered aggregate, especially in thin (0001) flakes enveloping the coke pores (fig. 321). During coking, it forms relatively quickly from durite and vitrite, while fusite changes into graphite only quite slowly. The industrial retort graphite, and that of analogous origin, which is formed accidentally in fractures in coke oven linings, consists of graphite in exceptional concentric scaley, botryoidal masses, in which (0001) is always // to the surface of the orb (fig. 322),

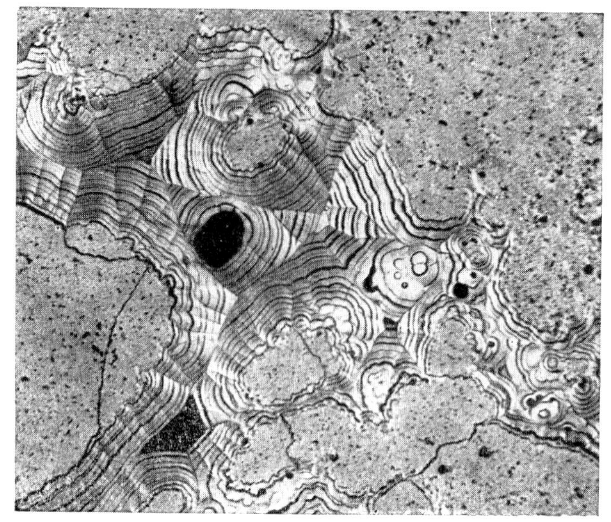

Fig. 322 125 ×, one nic. RAMDOHR

Graphite in "retort graphite". Concentrically layered, formation due to the decomposition of a gaseous phase; shows exceptionally well the optical orientation due to its pleochroism. (0001) is always // to the botryoidal surface

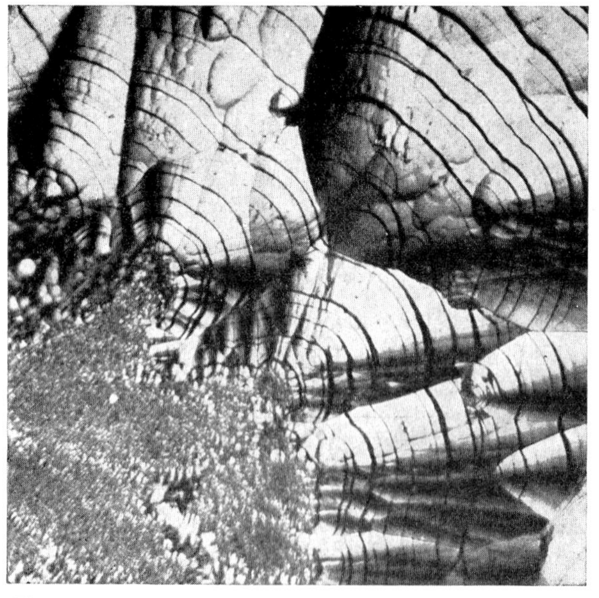

Fig. 323 450 ×, imm., + nic. RAMDOHR

Graphite in "retort graphite". Details of the above section. "Spherulitic cross"

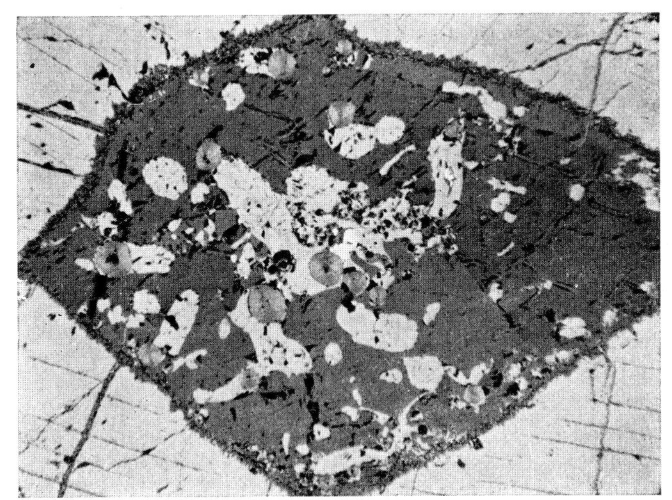

Fig. 324 75 × RAMDOHR
Lake Sanford, New York

Gangue inclusion (dark grey) in *titanomagnetite* (light grey). In the inclusion is *graphite* in the form of small radial spheres which show a vertical dark beam. Sulphides are pure white

Fig. 325 250 ×, imm. RAMDOHR
Lake Sanford, New York

Details of the previous figure. The structure of the graphite spheres is very easily recognized

and which exhibit nice spherulitic crosses between + nic. (fig. 323). More information about coke and retort graphites is given by the writer (1928c) and in many more recent works in the field of coal petrography.

The peculiar intermediate members between graphite and anisotropic coals have sometimes been called "pregraphites". Crystallochemically they appear to be explained by partial-ordering, which requires the exact graphite relationships within the basal plane, while normal to that, there are varying lattice spacings and also inclusions. The (0001) cleavage is less distinct, and the anisotropy is not quite as great. In the coal from Doberlug, which has been subjected to slight contact alteration, they form beautiful palmleaf-like structures.

X. Investigated occurrences (small selection). Liquid magmatic differentiates: Raana, Ofotenfjord; nickeliferous pyrrhotite rocks from the Bushveld, Transvaal; Lake Sanford, N.Y.; Lac de la Blache, Quebec.

Pegmatitic–pneumatolytic, contact-metasomatic: Huelva, Spain; Norberg, Sweden, Björkåsen, Ofotenfjord; Aliber mine on Lake Baikal; Ragedara, Ceylon; St. Remi D'Amherst, Quebec; Otjimbojo, South-West-Africa.

Contact-metamorphic: Radautal, Harz; Pargas, Finland, Kalkfeld, South-West-Africa; Franklin, N.J.

In basalts from Umanak, Upernivik and Blåfjeld, Greenland.

Natural cokes by eruptive rocks: e.g., Coly Hill Dyke, Newcastle, England; a "pregraphite" from Doberlug.

Dislocation metamorphic: Many graphitic slates, mica schists, gneisses and marbles, especially fine at Skaland, Norway; Outukumpu, Finland.

Artificial: Normal cokes and special cokes; retort graphite.

XI. Literature. Up to present, only RAMDOHR (1928c) has published a comprehensive paper on the microscopic properties of graphite; on the optical properties and the occurrence exist many single data; E. RYSCHKEWITSCH has compiled most of the information from the older reports.

XII. Powder diagram (B. & TH.). (10) 3·36, (5) 2·03, (8) 1·678, (6) 1·541, (9) 1·230, 9) 1·158 Å.

Carlsbergite, CrN, and **Osbornite**, TiN, are two cubic nitrides from enstate-rich meteorites. Both metallic properties. Carlsbergite with rose violet tinge, osbornite, when pure nearly gold-yellow; small changes in composition (C) alter the color quickly. For carlsbergite BUCHWALD & SCOTT give

R nm			
481	546	590	644
49.5	41.5	41.0	40.0

ALLOY-LIKE COMPOUNDS AND TELLURIDES

WHITNEYITE, ALGODONITE, DOMEYKITE

I. General data. The above three "minerals" are mostly *mixtures*, namely of "*whitneyite*" which represents a Cu-mixed crystal saturated with As, but differs from common copper, both in appearance and in microscopic properties, so widely that its separate discussion is justified; of *algodonite*, to which different formulas have been assigned (Cu_5As, Cu_6As, Cu_7As) the last of which probably giving the best accordance for the pressures of ore formation; of "*domeykite*" which is to be divided into isometric and hexagonal Cu_3As, the former actually better $Cu_{15}As_4$.

These are often joined by ordinary native copper, and that by no means only in the members poorest in As; further by niccolite, by silver, and may be by silver arsenide a.o. This does not yet exhaust the multiformity. The cases hitherto discussed are mostly of comparatively high temperature (roughly estimated $\gtrsim$ 220°). There are, however, formations of considerably lower temperature, as, e.g., those of Corocoro and Mechernich, which cannot be fitted into this scheme.

The study of the literature has been made extraordinarily difficult by the fact that, at least, four writers speak of α-, β-, γ-, δ-forms, modifications, varieties which time and again mean something different or partly different!

I attempt to bring my exposition especially in harmony with the roentgenographic results of PADĚRA's work (1951) which I consider to be exceptionally thorough. The writer has hitherto underestimated the importance of the hexagonal Cu_3As, and, owing to misstatements in analyses as well as faulty designations on labels, often also interchanged (α) domeykite and algodonite.

"**Whitneyite**", Cu with dissolved As. The analyses vary; at about "Cu_9As" the limit of solid solubility seems to be reached. Massive; fracture partly conchoidal, partly already hackly and nearly malleable. a_0 a bit larger than Cu s.str., H = $3^1/_2$, D > 8·5. Opaque, on fresh fracture very high metallic lustre deep cream-yellow. It tarnishes the most rapidly of all the members, and as a matter of fact also much more strongly than native copper.

"**Algodonite,** Cu_7As, orthorhombic, but decidedly pseudohexagonal: a_0 = 2·59, b_0 = 4·55, c_0 = 4·23 Å. Very near to hexagonal closest packing of the artificial E-Phase and corresponding properties: conchoidal fracture, no cleavage. D = 8·6, H = 4. Opaque, high metallic lustre, deep cream toward brass-yellow. Tarnishes like domeykite, but somewhat quicker.

Domeykite, Cu_3As, in the narrowest sense (α-D accord. to PADĚRA) is isometric (T_d^6) a_0 = 9·60 Å. Fracture conchoidal, H = 3–$3^1/_2$, D = 7·7, which is appreciably less than for algodonite and whitneyite. Opaque, high yellowish white metallic lustre; tarnishes rapidly iridescent, soon to be covered with a brown mouldlike coating.

Hexagonal Cu_3As (β-D accord. to PADĚRA) often accompanies or substitutes common domeykite which it closely resembles in properties. a_0 = 7·13 Å, c_0 = 7·30 Å. According to HEYDING & DESPAULT (1960) the formula $Cu_{15}As_4$ would be better.

II. Polishing properties. All members can be easily polished very well, but remain somewhat scratched, especially whitneyite. Abrasion hardness for all medium, higher than chalcocite, lower than breithauptite. The polished surface must be examined without delay because all the members tarnish rapidly, especially when other ore minerals are present. Talmage hardness: everywhere C^-.

III. Reflection behavior. *Color and reflection impression.* First and foremost quite

	in air		
	whitneyite	algodonite	domeykite (α and β)
general	white-cream, very strongly metallic	bright cream-white strongly metallic	yellowish white, towards very light grey, highly metallic
in comparison	with algodonite very similar, somewhat more yellow with domeykite somewhat brighter	with whitneyite very similar, somewhat whiter with domeykite somewhat brighter and whiter	with algodonite somewhat darker, somewhat greyish yellow with whitneyite somewhat darker
	differences very slight		

Reflectivity curves of domeykite and algonidote see PICOT, SAINFELD, VERNET. Bull. Soc. fr. Min. **89**, 1966, 259–261.

	in oil		
	whitneyite	algodonite	domeykite (α and β)
general	not appreciably different, very little pinkish	not appreciably different	lowered just perceptibly; slight change to greyish
in comparison	with algodonite very similar, yellower and a trace brighter with domeykite brighter and whiter	with whitneyite similar, somewhat whiter and a trace darker with domeykite brighter and whiter	with algodonite darker and greyish with whitneyite darker and yellowish grey
	differences considerably increased, but yet slight		

BESMERTNAYA — air (the differences seem to be rather high!)

nm	450	500	550	600	650	
Whitneyite	48	55	65	80	78	Mixed crystals,
Algdonite	54	58·5	62·5	62·5	60	values obviously
Domeykite	44·5	47·5	48·5	48·0	47	approximative!
Photocell (MOSES)			66·5		55·7	

α-domeykite and β-domeykite are of course very similar; in oil small differences.

α-domeykite next to β-domeykite is more yellowish, β-domeykite itself more bluish; only visible in freshly polished sections.

similar, cream-yellowish white with very high, almost metallic reflectivity. Only in oil are slight differences in color recognizable.

A part of the divergent statements about the brightness might only rest on the application of day-light filters or unfiltered light.

Bireflection and anisotropy-effects with +N. α-domeykite and whitneyite are isotropic and show no pleochroism. Algodonite and β-domeykite are anisotropic, the last one being against α-domeykite in oil a trace darker. Bireflection in oil is very weak in both cases, in oil at grain-broundaries distinct. Anisotropic effects with +N are low, color effects not bright. Pieces of the same locality and of different ones do not show absolutely the same behaviour.

IV. Etching. The statements of the literature indicate on the one hand that also the best experts are still undecided as to the demarcation of the minerals, on the other, that, obviously, the etch reactions have been affected by the accompanying minerals. It is expressedly warned to use diagnostic etching (see VII).

V. Physico chemistry. Many older examinations seem to be surpassed. MASKE & SKINNER (1971) studied experimentally the conditions on the Cu-rich side of the Cu–As diagram (fig. 325a). In spite of generally for natural occurrences a bit low looking temperatures, the diagram seems to be rather reliable, at least of the Lake Superior deposits, likewise those of Talmessi and the Cashin Mines. According to the diagram α-domeykite could be really Cu_3As, β-domeykite $Cu_{2.7}As$ and algodonite between $Cu_{5.2}$ till Cu_8As the last one at $\sim 300°$.

VI. Fabric. There are specimen of all members which contain one component only, but as a rule "domeykite" contains some algodonite, "algodonite" some domeykite or whitneyite, and "whitneyite" some algodonite. Whitneyite with domeykite is rare.

The relations of α-domeykite (cubic) and β-domeykite (hexagonal) are interesting. The "domeykites" of Ahmeek and other mines of the Calumet Peninsula, but likewise those of Talmessi and the Cashin Mine are obviously formed at first as a granular aggregate of a cubic domeykite which breaks down in all stages forming "restdomeykite" and (111) lamellae of β-domeykite. Especially "domeykite" of Talmessi shows, that the large lamellae of β-domeykite // (111) cannot be oriented with // (111) but have an oblique extinction. Often they are twinned and the twinning plain is really // (111). This observation is actually not astonishing regarding the distinct difference of c_0 of β-domeykite (7·33 Å) and a $\sqrt{3}$ of α-domeykite (16·57 Å). It remains remarkable that in spite of that the unmixing follows strictly (111). I think, one should consider intermediate phases.

With further progress of the α–β-alteration the fine β-lamellae form very thin films, on the larger grain boundaries of the old α-grains a bit coarser strings of pearls composed of algodonite. By that observation alone is shown that α- and β-domeykite must have different composition and β-domeykite must have higher As-content.

Besides that fine lamella–breakdown of α-domeykite into spindle or oleanderleaf-formed intergrowths is not rare. Sometimes small spindles are enclosed in the coarse ones. Forms can appear again like breakdown of mixed crystals or similar to such ones of dyscrasite of Cobalt.

Remarkably is the fact that some "algodonites" in the centre are distinctly richer in Cu, in other words, show a zoning.

The "minerals" "ledouxite" (said to be Cu_4As), "keweenawite" $(Cu, Ni, Co)_2As$ and mohawkite $(Cu, Ni, Co)_3As$ are mixtures, partly complicated ones, of the series reviewed before with niccolite, rammelsbergite, maucherite and breithauptite (perhaps more).

The author could collect a lot of material in the Seneca Mine, Calumet, Mich., and got similar ones, from Cashin Mine, Southwestern Colorado and Talmessi, Iran. There the relations of both "domeykites" could be studied very well.

In Seneca we have a loose aggregate of grains of β-domeykite, showing a few twin lamellae, imbedded in a matrix of former α-domeykite, which, as far as it could be observed, is quantitatively altered into an intergrowth of fine, partly in themselves twinned lamellae of β-domeykite. An excellent octahedric network is formed. The observation is not easy and needs fresh polish and crossed nicols.

In Cashin and Talmessi the cubic form predominates and is mostly preserved, but every section shows beginning of transformation into β-lamellae. — In spite of perfect octahedral orientation of the β-lamellae no simple metric relation of both forms is known.

All those data do *not* belong to the "domeykites" and the "algodonites" obviously formed at distinctly lower temperatures from Coro coro, Mechernich, Zwickau a. o. Their physico chemical position and occurrence is not yet known.

The "domeykites" of Ahmeek and other mines of Calumet obviously are formed as cubic domeykite but altered in all stages into hexagonal β-domeykite. After complete alteration the β-lamellae are incrusted by fine films or, especially following old grain boundaries by strings of algodonite-pearls. That shows that the composition of α- and β-domeykite must be different, β-domeykite having a higher content in As.

Structures and textures. All minerals of that series form thick crusts or coarse solid masses, but practically never showing any crystal forms. All are aggregates of rounded, not interwoven grains of very various size, domeykite being often rather fine-grained (many exceptions!). Locally the minerals (or only one) can form the cement of breccias and sandstones (fig. 208).

VII. Diagnostic features. The immediate recognition of these minerals, one against the other and all against some other minerals with high reflectivity, is difficult. Whitneyite is remarkable though, as compared with the brittleness of domeykite-rich mixtures. Algodonite and β-domeykite are anisotropic, α-domeykite and whitneyite isotropic. The differences in reflectivity are no doubt distinct on freshly polished surfaces, but overlap in using different filters and become veiled very quickly by the variable tarnish. Powder diagrams are necessary but should be used cautiously because there can be very unexpected and complicated mixtures. — Etching tests are without any value and misleading. All these data do not refer to the so-called domeykite and algodonite of Corocoro, Mechernich, Zwickau etc. surely formed at definitely lower temperatures. Their paragenetic and physico-chemic position is until now unexplained and may remain so regarding the difficulty to get authentic specimens.

IX. Paragenetic position. All minerals occur in very similar, but rather rarely realized conditions. Mostly it seems to be a hydrothermal rearrangement of older assemblages of Cu-ores. Occasionnally the migration in reducing groundwater in deserts (red beds) may form algodonite. The quantities are in most instances very small and dispersed, but locally (Keweenaw Peninsula, Talmessi) there were rather large quantities.

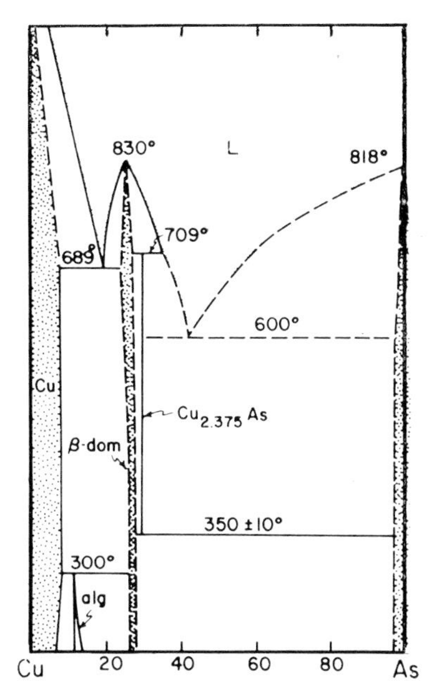

Fig. 325a
(accord. MASKE and SKINNER)

Fig. 325b 160 × RAMDOHR
Hellenic Mining Co., Cyprus

Oregonite (white) in crystals imbedded in *pyrrhotite* (light grey) which is intimately intergrown with xenoblastic *magnetite* (medium grey)

X. Investigated occurrences. For domeykite: "Lake Superior"; Cashin Mine (Utah); Coro-Coro (Bolivia — Buen Pastor Mine and others); Talmessi (Southern Iran). For algodonite: Mohawk Mine and many others in Michigan; Paracatas, Cigazula (Mexico); Mechernich and Brückenberg-Shaft near Zwickau (Germany). For whitneyite: "Lake Superior".

XI. Literature. A report on the older, partly strongly misleading literature is given by PADĚRA (1951). Very many older data generalize too much. After some good preparatory work of MACHATSCHKI (1929) and RAMSDELL (1929), MICHEEV (1957) and PADĚRA (1951) worked especially successfully, as did most recently HEYDING and DESPAULT (1960). The paper of BUTLER & BURBANK (1929) and the data of HARCOURT (1942), too can be used only with (much care or not at all). MASKE & SKINNER (1971) did good experimental work.

XII. Powder diagram (accord. to PADĚRA and B. & TH.). The intensities vary rather strongly with different techniques.

α-Domeykite: (4) 3·04, (4) 2·15, (10) 2·05, (5) 1·964, (7) 1·887 Å, some rather intense lines at 1·38, 1·22, 1·18 Å.

β-Domeykite. (4) 3·62, (3) 2·37, (6) 2·23, (10) 2·02, (7) 1·78, (9) 1·44 Å.

Algodonite. (2) 2·251, (4) 2·114, (10) 1·988, (2) 1·194 Å.
HARCOURT's values are misleading.

CUPROSTIBITE

I. *Chem.*, $Cu_2(Sb, Te)$. Hardly known. *Cryst.* tetragonal. $a_0\, c_0 = 3·99,\ 6·09$ Å Z = 2 (similar to synthetic Cu_2Sb). ⌗ (001), D = 8·42.

III. Steelgrey with violet-red tint; Rpl. strong in pink-violet tints strong anisotropism. — Accord. to the reflection values it changes from optically + to — at approx. 637 mm.

Photocell (in air) nm	460	540	580	660
R_E	56·3	46·8	43·3	42·9
R_O	51·8	40·7	40·8	51·4

IX. Until now only in traces in sodalite syenite from Ilimaussag, Greenland.

XI. Described by SÖRENSEN et al. (1969); from these all data.

XII. Powder diagram: (4) 2·82, (5) 2·56, (10) 2·07, (4) 1·993, (3) 1·424 Å.

Kutinaite

I. *Chem.* $(Cu_{2.07}Ag_{0.84})$ As; in spite of this complicated composition in narrow limits synthesizable. Cryst. cubic, a_0 = 11.76 Å. D = 8·38, malleable.

III. Isotropic, grey white with distinct blue tint.

R in air nm	420	620
	47·2	39·3

Less white than novakite.

IV. Etched by HNO_3 and $FeCl_3$.

IX. A rare component with As-rich paragenesis (Černy Dul): Novakite, koutekite, paxite, arsenolamprite, löllingite.

XI. All data by Hak et al. (1970).

XII. P.D.: (9) 2·702, (8) 2.398, (10) 2·259, (10) 2·078, (7) 1·991 Å.

Koutekite

I. *Chem.* — Cu_5As_2. *Cryst.* suppossedly hexagonal, a_0 c_0 = 11·51, 14·54 Å, Z = 18. Tiny crystals with twin lamellation upgrown on safflorite. H = 4. Observed till now only from Cerny Dul (Schwarzental), Czechoslowakia, in calcite veins with löllingite and safflorite. Seems to be abundant in Talmessi.

III. R 43%, compared with safflorite light bluish grey. Bireflection visible only at grain boundaries, anisotropic effects with + Nic. very weak, but colored, orange grey to bluish grey. — Twin lamellation in one direction (that does not fit to hexagonal system! Ra.).

XII. P.D. (Picot & Vernet): (7) 2·446, (10) 2·078, (10) 2·024, (6) 1·994, (6) 1·785 (8) 1·328 Å.

All data by Z. Johán (1960), Chemie der Erde 20, 217. Other data by Besmertnaja.*

Novakite

I. General data. *Chem.* $(Cu, Ag)_5$ As_3, possibly $(Cu, Ag)_{11}As_5$. *Cryst.* Tetragonal a_0 = 8·20, c_0 = 11·88 Å (no relation to maucherite). H = 3½, D = 6·7. Tarnishes steel blue. —

III. Reflectivity high, no numerical values. No bireflection, medium strong anisotropy. No internal reflections.

IV. Etching. *Positive*: HNO_3, HCl, $FeCl_3$; *negative*: $HgCl_2$ (partly positive), KOH, KCN.

XI. Literature. This mineral was described by Johán & Hak (1959).

XII. Powder diagram (Johán & Hak). (9) 1·998, (9) 1·957, (7) 1·910, (10) 1·870, (10) 1·182 Å.

Horsfordite

Chem. "Cu_5Sb", perhaps corresponding with the synthetic $Cu_{11}Sb_2$. *Cryst.* not known, only massive. — H = 4, D = 8·81, brittle. — Silvery white, strongly metallic, quickly tarnishing.

On ore-microscopical behavior short data of Murdoch only. The writer was not able to secure any material. — Murdoch reports: Takes good polish, compared to galena a little cream-colored, to silver blue-greyish white. — *Etching*: *positive*: HNO_3, blackens, efferv.; KOH, slowly tarnishing; *negative*: KCN, HCl, $FeCl_3$.

Dienerite

Chem. Ni_3As. *Cryst.* cubic. — The mineral was described very poorly; it one single crystal only. No microscopical data are available, the original specimen is lost! — *Oregonite*, described by the writer, is surely different from "dienerite" (p. 399).

Orcelite

I. *Chem.* Ni_2As, from the similar oregonite very different in the powder diagram and properties. D 6.5. — Polish good.

II. Distinctly rose to bronce. Weak pleochroism, anisotropic.

	green,	yellow,	red	
R_g nm	48	54	58	following
R_p	44	52	56	Caillére

* Neither koutekite nor novakite seem to be identical with the artificial $Cu_{5-x}As_2$ of Heyding and Despault

VI. Rare inclusions of a pentlandite-nodule in New Caledonia, in some awaruite nodules in Josephine Co., Oregon.

XII. Powder diagram. (4) 2.109, (10) 1.977, (10) 1.918, (4) 1.810, (4) 1.737 Å. Most data going back to CAILLÈRE et al. (1959).

Oregonite

I. *Chem.* Ni_2FeS_2. *Cryst.* Hexagonal, $a_0c_0 = 6.083, 7.130$ Å, $Z = 3$; $D = 6.92$, $H = 5$. Metallic white, break and filing metallic, a bit plastic; good *polish*.

III. Reflection: high, ~ pyrite, a bit too rose.

R (accord. PICOT & JOHÁN)						
nm	440	480	520	560	600	640
	44.3	46.1	44.2	48.9	51.3	54.2

Bireflection not visible in air and in oil, anisotropism weak but distinct.

IV. In the artificial system Ni–As–S appears a phase Ni_2FeS_2 which gives the powder-diagram of orogonite.

VI. In aggregates hypidiomorphic polygonal grains till 1/4 mm.; around inclusions sometimes idiomorphic with face striations. Slight deformations in the outermost parts of the nuggets, where the tiny inclusions of mineral × are broken.

IX. Discovered as a member of the nugget association of Josephine Co. in Oregon, where it is associated with awaruite, native copper, molybdenite etc. and probably precipitated by hydrogen or H_3As, it is discovered not too rarely in Laxia tou Mavrou in Cyprus, here again in a serpentine in an included nest of sulfides (Fig. 325 b).

MAUCHERITE

(= Temiskamite, artificial "Nickelspeise")

I. General data. *Chem.* Formulae Ni_4As_3 and Ni_3As_2; the determination of the structure leads to $Z = 4 \cdot Ni_{11}As_8$, an intermediate value. *Cryst.* Tetragonal, trapezoedric, D_4^4, $a_0 = 6{\cdot}84$, $c_0 = 21{\cdot}83$ Å, $c/a = 3{\cdot}19$. Plates // (001), e.g., in Ni-matte; rarely as aggregates and needles // (001). Cleavage was not observed anywhere. — $H = 5$, $D = 7{\cdot}80$. Opaque; silvery greyish white to distinctly yellowish pink.

	in air	in oil
general	high, white to pinkish	no essential change
in comparison with: niccolite	a little darker, light grey	analogous, in immediate comparison a little "bluish-violet-grey"
the white arsenides	a little darker, the pink tint is distinct in contrast	analogous

Photocell (BURKE)	nm	470	456	589	650
		46·3	48·0	50·6	54·2

VJALSOV's values see in his tabella.

II. Polishing properties. Excellent, polished surfaces remain fresh for a long time, broken ones tarnish quickly. Hardness is a little higher than for chalcopyrite and sphalerite, very similar to niccolite, partly higher, partly lower than smaltite, distinctly lower than safflorite, rammelsbergite, löllingite.

III. Reflection behavior. *Impression of color and reflectivity.* Highly reflecting, color without a mineral for comparison white, close to white minerals yellow to pink, compared with niccolite light grey.

Reflection pleochroism not visible.

Anisotropy with + nic. is poorly visible in air, in oil it becomes distinct, using low power objectives and strong illumination; the effects are sufficient for the determination of fabrics. — Some other records in literature may pertain more to bad observation than to real differences in the behavior.

IV. Etching (accord. to FLÖRKE & MURDOCH (1923)). *Positive*: HNO_3, efferv., immediately blackening, cleaned surface strongly pitted; $FeCl_3$, faint brown tarnish (not always!). Positive are further: $KMnO_4 + H_2SO_4$, $H_2SO_4 + H_2O_2$. *Negative*: HCl, KOH, NH_4OH, $HgCl_2$, KCN, NaOCl.

For *structural etching* FLÖRKE (1923) obtained good results with $KMnO_4 + H_2SO_4$ or $H_2O_2 + H_2SO_4$; MEYER (1926) recommends HNO_3 1:1 applied for a short time only.

VI. Fabric. *Internal properties of grains*: Twinning // (106) (?) is mentioned by PEACOCK & BERRY (139). Massive specimens of the "Mansfelder Rücken" show all a very characteristic fabric, subparallel to radiating tabular, and perhaps some fine twinning, other deposits contain well developed crystals (e.g., Schladming). Zoning can be seen by different lamellar striation.

Structure and textures, Grain shape. Mostly there are aggregates of xenomorphic grains in long "fibres" (sections of plates — fig. 326). Sometimes idiomorphic crystals, mostly tabular (very good, e.g., in the niccolite of Elk Lake), occasionally long prismatic, but very small, are randomly oriented (20–50 microns). Aggregates may reach the size of a fist.

VII. Special fabrics. Replacement of and by maucherite is known in many cases and with many minerals. Niccolite of Los Jarales and Gallega near Malaga (Spain) and Bou Azzer (Morocco) is replaced by maucherite in all stages starting at the border; this process is surely hypogene. Some aggregates of Mansfeld probably are formed likewise by replacement of niccolite.

Specimens of Schladming etc. show the reverse effect. Idiomorphic plates of maucherite are replaced by very fine grained randomly oriented niccolite (fig. 326). Often the interior part of the plates — perhaps caused by chemical zoning — remains rather well preserved. — More important is the fact, experimentally examined by CH. PALMER & BASTIN (1913), that maucherite can be replaced very easily in supergene, but perhaps also in hypogene processes, by solutions containing silver salts; 17 atoms of silver are assumed to be precipitated by one mole of Ni_4As_3, a ratio which may have been slightly overestimated. Actually this process is probably important only near the surface.

VIII. Diagnostic features. Close to niccolite maucherite can be overlooked, which explains its late discovery (so, e.g., its presence in Schladming, Bou Azzer, Los Jarales etc. was unknown before). In comparison with niccolite the very low anisotropy and the lack of bireflection are characteristic, not less the more dilute color. — Rammelsbergite, attacked much less by HNO_3, is whiter, harder and distinctly anisotropic. Safflorite, affected by HNO_3, is whiter, harder and likewise anisotropic. Some members and zones of the smaltite-chloanthite group can be, in their different hardnesses and colors, very similar to maucherite. Mostly they are whiter, but sometimes doubts can be avoided only by etching with HNO_3 developing zoning in smaltite, tabular structure in maucherite (fig. 326).

Fig. 326 55 ×, imm. etched with HNO_3, Eisleben RAMDOHR-EHRENBERG

Maucherite, etched (in the centre of photograph), shows well the fibrous-tabular texture and twinning. The surrounding, not distinctly etched material is composed of "white arsenides", mainly *rammelsbergite*

IX. Paragenetic position. Maucherite is present in the deposits of Los Jarales and Gallega, originally certainly magmatic differentiations of chromite and niccolite. Here, maucherite surely belongs to later stages. The occurrences in many mines of the Sudbury field are already hydrothermal.

All other occurrences are typically hydrothermal: Most essential are the deposits of the Co-, Ni-, Ag-, As-formations of the "Mansfelder Rücken", the veins of Cobalt, Bou Azzer and so on. All contain smaltite, niccolite, safflorite, mostly in larger amounts, and often native silver, argentite, ruby silver and some copper ores. Recently much maucherite was discovered by KULLERUD (priv. comm.) in the copper arsenides of the Keweenaw Peninsula.

X. Investigated occurrences. "Mansfelder Rücken" (many mines), Schladming, Styria; Los Jarales and Gallega near Malaga: Bou Azzer, Morocco; Cobalt and Elk Lake, Ontario; Frood and Falconbridge Mine, Sudbury; Seneca Mine, Michigan.

XI. Literature. Rather early records of maucherite can be found at CH. PALMER & BASTIN (1913), E. S. BASTIN (1939) etc. Some data by J. ORCEL (1928), J. ORCEL & G. R. PLAZA (1928), E. MEYER (1926), and HAWLEY (1962).

XII. Powder diagram (B. & TH.). (10) 2·70, (10) 2·02, (10) 1·175, (5) 1·430 Å.

HAUCHECORNITE

I. General data. *Chem.* Mostly it is referred as $(Ni, Co)_7(S, Sb, Bi)_8$.* *Cryst.* Tetragonal, c/a = 1·05; *lattice* (45° turned!) D^1_{4h} $a_0 = 7{\cdot}28$, $c_0 = 5{\cdot}39$ kX ($a_0 \sim a_0$ of maucherite, $c_0 \sim 1/4\ c_0$ of it). — Cleavage not observed, very brittle. H = 5, D = 6·4. — Opaque; color light bronze-yellow, metallic, darkens fast through tarnish.

II. Polishing properties. Polishes very well, but, influenced by the porosity of the surrounding minerals, scratches can be avoided only by careful impregnation; generally a little worse than millerite. Hardness intermediate, a little higher than millerite.
Talmage hardness: E (?).

III. Reflection behavior. Light bronze-yellow, high reflectivity.

	in air				in oil
general	light brown yellow				distinctly lower
in comparison with millerite	a distinct tinge to brownish-pink, *R* a little lower				the tinge is now a little more olive
Photometer ocular					
green	40·5				32
orange	42·5				34
red	44·5				39
Photocell nm	470	545	589	650	
R_g	43·0	47·1	49·2	51·6	the values of westphalian deposits are always 1–3% lower
R_p	41·6	46·2	48·2	50·8	

Reflection pleochroism in air very weak, hardly visible; in oil much more distinct. The darkest position links the brightest of pyrrhotite, the brightest is about the same as the darker of millerite. Very probably (due to the varying, tabular or prismatic forms, this is not certain) O has the darker, E the brighter color.

Anisotropy between + nic. is distinct in air and in oil likewise, but the colors are not distinct.

IV. Etching (MURDOCH). *Positive*: HNO_3, slightly brown, rubs easily off, leaves surface slightly etched. $HgCl_2$, a slightly brown tarnish, rubs clean. *Negative*: KCN, HCl, $FeCl_3$, KOH.

VI. Fabric. Lobate aggregates, at the border mostly surrounded by millerite. Rarely thick tabular or prismatic crystals in vugs. The intergrowth with millerite is most probably a replacement by the latter. — Itself, it replaces grains and crystals of ullmannite, similarly to millerite.

* For the Sudbury occurrence $(Ni, Co, Fe)_9(Bi, As, Sb)_2S_8$.

VIII. Diagnostic features. Can be overlooked very easily in association with millerite, against the hard white minerals color is striking. It is very similar to pyrrhotite, but the latter is much more anisotropic.

IX. Paragenetic position. The occurrences of hauchecornite are much more numerous than was previously assumed. Approved data on the genetical relations cannot be made. The classical locality of Mine Friedrich (Wissen at the Sieg, present in many collections) is surely a product of the cementation zone, originated from the minerals of the linneite-group. It contains further millerite, linneite, ? ullmannite, bravoite, sphalerite etc. A similar occurrence is Vermilion Mine, Sudbury. Another mineral perhaps pertaining to it, is a bit more brownish in the doubtlessly ascendent complex ores from Mte. Narba, beside ullmannite, breithauptite etc. Siegerland ores contain it perhaps of primary origin (Mine Grüne Aue).

XII. Powder diagram (B. & Th.). (5) 4·34, (10) 2·79, (6) 2·39, (6) 2·32 Å; very many weak lines.

PARKERITE

I. General data. *Chem.* $Ni_3Bi_2S_2$, perhaps a little Fe. *Cryst.* Orthorhombic, grains and cleavage splinters. C_{2v}^2, $a_0 = 4{\cdot}02$, $b_0 = 5{\cdot}52$, $c_0 = 5{\cdot}72$ Å, Z = 1. — Cleavage (001), parting (111) nearly perfect. Brittle, H < 3, D = 8·4. — Opaque, highly metallic lustre, light bronze colored, streak black.

II. Polishing properties. Excellent; hardness ~ galena. In polished section mostly only the cleavage (001) can be observed. Talmage hardness: B +.

III. Reflection behavior. Color and reflectivity are similar to galena but with a distinct tinge to cream. Reflectivity (estimated) in air ~ 45%.

Air nm	470	546	589	650
R_g	46·0	47·1	48·1	49·0
R_p	43·3	44·8	45·2	46·4

Reflection pleochroism in air is distinct, in oil strong, cream white: grey-creamish white. The twinning is made visible by the bireflection.

The *anisotropy effects with + nic.* are very strong in air and in oil, and show excellently the twinning. The color effects are distinct, but very sensitive concerning exact 90° position.

IV. Etching. (Michener & Peacock, in part Scholtz). *Positive*: HNO_3, instant blackening; $FeCl_3$ the same; $HgCl_2$ tarnishes iridescent. *Negative*: HCl, KCN, KOH.

V. Physico chemistry. $Ni_3Bi_2S_2$ is in the pure system Ni–Bi–S the only ternary compound and can be made easily; a pure analogue of ullmannite with Bi instead of Sb does not exist. — But lead can enter for Bi in an appreciable amount.

VI. Fabric. *Twinning.* Both known natural occurrences and the synthetic material likewise, show a very distinct twin lamellation after (111), with one set often very surpassing or with a complicated checker board pattern. Lamellae are lath-like or slender spindle-shaped.

The *grain-shape* is xenomorphic in polymineralic assemblages, in synthetic material it is a mosaic. The largest grains both in Insizwa and Frood reach 1–2 mm. — In artificial products without complete reaction $Ni_3Bi_2S_2$ can be idiomorphic against the interstices; the grains are here short prismatic, and show twins of pseudohexagonal, aragonite-like crossections.

VIII. Diagnostic features. In the assemblage of Ni–Fe–S deposits hardness, color and reflectivity are rather similar to galena or bismuthinite. But, the first one is isotropic and greyish bluewhite, especially in immediate comparison, bismuthinite is not twinned and often idiomorphic.

IX. Paragenetic position. Undoubted parkerite was observed till now only in pegmatitic or marginal parts of the magmatic nickeliferous pyrrhotite deposits and here it is connected with pentlandite, chalcopyrite, cubanite, galena, Bi-minerals, tellurides, sperrylite, gold, and many others. In weathered specimens it is more resistant than many of these minerals.

X. Investigated occurrences. The writer could investigate many sections of Frood Mine, Insizwa, and artificial material.

XI. Literature. First, parkerite has been isolated and described by SCHOLTZ (1936), but misinterpreted in regard to its chemical and crystallographic properties. MICHENER & PEACOCK (1943) defined it correctly; practically all their data agree with mine. HAWLEY (1962) mentioned it more recently again from Sudbury.

XII. Powder diagram (MICHENER & PEACOCK). (7) 4·01, (10) 2·85, (9) 2·33, (7) 1·65 Å and many others.

SHANDITE

I. General data. *Chem.* $Ni_3Pb_2S_2$. *Cryst.* Rhombohedral with a_{rh} = 7·89 Å, α_{rh} = 90°, Z = 4. Because of the angle α being exactly 90° some extraordinary relations can be expected. H = 4. Lustre metallic, white. Polishes very well.

III. Reflection behavior. Bright white with a faint tinge to cream. Exactly like heazlewoodite. — Strong reflection pleochroism, O is very distinctly darker and bluish grey (in oil!). The effects between crossed nicols are strong, but the colors not very variegated. — Opaque.

IV. Etching has not been investigated systematically.

V. Physico chemical. Shandite could be made artificially by KULLERUD up to 556°, i.e. that this is a bit low for a proper magmatic formation.

VI. Fabric. Practically always in oriented intergrowth with heazlewoodite with continuous cleavage in both minerals. Outlines mostly rounded.

VII. Special fabrics. Occasionally // (0001) most tiny, oriented inclusions of tabular sphalerite.

VIII. Diagnostic features. In a given assemblage, reflection pleochroism and anisotropic effect are very striking; actually there is no possibility of a mistake against other Ni minerals. The closely related parkerite is known from Ni-pyrrhotite deposits only.

IX–XI. Shandite is observed till now in minute quantities connected with heazlewoodite and sphalerite in a serpentine rock containing chromite and magnetite of Trial Harbour in Tasmania only (RAMDOHR, 1949). It could be overlooked in similar localities. — PEACOCK & MCANDREW (1950) investigated the cryst. structure and could confirm the data of RAMDOHR.

XII. Powder diagram (RAMDOHR). (9) 3·92, (10) 2·78, (8) 2·27, and then, very characteristic (6) 1·968, (5) 1·758, (5) 1·607 Å giving a sequence of four strong lines in very similar angle differences.

HEAZLEWOODITE

I. General data. *Chem.* Ni_3S_2. *Cryst.* Trigonal-trapezohedral; crystals of tabular or cube-like habit. D_3; a_0 = 5·73 Å, c_0 = 7·13 Å. (a_{rh} = 4·07 Å, α_{rh} = 89° 25'); Z = 1. H = 4, D = 5·82. Lustre metallic; light-colored; white to pale bronze. Non-magnetic.

II. Polish very good, easily obtainable. Cleavage // $(10\bar{1}1)$ frequently well observed; parting // (0001) develops from transformation into millerite. Hardness less than in pentlandite and millerite. Talmage hardness: C^+

III. Reflectivity very high, similar to pyrite, but more whitish; in air: white to cream-yellow". In oil little changed. — Reflectivity estimated to be at least 55%.

Reflection pleochroism very weak, hardly noticeable even in oil.

Anisotropy moderate with rather typical color effects. Extinction positions well marked revealing the crystallographic orientation of the pseudocubic mineral.

IV. Etching (PEACOCK). *Positive*: HNO_3, HCl (weak), $FeCl_3$, KOH, $HgCl_2$. *Negative*: KCN.

V. Physico chemical. Heazlewoodite could be made artificially by KULLERUD up to 556°, i.e. that this is a bit low for a proper magmatic formation.

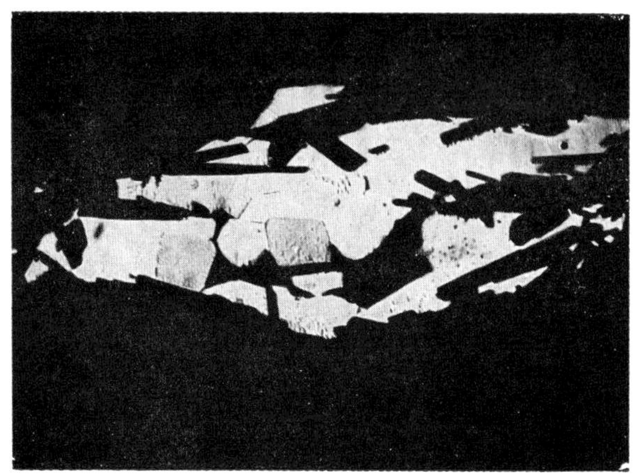

Fig. 327 120 ×, imm. + nic. RAMDOHR
Selva near Poschiavo, Graubünden, Switzerland

Heazlewoodite in a granular aggregate and in part cube-like crystals surrounded by a mixture of *serpentine* and *tremolite* (black). Very hard copying made visible the very faint anisotropism. (Internal reflections of silicates are blackened artificially

VI. Fabric. The shapes of the various grains and grain aggregates highly variable. At Heazlewood, the mineral occurs in granular, mosaic-like aggregates and intergrowth textures with pentlandite (see below); at Trial Harbour, mostly in rounded, isolated grains, besides younger intergrowth textures with pentlandite. At Poschiavo, loose aggregates of cube-like rhombohedrons (Fig. 327); at Hirt, isolated crystals.

Twin lamellae parallel a steep rhombohedron $(02\bar{2}1)$? not rare, probably due to secondary twinning.

Mosaic-like twin textures perhaps of mimetic origin, such as described by PEACOCK, were not observed by the writer.

VII. Special fabrics. Occasionally, heazlewoodite forms curved, bead-like and rootlike aggregates in pentlandite, apparently replacing the latter.

Veining by magnetite appears to be of genetic significance and could be interpreted as a hydrothermal oxidation of pentlandite, perhaps as follows: $Ni_6Fe_3S_8 + 12\ O = 2\ Ni_3S_2 + Fe_3O_4 + 4\ SO_2$.

Heazlewoodite is replaced by millerite, perhaps under hypogene conditions, with millerite not showing its hexagonal, columnar habit, but a tabular one // (0001). A rim of bravoite and vaesite is probably of supergene origin. At Trial Harbour, oriented intergrowth with shandite is common. Cleavage cracks // $(10\bar{1}1)$ pass through both minerals without deviation. In Canala, New Caledonia, heazlewoodite is pseudomorphed into an oxidation product having about the reflectivity of magnetite, at the same time retaining the crystallographic orientation, cleavage cracks and twin lamellae.

Heazlewoodite from Canala exhibits lenticular inclusions of pentlandite, probably exsolution blebs, // (0001). — Idiomorphic granules of an isotropic mineral of irregular arrangement could not be identified. At Poschiavo, cube-like rhombohedrons of heazlewoodite include tiny cubes of awaruite so that c_{Hea}/a_{Aw} (c_0 of heazlewoodite is 7·13 Å, 2 a_0 of awaruite is 7·18 Å).

VIII. Diagnostic features. The identification is difficult, but facilitated by paragenetic features. Reflectivity is similar to that of pyrite, but the hardness is much lower, and the color somewhat more whitish. The anisotropy may easily be overlooked, and this together with a (pseudo-)cubic cleavage could suggest one of the cubic arsenide minerals. Very tiny grains in serpentines may be mistaken for awaruite or iron.

X. Paragenetic position. Insufficiently understood. Heazlewoodite, so far is known only from serpentinized peridotites where it is associated with awaruite, pentlandite, shandite and magnetite, and could be of "hydrothermal" origin. Orthomagmatic mode of formation appears improbable in spite of its association with pentlandite. Sparsely disseminated in many serpentinites. — Material investigated was from Heazlewood and Trial Harbour, Tasmania; Tschagal near Rabad Sefid, Iran (very fine specimens); two occurrences in New Caledonia; Poschiavo in Graubünden, Switzerland; Hirt near Friesach, Carinthia, Austria; probably also from Josephine Co., Oregon, U.S.A.

XI. Literature. The identity of the mineral which was first described by Petterd (1896) was subject to doubt until it was newly confirmed by Peacock (1947) who gave a detailed description of it. The writer (1949) was able to greatly expand this knowledge on the basis of sample material which points to the frequent occurrence of the mineral.

XII. Powder diagram (Peacock). (5) 4·10, (9) 2·88, (4) 2·38, (5) 2·03, (10) 1·82, (8) 1·66 Å. — The six strong lines close to the primary ray are particularly characteristic.

DYSCRASITE AND RELATIVES

(The spelling discrasite is wrong)

I. General data. *Chem.* The system Ag-Sb is rather complicated and not yet clarified completely, especially not the question, how far the often occurring contents of Hg are of influence. Besides that, uncomplete exsolution, anisotropism of components which should be isotropic, complicated intergrowths etc. render the problem more difficult.

Accord. to Petruk et al. (1970) we have to treat Ag itself, bearing only tiny Sb-contents. — Phase I, not strongly separated from Ag, but forming often light lamellae (figs.!), which has up to 5% Sb, in the Cobalt area 8% Hg and a lattice constant of about 4.113 Å. — Phase II, which the author in agreement with Peacock previously held to be ε-Ag, as it is remarkably anisotropic. The author named it first "Allargentum". But since the author emphasized for that name the identity with ε-Ag, the definition must now be changed. Really phase II is cubic, Sb-rich, but poor in Hg ($a_0 \sim 4{\cdot}12$). The cause of the anisotropism is not clear, but the reason seems to be, that II and III together were originally hexagonal and later unmixed.

The inclusions of phase I-lamellae can hardly be explained otherwise. — Phase III, forming darker oleander-leaf-like lamellae in II, is really the ε-phase, therefore "Allargentum" redefined. (a_0 c_0 = 2·952, 4·773 Å, D = 10·0. — It follows the classic dyscrasite, Ag_3Sb, o'rhomb. — Nat. Antimony itself does not contain Ag.

Since often several phases occur side by side, a completely separate discussion is not recommendable.

Cryst. **Dyscrasite proper** is orthorhombic, occurring always in pseudohexagonal twins. C^1_{2v}, $a_0 = 2{\cdot}99$ Å, $b_0 = 5{\cdot}29$ Å, $c_0 = 4{\cdot}82$ Å; Z = Ag_3Sb. — Cleavage // (001) and (011) distinct, (110) only in particularly homogeneous material. H = $3^1/_2$–4; sectile,

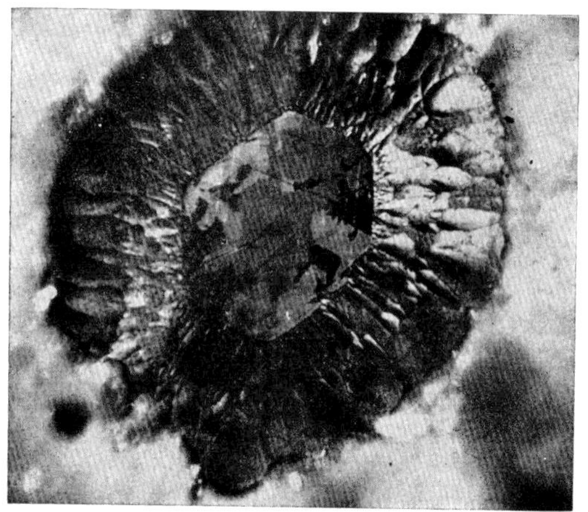

Fig. 328 RAMDOHR
40 ×, imm., nic. almost crossed
Andreasberg, Harz, Germany

Dyscrasite with a rim of native *arsenic*. Between + nic. and in basal sections, twinning according to the aragonite law is easily seen

	in air	in oil
general	very high, yellowish-white	hardly changed or diminished. A slight difference is noticeable between dyscrasite proper (a greyish tinge) and the cubic mixed crystals. If "allargentum" is present, its reflectivity is below that of the two other components mentioned above
contrasted with antimony	a little more yellowish; the difference is more marked in "old" preparations than in freshly polished ones reflectivity little or hardly lower than in antimony, and much higher than in arsenic	
Photometer ocular (loc.: Andreasberg)		
green	66	57
orange	62·5	57
red	61	55

Photocell air nm	470	546	589	660
accord PETRUK et al.	63·4	65·0	65·7	64·3
Allargentum	65·4	67·5	68·7	71·3

but brittle. D = 9·4–9·9. — Opaque; silvery-white to cream, strong metallic lustre. The cubic mixed crystals of Ag saturated with Sb have similar properties but, like "allargentum", are less brittle.

II. Polishing properties. Most perfect polish obtainable, though not free of scratches. Cleavage rarely visible. Abrasion hardness low, $\lesssim$ antimony. Talmage hardness: C.

III. Reflection behavior and color. Color white, very high metallic reflectivity, but distinctly less than in the commonly associated native Ag and Sb. At a first glance, in air the three components called "dyscrasite" are of almost identical appearance.

The *reflection pleochroism* of dyscrasite in air is very weak, a little more distinct in oil, and observable particularly along grain boundaries. In sections // (001) the twin boundaries are sometimes just noticeable.

Photocell air nm	470	546	589	560
accord. Petruk et al.	63.4	65.0	65.7	64.3
Allargentum	65.4	67.5	68.7	71.3

Anisotropy between + nic. is very weak in air, but more distinct in oil, particularly if the nicols are not perfectly crossed, and then sufficient for the distinction of dyscrasite proper from the (isotropic) Ag–Sb mixed crystals.

IV. Etching. All data are contradictory! Particularly those published before the composite nature of the material from Cobalt was recognized. The best will be to omit the data "diagnostic" at all.

For structure etching Van der Veen recommends HI which etches the mixed crystals in about 2 seconds, thus making dyscrasite proper stand out from the intergrown material. Schwartz obtained similar results by using HNO_3, $CuCl_2$, KCN, $KMnO_4$.

In air, dyscrasite gradually tarnishes, but much more slowly than pure native silver. No light-etching occurs.

V. Physicochem. In natural occurrences, where accord. Stumpfl (1967), some Hg is present, obviously the data of I vary a little, but otherwise microscopically not much is changed (see p. 407).

VI. Fabric. *Internal properties of grains, twinning.* Crystals of dyscrasite proper as found at Andreasberg are all pseudo-hexagonal triplets as revealed by the extinction patterns in basal sections observed between + nic. (fig. 328). Sub-individuals are also observed. — *Zoning* is not observed but deformations occur sometimes in crystals from Andreasberg.

Exsolutions. Dyscrasite proper itself is not rare as an exsolution in the high temperature galena of Broken Hill, N.S.W. The compact dyscrasite of Andreasberg, does not contain unmixing. On the left side of the Ag–Ag_3Sb-diagram as shown in figs. 329, 330, 331, 313E — all of Cobalt — the unmixing and breakdown structures are rather complicated. First are visible lancet- or oleander-leaf exsolutions in silver II or (rarely) silver I. It is "allargentum"* (ε-silver) originated by breakdown of a "silver III" into

* The name "allargentum" given by the author for a rather frequent component in the Ag–Sb-system was proposed according to "allopalladium" formerly regarded (wrongly!) as a hexagonal modification of Pd.

the silver II and allargentum. Such oleanderleaves" occur in *Silver Islet*, Chañarcillo, Colquechaca and Harmsarvet. — Unmixing lamellae of "silver I" in form of thin tabular sets — oriented *only* // the originally hexagonal "Silver III" are observed in Cobalt only (see later!).

Since per definitionem the ε-phase should represent "allargentum" that name must be transfered to the lamellae! Unexplicable remains then first the obvious (0001) lamellation of the latest silver-lamellae (silver I) and then the anisotropism of silver II, which cannot be missed. Spoken with all care: The most probable explanation is that the lamellae of ε-phase and silver II formed first one single phase of hexagonal symmetry into which "Silver I" lamellae // (0001) have immigrated; immediately later Silver II and allargentum separated. But then again remains enigmatic, that in the breakdown process the allargentum spindels formed in very probably octahedral arrangement in Silver II. Obviously that could be interpreted by an orientation // the hexagonal (0001) and an octahedron-like pyramid $(h0\bar{h}1)$ or $(hh\overline{2h}1)$ — but then further sets of lamellae should be observed; perhaps it happens sometimes really (fig. 329 and 331).

All this refers to phases Ag-richer than Ag_3Sb.

Structure and texture. Dyscrasite proper, formed on its own, exhibits crystal forms mostly of columnar habit, less commonly equal and even tabular. The "mixed-crystals" mostly occur in allotriomorphic granular masses, with hardly distinguishable grain shapes.

VII. Special fabrics. Dyscrasite is hardly ever affected by replacement phenomena, but similar to native silver, it may itself replace a variety of "older" ore minerals, such as niccolite and maucherite at Cobalt, Ontario. In particular circumstances, perhaps already in the course of a normal weathering process, Sb may be leached from the marginal parts of the brittle, creamish-yellow mixed crystals, occasionally originating from pyrostilpnite. The mixed crystals themselves are then rimmed by the brighter non-brittle native silver.

Intergrowth with native arsenic is not uncommon in dyscrasite from Andreasberg. Again, a thick rim of arsenic is intergrown with vermicular native antimony. Intergrowths of this kind (fig. 328) at least in part, correspond to what has been called "arsensilber" in the literature.

VIII. Diagnostic features. "Dyscrasite", or all of its four components, respectively, because of its very high reflectivity and cream-yellow color, is easily mistaken for either native antimony, silver, and bismuth, and, may be, for some of the very rare bismuth-silver minerals or similar compounds.

In contrast to antimony, it tarnishes already within a few days. Different from silver which tarnishes and exhibits various colors even more rapidly, it is characterized by its anisotropy, its slithly weaker reflectivity, and a slightly more yellowish hue. Bismuth is of much lower hardness and of lamellar twinning. The very rare tapalpite is very similar, but of much stronger anisotropy. The orthorhombic and hexagonal component on the one side, and the cubic mixed crystals on the other, are distinguished by their optical behavior between $+$ nic.

IX. Paragenetic position. Dyscrasite in larger quantity is found only in a few ore deposits of hydrothermal origin connected with intrusives or pseudo-intrusives but

Fig. 329 500 ×, imm. RAMDOHR
Comagas Mine, Cobalt, Ontario

"*Dyscrasite*": i.e. an intergrowth of the following: *dyscrasite* proper (grey) exsolved in spindleshaped lamellae out of a cubic (Ag, Sb) mixed crystal, *residual mixed crystal* of (Ag, Sb) stable at low temperature (light grey), *another mixed-crystal* similar to the above but without exsolution blebs (left-hand upper corner), and recrystallized *pure native silver* (white, right upper corner). Note the "reaction rim" of dyscrasite. *Chloanthite*, in part idiomorphic, is dark grey

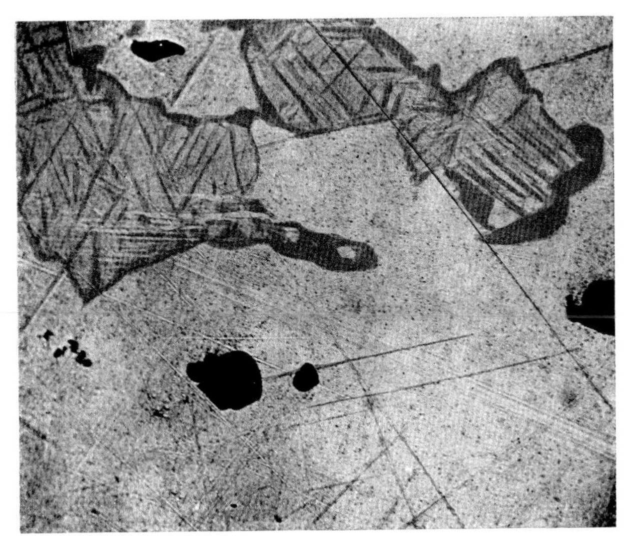

Fig. 330 250 ×, imm. RAMDOHR
Beaver Mine, Cobalt, Ontario, Canada

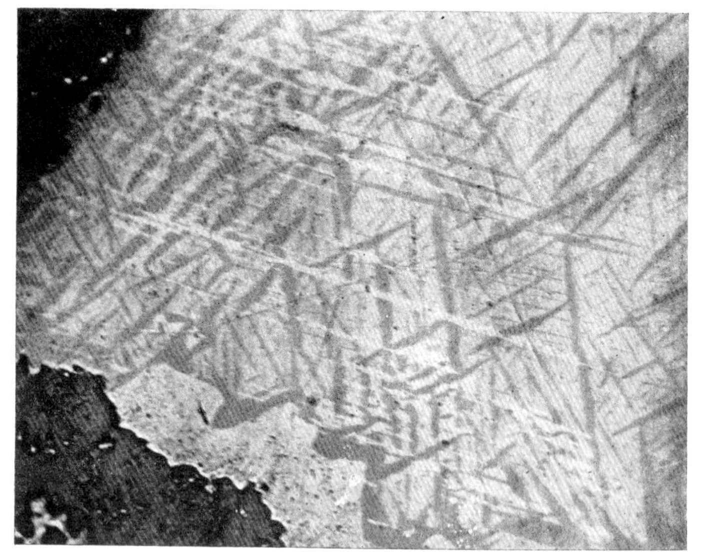

Fig. 331 500 ×, imm. Ramdohr
Beaver Mine, Cobalt, Ontario

"*Dyscrasite*": Originally a (Ag, Sb) mixed crystal, exsolved to *spindle-shaped* with a groundmass representing (Ag, Sb), but which cannot be cubic, since lamellae of native silver (white) are intergrown with it only in one direcion. This is the ε-modification of the (Ag, Sb) mixed crystals called "allargentum"

Fig. 331E 600 ×, imm. Ramdohr
Beaver Mine, Cobalt, Ontario

"*Allargentum*" with inclusions (following 0001) of As-rich *silver* (whitest). The original allargentum is now broken down into greyish white allargentum s.str. and grey *dyscrasite*. The dark grey mineral is *rammelsbergite*, black is *quartz*

may occur in great abundance and then constitute an important silver ore. According to the above mentioned textural relations dyscrasite intergrown with an Ag–Sb mixed crystal must be interpreted as being of hypogene origin. It appears that occasionally it can also be formed in the zone of secondary enrichment and then without the usual association, as reported by WHITEHEAD from Chañarcillo.

Investigations by the writer and DOLMAGE have confirmed that dyscrasite in the form of minute granules occurs as a silver-bearing mineral in numerous occurrences of bornite and galena. In most cases, these granules are too small for their certain identification.

X. Investigated occurrences. Andreasberg, Harz (numerous samples); Wolfach, Baden; Kongsberg and Sulitelma, Norway; Cobalt, Ontario (very numerous samples); Harmsarvet, near Gryksbo, Sweden; Chañarcillo, Chile; Colquechaca, Bolivia; Broken Hill, N.S.W.

XI. Literature. Dyscrasite is dealt with in a considerable number of publications and minor communications. In particular, the so-called "dyscrasite" from Cobalt has attracted the attention of various investigators, e.g., T. L. WALKER, E. THOMPSON (1930), F. N. GUILD, (1917) and others. The writer (1924) gave some details on the dyscrasite from Andreasberg, an X-ray analysis of which was carried out by F. MACHATSCHKI. Further details are by ÖDMAN (personal communication) and by PEACOCK (1940). The writer is obliged to Professors LAVES and PEACOCK for discussions regarding allargentum first found by him. A. B. EDWARDS (1954) already commented on the possible existence of the ε-Ag. MARKHAM & LAWRENCE (1962) observed allargentum with excellent unmixing networks in ores of the Consols lode near Broken Hill, N.S.W.

Essential is further a paper of STUMPFL (1967). The most important seems to be the work of PETRUK and CABRI (1969).

XII. Powder diagram (ASTM). (6) 2·58, (6) 2·40, (10) 2·28, (6) 1·765, (6) 1·500 Å.
Allargentum (PETRUK) (4) 2·548, (10) 2·370, (6) 2·52, (4) 1·353 Å.
A doubtful alloy Ag with 8% Bi has been called **"Chilenite"**

DYSCRASITE OF COBALT

The occurrences of Cobalt are after new observations at first glance only similar to many others but differ really very distinctly. Dyscrasite proper seems to be in Cobalt rare, if present at all.

In abundant material from Comagas and Beaver Mines the writer found the following textures in obvious breakdown-structures: oleander-leaf lamellae embedded in a matrix which contained often *one* system of unmixing lamellae of isotropic and *bright* silver I, myself, knowing that the ε-modification of AgSb is hexagonal, explained the weakly but distinctly anisotropic matrix as ε-phase, naming it allargentum, and regarded the oleander-leaves as dyscrasite proper.

My explanation was approved by PEACOCK and others. The problem was only the octahedrally looking arrangement of the oleander-leafs; but that could be after a hexagonal base and trigonal pyramide. This explanation was wrong.

PETRUK & CABRI (1969) found after careful examination with the microprobe that actually the spindle-lamellae are the ε-phase and the anisotropic matrix silver II. The bright white silver lamellae go obviously back to a time when the oleander-leaves and silver II were still hexagonal and the exsolution of the spindle-lamellae to a time when silver II had become cubic. The problem remains now, that a phase containing silver II + ε-silver has never been observed before.

For further informations compare the explanation of the figures 329-331 E.

STIBIOPALLADINITE

I. General Data. *Chem.* Pd_3Sb (DESBOROUGH means Pd_5Sb_2). *Cryst.* Data on symmetry vary B. F. LEONARD (priv. comm.) gives a_0 b_0 c_0 = 12.80, 15.04, 11.36 Å. RAMDOHR & SCHMITT found o'rhombic too but other cell-size. Against that GENKIN considers probably hexagonal. Mostly without x-forms. GENKIN reproduces barrel formed hexagonal crystals. — H = 4, D = 9.5. White metallic with bronce tint. The meanwhile asserted identity with "*allopalladium*" markers hexagonal very probable.

The formula is not approved, sometimes given as Pd_5Sb_2 or Pd_8Sb_3. GENKIN considers hexagonal a_0 = 7.59, c_0 = 28,11 Å.

II. Polish excellent, hardness a bit lower than platinum. Traces of basal cleavage in weathered sections.

III. Reflection behavior. *Color and anisotropy. Reflectivity high, nearly metallic,* hardly lowered in oil. Bright white, with a yellowish hue, and faintly pinkish in oil, particularly when compared with pentlandite in the presence of chalcopyrite. *Bireflection* in air and oil mostly not detectable, or only in traces. *Anisotropy between* + nic. very weak and easily overlooked at high magnifications; more distinct along grain boundaries when using medium-power oil immersion.

		air	oil
FRICK	green	51·5	50·0
	orange	54·0	53·5
	red	57·0	52·0 (?)
GENKIN	460 nm	48·0	
	540 nm	54·5	
	580 nm	55·4	
	660 nm	57·8	

IV. Etching. (accord. to SCHNEIDERHÖHN). *Positive*: aqua regia, structure etching, not very distinct. HCl (conc.) + $KClO_3$ produces superficial and grain-boundary etching in 3 to 5 seconds, with light-yellow to light-brown staining. All other reagent negative, particularly conc. HCl and HNO_3.

V. Physicochemistry. Apparently not yet sufficiently investigated. Natural weathering quite easily produces native Pd besides a little PdO.

VI. Fabric. Mostly coarse-grained aggregates of allotriomorphic grains. Etching reveals various textures which are incompatible with cubic symmetry. Grains of a stoutly columnar to thick tabular habit elongated towards the centre of grain aggregates may be made visible by etching. In polished sections from Rietfontein minute isolated crystals perched on cauliflower-shaped aggregates of Pt and hematite. The habit of these crystals is thick-tabular, with hexagonal cross-section; these show between + nic. an aragonite triplet pattern, from which an *orthorhombic symmetry* might be inferred, without excluding the possibility of a still lower symmetry.

VII. Special fabrics. Stibiopalladinite is frequently replaced by sperrylite which either encloses the former or penetrates into it as idioblasts (fig. 332). Weathering produces limonite-like palladium oxide (± antimony oxides) in rhythmic layers arising from grain boundaries or cleavage cracks, and mixed with powdery, sometimes massive, palladium.

VIII. Diagnostic features. Often difficult, particularly in view of misleading published data. Stibiopalladinite, among the ore forming platinoids seems to be one of the most frequently occurring mineral (with sperrylite being the most common one). Discriminatory characters are its excellent polish, the pinkish-yellow color, particularly noticed in oil immersion, it its almost unnoticeable anisotropy, particularly when contrasted with niccolite and the recently described PtAs, and the relatively easy weathering. The previous table gives a synopsis of the discriminatory characters of the minerals of platinum ores, with the exception of the easily-recognized iridosmium, and the little known potarite.

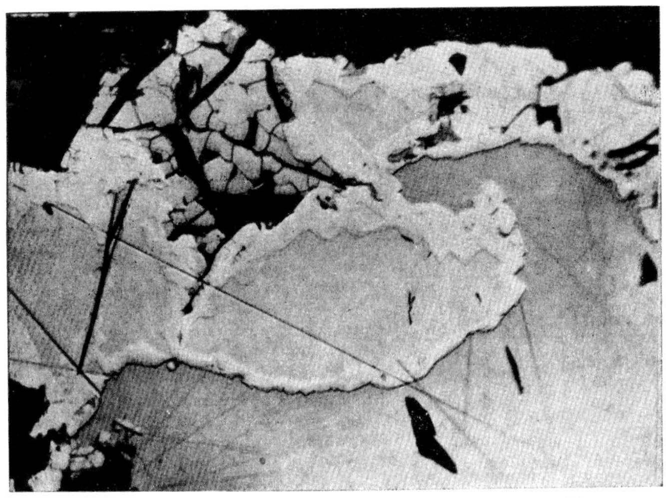

Fig. 332 500 ×, imm. RAMDOHR

Sandsloot Central Mine, Potgietersrust, Transvaal. *Stibiopalladinite* (greyish-white, smooth) with reaction rim of *sperrylite* (strong relief) and a little *native platinum* (white). Silicates and cavities black

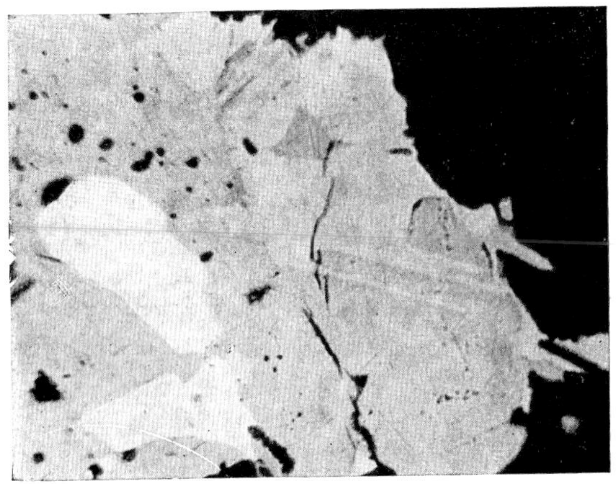

Fig. 332a 150 ×, imm. TISCHENDORF
Tilkerode, Harz

Allopalladium (greyish white, tattered skeletons) in *clausthalite* (grey, trace of cleavability), next is *gold* (white; black is gangue)

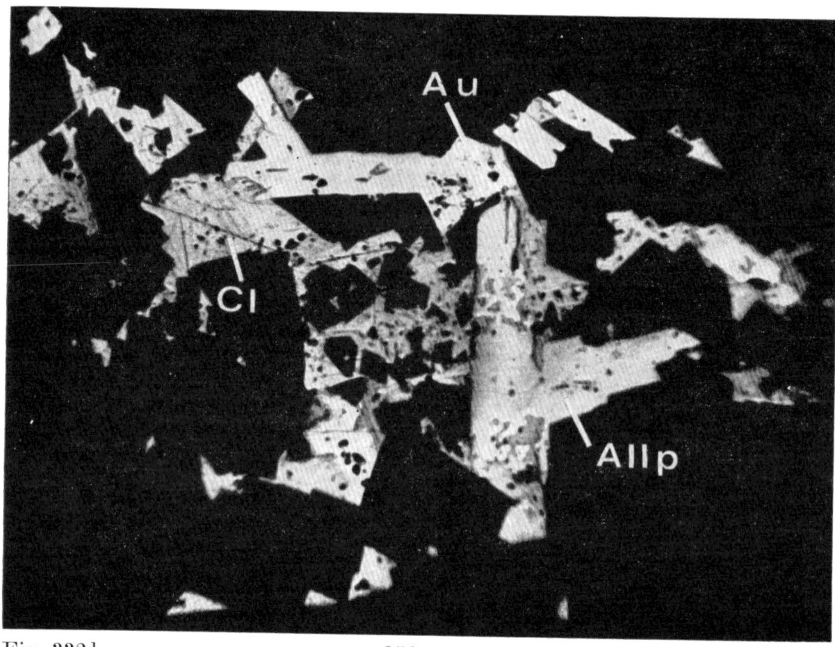

Fig. 332 b 250 × RAMDOHR
Tilkerode, Harz

"*Allopalladium*" (Allp.) lath-like crosscuts with *gold* (Au), a little brighter and softer, and *clausthalite* (Cl), soft and greyish, in quartz, dark.

IX. Paragenetic position. Because of its predominant occurrence in very small aggregates stibiopalladinite has been incompletely investigated. However, it seem to be of widespread occurrence and the principal palladium mineral in the related platinum deposits in the Bushveld complex which in part are of a contact-metasomatic nature. In these deposits it is associated with native platinum, sperrylite, cooperite, braggite, besides predominant pyrrhotite, chalcopyrite, pentlandite, etc. It was found in abundance and almost without the above associates, in graphic pegmatites at Tweefontein in the Potgietersrust district, Transvaal, where it is perhaps of a lateral-secretionary origin. Later, the writer observed tiny crystals of stibiopalladinite in rich ores of the platinum-quartz veins from the Elephantwinze near Rietfontain, Waterberg, Transvaal. It is also found in platinum nuggets from Choco, Colombia, South America. In each case, ample material was at the writer's disposal. In Sudbury, stibiopalladinite is missing.

XI. Literature. Stibiopalladinite was first recognized by REUNIG, named by P. A. WAGNER (1925) and described by H. R. ADAMS. A more detailed description, by SCHNEIDERHÖHN (1929) could be considerably augmented and in part corrected by the writer.

XII. Powder diagram (SCHMITT, Heidelberg). (10) 230, (8) 225, (7) 216, (5) 159, (4) 132, (4) 128 Å.

RICKARDITE

I. General data. *Chem.* Approximately Cu_4Te_3 (accord. to PATZAK: Cu_7Te_5), or $Cu_{4-x}Te_2$ where $x \sim 4/3$. *Cryst.* Tetragonal, D^7_{4h}; $a_0 = 3{\cdot}97$ kX, $c_0 = 6{\cdot}11$ kX (PEACOCK). Optic uniaxiality was predicted by the writer. Cleavage not observed. — $H = 3$, $D = 7{\cdot}6$. — Opaque. On fresh fracture surfaces a characteristic purplish-red, somewhat lighter and less tarnishing than umangite with which it bears similarity. When kept in collections for a long period, umangite darkens more intensively than rickardite.

Table for the identification of the more important *platinoid minerals*

Mineral	Hardness	Color in oil	Reflectivity in oil	Anisotropy in oil 1 nic.	+ nic	Remarks
Platinum	—	white to bluish	extremely high	isotropic	—	—
Palladium	$<$ platinum	white to yellowish	extremely high	isotropic	—	often secondary
Iridium	$\gtrsim$ platinum	white, similar to Pt	extremely high	isotropic	—	—
Stibiopalladi-nite	$\gtrsim$ platinum	pinkish-yellow, near chalcopyrite: distinctly pink	high ($>$ pentlandite) $\sim R_0$ of niccolite; compared with Pt a little brownish	rarely visible in traces	very weak, but often noticeable	cleav, basal rarely visible
Cooperite	$<$ platinum	coffeebrown to olive leather brown	moderate, $\sim$ pyrrhotite	—	almost unnoticeable (distinct. from pyrrhotite)	basal cleav.
Braggite	$>$ platinum and stibiopalladinite	bluish-grey-white, or brownish-white	high; $>$, but near to stibiopalladinite	recognizable, bluish-grey or brownish-grey	very distinct	—
Sperrylite	$>$ braggite $<$ laurite	white to distinctly bluish	high	isotropic	—	the recently discovered geversite is very similar
Laurite	$>$ sperrylite	white, somewhat to bluish	high	isotropic	—	

This tabella on properties of platinoid-minerals is at this moment very strongly outdated. During the last 20 years surely 60 new members have been described — more or less reliably — that it is impossible to look over all possibilities of similarities or errors. — In the meantime (1975) about 20 further ones have been described from new deposits in China. — Since there is some probability that the most frequent ones were first recognized the tabella may remain.

II. Polish. In spite of its considerable porosity rickardite without difficulty takes a good polish.

III. Color and reflection behavior. Highly characteristic, mottled red and greyishpurple colors visible without nicols. Moderate reflectivity. SINDEEVA gives (for white) $R_0 = 20\%$, $R_E = 14\%$.

Anisotropy: Between crossed nicols, effects of anisotropism are of an exceptional intensity as matched or surpassed only by very few other minerals: very vivid colors and hues, e.g., canary-yellow and a saturated brownish-red. Rickardite behaves as an optically uniaxially mineral, i.e., basal sections do not exhibit pleochroism and remain dark between + nic.

IV. Etching (accord. to MURDOCH and SHORT). *Positive*: HNO_3, immediate effervescence and blackening; HCl, greyish-brown; with KCN slowly bleaching to grey; $FeCl_3$ bleaching to greyish-blue; $HgCl_2$, purplish-red coating (SHORT). – Discoloration to bluish-green (DAVY & FARNHAM); KOH shiny coating.

V. Physico chemistry. Rickardite can be easily synthesized by several methods which in part may be similar to natural processes.

VI. Fabric. A specimen from Gunnison Co. exhibited a peculiar porosity and consisted of very small grains of random orientation, or of botryoidal aggregates of regularly arranged grains. The latter texture seems to indicate that the aggregate may have been formed as a pseudomorph after an originally cubic mineral, perhaps α-weissite, the high-temperature polymorph of Cu_2Te. A white, anisotropic and tabular mineral (perhaps weissite) was intergrown corresponding to the cubic orientation. Here, occasionally, recrystallized rickardite of larger grain size was observed.

In a sample from Hinokizawa rickardite replaced native tellurium either in the form of a network or by forming a broad rim. In the same sample, it is itself replaced by weissite.

In bornite and chalcocite ores from Namaqualand rickardite is formed as a supergene secondary mineral from melonite and chalcocite. Artificially produced crystals are of a columnar habit and a pronounced zoning, with a deep-red core and a purplish-grey outer shell. These crystals exhibit two types of twin lamellae: 1. a few isolated, sharp lamellae, probably // to a pyramid plane; 2. a system of fine lamellae permeating the whole section.

	in air	in oil
general	light purplish-red, somewhat lighter than umangite; many sections also pronounced violet	colors even more vivid; reflectivity is hardly diminished
contrasted with umangite, lighter, particularly in oil; marked differences from all other opaque minerals		
reflection pleochroism	striking effects, very pronounced O: crimson-red E: purplish-grey, considerably darker	intensified O: lighter, bright-red to salmon-pink E: darker, diffused purplish-grey

VIII. Diagnostic features. Cannot be mistaken for any other opaque mineral.

IX. Paragenetic position. The occurrences from the Vulcan mine, Gunnison Co., Colorado, U.S.A., from the Teine mine, Hokkaido, and from Hinokizawa, Rendaizi, Sizuoka, Japan, are all veins rich in tellurium ores of the subvolcanic type. – Ookiep is a bornite deposit of high temperature of formation. Samples from all of the above deposits were at hand as well as synthetic material.

XI. Literature. Brief data are found in the works by MURDOCH, DAVY & FARNHAM, and SHORT. A detailed description by STILLWELL (1931, 1953) pertains to weissite. More complete descriptions may be found in difficultly obtainable Japanese publications, particularly by M. WATANABE (1934).

XII. Powder diagram (THOMPSON in Å). (6) 3·35, (4) 2·54, (10) 2·07.

WEISSITE

I. General data. *Chem.* Cu_2Te, with slight excess of Te. *Cryst.* (pseudo-) cubic, a_0 = 7·23 Å, Z = 8. Color white, tarnishing to intensive blue. Opaque. Dissolves in H_2SO_4 with deep-red color. With the exception of its color, very similar to rickardite in most properties and reactions. Synthetic Cu_2Te is hexagonal, a_0 c_0 = 4·24, 7·27 Å.

Accord. to M. WATANABE (1934) weissite closely corresponds to synthetically produced Cu_2Te which takes up to 5% excess Te, very similar to the composition of weissite. The system Cu–Te was already investigated by M. CHIKASHIGE a long time ago, before the natural occurrence of corresponding Cu–Te minerals was known. More recently it was worked out thorougly by R. MOLÉ (1954) and I. PATZAK (1956).

WATANABE gives the following data referring to both the natural ("n") and synthetic ("s") material. *Polish* excellent. Talmage hardness: B.

III. Reflection behavior. Light-grey, similar to common "normal" chalcocite (accord. to SHORT), but creamish-white to dark blue accord. to WATANABE: photoelectric cell (accord. to MOSES) R = 31·8. Distinctly anisotropic, reflection pleochroism weak, but distinct.

IV. Etching (accord. to SHORT and WATANABE). *Positive* all normal etching reagents. HNO_3, conc., immediate effervescence (n) and brown stain: HCl, $FeCl_3$, $HgCl_2$, KCN, KOH stain brownish, but with different velocity.

V. Physicochemistry. Cu_2Te may take up to 5% Te in excess of formula (see above); above 351° it occurs in a cubic, high-temperature modification, below 351° it changes into the ordinary pseudo-tetragonal modification.

VI. Fabric. Little known. Occasionally in lamellar form in rickardite from Gunnison Co., the latter being developed as a paramorph. SCHERBINA (1941) observed weissite as a rim around pseudomorphosed nagyagite.

VIII. Diagnostic features. Not yet fully reliable; the blue tarnish developed in air after about seven days may be taken as characteristic.

X. Investigated occurrences. Good Hope mine near Vulcan, Gunnison Co., Colorado, U.S.A.; Teine mine, Hokkaido, Japan. – Synthetic material. The writer is obliged to Professor T. WATANABE for translating the paper by M. WATANABE from the Japanese original.

XII. Powder diagram (THOMPSON). (10) 3·65, (7) 3·22, (4) 2·09, (5) 1·99, (3) 1·45 Å.

VULCANITE

I. General data. *Chem.* CuTe. *Cryst.* O'rhombic, a_0 b_0 c_0 = 4·09, 6·95, 3·15 Å, Z = 2. Aggregates of prismatic or platy, but more often irregular grains. Two pinacoidal cleavages, H = 1–2. Light bronze to yellow. *Polish* is, in spite of the low hardness, very good.

III. Reflection behavior. Vulcanite has *very* peculiar optical properties! The reflection pleochroism is so strong, that it can be observed without nicols, and with the naked eye.

The reflectivity varies in different direction strongly (in white acc. C. & TH. ~ 58% to ~ 36%).

The bireflection is extremely high and increases in oil immersion. One direction (showing the best cleavage!) is bright yellow, like millerite or chalcopyrite, the others blue grey and ink blue resp.

The anisotropism is extreme, likewise, in air with yellow white, yellow orange and grey colors. — Straight extinction. — No internal reflections.

IV. Etching. (accord. to CAMERON & THREADGOLD): *Positive* HNO_3, HCl, $FeCl_3$, KCN. *Negative*: $HgCl_2$ and KOH.

VI. Fabric. Aggregates of small (1–10 mm) elongated grains intergrown with rickardite. Twinning according to a prism in two directions symmetric to a pinacoid is actually always present.

Probably exsolutions of rickardite.

IX. Occurrences. From associations rich in tellurium and copper together with weissite, tellurium, rickardite, and an unknown mineral, e.g. Teine Mine, Japan and Hood Hope Mine, Vulcan, Colorado, but surely more widespread.
Can be made artificially without difficulty.

XI. Literature. A mineral with the probable composition CuTe has been observed and in its exceptional optical properties described shortly by the present author in the ore of Teine, Japan (former editions of this book). CAMERON & THREADGOLD (1961) named it and gave full data (mostly repeated here). — Artificially it has been prepared as early as 1907 by M. CHIKASHIGE and M. WATANABE (1934).

XII. Powder diagram (C. & TH.). (6) 3·52, (7) 2·86, (10) 2·03.

CAMERON, E. N. & THREADGOLD, I. M., Vulcanite, a new telluride from Colorado. Amer. Min. 46, 258–268, 1961.

MELONITE

I. General data. *Chem.* Member of a solid solution series NiTe–$NiTe_2$, often close to $NiTe_2$. *Cryst.* hexagonal, c/a = 1·35 (NiAs structure, with systematic absence of layers). a_0 = 3·83 Å, c_0 = 5·25 Å, D^8_{3d} , Z = 1 × $NiTe_2$. Cleav. // (0001), differently perfect. — H = 1 to $1^1/_2$; Ni-rich till 4, D = 7·73. — Tin-white to pinkish. Strongly reflecting like bismuth, tarnishing to yellowish-brown. Data for both macroscopic and microscopic observation vary distinctly with the composition. Mixtures high in Ni are characterized by a less perfect cleavage, greater hardness and lesser flexibility.

II. Polishing properties. Good polish not always easy to obtain. Specimens bearing evidence of tectonic deformation, may retain a pitted and scratched surface. Polishing hardness quite variable, apparently increasing with increasing Ni-content and may (accord. to STILLWELL) occasionally be $\lesssim$ chalcopyrite, obviously > gold and krennerite; in other cases, e.g., in the Robb-Montbray mine, considerably softer than gold and montbrayite, approximately the same as tellurobismutite. — Cleavage // (0001) often visible in polished section. Sections approximately // (0001) may exhibit peculiar cleavage pits, perhaps due to translation and described as "Doppelwinkel" ("crosses"), similar to the percussion figures in calcite.
Talmage hardness: B.

III. Reflection behavior. White to faintly pinkish; high reflectivity, brighter than krennerite, and less lowered in oil than for associated minerals. SINDEEVA gives semiquantitatively 60%. Bireflection exceptionally weak, distinctly noticeable in oil only. O slightly brighter, and more creamish-white, E slightly more pinkish; these differences are particularly visible when contrasted with native tellurium. Between + nic., *anisotropy* weak, but distinctly visible when using medium-power oil immersion; extinction straight. Basal sections may exhibit a faint anisotropy if bent by deformation.

Variation in R can be expected regarding the different composition. The very high bireflection mentioned by *Besmertnaja* could never be observed.

IV. Etching (accord. to MURDOCH and SHORT). *Positive*: HNO_3, immediate effervescence and black stain; $FeCl_3$, weak brownish stain. *Negative*: KCN, HCl, KOH, $HgCl_2$; HCl sometimes gives light-brownish stain.

VI. Fabric. Little known. In a sample from Kalgoorlie, kindly provided by Dr. STILLWELL, melonite occurs enclosed in krennerite as thick-tabular crystals rich in crystal faces. Accord. to STILLWELL it also occurs together with calaverite, coloradoite and tetrahedrite. At Ookiep and other deposits, it occurs as minute tabular crystals in chalcopyrite, sometimes twinned with obtruse angles. Melonite in ores from the Robb-Montbray mine is of a different appearance, generally xenomorphic, but lamellar to coarse-myrmekitic when intergrown with tellurobismutite, having the basal planes in common. Peculiar differences in hardness occurring along irrgular zones are characteristic for the latter occurrence (see above).

In ores from Ookiep, crystals of melonite intergrown with chalcopyrite are weathered and have produced rickardite by chemical interaction.

A slightly bluish-white as well as a light-greyish mineral may occasionally be formed as rims by decomposition, and may be interpreted as native tellurium and weissite, respectively.

VIII. Diagnostic features. Diagnosis is made difficult because of variability of hardness. In a paragenesis of nickel-bearing minerals with tellurides the presence of melonite is to be considered as possible.

IX. Paragenetic position. Associated with other Te-minerals in Western Australia and Calaveras, California, U.S.A.; in nickel-bearing high temperature copper deposits of Namaqualand, S. Africa. The mineral N described by SCHOLTZ (1936) as occurring in Insizawa probably is melonite. PEACOCK & BERRY (1940) identified the structure of the natural mineral, at the same time describing two new occurrences.

X. Investigated occurrences. Kalgoorlie, Western Australia; Concordia mine, Spectakel mine, Springbockfontein mine, all near Ookiep, Namaqualand, S. Africa; Robb-Montbray mine, Quebec, Canada.

XII. Powder diagram (THOMPSON, in Å). (10) 2·81, (5) 2·05, (5) 1·912, (6) 1·544.

KITKAITE

I. *Chem.* NiTeSe. *Cryst.* hexagonal, a_0 c_0 = 3·72, 5·13 Å, D = 7·19. Pale yellowish.

III. Reflectivity. R_0 in air: green 56, yellow 65, red 62%, in oil 50, 56, 54% resp. Bireflection high, O pale yellowish grey, E reddish.

IV. Etching. *Positive*: HNO_3 strongly, $HgCl_2$ faintly positive.

IX. Occurrence. Observed by HÄKLI et al. from Kuusamo, Finland, in carbonate-albite-veins together with sederholmite β-NiSe, blockite ($NiSe_2$), clausthalite, polydymite, trüstedtite (Ni_3Se_4), paraguanajuatite (Bi_2Se_3), gold.

XII. Powder diagram. (HÄKLI): (10) 2·729, (5) 2·01, (3) 1·860, (2) 1·535, (4) 1·510 Å. All data from HÄKLI et al. (1965). Similar to melonite.

Imgreite, NiTe, from Kuusamo is distinctly different from kitkaite and melonite. Powder diagram: (5) 3·21, (10) 2·88, (7) 2·31, (5) 1·964, (5) 1·588 Å.

HESSITE

I. General data. *Chem.* Ag_2Te. *Cryst.* Macroscopically often in "cubes" (see below) and in fine (pseudo ?) rhombic crystals in part highly distorted. Microscopically recognized as anisotropic, accord. to ROWLAND & BERRY (1951) orthorhombic (cf. as contrasted to TOKODY). $a_0 = 16{\cdot}27$ Å, $b_0 = 2{\cdot}628$ Å, $c_0 = 7{\cdot}55$ Å; $Z = 48$; the high-temperature cubic modification has $a_0 = 6{\cdot}64$ Å, with $Z = 4$. Acc. to FRUEH it is monoclinic (C_{2h}^5), $a_0\, b_0\, c_0 = 8{\cdot}13,\ 4{\cdot}48,\ 8{\cdot}09$ Å, $\beta = 112°\ 55'$, $Z = 4$. – Cleavage (100) indistinct. H = 2–3; brittle to sectile. D = 8·3–9, calculated 8·35. – Opaque, steelgrey, with pinkish hue on fresh fracture surface; weak metallic lustre.

Fig. 332 c 250 ×, imm. + Nic. McAlpine Mine, Col. RAMDOHR

Hessite, a bit scratched, with typical inversion lamellae. At the corners *quartz* with plentiful internal reflections

II. Polish. Too soft to take good polish, similar argentite with plenty of scratches. Cleavage not visible; but may become visible on heating above the inversion point (BORCHERT, 1930). Hardness low (nearer to 2 than to 3), < altaite and < gold.

Talmage hardness: A+.

III. Reflection behavior. *Color and reflection.* Reflectivity moderate, strongly lowered in oil. Color greyish-white, in oil changed to brownish (see table).

IV. Etching (accord. to DAVY & FARNHAM, and MURDOCH). *Positive*: HNO_3, immediate strain, dark-brown, rough surface. KCN, weak blackish stain and dissolution, very rough surface. $FeCl_3$, tarnish, rubs off. *Negative*: HCl (accord. to DAVY & FARNHAM, faint coating). KOH. Light-etching not noticeable after ten minutes, tarnishes, however, in full daylight after 2 to 3 days with varying intensity, preferably along grain boundaries. HNO_3 (accord. to BORCHERT) is an excellent agent for structure etching.

V. Physico-chemistry. Accord. to BORCHERT hessite changes into the cubic modification at 155°, with $a_0 = 6{\cdot}64$ Å, $Z = 4$. The lamellar texture exhibited by hessite from almost all occurrences may serve as an important geological thermometer and as evidence for hypogene origin.

VI. Fabric. *Internal properties of grains.* Between + nic. the lamellar texture resulting from inversion is very characteristic and almost always noticeable. The orientation of the lamellae in granular aggregates is not yet established with certainty.

Macrostructure and textures are little investigated. Occurs in crystals but more commonly in loose or compact aggregates often of polygonal grains which are little interlocked. Grain size in samples investigated was from $^1/_2$ to 1 mm.

STILLWELL (1931) observed hessite from Kalgoorlie in myrmekitic intergrowth with sylvanite. It may in part have formed from krennerite. Samples unusually rich in gold (21·4%), are actually a microscopic network of numerous minute veinlets of native gold.

Photocell	air				photometer ocular CISSARZ			
nm	460	540	580	640		green	orange	red
LOGINOVA R_g	40·7	40·7	41·3	43·0	in air	29	27	26
R_p	40·3	38·7	37·8	37·4	in oil	43	40	42
FASTRÉ & ORCEL	41·1	39·1	37·5					
reflection pleochroism	hardly visible			quite distinct, dull brownish-white; white to greyish-blue-purplish				
anisotropy between + nic.	quite distinct; in diagonal position dark-orange and dark slate-blue, respectively			very distinct; polarisation effects along scratches interfere; characteristic colors although of moderate intensity only; dark brownish-purple and weakly olive-yellow				

VIII. Diagnostic features. The imperfect polish, low hardness and its association with other Te-minerals are characteristic. Argentite has a much weaker anisotropy; in oil argentite acquires a greenish hue, hessite a brownish one.

IX. Paragenetic position. Associated with other tellurides (calaverite, sylvanite, etc.), native gold and native tellurium, mostly in subvolcanic auriferous and argentiferous veins, e.g., from Siebenbürgen, Roumania, and many other places. A different type of deposit is at Kalgoorlie where hessite is found together with large quantities of other tellurides in metalliferous veins. Its occurrence in the Altai is still little understood. LEHNER showed that hessite has a strong tendency to precipitate gold from its chloride solutions, a feature which could also be significant in its natural deposits. In tiny specks in nickeliferous pyrrhotite deposits (Sudbury, Insizwa, Abu Suwagel).

X. Investigated occurrences. Glava, Värmland, Sweden; Baita Bihorului and Săcăramb, Rumania; Sawodinsk, Altai, U.S.S.R.; Hinsdale Co., Colorado, U.S.A.; Frood mine, Sudbury, some of Quebec, Canada; Kalgoorlie, Western Australia.

XI. Literature. Data about etching were given by MURDOCH and by DAVY & FARNHAM; some microscopic data by VAN DER VEEN (1925). H. BORCHERT (1930) determined the inversion temperature in synthetic material, at the same time stressing its significance as geological thermometer.

XII. Powder diagram (ROWLAND & BERRY, in Å). (6) 3·01, (8) 2·87, (10) 2·31, (7) 2·25, (6) 2·14; many other lines!

PETZITE

I. General data. *Chem.* Ag_3AuTe_2. Relation to hessite not fully understood; but probably not close. — *Cryst.* O^8, a_0 = 10·38 Å, Z = 8 (but not isotropic!), — H = $2^1/_2$, D = 9·13. — Opaque; dark-grey, metallic lustre.

II. Polish. Good polish obtainable in spite of brittleness, decidely better than in hessite. Talmage hardness: A +.

Fig. 332 d 250 ×, imm. RAMDOHR
Stanija, Roumania

Petzite, looking like galena, forms the matrix of a bit darker grains of *hessite*. The reflectivity of both is very similar but distinct at grain boundaries. Irregular cracks are typical for hessite. Besides that the section contains idioblasts of pyrite and traces of a needle-formed mineral, probably nagyagite

III. Reflection behavior. Published data in pt. contradictory, probably due to faulty identification. ORCEL and FASTRÉ gave photocell values (465 nm = 42·2, 527 nm = 39·8, 589 nm = 38·2%) which are identical, in the range of accuracy of measurement, with those of hessite, whereas most observers — including the writer — have observed a higher reflectivity.

Light grey, with a light-pinkish hue which is more pronounced in oil. In oil, particularly when contrasted with galena, the color is distinctly brownish-white.

Anisotropy with + nic. Distinctly weaker than in hessite, but quite noticeable. HELKE observed the association of an anisotropic with an isotropic component, which at the same time was slightly more greyish-purple and of better polish. This he interpreted as the association of an α- or high-temperature modification with a β- or low-temperature modification. However, the evidence given seems insufficient for this interpretation, and the relationship between the isotropic and anisotropic material needs further clarification.

IV. Etching. (MURDOCH and BORCHERT). *Positive*: HNO_3, immediate effervescence, dark-brownish stain; irregularly pitted surface; considerably stronger than with coloradoite.

$FeCl_3$, instant formation of an iridescent coating which rubs off easily. *Negative*: KCN, HCl (which occasionally may both produce an iridescent coating); KOH.

More detailed data on etching by BORCHERT (1935) who stresses the difficulty of its diagnosis by etching.

VI. Fabric. The writer had little opportunity of studying the fabric. HELKE (1934) observed, in material from a deposit in Siebenbürgen, the intergrowth of petzite proper (weakly anisotropic, with coarse twin lamellae), — and a minor amount of an isotropic component, of greyish-purple color. In a sample labelled "petzite" from Sacaramb stützite (?) was observed in addition to another weakly anisotropic, but pure-grey mineral. Accord. to STILLWELL (1931) and EDWARDS (1953), petzite from Kalgoorlie is a product of the decomposition of krennerite, and, jointly with sylvanite and native gold, a product of the decomposition of calaverite. Similarly, SCHERBINA (1941) noticed petzite to be younger than krennerite. STILLWELL stated that petzite is a very rare mineral and that samples so labelled often consist of an intergrowth of hessite and calaverite.

VIII. Diagnostic features. Difficult, particularly in view of the difficulty of obtaining suitable reference material. Coloradoite may be quite similar, but is attacked by HNO_3 less rapidly.

IX. Paragenetic position. Observed in auriferous veins of both the intrusive and subvolcanic cycle.

X. Investigated occurrences. Examples examined by the writer and originating from Sacaramb, Roumania, St. Vincent in Minas Geraes, Brazil, and Glava in Värmland, Sweden, were mostly unsuitable for diagnostic purposes because of their minute grainsize or ntricate intergrowths.

XI. Literature. See references quoted above.

XII. Powder diagram (THOMPSON, in Å). (10) 2·77, (5) 2·11, (4) 2·02.

Note. The so-called "antamokite", accord. to STILLWELL, is nothing but an intimate intergrowth of altaite, petzite, and native gold. It was first described though very incompletely by ALOIR from an occurrence in the Philippine Islands.

Fischesserite

Ag_3AuSe_4, has recently been described from an uranium bearing Se-deposit (Predborica, Bohemia) by JOHÁN et al. (1971). Isotypic with petzite, $a_0 = 9{\cdot}97$ Å. — Reflectivity: pink, isotropic, the color against chalcopyrite looks about between freshly-polished bornite and enargite.

R. V nm	420	460	520	560	620	660
	33·4	33·1	30·1	29·9	32·6	34·7

Powder diagram (JOHÁN). (10) 2·662, (8) 2·229, (8) 2·035, (8) 1·821, (7) 1·266 Å.

Note: Fischesserite is up to now the only compound of Au and Se.

STÜTZITE and EMPRESSITE

I. General. *Chem.* Both minerals for some time were mixed up; they are really very similar and genetically related, in so far as the low temperature compound AgTe at 210° breaks down into $Ag_5Te_3 + Te_2$. Empressite is AgTe, stützite Ag_5Te_3, which for the last one would mean a formation $> 210°$. *Cryst.* Acc. to HONEA the long known stützite is Ag_5Te_3; hexagonal, $a_0\ c_0 = 13{\cdot}38,\ 8{\cdot}45$ Å, Z = 16, D = 7·61. BERRY & THOMPSON call stützite "empressite I", empressite "empressite II". — Mostly finegrained masses with conchoidal fracture, stützite rarely in good crystals with many faces. — Brittle, $H = 3^1/_2$. Lustre metallic, black like petzite; empressite has a faint reddish to bronze tint. — *Polish* good, and easily obtained.

III. Reflection behavior. Moderate; contrasted with galena, of bluish-white, color. *Reflection pleochroism* very strong: light-grey: creamwhite, contrast intensified in oil. Anisotropy very strong. THOMPSON and others mention: white-greenish-yellow and russet-brown to brownish dark blue as "polarization colors" (for Stützite).

IV. Etching. *Positive*: HNO_3, weak effervescence, iridescent coating. $FeCl_3$, iridescent permanent coating; $HgCl_2$, immediate, similar effect. *Negative*: HCl, KOH. Etching data regarding "stützite" given by MURDOCH could not be confirmed.

V. Physico chemistry. Ag and Te melted together in the ratio 1:1 give an aggregate of empressite and tellurium, but in the ratio 5 Ag:3 Te produce a homogeneous material; with increasing amount of Ag hessite is formed. The investigation of the structure by THOMPSON et al. has not yet established the correct formula. Hydrothermal synthesis, even of good crystals, is possible without difficulty.

VI. Textural properties. At the original locality a medium grained aggregate in part of empressite, in part of stützite is present. The conditions for the formation of well formed crystals of stützite are rarely given.

IX.–X. Occurrences of "empressite", mostly later proved to be stützite, have been described mostly from Colorado (Golden Fleece, May Day, Red Cloud and Empress mines) and Sacaramb, Roumania. Empressite s. str. is known from Empress Josephine Mine only. The association always shows altaite, pyrite, chalcopyrite, sphalerite.

XI. Literature. The paper of HONEA (1964) brought only recently some elucidation into the before rather contradictory data of THOMPSON et al. (1951), BERRY & THOMPSON (1962) and a lot of older ones.

XII. Powder diagram (HONEA). Stützite: (7) 3·03, (7) 2·62, (8) 2·55, (10) 2·16 Å; empressite: (6) 3·81, (6) 3·33, (10) 2·70, (4) 2·31, (8) 2·23, (4) 2·13 Å.

SYLVANITE

I. General data. *Chem.* $AuAgTe_4$, in part with slight excess of Au. *Cryst.* monoclinic, C_{2h}^4, $a_0 = 8{\cdot}94$, $b_0 = 4{\cdot}48$, $c_0 = 14{\cdot}59$ kX. $\beta = 45°\ 36'$, Z = 2. Dendritic to graphic twin aggregates. — Cleavage // (010) perfect; brittle. H = $1^1/_2$–2, D = 8·16. — Opaque; high metallic lustre; silvery-white with yellowish tinge.

II. Polishing properties. In view of its low hardness and its association with predominantly very hard minerals, use of the polishing machine is essential. Cleavage rarely visible. Hardness low, well below that of sphalerite, but a little higher than that of nagyagite.

III. Reflection behavior. Considerably higher than that of galena, but color cream-white.

Reflection pleochroism highly distinct, both with and without oil immersion, and particularly marked along the boundaries of grains and twin lamellae. *Reflection pleochroism* in air: bright cream-white to darker cream-white brown, in oil: light-cream-white; creamish-brown, or leather-brown. Crystallographic orientation of the bireflection unknown. *Anisotropy* very strong; no darkness position obtainable, but close to it: "dirty" greyish-brown, rapidly changing with further rotation; in diagonal position pinkish-white: light greyish-white; in intermediate positions brownish, yellowish and bluish tints. In oil similar tints.

IV. Etching (MURDOCH). *Positive*: HNO_3, instantly discoloration to light-brown, iridescent coating; after rubbing-off: deep-brown. Aqua regia weak effervescence, vapors produce tarnish. *Negative*: KCN, HCl, $FeCl_3$, KOH, $HgCl_2$.

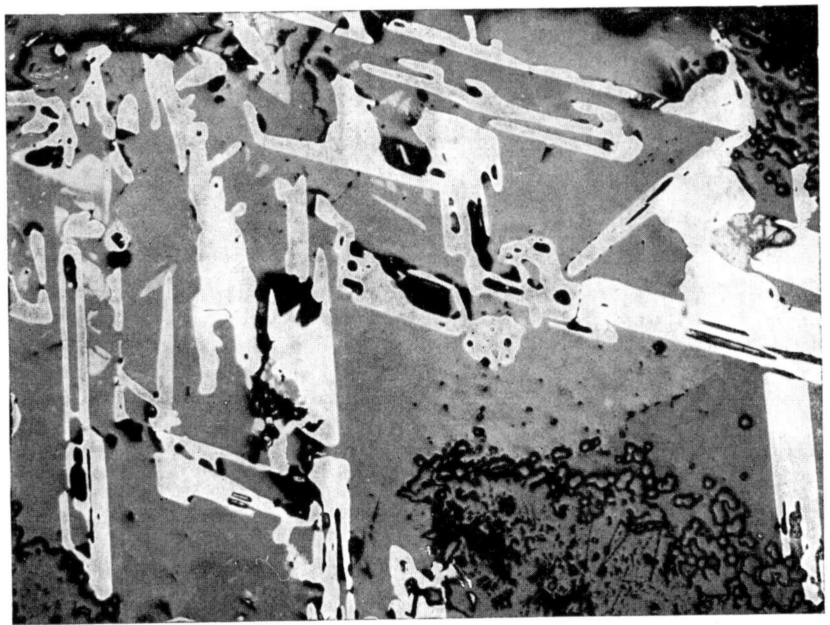

Fig. 333 15 × SCHNEIDERHÖHN

Offenbanja, Carpathians, Rumania

Skeletiform *sylvanite* crystals (white) in *calcite*; in right upper corner some *quartz*

	in air	in oil
general	quite high	strongly lowered, particularly for one direction
contrasted with		
tellurium	little more towards cream	ditto
galena	cream-white, lighter	
calaverite	darker	considerably darker
nagyagite	cream-white, lighter	lighter
altaite	darker	darker and more brown

	air				
VJALSOV	nm	460	540	580	650
	R_g	55·8	57·3	57·4	57·2
	R_p	41·9	44·8	45·5	44·6
FASTRÉ	R_g	60·0	61	60	These values are 10–20% higher than those of VJALSOV
	R_p	55	56	54	

Cleavage-etching // (010) is developed by HNO_3, as confirmed by BORCHERT (1930) who furthermore noticed a very indistinct cleavage // (001).

V. Phys.-chem. The relations of the Au- and Ag-tellurides are in spite of the careful examinations of SHERBINA & ZARYAN (1964) and MARKHAM (1960) not yet comprehensively explained and the data do not fit to all natural associations without difficulty. Very probably krennerite forms below 200° only, but near 200° it can contain higher Ag than calaverite appearing instead of the latter at ~ 200°. Sylvanite is rather exactly $AuAgTe_4$; higher excess in Au seems to signify a formation > 200°. Pure $AgTe_2$ does not occur; at high contents of Ag always appears the association sylvanite, empressite (? stuetzite) and nat. Te. — "Muthmannite" could not yet be synthesized.

VI. Fabric. Lamellar twinning very pronounced between + nic. Lamellae in part quite large, in other grains they are visible only in oil immersion. Twinning generally // (100), lamellae mostly oblique with regard to the extinction direction. Twins observed also in freely-grown crystals, as resulting from the growth process.

Interior parts of larger grains are often not twinned, where outermost parts are often full of twins. That may suggest wrongly two different minerals.

Fig. 333a 250 ×, imm. Vatukoula, Fiji RAMDOHR

A large grain of *sylvanite* is following an otherwise hardly visible cleavage altered to gold ("mustard-gold")

Structures and textures. Incompletely studied; well-developed crystals in vugs, often also skeleton-shaped ("schrifterz" = "graphic ore") (fig. 333). Also disseminated with a tendency towards idiomorphic development.

VII. Special fabrics. OEBBEKE & SCHWARZ describe native gold in skeletons in the core of sylvanite crystals. In the oxidation zone often decomposed, leaving gold in form of finest powder ("mustard gold"). Occasionally thereby an intermediate brownish material is produced, perhaps an independent mineral.

VIII. Diagnostic features. Not yet sufficiently investigated in its contrast to other minerals than auriferous tellurides. Easily distinguished from the latter by its marked reflection pleochroism, strong anisotropism effects and pronounced twinning. Nagyagite is darker, rather thin tabular, with pronounced cleavage. In contrast to other ore minerals, it is well characterized by its low hardness, twinning, and mineral association.

Its peculiar etching properties which distinguish it readily from calaverite and krennerite, have been described in detail by BORCHERT, but the optical phenomena may generally be sufficient for its diagnosis.

IX. Paragenetic position. Probably the most common of all gold tellurides. Known from many subvolcanic hydrothermal Ag–Au–Te-bearing metalliferous veins, and often the principal auriferous minerals in these veins. However, it is by no means restricted to this type of occurrences, but it is observed also in many intrusive-hydrothermal auriferous veins, sometimes in considerable quantities.

X. Investigated occurrences. Sacaramb, Faiïa Baţe, and other places in Siebenbürgen, Roumania; Smuggler Mine, Colorado, and Goldfield, Nevada, U.S.A.; Glava, Värmland, Sweden; Arakaka Goldfield, British Guyana; from some mines in Japan; Emperor Gold Mines, Fiji.

XI. Literature. Detailed and concised data provided by BORCHERT (1930), SCHERBINA (1941) and THOMPSON et al. (1951).

XII. Powder diagram (THOMPSON). (10) 3·04, (2) 2·97, (3) 2·25, (5) 2·14 Å.

VOLINSKYITE

(Spelling by different transcriptions variable)

I. General Data. *Chem.* $AgBiTe_2$. Cryst. o'rhombic, perhaps similarities with schapbachite and bogdanoviczite. H one good, two more indistinct. — *Polish* caused by softness high relief.

III. Reflection behavior. White with a tiny rose tint, in air and in oil not very different, tarnishing or darkening alike bismuth not observed. — Reflectivity higher than galena, Rg = 55.2, Rp = 52.8 in air (VJALSOV). The bireflection is only in oil and with much care distinct at grain boundaries. Anisotropism between + N. in oil distinct, the extinction straight.

VI. Structure. Lamellae oriented // the best cleavage. Along the cleavage starts the weathering.

VIII. Diagnostic data. Missing care may cause a mistake for native bismuth (bad polish, negative relief). Since for an immediate determination data not yet sufficient, be careful!

XI. The paragenetic positions insofar noticeable as in spite of presence of native Au no gold-telluride has been formed. Besides tellurobismutite hessite, altaite etc. are accompanying.

X. Only the occurrence of Armenia, from which GENKIN kindly dedicated a specimen to the author, could be investigated.

XI. Literature. The mineral was first described by BESMERTNAYA and SOBOLEVA (1963).

XII. Powder diagram. (8) 3.21, (10) 3.09, (5) 2.21, (3) 2.15, (3) 1.82 Å.

KOSTOVITE

I. *Chem.* $CuAuTe_4$. Symmetry unknown, low. Strongly twinned. One cleavage.. Caused by tarnishing, no distinct metallic lustre in specimen.

III. Reflectivity high (R = 49·3–60·1%). Bireflection very distinct, anisotropism high (reddish grey to brown, greyish yellow resp.).

IV. Etching. All standard reagents negative.

IX. Observed. in Chelopech, Bulgaria, only.

XII. Powder diagram (Terziev 1966): (5) 4·96, (10) 3·01, (4) 2·34, (4) 2·24, (6) 2·10, (5) 1·859 Å.

KRENNERITE

(= Müllerine, Bunsenine)

1. General data. *Chem.* $AuTe_2$, similar to calaverite, but mostly with higher silver content (Au:Ag ~ 4:1). *Cryst.* C^4_{2v}; crystals rich in faces. a_0 = 16·51, b_0 = 8·80, c_0 = 4·45, all in Å, Z = 8. — Cleavage quite distinct in coarse-grained material, different from calaverite. Brittle. H = 2½, D = 8·62 (slightly less than calaverite). — Opaque. Silvery-white with yellowish tinge, very high metallic lustre.

II. Polishing properties. Easily takes an excellent polish, but cleavage // (001) often visible in polished sections. A second cleavage (very indistinct) occasionally observed.

Talmage hardness: C.

III. Reflection behavior. Cream-white, strongly reflecting. Hardly distinguishable from calaverite, perhaps slightly less reflecting, and of a lesser yellowish tinge.

		465 nm		527 nm		589 nm	
Photocell	R_g	72·0	70·0	71·0	70·0	75·0	76·0
Fastré & Orcel	R_p	65·0	63·2	64·0	63·7	68·0	69·0
	Moses 61·8 All values of Fastré & Orcel seem to be high!						
(U. + B.)	nm	480	540	580	640		
	R_m	64·8	71·9	74·6	76·0 which		
	does not regard the high bireflextion						

further, I think more probable values in the enclosed tabellae!

The *reflection pleochroism* is weaker than one would expect from the above figures, but moderately distinct along grain boundaries; hardly increased in oil.

Anisotropy between + nic. quite distinct when using intensive illumination, and stronger than in calaverite. In the darkest position (in non-pinacoidal sections) no complete extinction, but dark-brown colors.

IV. Etching (accord. to Short and Stillwell). *Positive*: HNO_3, effervescence, iridescent coating; $FeCl_3$ and KOH give brownish coating which, with the latter, may sometimes be strong, sometimes weaker. *Negative*: HCl, KCN, $HgCl_2$.

Texture etching: HNO_3 produces cleavage etching which, in contrast to calaverite, occurs in several directions perpendicular to one direction (distinction, from calaverite!). Etch reaction with HNO_3 is considerably quicker than with calaverite.

VI. Fabric. As stated by STILLWELL, krennerite appears to be of more common occurrence than previously assumed. In addition, much of the material previously labelled "sylvanite" may be krennerite. Further investigations of the textures involved are much needed. — Idiomorphic grains are very rare; mostly coarse-grained, little interlocked, and mostly untwinned.

Replacements: In many of its deposits, krennerite is the first-formed gold telluride and is often replaced by other tellurides, even by non-auriferous ones. STILLWELL (1931) describes replacements by native gold, sylvanite, hessite, petzite, and chalcopyrite. The writer was able to study its replacement by sylvanite in excellent specimens kindly placed at his disposal by Dr. STILLWELL. BORCHERT's view (1930) that krennerite might be a low-temperature polymorph of calaverite, is erroneous.

Krennerite replacing nagyagite was observed by SCHERBINA (1941) in material from Glava. In a specimen from Siebenbürgen, krennerite occurs as a rim around decomposed sylvanite, together with newly-formed hessite. The mode of attachment of krennerite to hessite retains something of the twin structure of the original sylvanite. — The writer also observed myrmekitic intergrowth with silver-rich tetrahedrite in ores from Nagyag, and with pyrrhotite from other deposits.

VIII. Diagnostic features. In earlier investigations krennerite was considered to be a "variety" of calaverite. X-ray analysis as well as the investigations by STILLWELL, however, establish krennerite as a species of its own.

The differences in the properties are not always sufficient for an unambiguous identification in polished sections: krennerite has the better cleavage and stronger anisotropy, exhibits cleavage etching, and reveals a higher silver content by microchemical tests. Sylvanite likewise is very similar, but, different from krennerite, it exhibits a distinct reflection pleochroism accentuated by its characteristic twin lamellae which do not occur in krennerite. — "Müllerine" or "Müllerite" and the so-called "lichtes Weißtellurerz" essentially consist of krennerite.

IX. Paragenetic position. Similar to the other gold tellurides.

X. Investigated occurrences. Săcăramb and Fata Baii, Siebenbürgen, Roumania; La Plata Mine, Durango, Colorado, U.S.A.; Bevcourt Mine, Louvricourt, Quebec, Canada; Kalgoorlie, Western Australia.

XI. Literature. Among the most important papers are the following: STILLWELL (1931), BORCHERT (1930, 1935), and TUNELL & KSANDA (1933) (structure determinations).

XII. Power diagram (THOMPSON). (10) 3·05, (4) 2·94, (4) 2·25, (5) 2·11 Å.

Muthmannite, $(Au, Ag)Te_{1-2}$ could not be approved in its independence. Elongated grains with cleavage in the elongation. $H = 2^1/_2$. The data on properties and powder diagram differ from petzite, krennerite, sylvanite, calaverite. — Not much known.

CALAVERITE

I. General data. *Chem.* $AuTe_2$, with small amounts of Au substituted by Ag. *Cryst.* monoclinic; occasionally crystals with numerous faces, difficultly identifiable. C_{2h}, $a_0 = 7{\cdot}18$, $b_0 = 4{\cdot}10$, $c_0 = 5{\cdot}07$, all in Å, $\beta = 90°$, $Z = 2$. Twinning according to several laws. Cleavage: none; fracture conchoidal. Brittle. $H = 2^1/_2\text{–}3$, $D = 9{\cdot}2$ to $9{\cdot}3$. — Opaque. Strong metallic lustre, white to yellowish.

II. Polishing properties. Good polish easily obtainable, developing strong relief against hard minerals. No cleavage visible.

Polishing hardness very low, slightly higher than pyrargyrite, similar to galena, but just a little softer than the latter; much softer than chalcopyrite and fahlore.

Talmage hardness: C.

III. Reflection behavior. White, with a light tinge to brownish-yellow, tarnishing a little. Very bright, but distinctly less bright than well polished bismuth.

Reflection pleochroism weak, more pronounced along grain boundaries. In prismatic crystals from Cripple Ck. sections // columnar axis are hardly pleochroic, those normal to that axis more distinctly so: light yellowish-brown // the short diagonal, white to yellowish-white (and brighter) // the longer diagonal of the rhombic gross sections. In oil very similar.

	in air		in oil
general	very high		distinctly lower
contrasted with pyrite	similar, little less bright, and more pinkish yellow (contrary to measurements)		generally brighter than pyrite
galena	little brighter		distinctly brighter
sylvanite	distinctly brighter, and a little more colored		distinctly brighter
Photocell	the differences between R_g and R_p appear somewhat too high, but in general the values given by ORCEL seem to be better than those obtained with the photometer ocular		
	ORCEL		
	R_g	R_p	
465 nm	62·0	55·4	
527 nm	66·0	58·0	
589 nm	65·0	59·0	
MOSES/FOLINSBEE	64·1/63·2		

Specific color phenomena. The intensity of the reflection and the colors vary within a wider range than should be expected from the chemical composition. Tarnish cannot be responsible for this variation.

Anisotropy between + nic. distinct, but quite weak. Colors in diagonal position pronounced only when nicols are not perfectly crossed.

IV. Etching (accord. to MURDOCH, SHORT, BORCHERT). *Positive*: HNO_3, quick light brown tarnish, then blackening, under weak effervescence. $FeCl_3$, brownish strain, mostly negative; KOH similar. *Negative*: KCN, HCl, NaOH, $HgCl_2$. — *Texture etching*: HNO_3 1:1 produces very fine cleavage etching.

VI. Fabric. Twinning mostly not visible in polished sections. Occasionally along grain boundaries pronounced twinning according to still unknown twin laws which may be due to microtectonic deformation. — Sometimes subparallel growth.

Structure and texture. Mostly idiomorphic judging from the limited material available for investigation. In material from Cripple Greek, stout prismatic crystals in a "groundmass" of predominant fluorite. Grain contours often lobate, somewhat like idioblasts. Xenomorphic when finely dispersed through country rock, and then probably formed together with pyrite, the iron content of which probably derived from original magnetite.

VII. Special fabrics. WILLEMSE described the replacement of gold by calaverite. The reverse process, gold replacing calaverite, is common in weathered portions of the deposits, occasionally resulting in beautiful pseudomorphs of spongy gold after leached-out calaverite.

VIII. Diagnostic features. Reliable data are scarce. Contrasted with sylvanite, calaverite is somewhat brighter, and of much weaker anisotropy. Krennerite is very similar, but exhibits a distinct cleavage, and its silver content as revealed microchemically is much higher. Calaverite is harder than krennerite and the mostly twinned sylvanite. — No doubt krennerite and calaverite are to be considered as different species! Distinction from many other minerals by the reactions for Au and Te.

IX. Paragenetic position. Quite common as one of the last-formed ore minerals in auriferous veins formed at greater depths, but rarely in large quantities. Occasionally in epithermal, subvolcanic auriferous veins (mostly of somewhat unusual types), where it may be the only auriferous mineral. (Cripple Creek, Colorado; Calaveras, California, U.S.A.). Its association, particularly in the former deposit, is characterized by a multiplicity of minerals. Besides ever-present quartz and pyrite many tellurides such as altaite, hessite, sylvanite, tellurobismutite, often native gold or electrum, with galena, sphalerite and fahlore as usual associates.

X. Investigated occurrences. Last Dollar Mine, Cripple Creek, and Smuggler Mine, Colorado. Lake View Consolidated, Kalgoorlie, Western Australia.

XI. Literature. Besides older, mostly obsolete descriptions the publications by BORCHERT (1930, 1935) and the data given by SHORT, STILLWELL and ORCEL are the most important ones.

XII. Powder diagram (THOMPSON, in Å): (10) 3·01, (4) 2·19, (8) 2·09, (3) 1·76.

NAGYAGITE

I. General data. *Chem.* Formula only recently established with some reliability: $Pb_5Au(Te, Sb)_4S_{5-8}$; probably not a sulphide proper, but similar to intermetallic compounds. *Cryst.* Monoclinic, distinctly pseudotetragonal. *Lattice* constants in tetragonal interpretation: $a_0 = 4{\cdot}15$, $c_0 = 30{\cdot}25$ Å. Crystals thin tabular // (001), rarely granular, often bent. — Cleavage // (001) perfect, $H = 1^1/_2$, $D = 7{\cdot}5$. — Opaque; metallic lustre, greyish-white.

II. Polishing properties. Polish of moderate quality obtainable in spite of low hardness, but easily results in developing a relief. Cleavage // (001) mostly visible in polished sections, and similar to molybdenite. Hardness a little less than in sylvanite.

Talmage hardness: B^+.

III. Reflection behavior. Greyish-white, similar to galena.

Reflection pleochroism weak both in air and in oil, quite different from molybdenite; reflection // (001) distinctly higher than ⊥ (001).

Anisotropy berween + nic. weak, but distinct. Extinction nearly straight, but often undulating. No distinct deviation from tetragonal behavior.

	in air		in oil
general	high		very little lowered
contrasted with			
sylvanite	darker		considerably darker
hessite	light brownish-white ("white-coffee")		similar
krennerite	grey (to cream)		considerably darker
galena	about equal		similar
Photometer ocular (locality: Nagybanya)			
green	43		27·5
orange	35		24
red	34		24
Photocell			
Fastré (1933)	R_g	R_p	(in view of the weak bireflection these figures appear to be somewhat too high)
465 nm	49·0	43·8	
527 nm	47·0	41·6	
589 nm	40·7	38·1	
Moses	40·0	41·2	

IV. Etching (accord. to Murdoch). *Positive*: HNO_3, iridescent coating, after rubbing-off: dark-grey. *Negative*: KCN, HCl, $FeCl_3$, KOH. A suitable reagent for texture-etching is not yet known, but not essential for the recognition of textural features.

VI. Fabric. *Structure and texture.* Mostly thin-tabular crystals, often bent, and often rimmed by bright-white altaite. Helke identified small amounts of sylvanite and petzite in these reaction rims, as well as two other metalliferous minerals, one of which was identified by the writer as hessite. Careful observations reveal that quite often lamellae occur // (001) which could be interpreted as mimetic twinning on (100). Sections // (001) exhibit crossed twin lamellae, perhaps according to similar twin laws. (001), furthermore, seems to be a gliding plane.

Replacements. Besides the above mentioned reaction rims replacement phenomena have not been frequently observed. At Glava, Värmland, nagyagite is almost completely pseudomorphosed by krennerite, or even by an aggregate of up to six ore minerals. Fig. 255 could be interpreted as a pseudomorph after nagyagite.

VIII. Diagnostic features. In cases where the presence of tellurides is indicated, the diagnosis of nagyagite is not difficult. The low hardness, distinct cleavage, greyish-white color and its often indistinct and undulating extinction combined with a weak reflection pleochroism, are diagnostic criteria.

IX. Paragenetic position. Similar to most other tellurides, but generally less common in spite of its abundance in a few deposits. Particularly common in the epithermal mineral veins of Nagyag, but also at Kalgoorlie; Tararu Creek, New Zealand; Tavua, Fiji Islands. At Nagyag, it is associated with native gold, electrum, tellurides of Au and Ag, altaite, native tellurium, pyrite, alabandite, etc.

X. Investigated occurrences. Specimens from only three occurrences were at the disposal fo the writer: Nagyag (now Săcăramb) Rumania; Glava, Värmland, Sweden; Kalgoorlie, Western Australia.

XI. Literature. Microscopic investigations on nagyagite are described in the publications by BORCHERT (1930), STILLWELL (1931), HELKE (1939), SCHERBINA (1941).

XII. Powder diagram (B. & TH.). (10) 3·02), (6) 2·81, (4) 2·43, (3) 2·08, (6) 1·506 Å

MONTBRAYITE

I. General data. *Chem.* Au_2Te_3, perhaps with a little Pb and Bi. *Cryst.* Triclinic, $a_0 = 12{\cdot}08$, $b_0 = 13{\cdot}43$, $c_0 = 10{\cdot}78$ Å. $\alpha = 104° 30'$, $\beta = 97° 34'$, $\gamma = 107° 53'$; Z = 12. — Cleavage $(1\bar{1}0)$, $(0\bar{1}1)$, $(1\bar{1}1)$. H = $2^1/_2$, very brittle; D = 9·94. Strong metallic lustre, yellowish-white.

II. Polish excellent; a little, but distinctly harder than tellurobismuthite Cleavage distinct in several directions; quite brittle; cleavage pits similar to those of galena.

Talmage hardness: C.

III. Reflection behaviour. Very high, similar to krennerite. Color similar to that of krennerite and calaverite, perhaps a little more cream-pink (in oil). Contrasted with altaite, the latter is distinctly more white to bluish R ~ 60% acc. to SINDEEVA.

Reflection pleochroism extremely weak, just visible in oil only along grain boundaries when these are distinguishable by other reasons.

R. nm	480	540	580	640
	55·8	63·5	66·1	67·3

Anisotropy weak, but color effects quite distinct: light-grey, yellowish-brown, bluish-grey. Material from the Toburu mine exhibits a comparatively strong anisotropy as well as a marked dispersion of the extinction position. Its color (a light pink) is accentuated by contrast with its rim of altaite.

IV. Etching. (PEACOCK & THOMPSON). *Positive*: HNO_3, strong effervescence, light-yellowish tarnish. Peculiar etching pattern, similar to "alligator skin". — Stronger acid produces effervescence, and directional preference is noticeable in the etching pattern. Conc. HNO_3 produces only a weak etching with a homogeneous greyish brown tarnish. — *Negative*: HCl, KCN, $FeCl_3$, $HgCl_2$, KOH.

VI. Fabric. Relatively coarse-grained (∅ > 1 mm). Grains generally rounded. Additional investigations necessary.

Material at the disposal of the writer contained one single case of typical twinning. Age relations with regard to associated minerals insufficiently investigated.

VIII. Diagnostic features. Distinction from calaverite and krennerite is difficult; sylvanite has a much stronger reflection pleochroism. The cleavage seems to be a little more pronounced. Taken as a whole, its microscopic identification presents considerable difficulties.

X. Paragenetic position. A single sample, from the first only known occurrence was kindly placed at the disposal of the writer by Dr. FROHBERG (Robb Montbray Mine, Montbray township, Abitibi Co., Quebec, Canada). The writer's observations closely corresponds to those by PEACOCK & THOMPSON (1946). The following associated minerals were observed: native gold, chalcopyrite, frohbergite, "tellurbismut", melonite, coloradoite, etc., Toburn mine.

XII. Powder diagram (PEACOCK & THOMPSON, in Å). (8) 2·97, (8) 2·92, (10) 2·08. Many additional lines, mostly very weak.

TETRADYMITE AND TELLUROBISMUTITE

[*Tellurwismut*]

I. General data. Material formerly labelled as "tellurwismut" has in the meantime been found to represent a diversity of mineral species which are not yet sufficiently identified and distinguished from one another. Closer relations exist between the following which have nearly identical structures: *Tetradymite* Bi_2Te_2S, the most common of all, and *tellurobismutite*, Bi_2Te_3; *gruenlingite* Bi_4TeS_3 and *joseite* $Bi_4(Te, S)_3$; *ikunolite* $Bi_4(S, Se)_3$ and *pilsenite* Bi_3Te_2 and "*wehrlite*" of which the latter two may be identical. Only the first two of the above minerals have been appropriately investigated. The recently named *laitakarite* and *hedleyite* (Bi_7Te_3) may be identical, "*selenjoseite*"(Bi_4SeS_2) is similar to *ikunolite*.

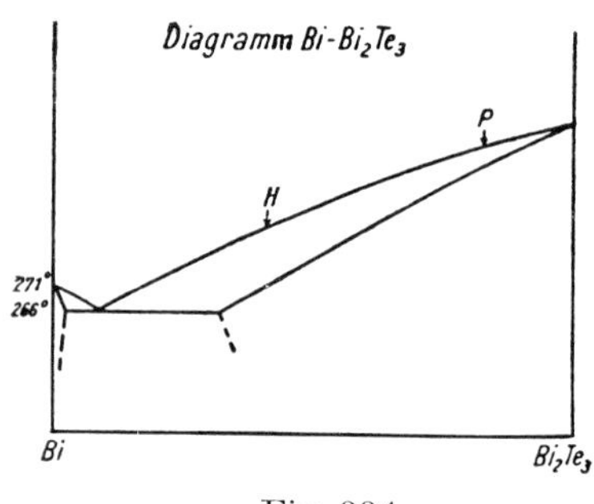

Fig. 334

Tetradymite and tellurobismutite are rhombohedral with typical layer structures, the former with $a_{rh} = 10{\cdot}13$ Å, $\alpha_{rh} = 24° 10'$, $Z = 1$ ($a_0 = 4{\cdot}22$, $c_0 = 30{\cdot}01$ Å), C_{3i}^2; tellurobismutite perhaps D_{3d}^5, $a_{rh} = 10{\cdot}51$ Å, $\alpha F = 24° 02'$ ($a_0 = 4{\cdot}37$, $c_0 = 30{\cdot}45$ Å); $c/a = 6{\cdot}96$ and $6{\cdot}99$ respectively. — Cleavage // (0001) very perfect, // (10$\bar{1}$0) indistinct. — H = $1^1/_2$–2, D = 7·3–7·8. — Opaque. Strong metallic lustre, white, on cleavage faces.

II. Polishing properties. Gives good polish, but rarely free of scratches when in association with harder gangues; sections // (0001) most difficult to polish. Cleavage // (0001) often visible in polished sections. Hardness low, < bismuthinite and galena

Talmage hardness: B^+.

III. Reflection behavior. White, with faint yellowish tinge. Reflectivity very high tellurobismutite still higher).

	in air	in oil
general	very bright, white to cream	not noticeably changed
contrasted with chalcopyrite	considerably brighter, and more white	
wehrlite	very little darker	similar
pyrite	similar, less yellowish	
Photocell	(tetradymite)	
green	48·5	44 } probably too low
orange	48·0	37 } probably too low
red	47·5	38 } probably too low

data of different authors vary so strongly, that none looks useful.

Moses	44 (probably too low)	
Folinsbee	56·9 (high, belongs probably to tellurobismutite)	

Bireflection: In air very weak, actually not visible; in oil very weak, but visible at grain boundaries, O brigther than E.

Anisotropy between + nic. distinct, with low power oil immersion more pronounced (in diagonal position brownish colors!).

IV. Etching. Published data are in part contradictory, those by DAVY & FARNHAM seem to be the most reliable ones. — *Positive*: HNO_3, rapid deep-brown stain under strong effervescence; reaction starts sometimes slowly. $FeCl_3$, weak coating, reaction along scratch marks. *Negative*: HCl (in pt., positive), $HgCl_2$ (in pt., positive), KCN, KOH.

V. Physico chemistry. The system Bi–Te–S has only been incompletely investigated. Apparently there exists an extended mixed-crystal series with Bi_2S_3 (and Bi_2Se_3) at high temperatures, and a complete mixed-crystal series between Bi_2Te_2S and Bi_2Te_3. At lower temperatures Bi_2S_3 forms by exsolution, and also the mixed crystals of Bi_2Te_2S–Bi_2Te_3 decompose into tetradymite and tellurobismutite.

VI. Fabric. *Grain shapes.* Tetradymite grown in open vugs mostly forms fourlings, otherwise thick-tabular or granular aggregates. At Oravița native bismuth in myrmekitic intergrowth with galena replaces tetradymite; parallel intergrowth with bismuthinite is also observed. Exsolution of (brighter) tellurobismutite from tetradymite is observed in the form of thin lenticular blebs // (0001) of the latter; exsolution of bismuthinite is observed in specimens from Boliden.

Besides the above exsolution phenomena, similar myrmekitic textures may result from the addition of sulphur and in part, lead.

Thus the brighter tellurobismutite may be rimmed by a mixture of tetradymite and galena.

In turn, tetradymite often occurs as an exsolution mineral in galena, its (0001) planes being parallel to (111) of the latter, thus giving rise to a pseudocleavage (parting) // (111) of galena. More coarse-grained intergrowth textures of the same kind may also be of primary origin, e.g., at Boliden. It is found in similar exsolution blebs in certain hypogene bornites (e.g., from Ookiep).

Peculiar oriented intergrowth textures with melonite, in part coarsely lamellar, in part myrmekitic, occur in the Robb Montbray Mine.

In some cases, the crystals are strongly exfoliated or bent, probably owing to tectonic deformation.

VIII. Diagnostic features. Identification is without difficulty, provided only that other criteria suggest the possibility of its presence, and that the mineral exhibits phenomena indicative of its layer structure. In contrast with all other layer-structure minerals, the almost complete absence of reflection pleochroism is highly discriminatory in combination with its high reflectivity.

IX. Paragenetic position. Tetradymite as well as tellurobismutite which is hardly distinguishable from the former, are generally, though wrongly, considered to be of very rare occurrence. However, they are of widespread occurrence although in small quantities, particularly so in hypothermal to mesothermal auriferous quartz veins, in metasomatic Bi-deposits, as well as in chalcopyrite–bornite deposits, and chalcocite–quartz veins. They are frequently associated with bismuthinite. They also occur besides other tellurides in subvolcanic metalliferous veins, in contact deposits and high-temperature complex pyrite deposits. In contact deposits and in some auriferous quartz veins, tetradymite and tellurobismutite are associates of native gold, other associated minerals being quartz, calcite, pyrite, galena, chalcopyrite, bismuthinite. Occasionally (e.g., at Highland, Montana), a mineral very similar to tetradymite is observed in the paragenesis, being perfectly cleavable, but strongly anisotropic, and of strong reflection pleochroism.

X. Investigated occurrences. Although numerous samples of tetradymite and tellurobismutite were at the disposal of the writer, their identification without doubt was not always possible. Investigated occurrences were: Boliden, Northern Sweden (particularly rich material); Baita Bihorului and Oravița, Banat, Rumania; "Deutsch Pilsen", Tatra, Č.S.S.R.; Arakaka goldfield, Guyana; several occurrences in Canada and Australia; Holkol, Korea; Glava, Värmland, Sweden.

XI. Literature. Publications on tetradymite before 1935 are scarce and not always dealing with correctly identified material. More recent references are in particular by T. WATANABE (1933), ÖDMAN (1941), WARREN & PEACOCK (1945), K. PADĚRA, SZTROKAY (1940), R. M. THOMPSON.

XII. Powder diagram: For tetradymite: THOMPSON (10) 3·11, (5) 2·28, (3) 2·11, (4) 1·963, (5) 1·207 Å. HARCOURT (1) 3·10, (4) 2·35, (3) 2·16, (3) 1·99 Å. For tellurobismutite: (B. & T.) (10) 3·22, (8) 2·37, (4) 2·20, (4) 2·04, (5) 1·486 Å. (The HARCOURT values, supposed to be for tetradymite, refer probably to an intermediate member!)

Kawazulite, Bi_3TeSe, has nearly the same properties.

OTHER TELLURIUM–BISMUTH COMPOUNDS

Besides tetradymite and tellurobismutite, during the last few decades a considerable number of similar "compounds" of Bi and Te, with or without S, sometimes with Pb have been described. Only more recently PEACOCK and co-workers, as well as SZTROKAY were able to introduce some order and clarity into this difficult field of study, All mineral varieties involved seem to be of very similar structures and could probably be interpreted as various members of a complex mixed-crystal series. In the following, the members free of sulphur (pilsenite = wehrlite), hedleyite, and, described more recently, laitakarite are distinguished from the sulphur-bearing compounds, i.e., joseite, grünlingite, and oruetite.

They all have close relations to tetradymite and tellurobismutite, as regards both structure and physical/chemical properties.

PILSENITE (= WEHRLITE) AND HEDLEYITE

I. General data. The melting equilibria between bismuth and bismuth-tri-telluride as represented in fig. 334 exhibit a binary eutectic point of two series of mixed crystals. Minerals representing chemical compositions between that of Bi_2Te_3 and that at the eutectic point have been repeatedly described and do, at least in part, exist as minerals. Regarding the best-known member of that series pilsenite (= wehrlite), wherefrom the element tellurium was discovered!), SZTROKAY (1946) has shown that at the type locality there occurs an aggregate of replacing minerals with tellurobismutite as the oldest, tetradymite, Bi_2TeS_2, ± bismuthinite, Bi_2S_3. There is no doubt that at many places elsewhere there exist minerals corresponding to the above formula of "pilsenite", and for this reason the writer proposes to retain the corresponding species or variety names. Points P and H in the diagram would approximately correspond to the compositions of pilsenite (~ BiTe) and hedleyite (~ $Bi_{14}Te_6$), respectively. *Cryst.* Both members have rhombohedral layer lattices, with a pseudocell of a_{hex} = 4·42 or 4·46 Å, respectively, and c_{hex} = 5·97 or 5·94 Å, respectively. Z = 1/2 (BiTe) in the rhombohedral cell with a_{rh} = 3.24 Å, and α = 86° 8½′ or 86° 42′, respectively. In both cases, the true cell is very elongated-prismatic or acute-rhombohedral, respectively, but may not always be exactly definable. In the case of hedleyite, e.g., a_{rh} = 39·68 Å, α_{rh} = 6° 26½′, with Z = 1 ($Bi_{14}Te_6$). Massive aggregates, basal cleavage perfect, cleavage flakes somewhat flexible. H = 2, D = 8·4–8·9 with increasing Bi-content. Metallic lustre, tin-white, in part tarnishing.

II. Polish similar to tetradymite, perhaps retaining more scratch marks.

III. Reflection behavior. Earlier descriptions to be discredited since the investigation by SZTROKAY. Not distinguishable from tellurobismutite. Very "bright-white". Very weak anisotropy, visible only along grain boundaries, in oil.

IV. Etching. *Positive*: HNO_3, rapid and strong reaction: $FeCl_3$, brown tarnish; H_2O_2. *Negative*: HCl, KCN, KOH, $HgCl_2$, all tests being practically identical for pilsenite and hedleyite.

VI. Fabric. The sample investigated, a specimen of homogeneous appearance and labelled "pilsenite" according to the original definition for material from "Deutsch Pilsen", proved to be of a complex composition, containing tellurobismutite, tetradymite, hessite, native bismuth, sphalerite, galena, bournonite, bismuthinite, pyrite, arsenopyrite. From the composition of that paragenesis one would expect that at least some members of that association had combined with one another. Similarly, hedleyite originates in a paragenesis of numerous minerals, but generally low in sulphur.

With the nicols perfectly crossed, and extra strong illumination, a very fine mosaic-like texture becomes faintly visible.

XI. Literature. Some microscopic data (partly doubtful) contained in a paper by v. PAPP. Hedleyite was described by WARREN & PEACOCK (1945) whose work makes use of further publications.

XII. Powder diagram: Pilsenite: (10) 3·22, (7) 2·36, (5) 2·21 (THOMPSON); hedleyite: (10) 3·25, (5) 2·36, (4) 2·23 Å (accord. to WARREN & PEACOCK).

JOSEITE, GRÜNLINGITE, ORUETITE

I. General data. According to the original definitions and calculation of analyses (the latter not being free of arbitrary assumptions), — there should be: Joseite = Bi_3TeS, grünlingite = Bi_4TeS_3, and oruetite = Bi_8TeS_4. The properties and characteristics found in this group are so similar to one another, and again to properties found in the tetradymite group that closer relationships between all these species were long suspected to exist. These are confirmed by X-ray analysis. — *Cryst.* rhombohedral, with layer lattices; minerals occurring mostly in aggregates of tabular grains. *Lattice* identical for the above species: D^5_{3d} with a_{hex} = 4·24, c_{hex} = 39·69 Å, with some occurrences of higher c_0 values, up to 42·1 Å. Alternatively, a_{rh} 13·45 Å, α_{rh} = 18° 08′, Z = Bi_4TeS. — *Chem.* The structure corresponds to the formula: $Bi_{4+x}Te_{1+x}$ or $Bi_{4+x}(Te, S)_{2+x}$ where x = 0 to 0·3. The latter formula does not readily comply with part of the analyses. — H = 2, D = 8·0–8·15 (> tellurobismutite < pilsenite ~ hedleyite). — Strong metallic lustre; silvery-white to greyish-white, but tarnishing.

II. Polishing properties. Similar to tetradymite; abrasion hardness slightly lower.

III. Reflection behavior (accord. to PEACOCK, 1941). White, brighter than galena. Basal sections isotropic, others anisotropic, with greenish-grey colors in diagonal positions. Not distinguishable with certainty from tetradymite.

BESMERTNAJA gives for joseite (a and b) rather different properties, joseite (a) in average 7–8 units lower. The VJALSOV values seem to be better but he gives only joseite (a).

IV. Etching. Not noticeably different for various members of the group: *Positive*: HNO_3, strong attack; HCl, light-grey coating. $FeCl_3$, strong coating. *Negative*: KCN, KOH, $HgCl_2$.

VI.–VIII. No data available.

IX. and X. The specimens investigated by PEACOCK originated in hydrothermal deposits of high temperature of formation, mostly auriferous veins.

XI. Literature. The above data mostly by PEACOCK (1941).

XII. Power diagram. The strongest lines of the powder diffraction diagram for "joseite I" are (in Å): (10) 3·07, (5) 2·24, (5) 2·11.

Temagamite, Pd_3HgTe_3, known only from a specimen from Temagami, Ontario, forms an inclusion in chalcopyrite. It is light greyish white and weakly anisotropic. After the given R-values it should be distinctly brownish.

OTHER TITLES IN THE INTERNATIONAL SERIES IN EARTH SCIENCES

Vol.	Author	Title
Vol. 1	BENIOFF et al.	Contributions in Geophysics in honor of Beno Gutenberg
Vol. 2	SWINEFORD	Clays and Clay Minerals
Vol. 3	GINZBURG	Principles of Geochemical Prospecting
Vol. 4	WAIT	Overvoltage Research and Geophysical Applications
Vol. 5	SWINEFORD	Seventh National Conference on Clays and Clay Minerals
Vol. 6	TYUTYUNOV	Introduction to the Theory of the Formation of Frozen Rocks
Vol. 7	KRINOV	Principles of Meteorites
Vol. 8	NALIVKIN	The Geology of the USSR
Vol. 9	SWINEFORD	Eighth National Conference on Clays and Clay Minerals
Vol. 10	POKORNY	Principles of Zoological Micropalaeontology Vol. 1
Vol. 11	SWINEFORD	Ninth National Conference on Clays and Clay Minerals
Vol. 12	SWINEFORD	Tenth National Conference on Clays and Clay Minerals
Vol. 13	BRADLEY	Eleventh National Conference on Clays and Clay Minerals
Vol. 14	ROSENQVIST & GRAFF-PETERSON	International Clay Conference Vol. 1
Vol. 15	COLOMBO & HOBSON	Advances in Organic Geochemistry 1962
Vol. 16	BREGER	Organic Geochemistry
Vol. 17	HELGESON	Complexing and Hydrothermal Ore Deposition
Vol. 18	BATTEY & TOMKEIEFF	Aspects of Theoretical Mineralogy in the USSR
Vol. 19	BRADLEY	Twelfth National Conference on Clays and Clay Minerals
Vol. 20	POKORNY	Principles of Zoological Micropalaeontology Vol. 2
Vol. 21	ROSENQVIST & GRAFF-PETERSON	International Clay Conference Vol. 2
Vol. 22	YERMAKOV et al.	Research on the Nature of Mineral-forming Solutions
Vol. 23	BRIDGE & BRIDGE	Clays and Clay Minerals
Vol. 24	HOBSON & LOUIS	Advances in Organic Geochemistry 1964
Vol. 25	BRADLEY	Thirteenth National Conference on Clays and Clay Minerals
Vol. 26	BAILEY	Fourteenth National Conference on Clays and Clay Minerals
Vol. 27	BAILEY	Fifteenth National Conference on Clays and Clay Minerals
Vol. 28	ANDREEV et al.	Transformation of Petroleum in Nature
Vol. 29	RAMDOHR	The Ore Minerals and their Intergrowths
Vol. 30	AHRENS	The Origin and Distribution of the Elements
Vol. 31	SCHENK & HAVENAAR	Advances in Organic Geochemistry 1968, 4th International Meeting
Vol. 32	HOBSON	Advances in Organic Geochemistry, 3rd International Meeting
Vol. 33	VON GAERTNER & WEHNER	Advances in Organic Geochemistry 1971, 5th International Meeting
Vol. 34	AHRENS	Origin and Distribution of the Elements, 2nd Symposium